AF410818

Ecosystem Service and Land-Use Changes in Asia: Implications for Regional Sustainability

Ecosystem Service and Land-Use Changes in Asia: Implications for Regional Sustainability

Editors

Kikuko Shoyama
Rajarshi Dasgupta
Ronald C. Estoque

MDPI • Basel • Beijing • Wuhan • Barcelona • Belgrade • Manchester • Tokyo • Cluj • Tianjin

Editors
Kikuko Shoyama
Department of Regional and
Comprehensive Agriculture
College of Agriculture
Ibaraki University
Ami
Japan

Rajarshi Dasgupta
Integrated Sustainability
Centre
Institute for Global
Environmental Strategies
Hayama
Japan

Ronald C. Estoque
Center for Biodiversity and
Climate Change
Forestry and Forest Products
Research Institute
Tsukuba
Japan

Editorial Office
MDPI
St. Alban-Anlage 66
4052 Basel, Switzerland

This is a reprint of articles from the Special Issue published online in the open access journal *Sustainability* (ISSN 2071-1050) (available at: www.mdpi.com/journal/sustainability/special_issues/ecosystem_service_land).

For citation purposes, cite each article independently as indicated on the article page online and as indicated below:

LastName, A.A.; LastName, B.B.; LastName, C.C. Article Title. *Journal Name* **Year**, *Volume Number*, Page Range.

ISBN 978-3-0365-5856-1 (Hbk)
ISBN 978-3-0365-5855-4 (PDF)

Cover image courtesy of Kikuko Shoyama

Contents

About the Editors

Kikuko Shoyama

Dr. Kikuko Shoyama is an Associate Professor at the Department of Regional and Comprehensive Agriculture, College of Agriculture, Ibaraki University, Japan. Her research interests are in the quantitative analysis of socio-ecological systems related to land use change, ecosystem services, and disaster resilience, in particular, the role of scenario modeling in decision-making under uncertain and changing environmentd. She contributed to the Intergovernmental Science-Policy Platform on Biodiversity and Ecosystem Services (IPBES) as a Lead Author for the Methodological Assessment on Scenarios and Models of Biodiversity and Ecosystem Services.

Rajarshi Dasgupta

Dr. Rajarshi Dasgupta is a Senior Policy Researcher in the Integrated Sustainability Center, Institute for Global Environmental Strategies. His research interests are risk-sensitive land use planning, ecosystem-based disaster risk reduction (Eco-DRR), socio-environmental scenario planning, spatial quantification of ecosystem services, community-based conservation, and disaster risk management. He served as a Lead Author (LA) for the Intergovernmental Science-Policy Platform for Biodiversity and Ecosystem Services (IPBES) Asia-Pacific Regional Assessment Report and Assessment of Sustainable Use of Wild Species, and a Chapter Scientist and contributing author for the Intergovernmental Panel on Climate Change's (IPCC) sixth assessment report (WG II).

Ronald C. Estoque

Dr. Ronald C. Estoque is a Senior Researcher at the Center for Biodiversity and Climate Change, Forestry and Forest Products Research Institute, Japan. He is, by profession, a forester and geo-environmental scientist. His research interests are at the interface between nature and society (sustainability; land change; ecosystem services; forest cover change; urbanization; and climate change mitigation, vulnerability, risk, and adaptation), with the application of various tools and techniques including remote sensing and GIS. He was a Contributing Author (CA) to the Intergovernmental Panel on Climate Change's (IPCC) sixth assessment report (WG II, Chapter 10). Currently, he serves as a Review Editor (RE) for the Intergovernmental Science-Policy Platform for Biodiversity and Ecosystem Services' (IPBES) nexus assessment (Chapter 2).

Preface to "Ecosystem Service and Land-Use Changes in Asia: Implications for Regional Sustainability"

The aim of this Special Issue (SI) of *Sustainability* is to provide insight into how land use, with its interaction with historical land management, has configured the services provided by ecosystems. This SI presents case studies that explore the impacts of direct and indirect drivers affecting the provision of ecosystem services in Asian countries, including China, India, Mongolia, Sri Lanka, and Vietnam. Findings from these empirical studies contribute to developing sustainability in Asia at both local and regional scales.

The Guest Editors are grateful to the authors for their contributions and, specifically, to the MDPI team for their effective assistance and cooperation.

Kikuko Shoyama, Rajarshi Dasgupta, and Ronald C. Estoque
Editors

Editorial

Ecosystem Service and Land-Use Changes in Asia: Implications for Regional Sustainability

Kikuko Shoyama [1,*], Rajarshi Dasgupta [2] and Ronald C. Estoque [3]

1. Department of Regional and Comprehensive Agriculture, College of Agriculture, Ibaraki University, Ami 300-0393, Japan
2. Integrated Sustainability Centre, Institute for Global Environmental Strategies, Hayama 240-0115, Japan
3. Forestry and Forest Products Research Institute, Tsukuba 305-8687, Japan
* Correspondence: kikuko.shoyama.sx68@vc.ibaraki.ac.jp

Citation: Shoyama, K.; Dasgupta, R.; Estoque, R.C. Ecosystem Service and Land-Use Changes in Asia: Implications for Regional Sustainability. *Sustainability* 2022, 14, 14263. https://doi.org/10.3390/su142114263

Received: 26 October 2022
Accepted: 28 October 2022
Published: 1 November 2022

Publisher's Note: MDPI stays neutral with regard to jurisdictional claims in published maps and institutional affiliations.

This Special Issue focuses on qualitative and quantitative analyses of ecosystem services (ESs) specifically toward sustainability in Asia. Asia is expected to experience population growth, peaking around 2030–2060 [1], which will likely result in unpredictable socio-economic changes that will present new challenges for land management. Sustainable land and natural resource management will play a crucial role in addressing these issues in the region. In particular, assessing land-use change and its effects on ESs is necessary to foster regional sustainability.

In response to the trade-offs in multiple land use, the concept of ESs has been introduced to find synergies between nature conservation and other aspects of human wellbeing. In recent years, many studies have addressed the impacts of land-use change on bundles of ESs by considering the influences of direct and indirect factors, e.g., region-specific changes in population and other socio-economic statuses. These case studies provide insight on how land use, with the interaction of historical land management, has configured the services provided by ecosystems. Thus, the findings from such empirical studies contribute to developing sustainability in Asia at both local and regional scales.

This Special Issue includes fifteen research papers that explore the impacts of direct and indirect drivers affecting ES provision in Asian countries, including China, India, Mongolia, Sri Lanka, and Vietnam. In this editorial, we briefly describe the contributions of each paper and how the analysis revealed the changes in ESs and could contribute to the intervention for regional sustainability in Asia.

The driving factor most frequently discussed in the Special Issue is urban development. The world's urban population is expected to nearly double by 2050, making urbanization one of the most disruptive developments. Yang and Liu [2], in their study titled "Spatio-Temporal Evolution and Driving Factors of Ecosystem Service Value of Urban Agglomeration in Central Yunnan", examined the spatiotemporal pattern of urban land changes in central Yunnan and found a significant decline of ecosystem service values (ESVs). The study highlights the need to have a balance between ecological conservation and urban development.

In examining the impacts of land change, scenario analysis is an effective method for providing implications for future land use planning and development. Yang and Su [3], in their study titled "Multi-Scenario Simulation of Ecosystem Service Values in the Guanzhong Plain Urban Agglomeration, China", investigated the response of ESVs to land use change in an urban agglomeration under different future scenarios and the trade-offs among various ESVs. Peng et al. [4], in their study titled "Evaluation of ESV Change under Urban Expansion Based on Ecological Sensitivity: A Case Study of Three Gorges Reservoir Area in China", applied the ecological sensitivity approach as a basis for predicting future urban expansion. These scenario assessments provide insight into ecosystem conservation under sustainable urbanization by predicting changes in urban land expansion that affect ESVs. In general, the scenario approach is applicable to other urban ecosystems. Athukorala et al. [5], in their study titled "Ecosystem Services Monitoring in the Muthurajawela Marsh and Negombo Lagoon, Sri Lanka, for Sustainable Landscape Planning", examined the impacts of urbanization on the natural landscape and ecosystem services of Muthurajawela Marsh and Negombo Lagoon (MMNL), an important wetland ecosystem in Sri Lanka. Between a business-as-usual scenario and an ecological protection scenario, the study suggests the latter as the more desirable future scenario because ecological protection policies can help flatten the MMNL's curve of continuous ecological degradation.

In contrast to these comprehensive scenario analyses, empirical approaches that focus on specific ecosystem services provide detailed evidence to practitioners in urban planning. Du et al. [6], in their study titled "The Impact of Impervious Surface Expansion on Soil Organic Carbon: A Case Study of 0–300 cm Soil Layer in Guangzhou City", showed that impervious surface expansion leads to soil organic carbon (SOC) loss and that retrofitting residential areas with low-rise buildings can significantly reduce SOC loss compared to the urbanization of agricultural land. For a large-scale development, Zhang and Hu [7] analyzed the effect of topography on ES in their study titled "Spatial Variation and Terrain Gradient Effect of Ecosystem Services in Heihe River Basin over the Past 20 Years". They showed that terrain gradient effects have a significant impact on ESs in inland watersheds, providing a scientific basis for optimizing local ecological patterns.

Two case studies in India are examples of analyses in which climate change scenarios were introduced in order to provide implications for climate change adaptation measures. Chandra et al. [8], in their study titled "Investigation of Spatio–Temporal Changes in Land Use and Heat Stress Indices over Jaipur City Using Geospatial Techniques", estimated the rise of thermal stress in urban areas in India based on land use and climate change scenarios. They applied physical indicators to assess urban conditions during various periods of thermal stress. This detailed spatial analysis of environmental changes in urban spaces is an important contribution to urban planning that takes into account urban ESs such as green space structure. Mitra et al. [9], in their study titled "Assessment of the Impacts of Spatial Water Resource Variability on Energy Planning in the Ganges River Basin under Climate Change Scenarios", simulated future water availability by applying climate change scenarios to examine the risks facing existing and planned power plants. The results provide development planners, energy planners, and investors with information on the spatial distribution of power plants that would be at risk, allowing them to make more accurate decisions regarding the siting of future power plants.

Asia also needs to address the vulnerability of arid ecosystems to climate change. Wang et al. [10], in their study titled "Evaluation of Qinghai-Tibet Plateau Wind Erosion Prevention Service Based on RWEQ Model", assessed the use of wind erosion control services as a way for improving the quality of the ecological environment. They identified factors governing the spatial differentiation of wind erosion control services and showed that the ES can be improved by reverting agricultural land in this area to grassland and controlling desertification.

Indirect drivers such as indigenous and local knowledge, technology, and financial assets play a major role in influencing the direct drivers of change in nature, nature's contribution to people, and quality of life at different spatial and temporal scales [11]. Ulziibaatar and Matsui [12], in their study titled "Herders' Perceptions about Rangeland Degradation and Herd Management: A Case among Traditional and Non-Traditional Herders in Khentii Province of Mongolia", focused on traditional and non-traditional pastoralists' land management. Herders play essential roles in sustaining Mongolia's economy and rangeland conditions. They found that pastoralists are willing to cooperate with local managers in rangeland management, providing implications for future management regimes.

Case studies exploring new systems in modern societies can provide insights into future natural resource management regimes in Asia. Pham et al. [13], in their study titled "Food Waste in Da Nang City of Vietnam: Trends, Challenges, and Perspectives toward Sustainable Resource Use", investigated the extent of food waste generation at the consumption stage, the eating habits of consumers, and the potential for reusing food waste as feed. They proposed both consumer waste prevention and waste management as effective measures for sustainable resource use. From a broader perspective, Morey et al. [14], in their study titled "Towards Circulating and Ecological Sphere in Urban Areas: An Indicator-Based Framework for Food-Energy-Water Security Assessment in Nagpur, India", proposed a new framework of indicators to address integrated food, energy, and water security in urban areas, based on the principles of a new concept called the Circular Ecosphere (CES). They concluded that food–energy–water nexus thinking can help establish the linkages between different resource management sectors and policies, thereby facilitating the application of CES.

As a method for exploring the nature of institutions aimed at conserving and restoring ecosystems, participatory approaches have an effective role in encouraging the participation of diverse stakeholders. Ratnayake et al. [15], in their study titled "Land Use-Based Participatory Assessment of Ecosystem Services for Ecological Restoration in Village Tank Cascade Systems of Sri Lanka", applied a participatory approach involving the integration of local knowledge, expert judgements and land use systems attribute data to assess the ESs. Chen et al. [16], in their study titled "Payments for Watershed Ecosystem Services in the Eyes of the Public, China", showed that the public had limited knowledge of the payment for ESs schemes. Therefore, improving public acceptance is essential for more effective payment schemes, and further research is needed on the impact of payment schemes on the public.

The integration of participatory methods and scientific simulation is also important in the context of coastal water resource management. Kumar et al. [17], in their study titled "Scenario-Based Hydrological Modeling for Designing Climate-Resilient Coastal Water Resource Management Measures: Lessons from Brahmani River, Odisha, Eastern India", applied a participatory modeling to evaluate current status and predict future conditions of river water quality, which is critical for people's livelihood in the region. The integration of participatory approach and computer simulation modeling, with the active participation of stakeholders, can enhance the science–policy interface for natural resource conservation and help co-generate future options.

In conclusion, this Special Issue presents case studies relating to the management of lands, natural resources and ESs in Asia. A variety of methods are employed, depending on the intended policy intervention, ranging from studies of individual services to more comprehensive ES assessments, studies that consider direct and indirect factors, and policy-oriented studies that explore with stakeholders how future institutions could be introduced into the society. It is noteworthy that continued efforts to address issues of ecosystem change and sustainability in Asia require not only monitoring of Asia's diverse ecosystems, but also recognition of the various values that can influence human engagement and policy decisions.

Author Contributions: Writing—original draft preparation, K.S.; writing—review and editing, R.D. and R.C.E. All authors have read and agreed to the published version of the manuscript.

Funding: This research received no external funding.

Institutional Review Board Statement: Not applicable.

Informed Consent Statement: Not applicable.

Conflicts of Interest: The authors declare no conflict of interest.

References

1. World Population Prospects. 2022. Available online: https://population.un.org/wpp/ (accessed on 21 October 2022).
2. Yang, L.; Liu, F. Spatio-Temporal Evolution and Driving Factors of Ecosystem Service Value of Urban Agglomeration in Central Yunnan. *Sustainability* **2022**, *14*, 10823. [CrossRef]
3. Yang, S.; Su, H. Multi-Scenario Simulation of Ecosystem Service Values in the Guanzhong Plain Urban Agglomeration, China. *Sustainability* **2022**, *14*, 8812. [CrossRef]

4.   Peng, H.; Hua, L.; Zhang, X.; Yuan, X.; Li, J. Evaluation of ESV Change under Urban Expansion Based on Ecological Sensitivity: A Case Study of Three Gorges Reservoir Area in China. *Sustainability* **2021**, *13*, 8490. [CrossRef]

5.   Athukorala, D.; Estoque, R.C.; Murayama, Y.; Matsushita, B. Ecosystem Services Monitoring in the Muthurajawela Marsh and Negombo Lagoon, Sri Lanka, for Sustainable Landscape Planning. *Sustainability* **2021**, *13*, 11463. [CrossRef]

6.   Du, J.; Yu, M.; Yan, J. The Impact of Impervious Surface Expansion on Soil Organic Carbon: A Case Study of 0–300 cm Soil Layer in Guangzhou City. *Sustainability* **2021**, *13*, 7901. [CrossRef]

7.   Zhang, L.; Hu, N. Spatial Variation and Terrain Gradient Effect of Ecosystem Services in Heihe River Basin over the Past 20 Years. *Sustainability* **2021**, *13*, 11271. [CrossRef]

8.   Chandra, S.; Dubey, S.K.; Sharma, D.; Mitra, B.K.; Dasgupta, R. Investigation of Spatio–Temporal Changes in Land Use and Heat Stress Indices over Jaipur City Using Geospatial Techniques. *Sustainability* **2022**, *14*, 9095. [CrossRef]

9.   Mitra, B.K.; Sharma, D.; Zhou, X.; Dasgupta, R. Assessment of the Impacts of Spatial Water Resource Variability on Energy Planning in the Ganges River Basin under Climate Change Scenarios. *Sustainability* **2021**, *13*, 7273. [CrossRef]

10. Wang, Y.; Xiao, Y.; Xie, G.; Xu, J.; Qin, K.; Liu, J.; Niu, Y.; Gan, S.; Huang, M.; Zhen, L. Evaluation of Qinghai-Tibet Plateau Wind Erosion Prevention Service Based on RWEQ Model. *Sustainability* **2022**, *14*, 4635. [CrossRef]

11. Intergovernmental Science-Policy Platform on Biodiversity and Ecosystem Services. *The IPBES Regional Assessment Report on Biodiversity and Ecosystem Services for Asia and the Pacific*; Karki, M., Senaratna Sellamuttu, S., Okayasu, S., Suzuki, W., Eds.; Secretariat of the Intergovernmental Science-Policy Platform on Biodiversity and Ecosystem Services: Bonn, Germany, 2018; p. 612.

12. Ulziibaatar, M.; Matsui, K. Herders' Perceptions about Rangeland Degradation and Herd Management: A Case among Traditional and Non-Traditional Herders in Khentii Province of Mongolia. *Sustainability* **2021**, *13*, 7896. [CrossRef]

13. Pham, N.-B.; Do, T.-N.; Tran, V.-Q.; Trinh, A.-D.; Liu, C.; Mao, C. Food Waste in Da Nang City of Vietnam: Trends, Challenges, and Perspectives toward Sustainable Resource Use. *Sustainability* **2021**, *13*, 7368. [CrossRef]

14. Morey, B.; Deshkar, S.; Sukhwani, V.; Mitra, P.; Shaw, R.; Mitra, B.K.; Sharma, D.; Rahman, M.A.; Dasgupta, R.; Das, A.K. Towards Circulating and Ecological Sphere in Urban Areas: An Indicator-Based Framework for Food-Energy-Water Security Assessment in Nagpur, India. *Sustainability* **2022**, *14*, 8123. [CrossRef]

15. Ratnayake, S.S.; Khan, A.; Reid, M.; Dharmasena, P.B.; Hunter, D.; Kumar, L.; Herath, K.; Kogo, B.; Kadupitiya, H.K.; Dammalage, T.; et al. Land Use-Based Participatory Assessment of Ecosystem Services for Ecological Restoration in Village Tank Cascade Systems of Sri Lanka. *Sustainability* **2022**, *14*, 10180. [CrossRef]

16. Chen, C.; He, G.; Lu, Y. Payments for Watershed Ecosystem Services in the Eyes of the Public, China. *Sustainability* **2022**, *14*, 9550. [CrossRef]

17. Kumar, P.; Dasgupta, R.; Dhyani, S.; Kadaverugu, R.; Johnson, B.A.; Hashimoto, S.; Sahu, N.; Avtar, R.; Saito, O.; Chakraborty, S.; et al. Scenario-Based Hydrological Modeling for Designing Climate-Resilient Coastal Water Resource Management Measures: Lessons from Brahmani River, Odisha, Eastern India. *Sustainability* **2021**, *13*, 6339. [CrossRef]

 *sustainability*

Article

Spatio-Temporal Evolution and Driving Factors of Ecosystem Service Value of Urban Agglomeration in Central Yunnan

Lei Yang and Fenglian Liu *

Institute of Land & Resources and Sustainable Development, Yunnan University of Finance and Economics, Kunming 650221, China
* Correspondence: zz2105@ynufe.edu.cn; Tel.: +86-137-2010-8008

Abstract: Urbanization and human activity have recently resulted in land use/cover change (LUCC), which has had a detrimental effect on the biological environment, on keeping the ecosystem's sustainable growth and on comprehending the ecosystem's quality and changes over the past 20 years in the central Yunnan urban agglomeration. The equivalent factor method and hotspot analysis were used to analyze the spatio-temporal changes in land use and ecosystem service value (ESV) in the urban agglomerations of central Yunnan province, and the effects of land use change on ESV were then examined. This study is based on the grid data of land use in 2000, 2005, 2010, 2015, and 2020. Finally, Geodetector was used to investigate the possible causes of ESV. The results showed that: (1) The urban agglomerations in central Yunnan's land-use structure and pattern clearly changed between 2000 and 2020, with continual declines in grassland, cultivated land, and woodland, and constant increases in construction land. There was significant growth in both speed and area. (2) The average ESV of the land decreased consistently, the hotspot areas shrank, and the cold-spot areas grew as the ecosystem service function declined and the total amount of ESV decreased by 1.517 billion Yuan. These events were mostly explained by an increase in construction land and a decrease in grassland, cultivated land, and woodland. (3) The synergistic effect of numerous factors is what causes the change in ESV in the urban agglomerations of central Yunnan. The key forces behind ESV change in the research area were land-use intensity, normalized difference vegetation index (NDVI), slope, and people density. The results can help decision makers establish policies for ecological conservation and land use.

Keywords: land-use change; ecosystem service value; hotspot analysis; Geodetector; central Yunnan urban agglomeration

Citation: Yang, L.; Liu, F. Spatio-Temporal Evolution and Driving Factors of Ecosystem Service Value of Urban Agglomeration in Central Yunnan. *Sustainability* **2022**, *14*, 10823. https://doi.org/10.3390/su141710823

Academic Editors: Ronald C. Estoque, Kikuko Shoyama and Rajarshi Dasgupta

Received: 29 June 2022
Accepted: 26 August 2022
Published: 30 August 2022

Publisher's Note: MDPI stays neutral with regard to jurisdictional claims in published maps and institutional affiliations.

1. Introduction

The level of social and economic development worldwide has altered substantially as urbanization has accelerated. Humans continue to seek development from nature during this phase. Numerous issues, including climatic warming, biodiversity loss, and aggravated soil erosion, have been brought on by the growing population, unsustainable land use patterns, and excessive demand for resources [1,2]. As a result, ecosystem functions have been severely compromised, endangering the long-term welfare of people and the sustainable development of ecosystems. In this setting, it is crucial for human survival and growth to control ecosystem quality and change, identify internal root causes of these issues, and implement appropriate solutions [3]. There are numerous indicators to track the dynamic changes in ecosystems, but few can accurately capture the value and expense of ecological deterioration [4–6]. Ecosystem service value may quantitatively evaluate ecosystem service value in monetary units, indicating the ecological consequences caused by diverse activities and effectively connecting ecosystem study with management decisions [7–11]. This concept has increasingly gained popularity in geography, ecology, and other related sciences.

Costanza et al. adopted the research method of economics, introduced "willingness to pay", and conservatively estimated the value of ecosystem services per unit area of different types of ecosystems and the economic value of global ecosystem services. Furthermore, they divided ecosystem services into 17 categories, which laid a theoretical foundation for the follow-up research [12]. The Millennium Ecosystem Assessment issued by the United Nations in 2005 formally established the concept of ecosystem services. They defined ecosystem services as the benefits that people directly or indirectly obtained from ecosystems and divided them into four categories: supply (food production, raw material supply, etc.), regulation (gas regulation, climate regulation, etc.), support (soil conservation, biodiversity maintenance, etc.) and culture (leisure and entertainment, aesthetic landscape, etc.) [13]. Since then, research on ecosystem service value has become more diversified, and research methods and framework systems have become more mature; The research objects include the forests [14,15], farmland [16], wetlands [17,18], grassland [19], river basin [20,21], city [22], island [23], coastal zone [24], etc. The research method is primarily quantitative research based on remote sensing technology [25]. The research content also extends from the detailed research of ecosystem service value to the correlation analysis with land use change, landscape pattern and landscape ecological risk, etc. [26–28] In addition, the existing research has gradually changed from evaluating and analyzing the total ecosystem service value of the research object to studying the value of a specific ecosystem or a single ecosystem service [29–31], from studying ESV itself to the comprehensive application of ESV (such as proposing an ecological compensation mechanism based on the ecosystem service value) [32].

In conclusion, the research on ESV is generally mature and productive; however, there are still two areas that require development. To begin with, the majority of prior research subjects were chosen from normal ecosystems, but there was a dearth of direction for the ecological preservation of administrative units. Especially in urban agglomerations with more complex structures and functions, the contradiction between social and economic development and ecosystem protection is more prominent. Secondly, the ecosystem is highly complex, and its changes resulting from the synergy of multiple factors. However, there are few studies on the changes in ecosystem structure and function of urban agglomerations, and the relationship between rapid urban development and ecosystem changes is still unclear. The potential driving elements of ESV have not been fully examined, especially the research on the influence degree and spatial differentiation of different types of driving factors on ESV is scarce. As a result, using the urban agglomerations in central Yunnan as the research object, this paper first examines the temporal and spatial evolution characteristics of land use and ecosystem service value, then assesses the effects of land use change on ecosystem service value, and finally, investigates the main driving forces of ecosystem service value. The research results can provide quantitative information for decision makers to grasp the ecological status of the urban agglomeration in central Yunnan to optimize the structure of land use and reduce the loss of ESV caused by unreasonable land use patterns. This will help to promote socioeconomic–ecological sustainable development in central Yunnan and urban agglomerations with similar conditions [33].

2. Materials and Methods

2.1. Study Area

The Central Yunnan Urban Agglomeration is one of the 19 state-level urban agglomerations located in the central part of Yunnan Province (101–104.5° E, 24–26.5° N), which is a typical plateau mountainous urban agglomeration with a topography dominated by mountains and intermountain basins (Figure 1). It is also the most economically developed, densely populated, and strongly developed region in Yunnan Province, with the highest concentration of intermountain basins. According to the Central Yunnan Urban Agglomeration Development Plan, which was published in July 2020, the region's scope includes the entire territory of Kunming, Qujing, Yuxi, and Chuxiong, as well as seven counties and cities in the north of Honghe Prefecture, totaling 49 counties (urban areas) with a combined area of 111,400 square kilometers, or 28.3% of Yunnan Province's total area. The urban agglomeration in central Yunnan has grown quickly in recent years, with a GDP of 1507.394 billion Yuan in 2020, 11.45 times that of 2000 (136.105 billion Yuan), and making up about 61.47 percent of the total GDP of Yunnan Province. The pace of urbanization has also accelerated significantly, rising from 30.62 percent in 2000 to 59.90 percent in 2020.

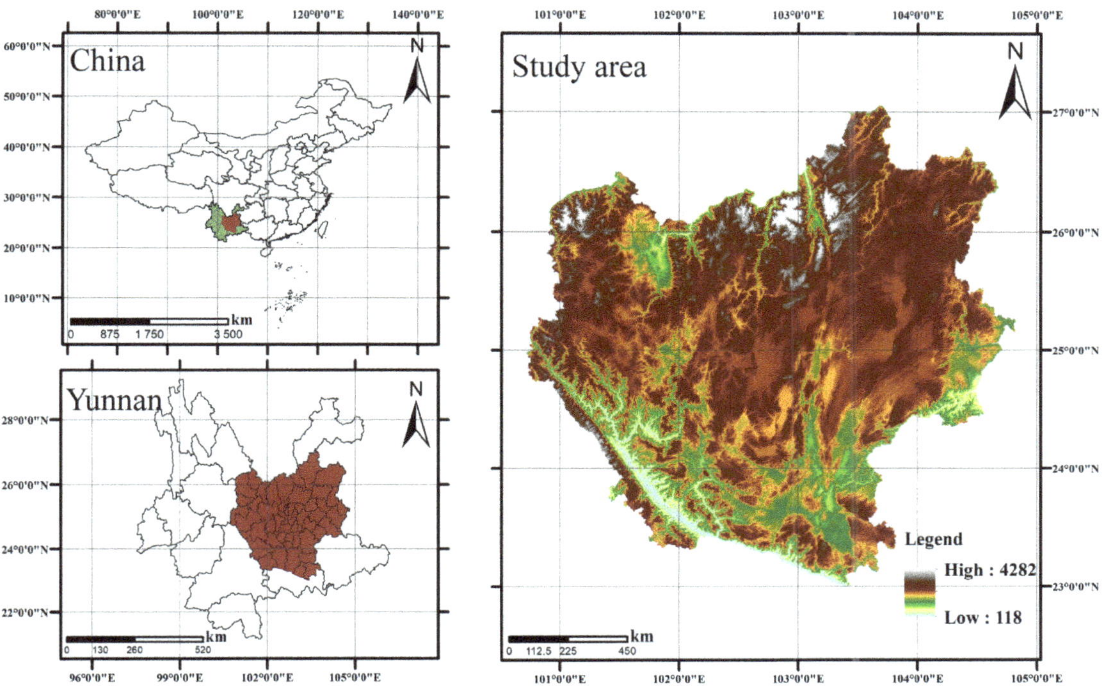

Figure 1. Overview of the study area.

2.2. Data Sources and Methods

2.2.1. Data Source

The five periods of land use/cover remote sensing monitoring data used in this study from 2000 to 2020 were downloaded from the Center of Resources and Environmental Science and Data, Chinese Academy of Sciences (http://www.resdc.cn, accessed on 25 May 2022), among which the land use grid data in 2000, 2005 and 2010 were interpreted by Landsat TM/ETM remote sensing image data, and the land use grid data in 2015 and 2020 were obtained by interpreting Landsat 8 remote sensing image data, with a spatial resolution of 30 m and an interpretation accuracy of over 90%, including 6 primary land types (cultivated land, woodland, grassland, water area, construction land, and unused land) and 25 secondary land types [34]. DEM (digital elevation model) is a branch of digital terrain model, whose purpose is to describe the terrain surface morphology and other ground elevation information in digital form [35]. The DEM images of urban agglomeration in central Yunnan in 2020 used in this study were downloaded from the geospatial data cloud (http://www.gscloud.cn, accessed on 25 May 2022), with a spatial resolution of 30 m, and were derived from the stitching and cropping of multiple remote sensing images. The grain yield and price data used to calculate the value of ecosystem services were obtained from the Yunnan Statistical Yearbook, the National Compilation of Costs and Benefits of Agricultural Products, and official government websites.

Considering the natural conditions and social development level of the urban agglomeration in central Yunnan, and combining it with the related research result [36], nine drivers were selected from the natural factors and human factors, including elevation, slope, average annual temperature, average annual precipitation, NDVI, soil erosion, population density, GDP, and land-use intensity in 2020. Elevation and slope data were extracted from DEM images; temperature, precipitation, NDVI and soil erosion data were downloaded from the Resource and Environment Science and Data Center of the Chinese Academy of Sciences (http://www.resdc.cn, accessed on 25 May 2022); population density and GDP data were obtained from the Yunnan Statistical Yearbook in 2020; and land-use intensity data were calculated by Equation (3).

2.2.2. Land-Use Dynamic Degree

The dynamic degree of single land use is an essential measure of the speed and magnitude of regional land-use change, which can effectively reflect the increase or decrease in the area of various land-use types in the study area in a certain period [37], the expression is:

$$K = \frac{U_b - U_a}{U_a} \times \frac{1}{T} \times 100\% \tag{1}$$

where K is the dynamic degree of a specific land-use type in the T period, which is used to measure the degree to which the change in the land-use type is disturbed by other factors, and the larger the value of K, the more unstable this land-use type is in the T period; U_a and U_b indicate the area of this type of land use at the beginning and end of the study period, respectively.

2.2.3. Land Use Transfer Matrix

The land use transfer matrix is a quantitative research method through systematic analysis, which is mainly used in related research on land-use change. The mathematical expression is [38]:

$$S_{ij} = \begin{bmatrix} S_{11} & S_{12} & \cdots & S_{1n} \\ S_{21} & S_{22} & \cdots & S_{2n} \\ \vdots & \vdots & \vdots & \vdots \\ S_{n1} & S_{n2} & \cdots & S_{nn} \end{bmatrix} \tag{2}$$

where S_{ij} represents the area change in a certain land-use type from the beginning to the end of the research period, and n represents the number of land-use types involved in the study.

2.2.4. Land-Use Intensity Index

The land-use intensity index reflects the degree of land-use change caused by human activities, and regional land development and utilization to a certain extent, and measures the depth and breadth of regional land use. The calculation formula is [39]:

$$L = 100 \times \sum_{i=1}^{n} \frac{A_i P_i}{A_t} \tag{3}$$

where L is the comprehensive land-use intensity index of the study area; n is the number of land-use types, A_i is the area of different land-use types, P represents the land-use intensity parameter, and A_t represents the total area of the study area. According to the previous research results, the unused land, woodland, grassland, water area, arable land, and construction land are divided into four grades and assigned values of 1–4; C_i is the area of the type i land-use type.

2.2.5. Calculation of Ecosystem Services Value

Costanza et al. proposed the equivalent factor method, which makes the assumption that the ecosystem service value per unit area of the same ecosystem type is constant; the total ecosystem service value can be obtained by multiplying this value by the area of each ecosystem type in the study area. Based on this, Chinese academics Xie et al. enhanced this method to determine the ecosystem service value equivalent per unit area acceptable for China [40,41]. Referring to the research results of Xie et al. and The Millennium Ecosystem Assessment [13], this study categorizes ecosystem services into four primary service kinds—supply, support, regulation, and culture—and 11 secondary service types. Combined with the situation of urban agglomeration in central Yunnan, the equivalent scale of ecosystem service value per unit area of ecosystem in the central Yunnan urban agglomeration is shown in Table 1, and the construction land equivalent coefficient is referred to in the study of Deng Shuhong [42]. The economic value of one ecosystem service value equivalent factor can be defined as 1/7th of the market value of the average grain yield per hectare of farmland in the study area under natural conditions without external interference [43]. The calculation equation is as follows:

$$E = \frac{1}{7} \times \sum_{a=1}^{n} \frac{S_a P_a Q_a}{S} \tag{4}$$

where E is the ecosystem service value of a standard equivalent factor, P_a is the average price of a grain in the study area during the study period, Q_a is the grain yield per unit area of a grain, S_a and S is the sum of the sown area of grain a and the sown area of three-grain crops. The value of ecosystem services for one common equivalent factor in the central Yunnan urban agglomeration was 1149.34 Yuan/hm^2. The formula for calculating the value of ecosystem services is as follows:

$$ESV = \sum_{q=1}^{n} S_q \times VC_q \tag{5}$$

$$ESV_k = \sum_{q=1}^{n} S_q \times VC_{qk} \tag{6}$$

$$AESV = \frac{\sum_{q=1}^{n} S_q \times VC_q}{\sum_{q=1}^{n} S_q} \tag{7}$$

where ESV, ESV_k and $AESV$ represent the total ecosystem service value, the individual ecosystem service value, and the average ecosystem service value, respectively; q is the land type q; S_q is the area of the land type q; VC_q, VC_{qk} represent the ecosystem service value per unit area of land type q and the value coefficient of the kth ecosystem service, respectively [26].

Table 1. Equivalent scale of ecosystem service value per unit area of ecosystem in central Yunnan urban agglomeration.

Primary Classification	Secondary Classification	Cultivated Land	Woodland	Grassland	Water Area	Construction Land	Unused Land
	Food production	0.96	0.26	0.21	0.80	0.00	0.00
Supply services	Raw material production	0.33	0.60	0.31	0.23	−7.51	0.00
	Water supply	−0.55	0.31	0.17	8.29	−2.42	0.00
	Gas regulation	0.77	1.97	1.08	0.77	0.00	0.02
Regulating services	Climate regulation	0.41	5.90	2.86	2.29	−2.46	0.00
	Purify the environment	0.12	1.72	0.95	5.55	0.00	0.10
	Hydrological regulation	0.80	3.79	2.10	102.24	0.02	0.03
	Soil conservation	0.81	2.40	1.32	0.93	0.00	0.02
Support services	Maintain nutrient cycling	0.14	0.18	0.10	0.07	0.34	0.00
	Biodiversity	0.15	2.19	1.20	2.55	0.01	0.02
Cultural services	Aesthetic landscapes	0.07	0.96	0.53	1.89	0.02	0.01

2.2.6. Hotspot Analysis

The hotspot analysis method can judge whether there is a clustering of high and low values of ecosystem service values in the study area and then identify the spatial distribution locations of the hotspot and cold-spot areas of ecosystem service values to reveal the spatial difference of ecosystem service capacity provided by various parts of the study area [44]. The formula is as follows:

$$G_i^* = \frac{\sum_{j=1}^{n} w_{ij}x_j - \overline{X}\sum_{j=1}^{n} w_{ij}}{S\sqrt{\dfrac{\left[n\sum_{j=1}^{n} w_{ij}^2 - \left(\sum_{j=1}^{n} w_{ij}\right)^2\right]}{n-1}}} \tag{8}$$

$$\overline{X} = \frac{1}{n}\sum_{i=1}^{n} x_i \tag{9}$$

$$S = \sqrt{\frac{1}{n}\sum_{i=1}^{n} x_i^2 - (X)^{-2}} \tag{10}$$

where G_i^* index is positive, it is a hotspot area with more ESV increase, and if the G_i^* index is negative, it is a cold-spot area with more ESV loss; n is the number of grids; x_i, x_j is the ecosystem service values of the *i*th and *j*th grids respectively.

2.2.7. Sensitivity Analysis

The sensitivity index is similar to the principle of the elasticity coefficient in economics. It is used to verify the reliability of the ESV results calculated using the ecosystem service value coefficient of different land-use types and then to analyze the impact of different land-use type changes on the value of ecosystem services [45]. The formula is as follows:

$$CS = \left| \frac{(ESV_b - ESV_a)/ESV_a \times 100\%}{\left(VC_{jk} - VC_{ik}\right)/VC_{ik}} \right| \tag{11}$$

where CS is the elasticity coefficient, which indicates the degree of influence of different land-use types on the ecosystem service value; ESV_b and ESV_a indicate the ecosystem service values at the end of the study period and the beginning of the study period; VC_{ik}, VC_{jk} represent the ecosystem service value coefficient of the K land-use type before and after adjustment, respectively.

2.2.8. Analysis of Driving Factors

Wang J.F.'s team first proposed Geodetector. It is a statistical analysis model, which mainly includes four parts: ecology, factor, risk, and interaction detectors. Each detector has different functions. The factor detector mainly indicates the explanatory strength of the independent variable to the dependent variable and the spatial heterogeneity of the dependent variable [46], The calculation method is as follows:

$$q = 1 - \frac{\sum_{h=1}^{L} N_h \sigma_h^2}{N\sigma^2} = 1 - \frac{SSW}{SST} \tag{12}$$

where q denotes the explanatory power of a factor on ESV, and the value is between 0 and 1. The larger the value of q, the stronger the explanatory power of the factor on ESV; h is the partition number of the independent variable; L is the total number of partitions N_h and N is the total number of samples in each partition, and the whole region, respectively; σ_h^2 and σ^2 are the variances of each partition and the variance of ESV in the whole region; SSW and SST are the sum of variance within the stratum and the total variance in the whole region, respectively.

3. Results

3.1. Analysis of Land Use Change

From 2000 to 2020, the most crucial land-use type in the central Yunnan urban agglomeration was woodland (Figure 2), accounting for about half of the total area, followed by grassland and cultivated land. During the study period, the land-use change showed "three increases and three decreases, that is, cultivated land, woodland, and grassland decreased, among which the grassland area decreased the most (a total decrease of 716.3 km^2); the area of construction land, water area, and unused land increased, among which the growth of construction land was more pronounced (1269.01 km^2)". As can be seen from Table 2, the dynamic degree of various land-use types in 2015–2020 was significantly higher than that of the previous three research periods, indicating that the land use in the study area changed drastically during this period. Overall, the land-use dynamics of woodland, cultivated land, and grassland from 2000 to 2020 were −0.03%, −0.10%, and −0.12%, respectively, indicating that the change rates of these three land-use types accelerated sequentially during the study period. The land-use dynamics of construction land, the water area, and unused land were all positive, with the land-use dynamics of construction land having changed relatively significantly, increasing from 1.42% in 2000–2005 to 7.40% in 2015–2020.

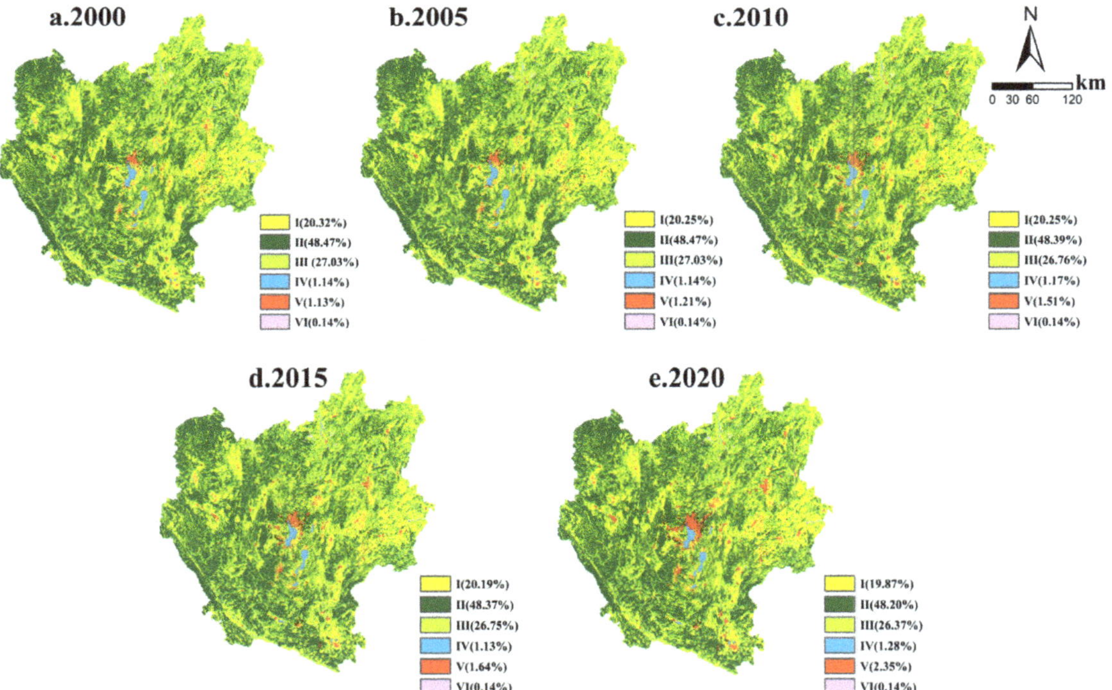

Figure 2. Spatial distribution map of land use. In the legend, I = cultivated land, II = woodland, III = grassland, IV = water area, V = construction land, and VI = unused land.

Table 2. Land use dynamics in central Yunnan urban agglomeration %.

Land Use Type	2000–2005	2005–2010	2010–2015	2015–2020	2000–2020
Cultivated land	−0.07	0.01	−0.06	−0.27	−0.10
Woodland	0.00	−0.03	−0.01	−0.06	−0.03
Grassland	0.00	−0.20	−0.01	−0.27	−0.12
Water area	−0.08	0.57	−0.73	2.91	0.65
Construction land	1.42	4.96	1.72	7.40	4.94
Unused land	0.02	0.74	0.00	−0.09	0.16

The land use transfer matrix can describe the change in land-use quantity and transfer direction from dynamic and static aspects, which is of great significance for analyzing the change in the ecosystem service value caused by land-use changes. Based on the land-use data, Table 3 and Figure 3 were obtained by processing with the raster calculator in ArcGIS to systematically reflect the quantitative characteristics and spatial distribution of land-use changes. Table 3 shows that the main characteristics of land-use transfer in the central Yunnan urban agglomeration are the outflow of grassland, cultivated land, and woodland, and the inflow of construction land. The outflow of grassland goes mainly to woodland (2187.90 km^2), followed by cultivated land and construction land. The cultivated land was mainly converted into woodland (1208.45 km^2), grassland (1195.59 km^2), construction land (936.21 km^2), and water areas (105.05 km^2), which is due to the project of returning farmland to forests, lakes, and grasslands implemented in Yunnan in recent years, as well as the rapid development of the city, resulting in a sharp increase in the demand for construction land. Another reason for the increase in the water area is the comprehensive implementation of the strategy of "strengthening Yunnan with water" and the construction of a series of water conservancy projects such as the "Central Yunnan Water Diversion Project", which has increased the land and water area for water conservancy facilities. Construction land inflow and outflow are quite different, and the primary source is cultivated land (936.21 km^2). From the spatial distribution of land use transfer, the conversion of arable land and woodland to construction land mainly occurred in Anning City, Chenggong District, Guandu District, Qilin District, Mengzi City, Hongta District, and Chuxiong City.

Table 3. Land-use transfer matrix of central Yunnan urban agglomeration from 2000 to 2020 km^2.

Land-Use Type	Grassland	Cultivated Land	Construction Land	Woodland	Water Area	Unused Land
Grassland	-	1393.76	400.04	2187.90	98.44	12.88
Cultivated land	1195.59	-	936.21	1208.45	105.05	6.72
Construction land	23.25	142.28	-	23.67	10.25	1.39
Woodland	2086.96	1325.68	272.82	-	116.23	4.73
Water area	40.13	70.31	31.35	24.36	-	2.37
Unused land	12.30	7.00	0.14	2.55	1.03	-

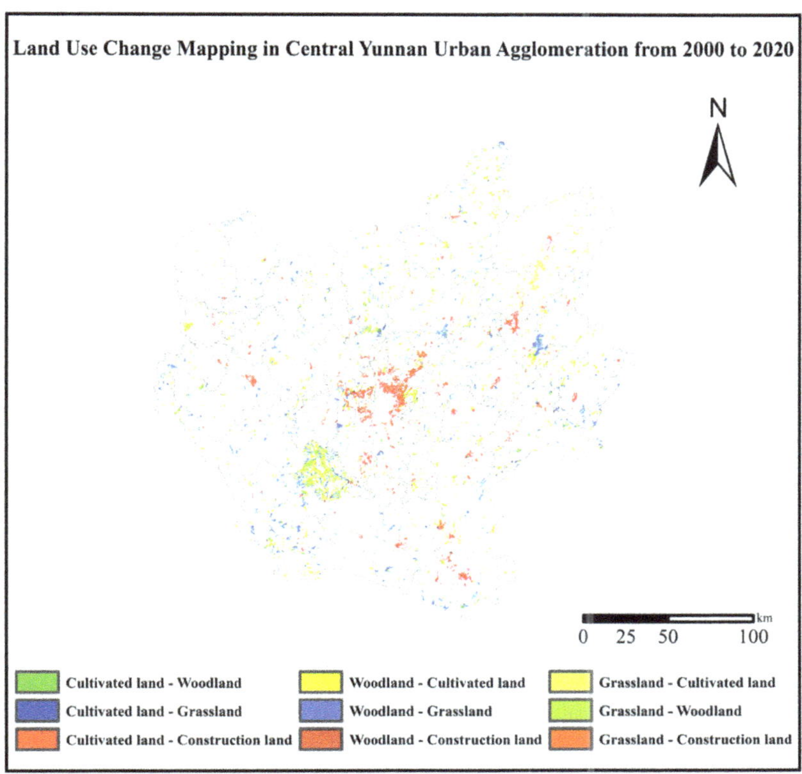

Figure 3. Map of land-use change in central Yunnan urban agglomeration.

3.2. Analysis of Ecosystem Service Value

3.2.1. Time Series Change in Ecosystem Service Value

Over the past 20 years, the central Yunnan urban agglomeration's ecosystem service value (ESV) has decreased by 1.517 billion Yuan (Table 4). During the study period, with the exception of 2015 to 2020, the total ESV continued to decline, especially from 2010 to 2015, when ESV decreased rapidly, and the ecosystem service value decreased by nearly 1×10^9 Yuan. The reason for the decrease in the total amount of ESV lies in the decrease in cultivated land, woodland, and grassland area and the increase in the construction land area, resulting in the loss of ESV of 2.37×10^8 Yuan, 7.74×10^8 Yuan, 9.39×10^8 Yuan, and 19×10^8 Yuan, respectively. From 2015 to 2020, although the ESV of other land types decreased, the water area increased, and its ecosystem service value coefficient was significant, which made the ecosystem service value of the study area rise in 2020.

Table 4. ESV of different types of land from 2000 to 2020 1×10^8 Yuan.

Land Use Type	2000		2005		2010		2015		2020	
	ESV	Proportion	ESV	Proportion	ESV	Proportion	ESV	Proportion	ESV	Proportion
Cultivated land	105.51	5.44%	105.12	5.43%	105.15	5.45%	104.85	5.46%	103.14	5.36%
Woodland	1281.34	66.12%	1281.20	66.20%	1279.28	66.27%	1278.73	66.59%	1273.60	66.24%
Grassland	381.83	19.70%	381.91	19.73%	378.06	19.58%	377.94	19.68%	372.44	19.37%
Water area	186.91	9.65%	186.19	9.62%	191.50	9.92%	184.50	9.61%	210.24	10.93%
Construction land	−17.71	−0.91%	−18.97	−0.98%	−23.67	−1.23%	−25.70	−1.34%	−36.71	−1.91%
Unused land	0.04	0.00%	0.04	0.00%	0.04	0.00%	0.04	0.00%	0.04	0.00%
Total	1937.92	100.00%	1935.48	100.00%	1930.36	100.00%	1920.35	100.00%	1922.75	100.00%

In terms of the value of individual ecosystem services (Table 5), from 2000 to 2020, the regulation service was dominant (67.74%), followed by the support service (22.15%), and their contribution rates to the total amount of ESV were close to 90%. Hydrological regulation and climate regulation were the most prominent types of secondary ecosystem services; followed by soil conservation, maintenance of biodiversity, gas regulation and environmental purification, at 218.58×10^8 Yuan, 186.77×10^8 Yuan, 182.13×10^8 Yuan, 153.10×10^8 Yuan at the end of the study period, respectively. Water supply and nutrient cycle maintenance were relatively weak. From the change in individual ecosystem service value, the ecosystem service value of raw material production and climate regulation decreased by 12.54×10^8 Yuan and 8.44×10^8 Yuan, respectively. During the study period, only the ecosystem service value of hydrological regulation and nutrient cycle maintenance increased.

Table 5. Value of individual ecosystem services from 2000 to 2020 1×10^8 Yuan.

Primary Classification	Secondary Classification	2000	2005	2010	2015	2020	ESV Variation
Supply services (5.77%)	Food production	50.46	50.36	50.30	50.17	49.75	−0.71
	Raw material production	46.95	46.13	43.03	41.70	34.41	−12.54
	Water supply	19.70	19.45	18.76	17.92	17.47	−2.23
Regulating services (67.74%)	Gas regulation	184.13	184.05	183.52	183.35	182.13	−2.00
	Climate regulation	484.17	483.84	481.41	480.64	475.73	−8.44
	Purify the environment	153.62	153.57	153.31	152.93	153.10	−0.52
	Hydrological regulation	486.35	485.67	488.90	483.03	501.63	15.28
Support services (22.15%)	Soil conservation	220.95	220.86	220.21	220.02	218.58	−2.37
	Maintain nutrient cycling	19.47	19.49	19.57	19.61	19.78	0.31
	Biodiversity	188.24	188.21	187.69	187.46	186.77	−1.47
Cultural services (4.34%)	Aesthetic landscapes	83.87	83.86	83.66	83.53	83.39	−0.48

3.2.2. Spatial Change in Ecosystem Service Value

Compared with the total ESV in the study area, the average local ESV can better reflect the regional ecological environment quality by excluding the influence of the administrative area. With the help of the average ESV calculation formula, the average ESV of 49 counties (cities, districts) in the central Yunnan urban agglomeration and its spatial distribution (Figure 4) was obtained, and then the spatial characteristics of ESV distribution were explored. On the whole, the ESV of the central Yunnan urban agglomeration shows the spatial distribution characteristics being "high in the west and low in the east." The low-value areas are concentrated in the eastern region, with an expanding trend; the lower value areas are mainly distributed in the northeast; the higher value areas are distributed in the southwest and northwest of the central Yunnan urban agglomeration; the high-value areas are fewer and mainly concentrated in the central region. Specifically, high-value areas include the Xishan District, Chenggong District in Kunming, Chengjiang County, and Jiangchuan District in Yuxi, which correspond to the spatial distribution of water areas. In 2000, the low-value areas included seven counties in Kunming, Qujing, and Honghe (Wuhua, Qilin, Fuyuan, Luliang county, Luoping, Shizong county, and Luxi county), and in 2020, three counties in Kunming (Anning, Songming, and Guandu) were added. In general, the quality of the ecological environment has shown a downward trend. For example, Guandu District has changed from a high-value area to a low-value area due to the surge in construction land. However, the quality of the ecological environment has also improved in some areas, among which Eshan and Luquan have changed from middle-value areas to higher-value areas, and the value of ecosystem services has increased.

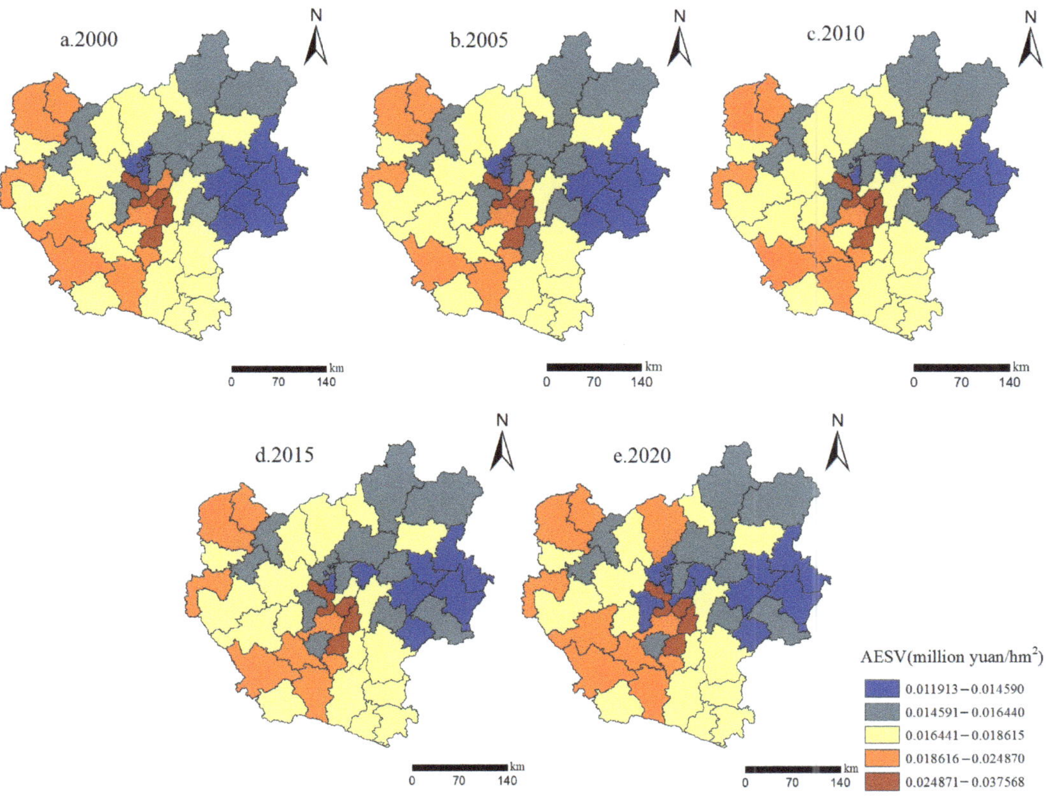

Figure 4. Spatial distribution of land average ESV of central Yunnan urban agglomeration.

To further explore the specific location of the spatial change in ESV, the hotspot analysis (Getis-Ord Gi*) in the ArcGIS spatial statistical tools was used to analyze the cold and hotspots of ESV in the urban agglomeration of central Yunnan in 2000, 2005, 2010, 2015, and 2020. The results are shown in Figure 5: the hotspots of ESV in the central Yunnan urban agglomeration are distributed in the western and northeastern regions, while the cold spots are concentrated in the central region. Specifically, the hotspots in 2000 included Dongchuan District, Xuanwei City, and Shuangbai County. The sub-hotspot area included Chuxiong City, Xinping Yi, Dai Autonomous County, and Yuanjiang Hani, Yi, and Dai Autonomous County. The cold-spot areas include eight counties and districts in Kunming City (Wuhua District, Panlong District, Songming County, Xishan District, Guandu District, Chenggong District, Yiliang County, and Jinning District) and Chengjiang County in Yuxi City. The sub-cold-spot areas involve Fumin County and Anning City in Kunming City and Hongta District, and Jiangchuan District in Yuxi City. By 2020, the number of hotspots decreased (Xuanwei City), and the area of cold spots expanded (Fumin County and Anning City), indicating the degradation of ecosystem service functions in the study area.

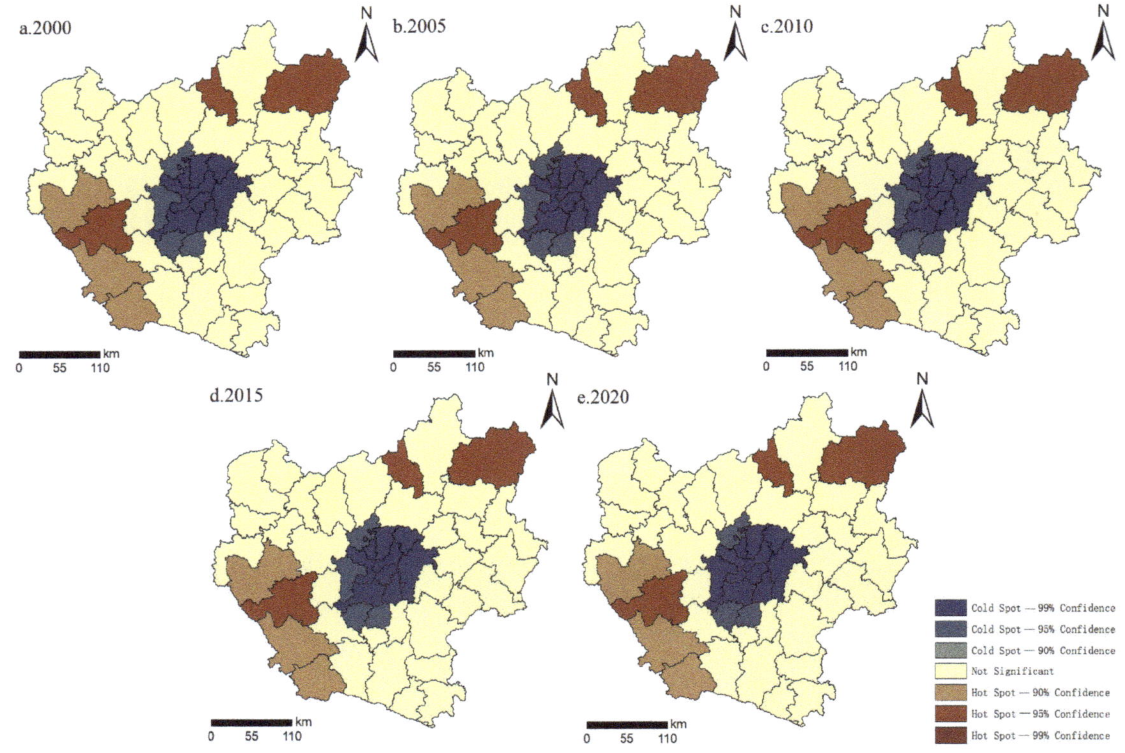

Figure 5. Analysis of ESV hotspots of central Yunnan urban agglomeration.

3.2.3. Sensitivity Analysis of Ecosystem Service Value

The sensitivity indices of different land-use types are different (Table 6). It is found that the sensitivity indices of each land-use type in the central Yunnan urban agglomeration are woodland, grassland, water area, cultivated land, unused land, and construction land in descending order, which is consistent with the contribution of each land-use type to the ESV. The sensitivity index of woodland was the highest, remaining above 0.66 during the study period, followed by grassland, with a sensitivity index close to 0.2. Different from the other land types, the sensitivity index of construction land was negative in different years, indicating that the increase in construction land would lead to a decrease in the ecosystem service value in the study area; the sensitivity index of unused land was shallow, indicating that the change in unused land had no significant effect on the change in ESV. In addition, the sensitivity indices of land-use types in each period were less than 1, indicating that ESV is inelastic to land-use changes. The study uses ecosystem service value coefficients of different land types suitable for the study area, and the ESV calculation results are reliable.

Table 6. Sensitivity coefficients of different land use types.

Sensitivity Index	2000	2005	2010	2015	2020
Cultivated land	0.054447	0.054311	0.054474	0.054601	0.053643
Woodland	0.661194	0.661954	0.662717	0.665880	0.662384
Grassland	0.197030	0.197320	0.195848	0.196808	0.193701
Water area	0.096451	0.096196	0.099203	0.096076	0.109345
Construction land	−0.009140	−0.009799	−0.012261	−0.013384	−0.019092
Unused land	0.000018	0.000018	0.000019	0.000019	0.000019

3.3. The Impact of Land Use Change on Ecosystem Service Value

The ESV profit and loss matrix (Table 7) is calculated based on the land-use transfer matrix combined with the ecosystem service value coefficients of different land-use types, which can reflect the impact of different land-use changes on the quantity change in ESVs. The land use pattern that led to the loss of ESV was the change from a high ESV land-use type to a low ESV land-use type. In particular, the conversion of woodland to cultivated land and grassland, and cultivated land to construction land contributed the most to the decrease in ESV, reducing the ESV by 24.84×10^6 Yuan, 22.65×10^6 Yuan, and 17.20×10^6 Yuan, respectively. The land-use patterns that contribute to the increase in ESV are grassland and cultivated land converted into woodland (23.75×10^6 Yuan, 22.64×10^6 Yuan), and other land types converted into water (43.51×10^6 Yuan).

Table 7. ESV profit and loss matrix of central Yunnan urban agglomeration from 2000 to 2020 10^6 Yuan.

Land Use Type	Grassland	Cultivated Land	Construction Land	Woodland	Water Area	Unused Land
Grassland	0.00	−10.98	−10.50	23.75	12.99	−0.16
Cultivated land	9.42	0.00	−17.20	22.64	14.68	−0.03
Construction land	0.61	2.61	0.00	0.88	1.62	0.02
Woodland	−22.65	−24.84	−10.12	0.00	14.07	−0.11
Water area	−5.29	−9.83	−4.96	−2.95	0.00	−0.34
Unused land	0.15	0.03	0.00	0.06	0.15	0.00

According to previous research results, land-use change has an important impact on ecosystem service value. However, few people have explored the impact of LUCC on ESV. In this study, the Pearson coefficient was used in the Origin2021 software (Origin 2021 edition was developed by OriginLab, USA, and downloaded from https://www.originlab.com/OriginProLearning.aspx, accessed on 25 May 2022) to analyze the relationship between land-use intensity and ESV and its change, and a correlation heat map was drawn (Figure 6) to discuss the impact of land-use intensity on both. The figure shows a significant negative correlation between land-use intensity and the total amount of ESV, and the correlation coefficient was between 0.68 and 0.72. The later the year, the stronger the correlation. There is also a negative correlation between land-use intensity and ESV variation. With the gradual increase in land-use intensity from 2000 to 2020, the influence on the ESV change was also gradually increasing.

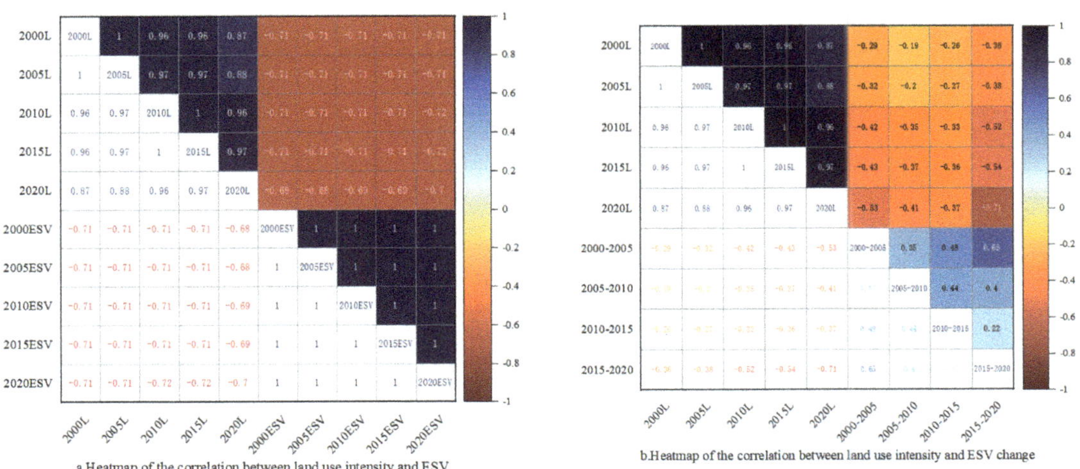

a.Heatmap of the correlation between land use intensity and ESV

b.Heatmap of the correlation between land use intensity and ESV change

Figure 6. Correlation heatmap. The 2000 L indicates the land-use intensity of the central Yunnan urban agglomeration in 2000; 2000 ESV represents the ecosystem service value in 2000; 2000–2005 denotes the amount of ESV change in central Yunnan urban agglomeration between 2000 and 2005.

3.4. Analysis of the Driving Factors of Ecosystem Service Value

The Geodetector can explore how different factors explain the value of ecosystem services, thereby identifying essential drivers of ESV changes. Kriging interpolation was used in the ArcGIS 10.3 software (Version: 10.3.0.4322; Founder: Environmental Systems Research Institute, Inc. American; Website: https://www.esri.com, accessed on 25 May 2022) to obtain a spatial distribution map of the nine drivers (Figure 7). The results of the factor detection are shown in Table 8. The explanatory strength of each driver in descending order is: land-use intensity (0.532), NDVI (0.497), soil erosion (0.314), slope (0.301), population density (0.29), temperature (0.233), GDP (0.214), elevation (0.21), and precipitation (0.134); the q value of each driving factor was more significant than 0.1, which indicates that these nine factors have an important impact on ESV. However, elevation, temperature, precipitation, soil erosion, and GDP failed the significance test of $p < 0.05$, so these five driving factors were excluded. The final result shows that the explanation of land-use intensity reached 53.2%, which is the most crucial driving factor leading to the change in ESV in the central Yunnan urban agglomeration, indicating that different land-use patterns and human disturbance will have a significant impact on ESV. The explanation of the NDVI to ESV is 49.7%, which indicates that the land cover has an important influence on the total amount and change in ESV. The reason for this is that the ESV coefficients of different vegetation types are significantly different, and the ability to provide ESV is also very different. Slope and population density are also important driving factors of ESV in the central Yunnan urban agglomeration. The greater the population density, the more intense the human activities, and the greater the impact on the ecosystem and its functions.

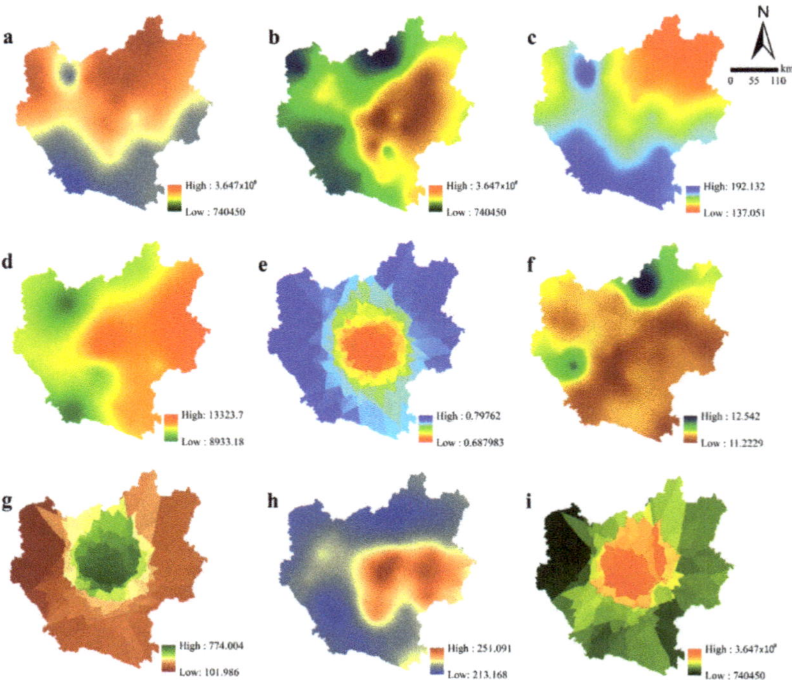

Figure 7. Spatial distribution of the main drivers of ESV. (**a**) Elevation; (**b**) Slope; (**c**) Temperature; (**d**) Precipitation; (**e**) NDVI; (**f**) Soil erosion; (**g**) Population density; (**h**) Land-use intensity; (**i**) GDP.

Table 8. Factor detection results.

Driving Factor	q Statistic	p Value	Degree of Impact
Elevation	0.21	0.099	8
Slope	0.301	0.015	4
Temperatures	0.233	0.063	6
Rainfall	0.134	0.357	9
NDVI	0.497	0.000	2
Soil erosion	0.314	0.072	3
Land-use intensity	0.532	0.000	1
Population density	0.29	0.004	5
GDP	0.214	0.131	7

The results of the interaction detection (Table 9) showed that the interaction between factors explained the dependent variable significantly more strongly than individual factors, involving both two-factor enhancement and non-linear enhancement, indicating that the spatial differentiation of ESV in the urban agglomeration of central Yunnan is the result of the synergistic effect of multiple factors. Specifically, the q value of land-use intensity ∩ NDVI has the largest value (0.785), indicating that the interaction between the two has the greatest explanation for ESV. In addition, the interaction types with q value of > 0.7 include precipitation ∩ temperature (0.743), temperature ∩ slope (0.740), NDVI ∩ temperature (0.711), land-use intensity ∩ soil erosion intensity (0.711), land-use intensity ∩ precipitation (0.705), soil erosion ∩ elevation (0.704), GDP ∩ slope (0.703), and GDP ∩ NDVI (0.702). The interaction among the remaining factors also had an enhancing effect on the spatial differentiation of ESV compared with an individual driver alone, especially the factor with a higher q value coupled and coordinated with other factors would amplify the effect on ESV.

Table 9. Interactive probe results.

Driving Factor	Elevation	Slope	Temperatures	Rainfall	NDVI	Soil Erosion	Land-Use Intensity	Population Density	GDP
Elevation	0.210								
Slope	0.487	0.301							
Temperatures	0.412	0.740	0.233						
Rainfall	0.445	0.581	0.743	0.134					
NDVI	0.681	0.565	0.711	0.685	0.497				
Soil erosion	0.704	0.518	0.697	0.540	0.677	0.314			
Land-use intensity	0.653	0.654	0.676	0.705	0.785	0.711	0.532		
Population density	0.444	0.533	0.504	0.393	0.599	0.560	0.640	0.290	
GDP	0.570	0.703	0.568	0.477	0.702	0.511	0.746	0.436	0.214

4. Discussion

This research employs land-use dynamic degree, transfer matrix, and hotspot analysis to methodically investigate the spatial and temporal dynamics of land-use and ecosystem service values. In order to support the sustainable development of urban agglomerations in terms of economy, ecology, and resources, we use this foundation to investigate the influence of LUCC on ESV by combining the ESV profit and loss matrix and Pearson correlation coefficient and further exploring other potential factors affecting ESV changes.

4.1. Land-Use Change

The results show that the land-use pattern of the central Yunnan urban agglomeration has changed dramatically in the last 20 years; cultivated land, woodland, and grassland has been significantly reduced, and the construction land, water area, and unused land have increased, which is consistent with the previous research results [47]. People destroyed forests and reclaimed land in the 1990s to pursue higher economic benefits, turning woodlands and grasslands into cultivated land. This trend was effectively stopped by the policy of returning farmland to forests and grasslands, which was fully implemented in early 2002 as the nation's economic strength and emphasis on environmental protection increased. The urbanization rate of the urban agglomeration in central Yunnan nearly doubled (from 30.62% to 59.90%) between 2000 and 2020 due to the rapid development of urbanization, and the demand for construction land soared, encroaching on a significant portion of high-quality agriculture. In the future, pertinent departments should concentrate on developing policies and procedures to enhance the protection of cultivated land, ensure its quantity and quality, and increase its productivity.

4.2. Changes in Ecosystem Service Value

In this paper, we used the equivalent factor method of Costanza et al. [12], referred to as the equivalent factor table of China's terrestrial ecosystem in 2015 by Xie G.D. et al. [41], combined with the data of the third land and resources survey in Yunnan Province, to calculate the value of ecosystem services in the central Yunnan urban agglomeration for the past 20 years. In the past, the construction land was not taken into account when calculating the total ESV because much research on the evaluation of ecosystem service value assumed that it served no purpose for ecosystem services [48,49]. However, this study believes that urban development has led to a substantial increase in construction land, which has a growing impact on the development pattern of land space and ecological security. On the one hand, certain tourist destinations provide excellent aesthetic and landscaping purposes (the Eiffel Tower, Forbidden City, Sydney Opera House, etc.). However, the production of raw materials, the availability of water resources, and other ecosystem services are all adversely impacted by construction land. Therefore, the ecosystem service function of construction land should not be neglected in ESV evaluation. Additionally, it is discovered that whether the ESV of construction land is considered or not and the selection of the equivalent coefficient for construction land have a significant effect on the estimation results of ESV. Even the ESV estimated using the same land use data may vary if the equivalent coefficients of the construction land used are quite different. Therefore, to enable cross-regional comparison, future study should establish a set of standard construction land equivalent coefficients.

4.3. Analysis of Driving Factors

There are many methods used to study the driving factors of ESV, such as the canonical correspondence analysis (CCA) model [50], decomposition analysis [29] and grey correlation analysis [51], etc. However, these methods are not thorough enough to analyze the spatial differentiation of factors. Geodetector, on the other hand, can not only quantitatively detect the interaction between different factors but also judge the significant differences in the influences of different driving factors on the research objects. According to the analysis results of driving factors, it is clear that land-use intensity is the most important driving factor for ESV change, and NDVI, slope, and population density also have significant effects on ESV, which has been verified in previous studies [47,52–55]. The type and quantity of data, the method of dispersion, and the carrier employed by the Geodetector may be factors in why elevation, temperature, precipitation, soil erosion, and GDP failed the significance test.

4.4. Innovation, Shortcomings, and Outlook

In order to examine the ESV changes and driving factors in the central Yunnan urban agglomeration, this study chose it as the research object. It also took into account the spatial heterogeneity of ESV and the interaction between the driving variables. The analysis was more thorough than the study on the impact of a single component on ESV, which helped improve the ecosystem service value. However, there are some restrictions. Firstly, the changed dynamic equivalent coefficients are not employed with the static ESV accounting method [56], which may cause a little inaccuracy in the evaluation of ESV. Secondly, the options for the driving factors are too limited, and the multicollinearity test was not performed beforehand, leaving only four components that passed the significance test. Additionally, the disparities between districts (counties) and years for ESV drivers were not examined. The next phase should use a more precise technique of ESV accounting in conjunction with the study region itself, gathering as much information as possible regarding the driving causes. To suggest tailored ecological protection plans, the driving variables of Geodetector should be studied for various administrative districts in various years [57,58].

5. Conclusions

This research reveals the temporal and spatial characteristics and laws of land use and ESV in the central Yunnan urban agglomeration in 2000, 2005, 2010, 2015, and 2020, then investigates the influence of LUCC on ESV, and lastly, examines the main driving factors of ESV changes using Geodetector, which can provide a valuable reference for optimizing regional territorial spatial patterns and ecological protection and restoration. The results show that woodland is the primary land use type in the urban agglomeration of central Yunnan, accounting for about half of the total area; the land use change is manifested as "three increases and three decreases," that is, cultivated land, woodland, and grassland decreased, while water, construction land, and unused land increased. Among them, the construction land increased the most and changed the fastest. From 2000 to 2020, the total ecosystem service value of the urban agglomeration in central Yunnan decreased by 1.517 billion Yuan. Among the ecosystem types, woodland, grassland, and water area provide more ecosystem services. Among the single ecosystem services, the ESV of climate and hydrological regulation was the highest. The land average ESV showed a spatial distribution characteristic of "high in the west and low in the east." The low-value areas were mainly concentrated in the eastern region and tended to increase with time, and the high-value areas were located in the central region, consistent with the spatial distribution of water. The results of hotspot analysis show that hotspots are mainly distributed in the west and northeast of the urban agglomeration in central Yunnan, while cold spots are concentrated in the central region, and the hotspots decrease, while the cold spots increase. Land-use intensity, NDVI, slope, and population density are the main driving factors of ESV changes. The interaction detection results showed that the effects of pairwise interactions between factors on ESV were significantly enhanced compared with single factors.

In recent years, under the influence of the national urban agglomeration policy, the urban agglomeration in central Yunnan has developed rapidly, and the land use structure has changed, significantly impacting the ecosystem. Therefore, preliminary suggestions are put forward according to the above research results. First of all, from 2000 to 2020, the area of construction land in the urban agglomeration of central Yunnan continued to increase, mainly encroaching on other types of land, primarily cultivated land. This trend is not conducive to food security. Therefore, the disorderly expansion of construction land should be strictly controlled and should strictly abide by the cultivated land "occupation–compensation balance." Secondly, woodland and grassland is important ecological land in the urban agglomeration of central Yunnan, with a large area. In the future, ecological restoration methods should be implemented according to local conditions to strengthen the ecosystem services of woodlands and grasslands. In addition, the unit area of the water area ESV was higher, and it is better to strengthen the comprehensive treatment of water pollution, and ecological and environmental protection and restoration. Finally, it is recommended to tap into the potential of land use, and improve the efficiency of land use by reorganizing and rehabilitating urban villages, "hollow villages," and abandoned industrial and mining land, in addition to adhering to both development and protection, and gradually improving and perfecting the system of paid use of resources and an ecological compensation mechanism. The land use data used in this study are only divided into six categories, eliminating wetlands with high ecosystem services, which will have an effect on the outcomes of ESV accounting. Additionally, it is relatively challenging to obtain data for multiple driving factors in a long time series, so this study did not explore the driving factors of ESV in different research periods, which had some limitations. The interpretation of remote sensing images will be used in the future to obtain data on different land use types, and attempts will be made to compile a long-term series of data on driving factors for further research.

Author Contributions: Conceptualization, writing—review and editing, funding acquisition, F.L.; methodology, software, validation, formal analysis, resources, data curation, writing—original draft, L.Y. All authors have read and agreed to the published version of the manuscript.

Funding: Supported by the Scientific Research Fund Project of Yunnan Education Department (Grant no. 2021J0592), and Graduate Student Innovation Fund Project of Yunnan University of Finance and Economics (Grant no. 2022YUFEYC098).

Institutional Review Board Statement: Not applicable.

Informed Consent Statement: Not applicable.

Data Availability Statement: Not applicable.

Conflicts of Interest: The authors declare no conflict of interest.

References

1. Bunker, D.E.; DeClerck, F.; Bradford, J.C.; Colwell, R.K.; Perfecto, I.; Phillips, O.L.; Sankaran, M.; Naeem, S. Species loss and aboveground carbon storage in a tropical forest. *Science* **2005**, *310*, 1029–1031. [CrossRef] [PubMed]
2. Carpenter, S.R.; Mooney, H.A.; Agard, J.; Capistrano, D.; DeFries, R.S.; Díaz, S.; Dietz, T.; Duraiappah, A.K.; Oteng-Yeboah, A.; Pereira, H.M. Science for managing ecosystem services: Beyond the Millennium Ecosystem Assessment. *Proc. Natl. Acad. Sci. USA* **2009**, *106*, 1305–1312. [CrossRef] [PubMed]
3. Costanza, R.; de Groot, R.; Sutton, P.; van der Ploeg, S.; Anderson, S.J.; Kubiszewski, I.; Farber, S.; Turner, R.K. Changes in the global value of ecosystem services. *Glob. Environ. Change* **2014**, *26*, 152–158. [CrossRef]
4. De Groot, R.; Brander, L.; van der Ploeg, S.; Costanza, R.; Bernard, F.; Braat, L.; Christie, M.; Crossman, N.; Ghermandi, A.; Hein, L.; et al. Global estimates of the value of ecosystems and their services in monetary units. *Ecosyst. Serv.* **2012**, *1*, 50–61. [CrossRef]
5. Blignaut, J.; Moolman, C. Quantifying the potential of restored natural capital to alleviate poverty and help conserve nature: A case study from South Africa. *J. Nat. Conserv.* **2006**, *14*, 237–248. [CrossRef]
6. Economics of Ecosystems and Biodiversity (TEEB). *Mainstreaming the Economics of Nature: A Synthesis of the Approach, Conclusions and Recommendations of TEEB*; Progress Press: Valletta, Malta, 2010.
7. Payne, C.; Sand, P. *Environmental Liability: Gulf War Reparations and the UN Compensation Commission*; Oxford University Press: London, UK; New York, NY, USA, 2011.
8. Farley, J. The role of prices in conserving critical natural capital. *Conserv. Biol.* **2008**, *22*, 1399–1408. [CrossRef] [PubMed]
9. Farley, J.; Costanza, R. Payments for ecosystem services: From local to global. *Ecol. Econ.* **2010**, *69*, 2060–2068. [CrossRef]
10. Leimona, B. *Fairly Efficient or Efficiently Fair: Success Factors and Constraints of Payment and Reward Schemes for Environmental Services in Asia*; Wageningen University and Research: Wageningen, Poland, 2011.
11. Crossman, N.D.; Bryan, B.A. Identifying cost-effective hotspots for restoring natural capital and enhancing landscape multifunctionality. *Ecol. Econ.* **2009**, *68*, 654–668. [CrossRef]
12. Costanza, R.; D'Arge, R.; De Groot, R.; Farber, S.; Grasso, M.; Hannon, B.; Limburg, K.; Naeem, S.; O'neill, R.V.; Paruelo, J. The value of the world's ecosystem services and natural capital. *Nature* **1997**, *387*, 253–260. [CrossRef]
13. Reid, W.V. *Millennium Ecosystem Assessment. Ecosystems and Human Well-Being: Synthesis*; Island Press: Washington, DC, USA, 2005.
14. Kibria, A.S.M.G.; Behie, A.; Costanza, R.; Groves, C.; Farrell, T. The value of ecosystem services obtained from the protected forest of Cambodia: The case of Veun Sai-Siem Pang National Park. *Ecosyst. Serv.* **2017**, *26*, 27–36. [CrossRef]
15. Ninan, K.N.; Kontoleon, A. Valuing Forest ecosystem services and disservices–Case study of a protected area in India. *Ecosyst. Serv.* **2016**, *20*, 1–14. [CrossRef]
16. Liao, J.; Yu, C.; Feng, Z.; Zhao, H.; Wu, K.; Ma, X. Spatial differentiation characteristics and driving factors of agricultural eco-efficiency in Chinese provinces from the perspective of ecosystem services. *J. Clean. Prod.* **2021**, *288*, 125466. [CrossRef]
17. Kozak, J.; Lant, C.; Shaikh, S.; Wang, G. The geography of ecosystem service value: The case of the Des Plaines and Cache River wetlands, Illinois. *Appl. Geogr.* **2011**, *31*, 303–311. [CrossRef]
18. Walters, D.; Kotze, D.; Rebelo, A.; Pretorius, L.; Job, N.; Lagesse, J.; Riddell, E.; Cowden, C. Validation of a rapid wetland ecosystem services assessment technique using the Delphi method. *Ecol. Indic.* **2021**, *125*, 107511. [CrossRef]
19. Gascoigne, W.R.; Hoag, D.; Koontz, L.; Tangen, B.A.; Shaffer, T.L.; Gleason, R.A. Valuing ecosystem and economic services across land-use scenarios in the Prairie Pothole Region of the Dakotas, USA. *Ecol. Econ.* **2011**, *70*, 1715–1725. [CrossRef]

20. Du, Y.; Li, X.; He, X.; Li, X.; Yang, G.; Li, D.; Xu, W.; Qiao, X.; Li, C.; Sui, L. Multi-Scenario Simulation and Trade-Off Analysis of Ecological Service Value in the Manas River Basin Based on Land Use Optimization in China. *Int. J. Environ. Res. Public Health* **2022**, *19*, 6216. [CrossRef]
21. Ericksen, P.; de Leeuw, J.; Said, M.; Silvestri, S.; Zaibet, L. Mapping ecosystem services in the Ewaso Ng'iro catchment. *Int. J. Biodivers. Sci. Ecosyst. Serv. Manag.* **2012**, *8*, 122–134. [CrossRef]
22. Kreuter, U.P.; Harris, H.G.; Matlock, M.D.; Lacey, R.E. Change in ecosystem service values in the San Antonio area, Texas. *Ecol. Econ.* **2001**, *39*, 333–346. [CrossRef]
23. Lin, T.; Xue, X.; Shi, L.; Gao, L. Urban spatial expansion and its impacts on island ecosystem services and landscape pattern: A case study of the island city of Xiamen, Southeast China. *Ocean. Coast. Manag.* **2013**, *81*, 90–96. [CrossRef]
24. Liu, Y.; Hou, X.; Li, X.; Song, B.; Wang, C. Assessing and predicting changes in ecosystem service values based on land use/cover change in the Bohai Rim coastal zone. *Ecol. Indic.* **2020**, *111*, 106004. [CrossRef]
25. Song, W.; Deng, X.; Yuan, Y.; Wang, Z.; Li, Z. Impacts of land-use change on valued ecosystem service in rapidly urbanized North China Plain. *Ecol. Model.* **2015**, *318*, 245–253. [CrossRef]
26. Long, H.; Liu, Y.; Hou, X.; Li, T.; Li, Y. Effects of land use transitions due to rapid urbanization on ecosystem services: Implications for urban planning in the new developing area of China. *Habitat Int.* **2014**, *44*, 536–544. [CrossRef]
27. Estoque, R.C.; Murayama, Y. Landscape pattern and ecosystem service value changes: Implications for environmental sustainability planning for the rapidly urbanizing summer capital of the Philippines. *Landsc. Urban Plan.* **2013**, *116*, 60–72. [CrossRef]
28. Jia, Y.; Tang, X.; Liu, W. Spatial–Temporal Evolution and Correlation Analysis of Ecosystem Service Value and Landscape Ecological Risk in Wuhu City. *Sustainability* **2020**, *12*, 2803. [CrossRef]
29. Song, F.; Su, F.; Mi, C.; Sun, D. Analysis of driving forces on wetland ecosystem services value change: A case in Northeast China. *Sci. Total Environ.* **2021**, *751*, 141778. [CrossRef]
30. Cheng, F.; Liu, S.; Hou, X.; Wu, X.; Dong, S.; Coxixo, A. The effects of urbanization on ecosystem services for biodiversity conservation in southernmost Yunnan Province, Southwest China. *J. Geogr. Sci.* **2019**, *29*, 1159–1178. [CrossRef]
31. Gaglio, M.; Aschonitis, V.; Castaldelli, G.; Fano, E.A. Land use intensification rather than land cover change affects regulating services in the mountainous Adige river basin (Italy). *Ecosyst. Serv.* **2020**, *45*, 101158. [CrossRef]
32. Gao, X.; Shen, J.; He, W.; Sun, F.; Zhang, Z.; Zhang, X.; Zhang, C.; Kong, Y.; An, M.; Yuan, L. Changes in ecosystem services value and establishment of watershed ecological compensation standards. *Int. J. Environ. Res. Public Health* **2019**, *16*, 2951. [CrossRef] [PubMed]
33. Lafortezza, R.; Sanesi, G.; Chen, J. Large-scale effects of forest management in Mediterranean landscapes of Europe. *Iforest* **2013**, *6*, 342. [CrossRef]
34. Jiyuan, L.; Zengxiang, Z.; Dafang, Z.; Yimou, W.; Wancun, Z. A study on the spatial-temporal dynamic changes of land-use and driving forces analyses of China in the 1990s. *Geogr. Res.* **2003**, *22*, 1–12.
35. Balasubramanian, A. *Digital Elevation Model (DEM) in GIS*; University of Mysore: Mysore, India, 2017.
36. Chen, W.; Chi, G.; Li, J. Ecosystem Services and Their Driving Forces in the Middle Reaches of the Yangtze River Urban Agglomerations, China. *Int. J. Environ. Res. Public Health* **2020**, *17*, 3717. [CrossRef]
37. Su, K.; Wei, D.-Z.; Lin, W.-X. Evaluation of ecosystem services value and its implications for policy making in China—A case study of Fujian province. *Ecol. Indic.* **2020**, *108*, 105752. [CrossRef]
38. Lin, X.; Xu, M.; Cao, C.; Singh, P.R.; Chen, W.; Ju, H. Land-use/land-cover changes and their influence on the ecosystem in Chengdu City, China during the period of 1992–2018. *Sustainability* **2018**, *10*, 3580. [CrossRef]
39. Hu, S.; Chen, L.; Li, L.; Wang, B.; Yuan, L.; Cheng, L.; Yu, Z.; Zhang, T. Spatiotemporal Dynamics of Ecosystem Service Value Determined by Land-Use Changes in the Urbanization of Anhui Province, China. *Int. J. Environ. Res. Public Health* **2019**, *16*, 5104. [CrossRef]
40. Xie, G.; Zhen, L.; Lu, C.; Xiao, Y.; Chen, C. Expert knowledge based valuation method of ecosystem services in China. *J. Nat. Resour.* **2008**, *23*, 911–919.
41. Xie, G.; Zhang, C.; Zhang, L.; Chen, W.; Li, S. Improvement of the Evaluation Method for Ecosystem Service Value Based on Per Unit Area. *J. Nat. Resour.* **2015**, *30*, 1243–1254.
42. Deng, S. *Dynamic Effects on Ecosystem Services Value with Regional Landuse Change*; Zhejiang University: Hangzhou, China, 2012.
43. Xie, G.; Xiao, Y.; Zhen, L.; Lu, C. Study on ecosystem services value of food production in China. *Chin. Acad. Sci.* **2005**, *13*, 10–13.
44. Li, Y.; Zhang, L.; Yan, J.; Wang, P.; Hu, N.; Cheng, W.; Fu, B. Mapping the hotspots and coldspots of ecosystem services in conservation priority setting. *J. Geogr. Sci.* **2017**, *27*, 681–696. [CrossRef]
45. Hasan, S.; Shi, W.; Zhu, X. Impact of land use land cover changes on ecosystem service value—A case study of Guangdong, Hong Kong, and Macao in South China. *PLoS ONE* **2020**, *15*, e0231259. [CrossRef]
46. Wang, J.; Xu, C. Geodetector: Principle and prospective. *Acta Geogr. Sin.* **2017**, *72*, 19.
47. Wang, R.; Bai, Y.; Alatalo, J.M.; Guo, G.; Yang, Z.; Yang, Z.; Yang, W. Impacts of Urbanization at City Cluster Scale on Ecosystem Services Along an Urban-Rural Gradient: A Case Study of Central Yunnan City Cluster, China. *Environ. Sci. Pollut. Res. Int.* **2022**. [CrossRef] [PubMed]
48. Zhang, P.; Geng, W.; Yang, D.; Li, Y.; Zhang, Y.; Qin, M. Spatial and temporal evolution of land use and ecosystem service values in the lower reaches of the Yellow River. *Trans. Chin. Soc. Agric. Eng* **2020**, *36*, 277–288.

49. Li, F.; Zhang, X.; Guo, H. Spatial and temporal variation characteristics of ecosystem service values in the Three Gorges reservoir area in the past 30 years based on land use. *Soil Water Conserv. Res.* **2021**, *28*, 309–318.
50. Hu, M.; Li, Z.; Wang, Y.; Jiao, M.; Li, M.; Xia, B. Spatio-temporal changes in ecosystem service value in response to land-use/cover changes in the Pearl River Delta. *Resour. Conserv. Recycl.* **2019**, *149*, 106–114. [CrossRef]
51. Zhang, D.; Lan, Z.; Wang, Q.; Wang, X.; Zhang, W.; Li, Z. In Study on the mangrove ecosystem services value change in Zhangjiang river estuary based on remote sensing and grey relational analysis-art. *Proc. SPIE* **2007**, *6790*, 79059.
52. Chillo, V.; Vázquez, D.P.; Amoroso, M.M.; Bennett, E.M.; Koricheva, J. Land-use intensity indirectly affects ecosystem services mainly through plant functional identity in a temperate forest. *Funct. Ecol.* **2018**, *32*, 1390–1399. [CrossRef]
53. Eigenbrod, F.; Bell, V.A.; Davies, H.N.; Heinemeyer, A.; Armsworth, P.R.; Gaston, K.J. The impact of projected increases in urbanization on ecosystem services. *Proc. Biol. Sci.* **2011**, *278*, 3201–3208. [CrossRef]
54. Fei, L.; Shuwen, Z.; Jiuchun, Y.; Kun, B.; Qing, W.; Junmei, T.; Liping, C. The effects of population density changes on ecosystem services value: A case study in Western Jilin, China. *Ecol. Indic.* **2016**, *61*, 328–337. [CrossRef]
55. Li, G.; Chen, W.; Zhang, X.; Yang, Z.; Bi, P.; Wang, Z. Ecosystem Service Values in the Dongting Lake Eco-Economic Zone and the Synergistic Impact of Its Driving Factors. *Int. J. Environ. Res. Public Health* **2022**, *19*, 3121. [CrossRef]
56. Yuan, K.; Li, F.; Yang, H.; Wang, Y. The Influence of Land Use Change on Ecosystem Service Value in Shangzhou District. *Int. J. Environ. Res. Public Health* **2019**, *16*, 1321. [CrossRef]
57. Luo, Q.; Zhou, J.; Li, Z.; Yu, B. Spatial differences of ecosystem services and their driving factors: A comparation analysis among three urban agglomerations in China's Yangtze River Economic Belt. *Sci. Total Environ.* **2020**, *725*, 138452. [CrossRef]
58. Lyu, R.; Clarke, K.C.; Zhang, J.; Feng, J.; Jia, X.; Li, J. Spatial correlations among ecosystem services and their socio-ecological driving factors: A case study in the city belt along the Yellow River in Ningxia, China. *Appl. Geogr.* **2019**, *108*, 64–73. [CrossRef]

 *sustainability*

Article

Multi-Scenario Simulation of Ecosystem Service Values in the Guanzhong Plain Urban Agglomeration, China

Shuo Yang and Hao Su *

School of Public Administration, Xi'an University of Architecture and Technology, Xi'an 710055, China; yangshuo@xauat.edu.cn
* Correspondence: suhao@xauat.edu.cn

Abstract: Rapid urbanization and human activities enhanced threats to the degradation of various ecosystem services in modern urban agglomerations. This study explored the response of ecosystem service values (ESVs) to land use changes and the trade-offs among various ESVs in urban agglomerations under different future development scenarios. The patch-general land use simulation (PLUS) model and ESV calculation method were used to simulate the ESVs of Guanzhong Plain Urban Agglomeration under the Business As Usual scenario (BAU), Ecological Conservation scenario (EC), and Economic Development scenario (ED) in 2030. Global and local Moran's I were used to detect the spatial distribution pattern, and correlation analysis was used to measure trade-offs among ecosystem services. The results showed that: (1) The simulated result of land use in Guanzhong Plain Urban Agglomeration showed high accuracy compared to the actual observed result of the same period, with a Kappa coefficient of 0.912. From 2000 to 2030, land use changes were significant, with the rapid decrease in farmland and an increase in construction land. The area of woodland increased significantly under the EC scenario, and the area of construction land increased rapidly under the ED scenario. (2) The decline of total ESV was CNY 218 million from 2000 to 2020, and ESVs remained the downward trend in the BAU and ED scenarios compared to 2020, decreasing by CNY 156 million and CNY 4731 million, respectively. An increasing trend of ESV showed under the EC scenario, with a growth of CNY 849 million. (3) Significant spatial autocorrelation showed in Guanzhong Plain Urban Agglomeration, as the Global Moran's I were all positive and the p-values were zero. The ESV grids mainly showed "High-High" clusters in the mountainous areas and "Low-Low" clusters in plain areas. Except for food production, a majority of ecosystem services exhibited positive synergistic relationships. In future planning and development, policymakers should focus on the coordinated development of the urbanization process and ecological preservation to build an ecological safety pattern.

Keywords: land use change; ecosystem service value; patch-general land use simulation (PLUS) model; Guanzhong Plain Urban Agglomeration

Citation: Yang, S.; Su, H. Multi-Scenario Simulation of Ecosystem Service Values in the Guanzhong Plain Urban Agglomeration, China. *Sustainability* **2022**, *14*, 8812. https://doi.org/10.3390/su14148812

Academic Editors: Kikuko Shoyama, Rajarshi Dasgupta and Ronald C. Estoque

Received: 17 May 2022
Accepted: 17 July 2022
Published: 19 July 2022

Publisher's Note: MDPI stays neutral with regard to jurisdictional claims in published maps and institutional affiliations.

1. Introduction

Ecosystem services are the direct or indirect contributions of ecosystems to human well-being, linking natural ecosystems to society and the economy through ecosystem functions [1]. Land, combining various ecosystems, natural and human factors, is a geographical entity [2]. Urbanization closely links regional land use type changes to changes in the ability to provide ecosystem services [3]. Urban agglomerations are important power sources and growth poles in China's urbanization process [4]. An urban agglomeration is of great significance for regional integration; coordinated development of large, medium, and small cities; and high-quality regional development [5]. Urban agglomerations have become the strategic core of China's national economic development and the main component of new urbanization. In recent years, there has been a gradual increase in the number of ecological studies with the perspective of urban agglomerations. Scholars have studied different urban agglomerations from the perspectives of quantification of ecosystem services, synergistic relationships, accounting for ecosystem service values (ESVs), and comprehensive evaluations [6–9].

The Guanzhong Plain Urban Agglomeration is a national urban agglomeration in northwest China, with an important position in transportation, tourism, and industrial production. In recent years, with the increase in the urbanization rate, ecosystems in the region are facing degradation risk along with the phenomenon of groundwater and haze pollution [10]. Ecosystem issues have gradually become hot topics in this study area. Yang analyzed the spatio-temporal variations of ecological footprints and ESV from 2005 to 2017 in the Guanzhong Plain Urban Agglomeration, and found that the area's consumption demand of natural resources was greater than the natural capital output [11]. Dong used the coupling coordination degree method to evaluate the relations between urbanization degree and ecological environment in the urban agglomeration, and further detected influencing factors of spatial divergence [12]. Peng estimated the future land use of Guanzhong Plain Urban Agglomeration in 2030 using the FLUS model, and simulated the supply and demand of ecosystem services [13]. His research was based on expert experience, which in part influenced judgments about overall ecosystem service budgets. Chen proposed an ecological pattern of the urban agglomeration by combining various ecological indicators and found ecological sources and corridors using the MCR method [5]. Current studies are fundamental for enriching the development patterns of ecosystem services in the Guanzhong Plain Urban Agglomeration. However, future ecological environment patterns had been a largely underexplored domain, especially under high-accuracy simulations.

Whether to pursue the ecosystem service value as ecological use or to pursue GDP as construction land is a point of conflict in land use planning. During the future development of urban agglomerations, different land use demands and development patterns induce changes in various ecosystem services. How to predict the future ecological degradation areas and develop land use planning scientifically has become the focus of research. In the context of natural resource value accounting by the Chinese government, by setting up three different development scenarios for the Guanzhong Plain Urban Agglomeration, it is possible to accurately predict future land use under different development patterns; thus, future urban development boundaries can be determined, and optimal management strategies for sustainable urban development and ecological security can be made. This paper attempts to construct three scenarios to simulate the trend, lower, and upper limits of future ESV changes and to assist in identifying changes in overall ecosystem services in the urban agglomeration.

In view of this, the specific research objectives of this paper are as follows: (1) to simulate the land use patterns of different scenarios in the Guanzhong Plain Urban Agglomeration in 2030, (2) to calculate the ESV changes from 2000 to 2020 and the differences in ESVs for the three scenarios, and (3) to analyze the spatial-temporal distributions and trends of ESVs and their trade-offs. This study provides significant support for securing land use planning under the ecological red line, a reference for effective allocation of land resources, and promotion of natural resource management under ecological protection.

2. Theoretical Background

2.1. Ecological Response to Land Use Changes

As human activities continue to intensify, rapid economic and social development has put increasing pressure on global natural resources and the ecological environment [14]. Human activities have altered the landscape patterns within their sphere of activity, affecting environmental diversity and further altering the geological and biological diversity of the region [15]. Various land use types are demonstrations of geodiversity [16]. The protection of geology is fundamental to the protection of the ecological environment [17]. In the process of rapid urbanization, urban sprawl continues to harm the ecological environment due to urbanization and industrialization, especially in China [18]. Since the reform and opening-up, China's land use pattern has changed dramatically, manifested by changes in the spatial distribution of ecological land and land degradation [19]. Urbanization encompasses multiple forms of transformation, with land use shifting from agricultural to construction land, economic models shifting from primary to secondary and tertiary industries, and residents' behavior patterns changing dramatically [20]. Correspondingly, land use conversion caused structure and function changes in the original ecosystems, concerning sustainable urban development [21]. Land use is the main form of response to the ecosystem service value [22–25]. Changes in land use types affect the structure, processes, and functions of ecosystems, which in turn affect the ecosystem service value [26]. Quantifying and analyzing changes in ecosystem service values (ESVs) is an important tool to raise ecological protection awareness [27,28]. Studying ESV response to land use/cover change (LUCC) has become a very popular research topic [29–32].

2.2. Land Use Simulation

Current land use simulation studies are mainly empirical studies, using GIS research tools combined with scientific theoretical methods to explore the characteristics of land use change and the driving factors [33]. Scholars obtained long-time-series land use data through remote sensing interpretation. Natural and socio-economic driving factors were spatialized by GIS software, and policy factors that cannot be quantified can be used as constraints for land use type conversion. By using different models, land use change rules were established using different driving mechanisms, and simulation results were generated [34]. The simulation results were compared with the actual land use status, and the accuracy was generally calibrated using Kappa coefficients or FoM coefficients. Further, multiple development scenarios were set up to predict land use under different development patterns.

Scholars implemented simulations of land use under different scenarios by adjusting the transfer cost matrices, neighborhood weights, and total future land use projections in the modeling [35,36]. In the different scenario settings of the simulation, scholars differed in the naming of future land use patterns, but their research designs were similar. The most common scenario settings were mainly three types, namely, the Business As Usual scenario (BAU), Ecological Conservation scenario (EC), and Economic Development scenario (ED) [35,37,38]. In some studies, Farmland Protection (FP) scenario was also a more common setting, while some scholars consider the conservation of farmland in the ED scenario [29,39]. Some researchers combined multiple scenarios with the UN's SDG development goals or the IPCC's climate development goals [40,41]. The common features of the multi-scenario simulations were that one scenario represented a non-interventionist development trend, one scenario focused on ecological protection, and one or more scenarios focused on economic development and its synergies [42,43]. Generally, the neighborhood weights of woodland, grassland, and water increased in the EC scenario and the scale of land use increased, while the neighborhood weight of construction land was generally the highest value in the ED scenario. Such scenario settings combined with changes in total land use can distinguish land use variations under multiple scenarios.

In the methodology choice of future land use simulation, most scholars adopted the CA-Markov model, CLUE-S model, FLUS model, etc., for the simulation [44–46]. The patch-general land use simulation (PLUS) model proposed by Liang et al. can better explore the causal factors of various types of land use changes and better simulate the changes at the patch level of multiple types of land use than other models. The PLUS model obtains the transition rules by analyzing the growing patches of each changed land use. A random forest classification algorithm is used to explore the relationships between the growth in each land use type and the multiple driving factors [47,48]. The model has been applied in the latest land use simulation studies [47,49–51].

2.3. Quantification of Ecosystem Service Values

Studies accounting for ecosystem service values can be broadly divided into two categories: methodologies based on unit service function values, or methodologies based on the unit area value equivalent factor [52]. The unit service function value approach, in which the total value is obtained based on the amount of ecosystem service function and the unit price of the functional volume, models the ecosystem service function of a small area by establishing a production equation between a single service function and local ecological variables [53]. The method has more input parameters, the calculation process is more complicated, and the evaluation method and parameter criteria for each service value are difficult to unify [54]. The unit area value equivalent factor approach is more intuitive and easy to use, requires fewer data, and is particularly suitable for the valuation of ecosystem services at regional and global scales [55]. The method combines land use types with different kinds of ecosystem services and generates a scale to assign values to each ecosystem service. In this way, the ESV of an area can be measured.

Costanza et al. first proposed an approach for accounting the ecosystem service value using the unit area value equivalent factor approach in 1997 [1]. Natural resources were viewed as a form of capital, and their capital stock contains flows of materials, energy, and information for ecosystem services. The value of each ecosystem service was estimated, assuming that the ecosystem service supply and demand curve is a vertical line. Xie et al. used the ecosystem service value table per unit area derived from the expert scoring method to account for the value of each ecosystem on the Tibetan Plateau [56]. Based on the method and the practical situation of the study area, scholars have studied the spatial and temporal changes in the ecosystem service values in different study areas, including different countries, provinces, cities, watersheds, and various ecological zones [57–61]. Most studies have been conducted on the response of past–present land use change to ecosystem service values, but studies on future land use and ESV changes are still to be further investigated [62,63].

3. Materials and Methods

3.1. Study Area

The Guanzhong Plain Urban Agglomeration (104°34′—112°34′ E, 33°34′—36°56′ N) is located in the inland area of northwest China, in the core area of the Wei River Basin, the first major tributary of the Yellow River, and the Fen River Basin, and the second major tributary of the Yellow River. The urban agglomeration covers an area of 107,000 km^2 and belongs to the warm temperate continental monsoon climate zone, with a resident population of 36,906,200 in 2020, and the regional GDP is CNY 1.91 trillion. The Guanzhong Plain Urban Agglomeration is an important fulcrum of the Asia–Europe Continental Bridge, and also the second-largest urban agglomeration in western China, connecting the China–Mongolia–Russia International Economic Cooperation Corridor in the north and the Chengdu-Chongqing Urban Agglomeration in the south [13]. The cities include Xi'an, Baoji, Xianyang, Tongchuan, Weinan, Shangluo City in Shaanxi Province; Yuncheng City (except Pinglu County and Quanqu County), and Linfen City (Yaodu District, Houma City, Xianfen County, Huozhou City, Quwo County, and Yicheng County) in Shanxi Province; Tianshui City (Hongdong County, Fushan County), Pingliang City (Kongdong District, Huating County, Jingchuan County, Chongxin County, and Lingtai County), and Qingyang City (Xifeng District) in Gansu Province. The study area is shown in Figure 1.

3.2. Data Sources

The land use data for 2000, 2010, and 2020 used in the paper were obtained from the Chinese Academy of Sciences, with a spatial resolution of 30 m. The data are based on the Landsat series of satellite images, generated through human–computer interaction and manual visual interpretation, with a comprehensive evaluation accuracy of more than 93% (https://www.resdc.cn/, accessed on 15 April 2022). The 30 m resolution elevation data were acquired from ASTER GDEM V2 datasets in the Geospatial Data Cloud (http://www.

gscloud.cn/, accessed on 15 April 2022). Furthermore, details of the used datasets are shown in Table 1. In the PLUS model, all the driving factors were resampled to a spatial resolution of 30 m and reprojected to the coordinate system of WGS_1984_UTM_zone_49N (Figure 2). Multi-year grain production and planted area in Shaanxi, Shanxi, and Gansu provinces were obtained from the statistical yearbooks of each province, and grain prices were obtained from the National Compilation of Costs and Benefits of Agricultural Products.

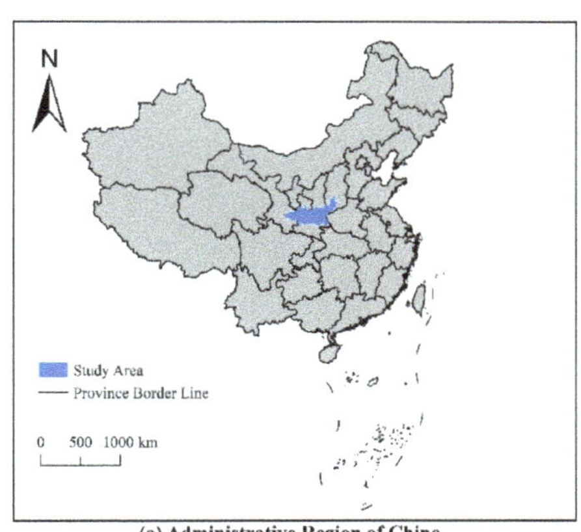

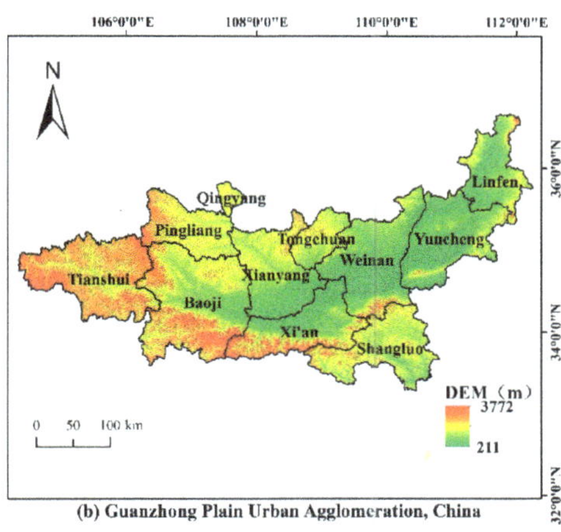

Figure 1. Location map of Guanzhong Plain Urban Agglomeration, China.

Table 1. Source of Datasets.

Data Type	Data Source	Website	Spatial Resolution
Slope	Calculated with DEM from ArcGIS	/	30 m
GDP grid			
Annual average precipitation	RESD	https://www.resdc.cn/, accessed on 15 April 2022	1000 m
Annual average temperature			
Railway			
State road			
Highway	API interface of AMap	https://lbs.amap.com/, accessed on 15 April 2022	Vector
Provincial road			
City main road			
County road			
Urban and rural settlement			
Built-up area	NCSGI	https://www.webmap.cn/, accessed on 15 April 2022	Vector
River			
Population density	WorldPop	http://www.worldpop.org, accessed on 15 April 2022	100 m
NPP-VIIRS nighttime light image	Earth Observation Group	https://eogdata.mines.edu/products/vnl, accessed on 15 April 2022	500 m

3.3. Methodology Flow

Integrating land use with ESV accounting, the methodology flow of this study can be divided into three parts: (1) Analysis of land use change characteristics from 2000 to 2020 in the Guanzhong Plain Urban Agglomeration. The change characteristics of each type of land use were obtained by overlapping analyses and constructing land use transfer matrices. (2) Establishment of multi-scenario simulations of land use in 2030 using PLUS model. The total land use, land transfer rules, and neighborhood weights under the BAU scenario, ED scenario, and EC scenario were set, and we obtained the high-accuracy simulation results under the three scenarios by the PLUS model. (3) Calculation of ESVs, spatial and temporal variations and their synergistic relationships. Based on the constructed regional ecosystem service value scale and the land use data of each year, we accounted for the ESV of each land type and each ecosystem service. Further, we used the Global and Local Moran's I for the spatial distribution characteristics and calculated the changes of correlation coefficients among each ecosystem service.

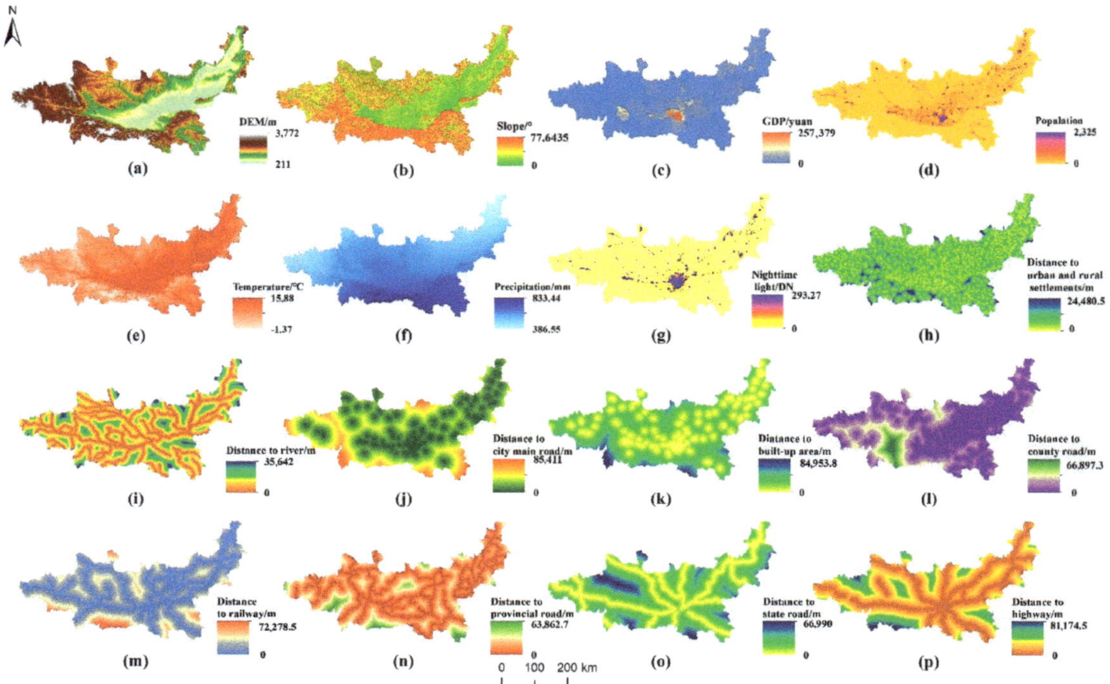

Figure 2. Driving factors of land use simulation in Guanzhong Plain Urban Agglomeration. (**a**) Digital elevation model (DEM), (**b**) slope, (**c**) GDP, (**d**) population density, (**e**) temperature, (**f**) precipitation, (**g**) nighttime light, (**h**) distance to urban and rural settlements, (**i**) distance to river, (**j**) distance to city main road, (**k**) distance to built-up area, (**l**) distance to county road, (**m**) distance to railway, (**n**) distance to provincial road, (**o**) distance to state road and (**p**) distance to highway.

3.4. Future Land Use Simulation

3.4.1. PLUS Model

The PLUS model is a future land use change simulation model that integrates a land expansion strategy analysis module and a metacellular automata model based on multi-class random patch seeds [64]. The rule mining method of the land expansion analysis strategy (LEAS) module extracts the part of each type of land use expansion between two periods of land use change and uses the random forest algorithm to mine the factors of each type of land use expansion and driving factors one by one to obtain the development probability of each type of land use, the driving factors' contribution to each type of land use in that time period, and the contribution of the drivers to the expansion of each type of land use in that time period. The PLUS model is better than the CLUE-S and CA-Markov models in terms of explaining the factors influencing land use change and the accuracy of the simulation results [47].

3.4.2. Multi-Scenario Simulation Settings

After referring to existing studies, consulting with relevant experts, and repeated adjustments, the probability of land use transfer under each scenario was set [50,65–67].

(1)   BAU scenario

The development pattern of land use in the Guanzhong Plain Urban Agglomeration from 2020 to 2030 was assumed to remain unchanged, i.e., the Markov chain model results from 2010 to 2020 were used to estimate the 2030 simulation results generated by the CA based on multi-type random patch seeds (CARS) module.

(2)   ED scenario

The land use in the economic development scenario is mainly referred to as the development plan of the Guanzhong Plain Urban Agglomeration (a guidance document issued by the Chinese government in 2018) and the overall land use plan of each city. "The 14th Five-Year Plan and the outline of the 2035 Vision" of China (official released in 2021) proposed to optimize the spatial layout of new urbanization and vigorously promote the construction of new urbanization with the county as the carrier to ensure food security, the supply of important agricultural products in the process of urban-rural integration, and development. Farmland is an important economic land type, so in the transfer condition matrix, it is set to not transfer to other land types except construction land, and construction land will not transfer to other land types. In the ED scenario in 2030, the transferring probability of farmland, woodland, grassland, and other land to construction land was set to increase by 50%, and the transferring probability of construction land to other land types except farmland was set to reduce by 50%.

(3)   EC scenario

The development plan of Guanzhong Plain Urban Agglomeration emphasizes ecological environmental protection as the task and prerequisite for the construction of the urban agglomeration, optimizing the ecological security pattern, and strengthening ecological protection and restoration. In the settings of land transfer rules, the transfer out of woodland and water was strictly restricted, and construction land was set to be transferable to woodland and grassland due to ecological remediation. In this study, the transferring probability of woodland and grassland to construction land was set to reduce by 50%, and the transferring probability of farmland, grassland, construction land, and other land to woodland was set to increase by 50%. The woodland and water areas were used as a restricted area, and the transfer of this type of land is prohibited. In the simulation of the EC scenario, ecological protection zones and development restriction zones will not be involved in land transfer. A buffer zone with 100 m around water systems was generated to limit the participation of land use transfer in the area [37]. The restricted zone of Guanzhong Plain Urban Agglomeration is shown in Figure 3.

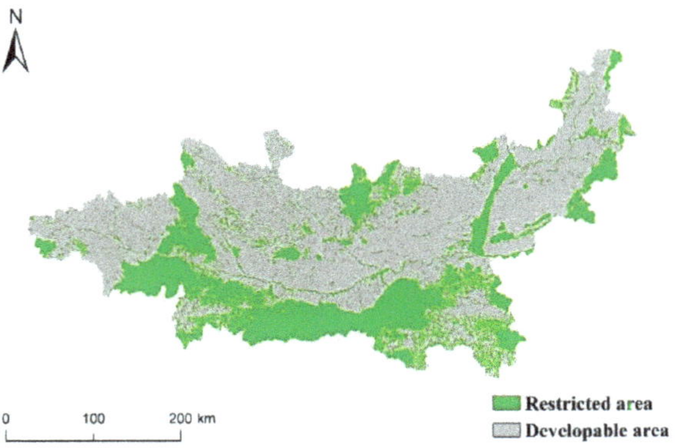

Figure 3. Restricted and developable area of Guanzhong Plain Urban Agglomeration.

3.5. Valuation of Ecosystem Services

The calculation of total ESV in Guanzhong Plain Urban Agglomeration was referred to Equation (1). The ESV per unit area of terrestrial ecosystem table proposed by Xie et al. set the food production function of farmland to 1, which presents the economic value of natural food production per unit area of farmland per year on average nationwide. The main food crops in the Guanzhong Plain Urban Agglomeration are wheat and corn. The value of different ecosystem services in the Guanzhong Plain Urban Agglomeration was corrected according to the value of one standard equivalent equal to 1/7 of the average grain yield market value in the region [56]. A provincial coefficient was used in the weighted average calculation between different provinces regarding Xie's study [68]. We also referred to Li's study to assign values to each coefficient of construction land [69]. The value of ecosystem services in Guanzhong Plain Urban Agglomeration was finally found to be CNY 1057.68/hm^2/a, as shown in Table 2. The exchange rates of CNY to USD and EUR are 6.698:1 and 7.058:1, respectively (obtained on 17 June 2022), same below.

$$ESV = \sum_{f=1}^{m} \sum_{i=1}^{n} (C_f \times E_{fi}) \tag{1}$$

where ESV is the total ecosystem service value (yuan) of the Guanzhong Plain Urban Agglomeration, C_f is the area of the f_{th} land type, E_{fi} is the service function value of the jth land use type, m is the number of land use types, and n is the number of ecosystem service categories.

Table 2. ESV per unit area for different LULC types in the Guanzhong Plain Urban Agglomeration (CNY/hm^2).

Primary Classification	Secondary Classification	Farmland	Woodland	Grassland	Water	Construction Land	Other Land
Provisioning services	Food production	1057.68	105.77	317.30	105.77	10.58	10.58
	Raw material	105.77	2749.97	52.88	10.58	0.00	0.00
	Gas regulation	528.84	3701.88	846.14	0.00	0.00	0.00
Regulating services	Climate regulation	941.34	2855.74	951.91	486.53	0.00	0.00
	Waste treatment	1734.60	1385.56	1385.56	19,228.62	−2601.89	10.58
	Water flow regulation	634.61	3384.58	846.14	21,555.52	−7943.18	31.73
Supporting services	Soil fertility maintenance	1544.21	4124.95	2062.48	10.58	21.15	21.15
	Biodiversity protection	750.95	3448.04	1152.87	2633.62	359.61	359.61
Cultural services	Recreation and culture	10.58	1353.83	42.31	4590.33	10.58	10.58

3.6. Spatial Autocorrelation Analysis

Spatial auto-correlation is an important indicator to test the correlated significance of the attribute value of an ecological index with the attribute value of its adjacent space [70]. The Global Moran's I index reflects the correlation of attribute values of adjacent spatial units [71]. The absolute value of Moran's I is close to 1, indicating a stronger spatial auto-correlation. The Global Moran's I can be calculated as follows:

$$\text{Global Moran's I} = \frac{N \sum_i \sum_{ij} w_{ij} (x_i - \bar{x})(x_j - \bar{x})}{\left(\sum_i \sum_{ij} w_{ij} \right) \sum_i (x_i - \bar{x})^2} \tag{2}$$

where w_{ij}, x_i, x_j, μ, and N indicate the normalized weights, value in the ith pixel, value in the jth pixel, mean value of the study area, and the total number of pixels, respectively. The Moran's I index is approximately +1 for places with complete correlation, while it is approximately −1 for places that are completely non-correlated.

Local Moran's I (LISA) index can effectively reflect the correlation between the ecological environment quality of each grid unit in the study area [29]. The calculation formula is as follows:

$$\text{Local Moran's I} = \frac{(x_i - \bar{x}) \sum_{ij} w_{ij} (x_j - \bar{x})}{\sum_i (x_i - \bar{x})^2} \tag{3}$$

where the calculation parameters are the same as the Moran's I index. LISA cluster map has five types of local spatial aggregation, namely High-High (H-H), Low-Low (L-L), Low-High (L-H), High-Low (H-L), and Not Significant.

4. Results

4.1. Land Use Change Characteristics

Figure 4 shows the land use types of the Guanzhong Plain Urban Agglomeration from 2000 to 2020. According to the land reclassification results, the land types of Guanzhong Plain Urban Agglomeration are mainly farmland, woodland, and grassland, which together account for 95% of the total area. Farmland accounts for about 45% of the total area of the region, woodland accounts for about 22% of the total area, and grassland accounts for about 27%. Between 2000 and 2020, the area of farmland and other land decreased, and the area of woodland, grassland, water, and construction land increased (Figure 5). The land use type transfer matrix is shown in Table 3. The total area of farmland was 49,593.83 km^2 in 2000, dropped to 48,655.17 km^2 in 2010, and reduced to 46,877.72 km^2 in 2020. The decreasing rate of farmland was about 0.27% per year. The most dramatic expansion of the land types is the construction land, rising from 4400.11 km^2 in 2000 to 6173.44km^2 in 2020, with an increase of 1773.33 km^2 in total. The total increasing rate of construction land reached 40.30%, almost 2.02% per year. The area of woodland, grassland, and water increased slightly, with a growth of 271.41 km^2, 628.05 km^2, and 44.04 km^2, respectively.

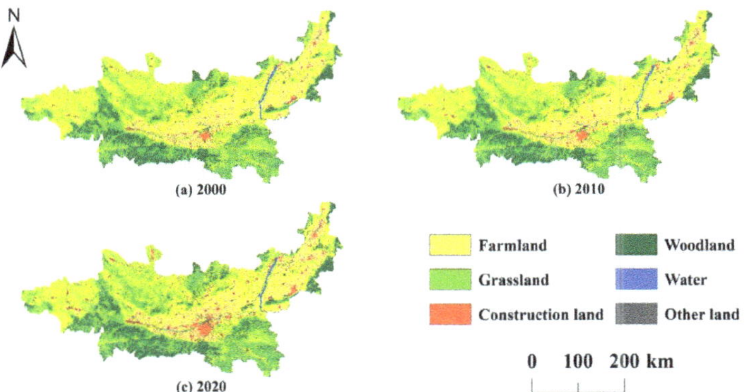

Figure 4. Land use types in Guanzhong Plain Urban Agglomeration in 2000, 2010, and 2020.

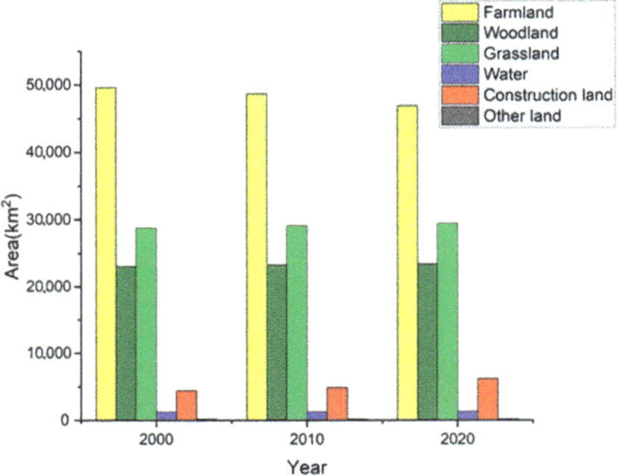

Figure 5. Land use area in Guanzhong Plain Urban Agglomeration in 2000, 2010, and 2020.

Table 3. Transfer matrices of land use type during 2000–2020 (km^2).

Period	Land Use Types	Farmland	Woodland	Grassland	Water	Construction Land	Other Land	Transfer Out
	Farmland	48,256.30	187.27	615.12	109.07	459.14	6.49	1377.09
	Woodland	51.05	22,921.05	69.72	6.40	17.48	1.56	146.22
	Grassland	256.46	129.32	28,361.06	20.46	21.08	1.56	428.88
2000~2010	Water	108.15	7.78	18.86	1113.46	2.58	1.47	138.84
	Construction land	13.51	1.17	3.68	0.55	4383.12	0.04	18.95
	Other land	8.29	4.17	10.66	0.04	0.62	136.67	23.79
	Transfer In	437.46	329.72	718.04	136.52	500.91	11.13	-
	Farmland	45,482.36	260.65	1305.63	122.51	1508.63	13.46	3210.89
	Woodland	144.11	22,738.86	315.51	8.33	34.06	7.89	509.89
	Grassland	889.81	325.14	27,748.15	26.04	73.41	15.36	1329.75
2010~2020	Water	67.10	3.75	31.07	1117.27	28.49	2.09	132.50
	Construction land	326.02	8.59	15.25	6.19	4527.64	0.33	356.38
	Other land	2.88	0.79	0.00	15.98	3.84	120.76	23.49
	Transfer In	1429.91	598.92	1667.46	179.05	1648.43	39.14	-

4.2. Multi-Scenario Land Use Simulation

Referring to existing studies and considering the current situation of the study area, the authors selected 16 natural and social factors such as elevation, slope, GDP, population, average annual precipitation, average annual temperature, distance to railroads, distance to roads (city main road, highway, state road, provincial road, county road), distance to rivers, distance to built-up areas, distance to urban and rural settlements, and nighttime light brightness as the driving factors of land use change in the study area [50,66,72]. Their driving forces were identified using the LEAS module, the number of random forest decision trees was set to 50, the sampling rate was 0.01, and the number of features in the training RF was 16 to obtain the suitability images for the six land use types. The contribution of each driving factor is shown in Figure 6.

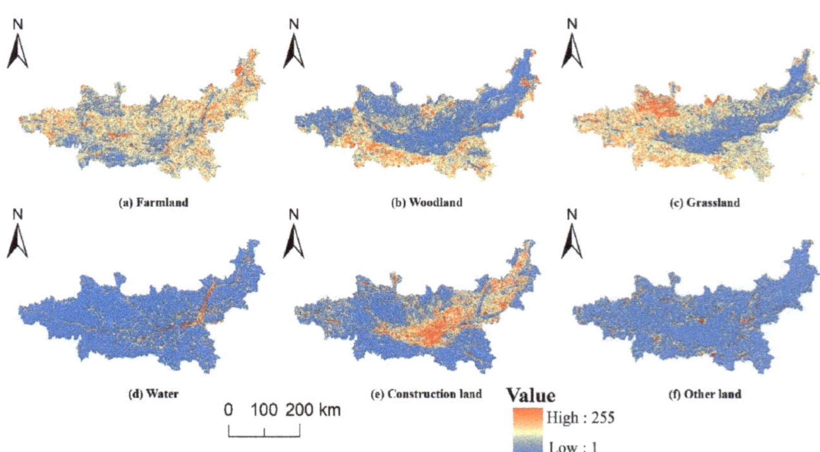

Figure 6. Development potentials of six land use types.

Compared with the actual situation in 2020, the Kappa coefficient of the simulation results was 0.912, greater than 0.8, meeting the simulation accuracy requirements. The three scenarios' conversion cost matrix and neighborhood weight matrix are presented in Tables 4 and 5, respectively. The land use images of the Guanzhong Plain Urban Agglomeration in 2030 under three scenarios were obtained by combining the development zones and restricted zones (Figure 7, Table 6). In the BAU scenario, the area of farmland in 2030 was 4533.58 km^2, which decreased by 1544.14 km^2 compared to 2020, which was the largest area transferred out. The area of woodland and grassland increased slightly by 67.37 km^2

and 296.81 km^2, respectively. The area of water rose to 1333.23 km^2, with an increase of 37.33 km^2, and the area of other land slightly increased from 159.82 km^2 to 169.62 km^2. Construction land was the largest transferred-in land type, compared to 2020, with a total growth of 1146.96 km^2.

Table 4. Conversion cost matrix for simulating different land use scenarios.

Different Scenarios		Farmland	Woodland	Grassland	Water	Construction Land	Other Land
BAU scenario	Farmland	1	1	1	1	1	1
	Woodland	1	1	1	1	1	1
	Grassland	1	1	1	1	1	1
	Water	1	1	1	1	1	1
	Construction land	0	0	0	0	1	0
	Other land	1	1	1	1	1	1
ED scenario	Farmland	1	0	0	0	1	0
	Woodland	1	1	1	1	1	1
	Grassland	1	1	1	1	1	1
	Water	1	1	1	1	1	1
	Construction land	0	0	0	0	1	0
	Other land	1	1	1	1	1	1
EC scenario	Farmland	1	1	1	1	1	1
	Woodland	0	1	0	0	0	0
	Grassland	0	1	1	1	0	0
	Water	0	1	1	1	0	0
	Construction land	0	1	1	1	1	0
	Other land	1	1	1	1	1	1

Table 5. Neighborhood weight for simulating different land use scenarios.

Different Scenarios	Farmland	Woodland	Grassland	Water	Construction Land	Other Land
BAU scenario	0.7	0.5	0.2	0.5	0.9	0.1
ED scenario	0.7	0.5	0.2	0.5	1	0.1
EC scenario	0.5	1	0.2	0.5	0.8	0.1

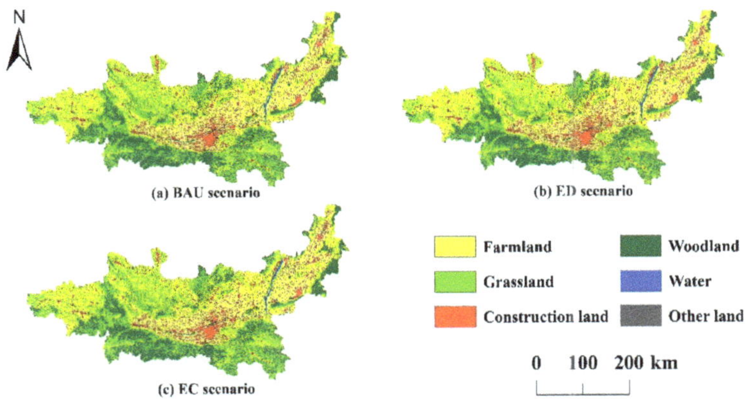

Figure 7. Land use simulation results under three scenarios.

Under the ED scenario, compared to 2020, the area of woodland, grassland, water, and other land decreased. Conversely, farmland and construction land area increased. Construction land, with the largest increase, increased by 1947.85 km^2, followed by 172.69 km^2 in farmland. Under this scenario, the urban agglomeration's construction land area expanded rapidly, with Xi'an as the primary core, and Baoji and Yuncheng as the secondary cores, developing rapidly along the Wei and Fen rivers. In this scenario, the high speed of farmland conversion was slowed down and the balance of farmland occupation was taken into account to ensure the important economic productivity.

Table 6. Simulated areas of three scenarios in Guanzhong Plain Urban Agglomeration (km^2).

Scenarios	Farmland	Woodland	Grassland	Water	Construction Land	Other Land
BAU scenario	45,333.58	23,381.52	29,670.14	1333.23	7320.40	169.62
ED scenario	47,050.41	22,849.78	27,898.48	1156.53	8121.29	132.01
EC scenario	45,233.17	23,694.21	32,243.73	1311.02	7257.18	169.18

In the EC scenario, the trends for each land type remained consistent with the BAU scenario. Farmland was still the main source of construction land, 1644.55 km^2 less than in 2020. The area of woodland, grassland, water, construction land, and unused land all increased to some extent. The area of ecological land types expanded significantly, as the area of woodland and grassland increased by 1.63% and 9.77%, respectively. The enhancement of the water area was rather reduced compared to the BAU scenario, which may be due to the transfer limitation of the buffer zone around the water systems. Compared with the BAU scenario, the growth rate of construction land was not significantly reduced, and the total area only decreased by 63.22 km^2.

4.3. Spatial and Temporal Variation of Ecosystem Service Values

4.3.1. Total ESVs and Variations of Each Ecosystem Service

Based on the land use situation of 2000, 2010, 2020, and three scenarios in 2030, we calculated the total ESVs and ESV of each classification in the Guanzhong Plain urban agglomeration (Tables 7 and 8). From 2000 to 2020, the overall ESV continuously reduced. In 2000, the total ESV was CNY 113.14 billion, and the number continuously dropped to CNY 110.68 billion in 2020. The changing rate from 2000 to 2010, 2010 to 2020, and 2000 to 2020 was −0.48%, −1.70%, and −2.18%, respectively. For different land types, the ESV of woodland, grassland, water, and other land enhanced, while farmland and construction land decreased. The most degradation was the construction land ESV change from 2000 to 2020, which was 40.31% lower than in 2000. The trend of ESV reduction in farmland continued to increase, with a total decline rate of −3.30% between 2000 and 2020, the largest reduction except for construction land. The total increase of water ESV was the highest in the last 20 years, with an increased rate of 3.52%. Among the different scenarios' simulations, ESVs showed different trends in comparison to 2020. In the BAU and ED scenario, the total ESVs decreased by CNY 156 million and CNY 4731 million, respectively, while in the EC scenario, the total ESV increased by CNY 849 million. The growth of construction land area was the main reason for the fluctuation of ESV under each scenario.

In the intercomparison of different ecosystem services, the ESVs of food production, climate regulation, waste treatment, water flow regulation, and soil fertility maintenance showed downward trends, while the ESVs of raw material, gas regulation, biodiversity protection, and recreation and culture increased from 2000 to 2020. The value of soil fertility maintenance was the highest of the nine ecosystem services, and the value of recreation and culture remained the lowest. In the BAU scenario, the ESV of raw material increased slightly, and the other eight ecosystem services all showed declining trends. In the ED scenario, the trends were generally consistent, with all ESVs showing decreasing trends. The EC scenario showed improvement in the situation of various ecosystem services. Raw material, gas regulation, climate regulation, soil fertility maintenance, biodiversity protection, and recreation and culture functions all increased to some extent, indicating the improvement of the ecological environment in the Guanzhong Plain Urban Agglomeration in this scenario.

Table 7. ESVs of different land use types in Guanzhong Plain Urban Agglomeration from 2000 to 2030.

ESV/ CNY 100 million	Year/Scenario	Farmland	Woodland	Grassland	Water	Construction Land	Other Land	Total
Changing rate/%	2000	362.46	532.53	220.12	60.87	−44.63	0.07	1131.42
	2010	355.60	536.77	222.33	60.75	−49.52	0.06	1125.99
	2020	342.61	538.80	224.93	63.01	−62.62	0.07	1106.80
	2030 BAU	331.32	540.35	227.20	64.82	−74.25	0.08	1089.52
	2030 ED	343.87	528.07	213.64	56.23	−82.38	0.06	1059.49
	2030 EC	330.59	547.58	246.91	63.74	−73.61	0.08	1115.29
	2000–2010	−1.89	0.80	1.00	−0.20	−10.96	−14.29	−0.48
	2010–2020	−3.65	0.38	1.17	3.72	−26.45	16.67	−1.70
	2000–2020	−5.48	1.18	2.19	3.52	−40.31	0.00	−2.18
	2020-BAU	−3.30	0.29	1.01	2.87	−18.57	14.29	−1.56
	2020-ED	1.26	−10.73	−11.29	−6.78	−19.76	−0.01	−47.31
	2020-EC	−12.02	8.78	21.98	0.73	−10.99	0.01	8.49

Table 8. ESVs of different ecosystem services in Guanzhong Plain Urban Agglomeration from 2000 to 2030 (CNY 100 million).

Ecosystem Service Type	2000	2010	2020	BAU	Variation	ED	Variation	EC	Variation
Food production	64.19	63.32	61.57	60.06	−1.51	61.24	−0.33	60.80	−0.77
Raw material	70.15	70.57	70.64	70.68	0.04	69.30	−1.34	71.66	1.02
Gas regulation	135.85	136.28	135.95	135.63	−0.32	133.08	−2.87	138.92	2.97
Climate regulation	140.46	140.37	139.30	138.34	−0.96	136.66	−2.64	141.58	2.28
Waste treatment	170.41	168.13	163.17	158.73	−4.44	153.04	−10.13	162.30	−0.87
Water flow regulation	125.82	122.21	112.41	103.61	−8.80	91.23	−21.18	106.80	−5.61
Soil fertility maintenance	231.03	230.94	229.29	227.82	−1.47	224.64	−4.65	234.26	4.97
Biodiversity protection	154.77	155.19	155.14	155.07	−0.07	152.30	−2.84	158.96	3.82
Recreation and culture	38.73	38.98	39.32	39.59	0.27	38.01	−1.31	40.02	0.70

4.3.2. Spatial Distribution and Trade-Offs of Different ESVs in Different Scenarios

To compare the local variation and spatial correlation, a 5 km × 5 km grid was generated by ArcGIS 10.8 to calculate the local ESVs. The spatial and temporal ESV variations of Guanzhong Plain Urban Agglomeration are shown in Figure 8. The high ESV grids were mainly distributed in the southeast, east, and north, and the low ESV grids were mainly distributed in the middle and northeast. The areas with ESVs below CNY 0 and 10 million were mainly urban built-up areas, and a significant expansion of areas with low ESV values can be observed in the three scenarios.

The Global Moran's I was calculated by ArcGIS software, and the index showed an increasing trend year by year during 2000–2020 (Table 9). The Moran's I was 0.6979 in 2000, 0.7049 in 2010, and 0.7057 in 2020. In the three scenarios, the highest Moran's I was 0.7122 in the EC scenario, and the lowest was 0.5755 in the ED scenario. All p-values were zero, indicating that the ESVs of the Guanzhong Plain Urban Agglomeration showed a high degree of spatial autocorrelation. The total ESV value of the Guanzhong Plain Urban Agglomeration showed an obvious "two-pole" pattern, i.e., it was mainly manifested as "High-High" agglomeration and "Low-Low" agglomeration (Figure 9). The "High-High" clusters were mainly concentrated in the Qinling Mountains in the south and in the hilly and ravine areas of the Loess Plateau in the north, while the "Low-Low" clusters were mainly located in the plains and other areas where human activities were frequent.

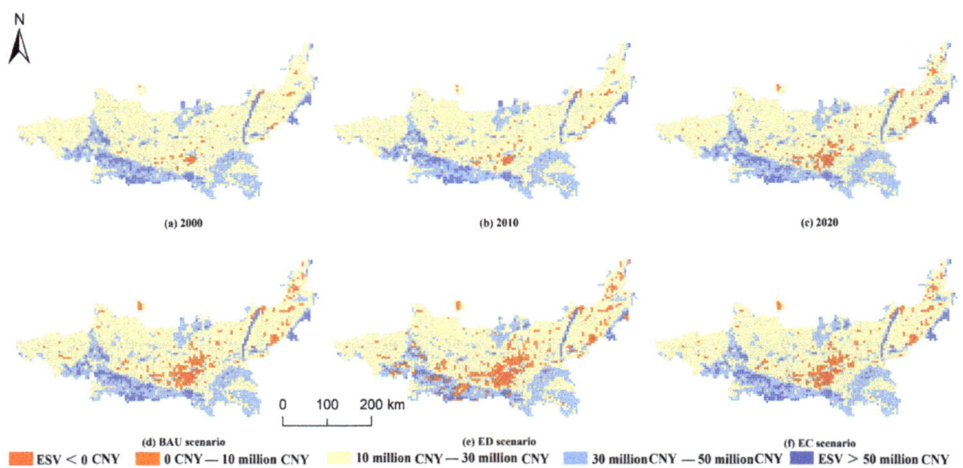

Figure 8. Spatial and temporal ESV variations of Guanzhong Plain Urban Agglomeration.

Table 9. Global Moran's I statistics.

Year	Moran's I	Z-score	*p*-Value
2000	0.6979	83.9632	0.0000
2010	0.7049	84.8010	0.0000
2020	0.7057	84.8883	0.0000
2030BAU	0.7087	85.2570	0.0000
2030ED	0.5755	69.2342	0.0000
2030EC	0.7122	85.6717	0.0000

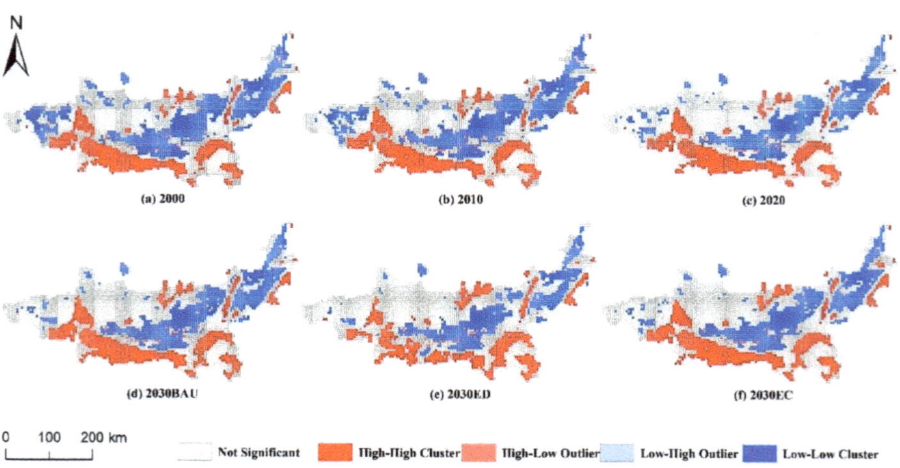

Figure 9. LISA map of three scenarios in Guanzhong Plain Urban Agglomeration.

Further, we calculated correlation coefficients for the ESVs in each secondary class in 2000, 2010, 2020, and three scenarios in 2030 to examine the trade-offs among the ecosystem services. The Pearson coefficient was calculated by SPSS 21 and visualized by MATLAB 2019a (Figure 10). Significant positive correlations showed in some ESV types. Except for FP and WT, all six ecosystem services showed positive synergistic relationships. In the ecosystem service of FP, synergistic features remained consistent across the six periods, with negative correlations with RM, GR, CR, WFR, SFM, BP, and RC. This may be due to the fact that the largest proportion of land type in the Guanzhong Plain Urban Agglomeration was farmland and the reduction in the area of farmland caused a weakening of the food production function. The correlation of the waste treatment function followed the same pattern of changes as the food production function. From 2000 to 2020, the negative correlation between FP and other services showed a gradual reduction. In the three scenarios in 2030, excellent synergistic relationships remained among the various ecosystem services, and in the EC scenario, the observation of the most significant synergies was detected.

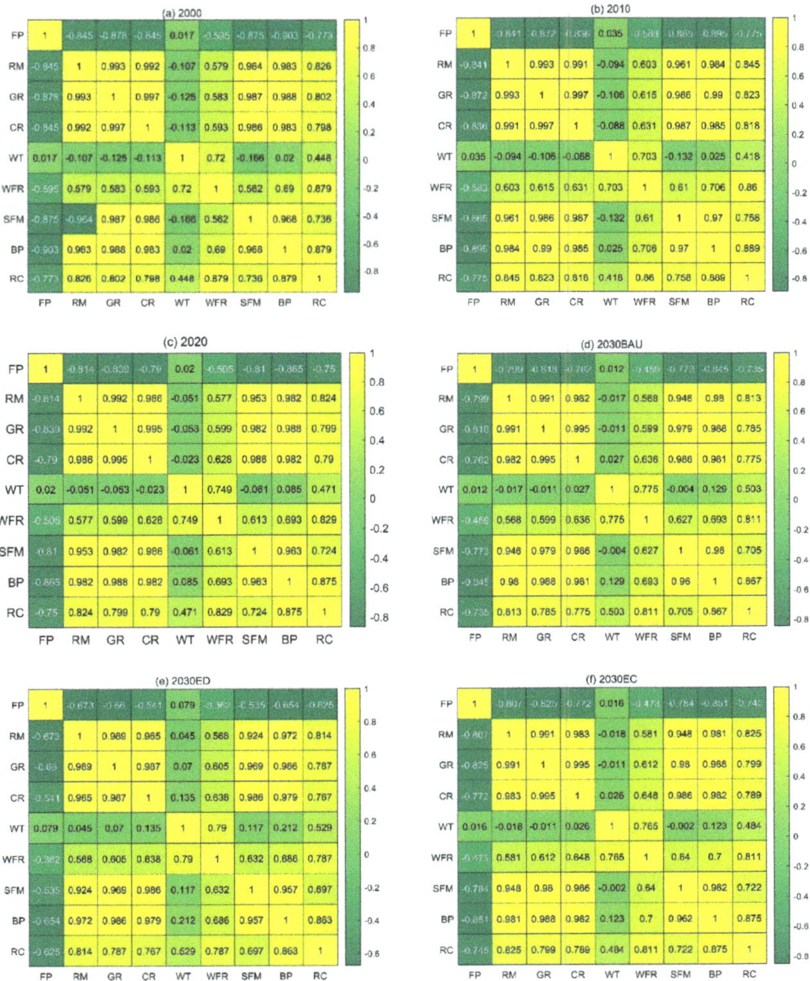

Figure 10. Correlation analysis of ESVs between each ecosystem service in Guanzhong Plain Urban Agglomeration (where FP, RM, GR, CR, WT, WFR, SFM, BP, and RC indicate the food production, raw material, gas regulation, climate regulation, waste treatment, water flow regulation, soil fertility maintenance, biodiversity protection, and recreation and culture, respectively).

5. Discussion

5.1. Feasibility of PLUS Model Application

The accuracy validation of simulation results is the main way to measure the land simulation model [73]. The simulated result of land use in Guanzhong Plain Urban Agglomeration was compared with the actual land use in 2020. The Kappa coefficient of the predicted image was 0.912 compared with the actual land use image. As the results of Landis' research, the Kappa statistic larger than 0.8 was recognized as almost perfect [74]. This indicates that the PLUS model has great advantages in terms of land use simulation accuracy [75]. Wang et al. tested the FLUS model and the PLUS model with land use simulations in western Beijing and found that the accuracy of the PLUS model was higher [65]. Guo et al. compared and analyzed the simulation results of the RNN-CA, the ANN-CA, and the PLUS models, and the PLUS model continued to perform the best [76]. To further verify the results of the PLUS model, we calculated the simulation accuracy for each of the six land use types. The simulation accuracies were 0.941, 0.975, 0.948, 0.862, 0.780, and 0.816 for farmland, woodland, grassland, water, construction land, and other land, respectively. The overall accuracy was 0.940, indicating that the prediction accuracy of the PLUS model is excellent and can be used as a method for future land use simulation.

5.2. Land Use Change Patterns and ESV Trade-Offs

The land use changes in the Guanzhong Plain Urban Agglomeration were the result of a combination of various factors. In general, the trend of land use changes showed that the area of farmland continued to decrease and the area of woodland, grassland, water, construction land, and other land increased. This feature was consistent with findings from similar study areas [66,77,78]. From 2000 to 2020, farmland was the most transferred out land type. The net transferred out area of farmland shifted mainly to construction land and grassland. On the one hand, the policy of returning farmland to woodland and grassland was being gradually implemented, and, on the other hand, the rapid development of the economy and society significantly increased the demand for urban construction land, and a large amount of farmland on the outskirts of cities had been occupied [79]. Under the three scenarios for 2030, land use characteristics continued to be characterized by a rapid turn-out of farmland. The most significant increase in woodland was observed in the EC scenario, while the expansion of construction land reached its greatest level in the ED scenario.

Multiple scenarios of future land use changes would trigger changes in the value of various ecosystem services, which in turn would affect their trade-off relationships [80]. ESV degradation areas are mainly distributed around urban built-up areas, which were the typical characteristics of land use type changes. In the high-altitude region, ESVs under BAU and EC scenarios, on the other hand, showed an increasing trend due to the effect of ecological remediation works. In the BAU scenario and the EC scenario, the ESV differences revealed by the grids were not significant, while the ESV changes in the ED scenario were more significant. It is noteworthy that the ESV in the southwestern mountains exhibited a rapid decline under the ED scenario if the transfer out of woodland and grassland was not constrained. With the expansion of urban and rural construction land, the "two-pole" pattern of ESV distribution in the Guanzhong Plain Urban Agglomeration had become more obvious.

Trade-offs between ecosystem services remain the focus of such studies [81]. Scholars calculated the correlation coefficients between ecosystem services, ecosystem service values in the region, or the spatial distribution patterns based on the characteristics of the study area [82,83]. However, the correlation coefficients between the various types of ESVs showed different characteristics due to the different study scopes (municipal, county, and grid) [29,84,85]. Since the land use change patterns in rapidly urbanizing regions are alike, the correlation pattern of ESVs we found was similar to that of Zhang et al. [86]. The synergistic relationships of the nine types of ecosystem services did not change significantly under the three simulated scenarios. This indicated that the trend of the Guanzhong Plain Urban Agglomeration is characterized by the development of coupled urbanization and ecological protection.

5.3. Policy Implications

By comparing the future land use simulation results of the Guanzhong Plain Urban Agglomeration under the three scenarios, we found that the expansion of urban construction land was rapid. The changes of ecological land such as woodland, grassland, and water varied under different development scenarios, which also caused the great differences of ESVs. Based on our findings, we have made policy suggestions for governments:

Scientifically delineate primary functional zones based on research results. China's current territorial spatial planning zoning is at the district and county levels. Each district and county is set as a key development zone, a major agricultural production zone, or an ecological function zone. In contrast, different districts and counties' natural and social conditions vary greatly and cannot be divided simply by living, production, or ecological space. Some major agricultural production zones are more suitable for new land for construction, while some key development zones contain important ecological source sites. The existing land use control requirements and policy convergence rules are also unclear, making it more challenging to match the spatial governance requirements of different scales. The prediction results of this study can refine future land use planning to the community level and help to allocate land development targets more scientifically. By aggregating the future land demand of each region through bottom-up statistics and preparing a new land use plan, we can ensure a reasonable allocation while meeting the land use laws.

Establish a long-term mechanism for balancing the ESVs from the perspective of urban agglomerations. The effectiveness of the current ecological project implementation is unstable and lacks systematic evaluation. Most projects have focused on the efficacy of single ecological projects, and the evaluation indexes are limited to the expected objectives of the original projects, such as the area of afforestation, the area of soil erosion control, or the number of dam systems projects, etc. Only a few plans have focused on the indicators of ecosystem service functions. The results of our research can provide new ideas for land use planning to plan the area of various land use types based on future ESV changes as a constraint to ensure the overall ESV of urban agglomerations is not degraded. The general land use target of the urban agglomeration is planned within limits using the ESV under the EC scenario as the upper limit and the ESV under the ED scenario as the lower limit.

Further consolidate the effectiveness of ecological restoration projects. There are numerous ecological restoration projects in government departments at this stage, but the project effects are still lacking. Economic forests are generally operated inefficiently and roughly: some plantation forests have a single structure, low ecological stability, and service functions; and some areas are close to the upper limit of regional water carrying capacity for vegetation restoration. The existing ecological funds come from a single source, still dominated by state input, lacking long-term stable input mechanisms and investment channels. Ecological conservation and restoration expenditures are still far below ESV losses. Since ecological optimization has powerful positive externalities, cities should negotiate the establishment of an ecological fund dedicated to ecological remediation. In inner-city space layouts, municipal engineering should increase the number of urban parks, green areas, green belts, artificial water systems, and lakes. In the ravine, the river bank, the slope area, and the idle vacant land, departments should grow green plants. Take traditional villages as units, greening and beautifying beside villages, houses, roadsides, and watersides, forming a combination of ecological restoration of points, lines, and surfaces. Within the mountains, natural restoration is the primary measure, supplemented by artificial restoration, to maximize the recovery of damaged land and destroyed vegetation to a zonal ecological landscape. Technical measures include hanging net guest soil spraying, fish scale pit planting (planting trough), planting bag, vine plant climbing, seedling replanting, etc.

5.4. Establishment of Smart and Sustainable Cities

Sustainability issues have received much attention from the United Nations Millennium Development Goals (MDGs) to the 2030 Sustainable Development Goals (SDGs) [87,88]. The realization of SDGs depends on harmonizing social, ecological, and natural resource elements. China was one of the fastest-growing countries in the SDGs' global score ranking, with a score that increased from 59.1 in 2016 to 73.89 in 2020 and a corresponding rise in ranking from 76th to 48th. Despite China's massive progress under the SDGs, rapidly growing population and resource pressures have strongly impacted ecosystem services in the context of rapid urbanization [89].

"Sustainable Cities and Communities" (SDG 11) is central to the achievement of all seventeen UN SDGs. Sustainable cities are based on the idea of sustainable development as a contemporary paradigm for building the ideal city of the future [90]. The concept was officially conceptualized by UN-Habitat at its second conference in 1996 as cities that are sustainable in three dimensions: environmental, social, and economic; that use resources at a sustainable level; and that are highly resilient to risk. As technology evolves, the understanding of sustainable cities is gradually gaining momentum. Subjects of compactness, sustainable transport, density, mixed land uses, habitat diversity, and greening were considered in the establishment of sustainable cities [91]. With the development of modern information technology, smart cities have gradually begun to attract attention [92]. However, since the concept of smart cities was proposed, due to the different starting points and focus, the connotation of smart cities has not yet formed a unified understanding [93]. Bibri synthesized the related research on smart and sustainable cities and proposed the research dimensions, technologies, and ideas for the future development of smart and sustainable cities [94]. Since then, studies on smart and sustainable cities have gradually emerged.

Our study attempts to propose a vision of smart and sustainable cities from the perspective of ecological security based on the current research results. It offers construction ideas from the ecological environment level to provide a paradigm for the high-quality development of the Guanzhong Plain Urban Agglomeration. The purpose of smart sustainable urban ecosystem management is to build a cyclic system that harmonizes human activities with natural development, to form and maintain a healthy ecosystem, to facilitate the exchange of material and information flows, to build landscapes with regional characteristics, and to enhance the quality of development with spatial control. The government should establish an ecological detection network. The network is based on remote sensing and drone images, combined with 5G transmission technology and AI recognition technology. Based on big data, the network enables real-time monitoring of changes in land cover, natural disasters such as wildfires and floods, and concentrations of PM2.5, PM10, and other information to be passed on to the relevant authorities. Ecological and environmental departments can make predictions on the current status and future trends of ecological quality based on existing high-precision data, making the protection of ecosystem services more scientific. At the same time, the government can incorporate more refined ecological and environment-related indicators into the vision, such as targets for total energy consumption, carbon emissions, forest accumulation, and green space per capita [95]. Each city in the Guanzhong Plain Urban Agglomeration should play a role in constructing smart and sustainable cities based on its characteristics and local conditions.

5.5. Limitations and Future Perspectives

The present study may have the following shortcomings. (1) This study cannot exhaust all future land use patterns and can only use three scenario representations. In the scenario settings of this paper, three scenarios were set up, namely the BAU scenario, ED scenario, and EC scenario. For example, we did not consider the Farmland Protection (FP) scenario in the scenario settings because the spatial distribution of permanently protected farmland is confidential government data that are not available. In this study, the protection of farmland was mainly reflected in the restriction of farmland to land types other than construction land in the ED scenario. (2) The selection of land use driving forces was based on data availability and accessibility, and therefore the driving factors of land use change could not be exhaustive. (3) The PLUS model can only simulate the future land use based on the changing pattern of existing land use types, and future policy factors were not considered. Policy factors are the main contributors to land use change in China. Policy uncertainty presents a variety of possibilities for future land use patterns.

This study provides methodological and data support for predicting the future ESVs in rapid urbanization regions. In our future studies on land use simulation and ESVs, we will improve the accuracy of regional ecosystem service value scales, collect natural and socio-economic data with higher accuracy, and conduct studies on larger-scale ecosystems based on land use simulation methods, focusing on the driving forces of land use transfers on the ecological environment.

6. Conclusions

Cities are spaces with a high concentration of population, resources, wealth, and human socio-economic activities. Due to the highly intensive nature of cities, ecological environment destruction, disordered development of resources, high population density, traffic congestion, declining quality of life, and other problems have become obstacles to the development of cities. Taking the Guanzhong Plain Urban Agglomeration as the study area, this paper analyzed scenario-based land use predictions and ESV responses. Through its spatial and temporal variation characteristics and synergistic patterns, we provided policy suggestions and data support for the coordinated and sustainable development of urban agglomerations.

The Guanzhong Plain Urban Agglomeration is a highly representative inland urban agglomeration among Chinese urban agglomerations. Its urbanization characteristics have promising implications for urban development in central and western China. The first feature is that farmland is the primary source of land transfer. In recent years, the national spatial planning has restricted the transfer-out of woodland, grassland, and water, and the policy of returning farmland to woodland and grassland has been gradually implemented. In addition to permanently protected farmland, a large amount of farmland has been converted to construction land. The massive withdrawal of farmland can trigger a dramatic degradation of ESVs. Such characteristics were also seen in other similar rapidly urbanizing regions. A regional ecosystem that wants to remain stable should follow the occupancy balance of ESV or at least ensure that no drastic changes occur.

The second feature is that in the single-core urban agglomeration, new construction land is mainly distributed around the main urban areas of the core city. In the simulation results of the three scenarios in 2030, the Guanzhong Plain Urban Agglomeration with Xi'an city as the core all exhibit a substantial expansion of the single-core. Due to policy-oriented factors and location conditions, this region is more prone to the urbanization process and accordingly triggers ecological degradation. Building multi-core urban agglomerations has become the key to solving this problem. In the process of new urbanization, on the one hand, the gravitational effect of core cities should be fully utilized to accelerate economic growth; on the other hand, the urban development boundary should be strictly observed in spatial planning and land use indexes should be reasonably allocated.

The third feature is that the spatial distribution pattern of ESVs shows significant spatial heterogeneity. Land use type differences are the most crucial cause of regional ESV changes and thus exhibit distinct spatial characteristics. The woodland is mainly distributed in the mountains with high altitudes and steep terrain, and most of them are located within the ecological protection red line, so the intensity of human exploitation is small. Grassland and farmland are mainly located in flat areas at lower elevations and are vulnerable to human activities. The plain area of the Guanzhong Plain Urban Agglomeration is dominated by farmland, grassland, and construction land, and shows L-L aggregation. The Qinling Mountains in the southern part and the Loess Hills in the northern region are concentrated areas of high ESVs, which shows H-H aggregation in the LISA results. Urban ecological security patterns are not static, which calls for governments to focus on synergistic relationships among ecosystem services. Constraint on the area of construction land and vigorous afforestation can effectively enhance the positive synergistic effect. Maintaining ESV supply and improving the quality of ecosystem services requires ongoing policy attention.

Author Contributions: Conceptualization, S.Y. and H.S.; methodology, S.Y. and H.S.; software, H.S.; validation, S.Y. and H.S.; formal analysis, S.Y. and H.S.; investigation, S.Y. and H.S.; resources, S.Y.; data curation, H.S.; writing—review and editing, S.Y. and H.S.; visualization, H.S.; supervision, S.Y. All authors have read and agreed to the published version of the manuscript.

Funding: This research was funded by the Annual Project of the Social Science Foundation of Shaanxi Province (2020D015), the Key Scientific Research Program of Shaanxi Provincial Education Department (20JT040), and the Soft Science Research Program Project of Xi'an (21RKYJ0053).

Institutional Review Board Statement: Not applicable.

Informed Consent Statement: Not applicable.

Data Availability Statement: Not applicable.

Conflicts of Interest: The authors declare no conflict of interest.

References

1. Costanza, R.; d'Arge, R.; de Groot, R.; Farber, S.; Grasso, M.; Hannon, B.; Limburg, K.; Naeem, S.; Oneill, R.V.; Paruelo, J.; et al. The value of the world's ecosystem services and natural capital. *Nature* **1997**, *387*, 253–260. [CrossRef]
2. Xie, H.L.; Zhang, Y.W.; Choi, Y.; Li, F.Q. A Scientometrics Review on Land Ecosystem Service Research. *Sustainability* **2020**, *12*, 2959. [CrossRef]
3. Cai, W.B.; Jiang, W.; Du, H.Y.; Chen, R.S.; Cai, Y.L. Assessing Ecosystem Services Supply-Demand (Mis)Matches for Differential City Management in the Yangtze River Delta Urban Agglomeration. *Int. J. Environ. Res. Public Health* **2021**, *18*, 8130. [CrossRef]
4. Fang, C. Chinas Urban Agglomeration and Metropolitan Area Construction under the New Development Pattern. *Econ. Geogr.* **2021**, *41*, 1–7.
5. Chen, J.; Wang, S.S.; Zou, Y.T. Construction of an ecological security pattern based on ecosystem sensitivity and the importance of ecological services: A case study of the Guanzhong Plain urban agglomeration, China. *Ecol. Indic.* **2022**, *136*, 108688. [CrossRef]
6. Lan, X.; Tang, H.P.; Liang, H.G. A theoretical framework for researching cultural ecosystem service flows in urban agglomerations. *Ecosyst. Serv.* **2017**, *28*, 95–104. [CrossRef]
7. Shao, M.; Wu, L.F.; Li, F.Z.; Lin, C.S. Spatiotemporal Dynamics of Ecosystem Services and the Driving Factors in Urban Agglomerations: Evidence from 12 National Urban Agglomerations in China. *Front. Ecol. Evol.* **2022**, *10*, 804969. [CrossRef]
8. Ouyang, X.; Tang, L.S.; Wei, X.; Li, Y.H. Spatial interaction between urbanization and ecosystem services in Chinese urban agglomerations. *Land Use Pol.* **2021**, *109*, 105587. [CrossRef]
9. Li, S.N.; He, Y.Y.; Xu, H.L.; Zhu, C.M.; Dong, B.Y.; Lin, Y.; Si, B.; Deng, J.S.; Wang, K. Impacts of Urban Expansion Forms on Ecosystem Services in Urban Agglomerations: A Case Study of Shanghai-Hangzhou Bay Urban Agglomeration. *Remote Sens.* **2021**, *13*, 1908. [CrossRef]
10. Guo, B.; Wang, X.X.; Pei, L.; Su, Y.; Zhang, D.M.; Wang, Y. Identifying the spatiotemporal dynamic of PM2.5 concentrations at multiple scales using geographically and temporally weighted regression model across China during 2015–2018. *Sci. Total Environ.* **2021**, *751*, 141765. [CrossRef]
11. Yang, Y.; Yang, H.; Cheng, Y. Why is it crucial to evaluate the fairness of natural capital consumption in urban agglomerations in terms of ecosystem services and economic contribution? *Sust. Cities Soc.* **2021**, *65*, 102644. [CrossRef]
12. Dong, L.Y.; Shang, J.; Ali, R.; Rehman, R.U. The Coupling Coordinated Relationship Between New-type Urbanization, Eco-Environment and its Driving Mechanism: A Case of Guanzhong, China. *Front. Environ. Sci.* **2021**, *9*, 638891. [CrossRef]
13. Peng, L.X.; Zhang, L.W.; Li, X.P.; Wang, Z.Z.; Wang, H.; Jiao, L. Spatial expansion effects on urban ecosystem services supply-demand mismatching in Guanzhong Plain Urban Agglomeration of China. *J. Geogr. Sci.* **2022**, *32*, 806–828. [CrossRef]
14. Foley, J.A.; De Fries, R.; Asner, G.P.; Barford, C.; Bonan, G.; Carpenter, S.R.; Chapin, F.S.; Coe, M.T.; Daily, G.C.; Gibbs, H.K.; et al. Global consequences of land use. *Science* **2005**, *309*, 570–574. [CrossRef] [PubMed]
15. Lugeri, F.R.; Farabollini, P. Discovering the Landscape by Cycling: A Geo-Touristic Experience through Italian Badlands. *Geosciences* **2018**, *8*, 291. [CrossRef]
16. Ruban, D.A. Quantification of geodiversity and its loss. *Proc. Geol. Assoc.* **2010**, *121*, 326–333. [CrossRef]
17. Brilha, J. Geoconservation and protected areas. *Environ. Conserv.* **2002**, *29*, 273–276. [CrossRef]
18. Monkkonen, P.; Comandon, A.; Escamilla, J.A.M.; Guerra, E. Urban sprawl and the growing geographic scale of segregation in Mexico, 1990–2010. *Habitat Int.* **2018**, *73*, 89–95. [CrossRef]
19. Kong, X.S.; Fu, M.X.; Zhao, X.; Wang, J.; Jiang, P. Ecological effects of land-use change on two sides of the Hu Huanyong Line in China. *Land Use Pol.* **2022**, *113*, 105895. [CrossRef]
20. Zhou, D.Y.; Tian, Y.Y.; Jiang, G.H. Spatio-temporal investigation of the interactive relationship between urbanization and ecosystem services: Case study of the Jingjinji urban agglomeration, China. *Ecol. Indic.* **2018**, *95*, 152–164. [CrossRef]

21. Cai, A.L.; Wang, J.; MacLachlan, I.; Zhu, L.K. Modeling the trade-offs between urban development and ecological process based on landscape multi-functionality and regional ecological networks. *J. Environ. Plan. Manag.* **2020**, *63*, 2357–2379. [CrossRef]

22. Song, W.; Deng, X.Z. Land-use/land-cover change and ecosystem service provision in China. *Sci. Total Environ.* **2017**, *576*, 705–719. [CrossRef] [PubMed]

23. Hao, F.H.; Lai, X.H.; Ouyang, W.; Xu, Y.M.; Wei, X.F.; Song, K.Y. Effects of Land Use Changes on the Ecosystem Service Values of a Reclamation Farm in Northeast China. *Environ. Manag.* **2012**, *50*, 888–899. [CrossRef] [PubMed]

24. Lambin, E.F.; Turner, B.L.; Geist, H.J.; Agbola, S.B.; Angelsen, A.; Bruce, J.W.; Coomes, O.T.; Dirzo, R.; Fischer, G.; Folke, C.; et al. The causes of land-use and land-cover change: Moving beyond the myths. *Glob. Environ. Chang.-Hum. Policy Dimens.* **2001**, *11*, 261–269. [CrossRef]

25. Li, X.M.; Wang, Y.; Song, Y. Unraveling land system vulnerability to rapid urbanization: An indicator-based vulnerability assessment for Wuhan, China. *Environ. Res.* **2022**, *211*, 112981. [CrossRef]

26. Jing, L.; Zhiyuan, R. Variations in Ecosystem Service Value in Response to Land use Changes in the Loess Plateau in Northern Shaanxi Province, China. *Int. J. Environ. Res.* **2011**, *5*, 109–118.

27. Kindu, M.; Schneider, T.; Teketay, D.; Knoke, T. Changes of ecosystem service values in response to land use/land cover dynamics in Munessa-Shashemene landscape of the Ethiopian highlands. *Sci. Total Environ.* **2016**, *547*, 137–147. [CrossRef]

28. Liu, S.A.; Costanza, R.; Troy, A.; D'Aagostino, J.; Mates, W. Valuing New Jersey's Ecosystem Services and Natural Capital: A Spatially Explicit Benefit Transfer Approach. *Environ. Manag.* **2010**, *45*, 1271–1285. [CrossRef]

29. Ji, Z.X.; Wei, H.J.; Xue, D.; Liu, M.X.; Cai, E.X.; Chen, W.Q.; Feng, X.W.; Li, J.W.; Lu, J.; Guo, Y.L. Trade-Off and Projecting Effects of Land Use Change on Ecosystem Services under Different Policies Scenarios: A Case Study in Central China. *Int. J. Environ. Res. Public Health* **2021**, *18*, 3552. [CrossRef]

30. Long, H.L.; Liu, Y.Q.; Hou, X.G.; Li, T.T.; Li, Y.R. Effects of land use transitions due to rapid urbanization on ecosystem services: Implications for urban planning in the new developing area of China. *Habitat Int.* **2014**, *44*, 536–544. [CrossRef]

31. Wu, K.Y.; Ye, X.Y.; Qi, Z.F.; Zhang, H. Impacts of land use/land cover change and socioeconomic development on regional ecosystem services: The case of fast-growing Hangzhou metropolitan area, China. *Cities* **2013**, *31*, 276–284. [CrossRef]

32. Cui, Y.; Lan, H.F.; Zhang, X.S.; He, Y. Confirmatory Analysis of the Effect of Socioeconomic Factors on Ecosystem Service Value Variation Based on the Structural Equation Model-A Case Study in Sichuan Province. *Land* **2022**, *11*, 483. [CrossRef]

33. Aburas, M.M.; Ahamad, M.S.S.; Omar, N.Q. Spatio-temporal simulation and prediction of land-use change using conventional and machine learning models: A review. *Environ. Monit. Assess.* **2019**, *191*, 1–28. [CrossRef] [PubMed]

34. Qiao, Z.; Jiang, Y.Y.; He, T.; Lu, Y.S.; Xv, X.L.; Yang, J. Land use change simulation: Progress, challenges, and prospects. *Acta Ecol. Sin.* **2022**, *42*, 1–12. [CrossRef]

35. Fu, Q.; Hou, Y.; Wang, B.; Bi, X.; Li, B.; Zhang, X.S. Scenario analysis of ecosystem service changes and interactions in a mountain-oasis-desert system: A case study in Altay Prefecture, China. *Sci. Rep.* **2018**, *8*, 1–13. [CrossRef]

36. Zhao, B.; Li, S.; Liu, Z. Multi-Scenario Simulation and Prediction of Regional Habitat Quality Based on a System Dynamic and Patch-Generating Land-Use Simulation Coupling Model: A Case Study of Jilin Province. *Sustainability* **2022**, *14*, 5303. [CrossRef]

37. Shi, M.J.; Wu, H.Q.; Fan, X.; Jia, H.T.; Dong, T.; He, P.X.; Baqa, M.F.; Jiang, P.G. Trade-Offs and Synergies of Multiple Ecosystem Services for Different Land Use Scenarios in the Yili River Valley, China. *Sustainability* **2021**, *13*, 1577. [CrossRef]

38. Domingo, D.; Palka, G.; Hersperger, A.M. Effect of Zoning Plans on Urban Land-Use Change: A Multi-Scenario Simulation for Supporting Sustainable Urban Growth. *Sust. Cities Soc.* **2021**, *69*, 102833. [CrossRef]

39. Gao, X.; Wang, J.; Li, C.X.; Shen, W.N.; Song, Z.Y.; Nie, C.J.; Zhang, X.R. Land use change simulation and spatial analysis of ecosystem service value in Shijiazhuang under multi-scenarios. *Environ. Sci. Pollut. Res.* **2021**, *28*, 31043–31058. [CrossRef]

40. Pan, Z.Z.; He, J.H.; Liu, D.F.; Wang, J.W. Predicting the joint effects of future climate and land use change on ecosystem health in the Middle Reaches of the Yangtze River Economic Belt, China. *Appl. Geogr.* **2020**, *124*, 102293. [CrossRef]

41. Peng, K.F.; Jiang, W.G.; Ling, Z.Y.; Hou, P.; Deng, Y.W. Evaluating the potential impacts of land use changes on ecosystem service value under multiple scenarios in support of SDG reporting: A case study of the Wuhan urban agglomeration. *J. Clean Prod.* **2021**, *307*, 127321. [CrossRef]

42. Chen, Y.H.; Wang, J.; Xiong, N.N.; Sun, L.; Xu, J.Q. Impacts of Land Use Changes on Net Primary Productivity in Urban Agglomerations under Multi-Scenarios Simulation. *Remote Sens.* **2022**, *14*, 1755. [CrossRef]

43. Sun, Q.; Qi, W.; Yu, X.Y. Impacts of land use change on ecosystem services in the intensive agricultural area of North China based on Multi-scenario analysis. *Alex. Eng. J.* **2021**, *60*, 1703–1716. [CrossRef]

44. Mitsova, D.; Shuster, W.; Wang, X.H. A cellular automata model of land cover change to integrate urban growth with open space conservation. *Landsc. Urban Plan.* **2011**, *99*, 141–153. [CrossRef]

45. Verburg, P.H.; Soepboer, W.; Veldkamp, A.; Limpiada, R.; Espaldon, V.; Mastura, S.S.A. Modeling the spatial dynamics of regional land use: The CLUE-S model. *Environ. Manag.* **2002**, *30*, 391–405. [CrossRef] [PubMed]

46. Liu, X.P.; Liang, X.; Li, X.; Xu, X.C.; Ou, J.P.; Chen, Y.M.; Li, S.Y.; Wang, S.J.; Pei, F.S. A future land use simulation model (FLUS) for simulating multiple land use scenarios by coupling human and natural effects. *Landsc. Urban Plan.* **2017**, *168*, 94–116. [CrossRef]

47. Liang, X.; Guan, Q.F.; Clarke, K.C.; Liu, S.S.; Wang, B.Y.; Yao, Y. Understanding the drivers of sustainable land expansion using a patch-generating land use simulation (PLUS) model: A case study in Wuhan, China. *Comput. Environ. Urban Syst.* **2021**, *85*, 101569. [CrossRef]

48. Zhang, S.; Yang, P.; Xia, J.; Wang, W.; Cai, W.; Chen, N.; Hu, S.; Luo, X.; Li, J.; Zhan, C. Land use/land cover prediction and analysis of the middle reaches of the Yangtze River under different scenarios. *Sci. Total Environ.* **2022**, *833*, 155238. [CrossRef]

49. Li, C.D.; Yang, M.Y.; Li, Z.B.; Wang, B.Q. How Will Rwandan Land Use/Land Cover Change under High Population Pressure and Changing Climate? *Appl. Sci.* **2021**, *11*, 5376. [CrossRef]

50. Li, C.; Wu, Y.M.; Gao, B.P.; Zheng, K.J.; Wu, Y.; Li, C. Multi-scenario simulation of ecosystem service value for optimization of land use in the Sichuan-Yunnan ecological barrier, China. *Ecol. Indic.* **2021**, *132*, 108328. [CrossRef]

51. Hu, Y.Y.; He, Y.; Li, Y.L. Urban Spatial Development Based on Multisource Data Analysis: A Case Study of Xianyang City's Integration into Xi'an International Metropolis. *Sustainability* **2022**, *14*, 4090. [CrossRef]

52. Xie, G.D.; Zhang, C.X.; Zhang, L.M.; Chen, W.H.; Li, S.M. Improvement of the Evaluation Method for Ecosystem Service Value Based on Per Unit Area. *J. Nat. Resour.* **2015**, *8*, 1243–1254.

53. Robertson, G.P.; Swinton, S.M. Reconciling agricultural productivity and environmental integrity: A grand challenge for agriculture. *Front. Ecol. Environ.* **2005**, *3*, 38–46. [CrossRef]

54. Zhang, B.A.; Li, W.H.; Xie, G.D. Ecosystem services research in China: Progress and perspective. *Ecol. Econ.* **2010**, *69*, 1389–1395. [CrossRef]

55. Costanza, R.; de Groot, R.; Sutton, P.; van der Ploeg, S.; Anderson, S.J.; Kubiszewski, I.; Farber, S.; Turner, R.K. Changes in the global value of ecosystem services. *Glob. Environ. Chang.-Hum. Policy Dimens.* **2014**, *26*, 152–158. [CrossRef]

56. Xie, G.D.; Lu, C.X.; Leng, Y.F.; Zheng, D.U.; Li, S.C. Ecological assets valuation of the Tibetan Plateau. *J. Nat. Resour.* **2003**, *18*, 189–196.

57. Akhtar, M.; Zhao, Y.Y.; Gao, G.L.; Gulzar, Q.; Hussain, A. Assessment of spatiotemporal variations of ecosystem service values and hotspots in a dryland: A case-study in Pakistan. *Land Degrad. Dev.* **2022**, *33*, 1383–1397. [CrossRef]

58. Liu, Z.Y.; Gan, X.Y.; Dai, W.N.; Huang, Y. Construction of an Ecological Security Pattern and the Evaluation of Corridor Priority Based on ESV and the "Importance-Connectivity" Index: A Case Study of Sichuan Province, China. *Sustainability* **2022**, *14*, 3985. [CrossRef]

59. Yang, N.; Mo, W.B.; Li, M.H.; Zhang, X.; Chen, M.; Li, F.; Gao, W.C. A Study on the Spatio-Temporal Land-Use Changes and Ecological Response of the Dongting Lake Catchment. *ISPRS Int. J. Geo-Inf.* **2021**, *10*, 716. [CrossRef]

60. Ling, H.B.; Yan, J.J.; Xu, H.L.; Guo, B.; Zhang, Q.Q. Estimates of shifts in ecosystem service values due to changes in key factors in the Manas River basin, northwest China. *Sci. Total Environ.* **2019**, *659*, 177–187. [CrossRef]

61. Peng, H.J.; Hua, L.; Zhang, X.S.; Yuan, X.Y.; Li, J.H. Evaluation of ESV Change under Urban Expansion Based on Ecological Sensitivity: A Case Study of Three Gorges Reservoir Area in China. *Sustainability* **2021**, *13*, 8490. [CrossRef]

62. Zhang, X.Y.; Zhang, X.; Li, D.H.; Lu, L.; Yu, H. Multi-Scenario Simulation of the Impact of Urban Land Use Change on Ecosystem Service Value in Shenzhen. *Acta Ecol. Sin.* **2022**, *42*, 2086–2097.

63. Jing, Y.D.; Chang, Y.Q.; Cheng, X.Y.; Wang, D. Land-use changes and ecosystem services under different scenarios in Nansi Lake Basin, China. *Environ. Monit. Assess.* **2021**, *193*, 14. [CrossRef] [PubMed]

64. Zhang, S.H.; Zhong, Q.L.; Cheng, D.L.; Xu, C.B.; Chang, Y.N.; Lin, Y.Y.; Li, B.Y. Landscape ecological risk projection based on the PLUS model under the localized shared socioeconomic pathways in the Fujian Delta region. *Ecol. Indic.* **2022**, *136*, 108642. [CrossRef]

65. Wang, J.; Zhang, J.P.; Xiong, N.N.; Liang, B.Y.; Wang, Z.; Cressey, E.L. Spatial and Temporal Variation, Simulation and Prediction of Land Use in Ecological Conservation Area of Western Beijing. *Remote Sens.* **2022**, *14*, 1452. [CrossRef]

66. Yang, C.; Wei, T.X.; Li, Y.R. Simulation and Spatio-Temporal Variation Characteristics of LULC in the Context of Urbanization Construction and Ecological Restoration in the Yellow River Basin. *Sustainability* **2022**, *14*, 789. [CrossRef]

67. Zhang, C.C.; Wang, P.; Xiong, P.S.; Li, C.H.; Quan, B. Spatial Pattern Simulation of Land Use Based on FLUS Model under Ecological Protection: A Case Study of Hengyang City. *Sustainability* **2021**, *13*, 10458. [CrossRef]

68. Xie, G.D.; Xiao, Y.; Zhen, L.; Lu, C. Study on ecosystem services value of food production in China. *Chin. J. Eco-Agric.* **2005**, *13*, 10–13.

69. Li, X.; Zhu, Y.; Zhao, L.; Tian, J.; Li, J. Ecosystem services value change in Qinglong County from dynamically adjusted value coefficients. *Chin. J. Eco-Agric.* **2015**, *23*, 373–381.

70. Li, L.M.; Tang, H.N.; Lei, J.R.; Song, X.Q. Spatial autocorrelation in land use type and ecosystem service value in Hainan Tropical Rain Forest National Park. *Ecol. Indic.* **2022**, *137*, 108727. [CrossRef]

71. Martin, D. An assessment of surface and zonal models of population. *Int. J. Geogr. Inf. Syst.* **1996**, *10*, 973–989. [CrossRef]

72. Li, C.; Gao, B.; Wu, Y.; Zheng, K.; Wu, Y. Dynamic Simulation of Landscape Ecological Risk in Mountain Towns Based on Plus Model. *J. Zhejiang AF Univer.* **2022**, *39*, 84–94. [CrossRef] [PubMed]

73. Li, J.Y.; Gong, J.; Guldmann, J.M.; Yang, J.X.; Zhang, Z. Simulation of Land-Use Spatiotemporal Changes under Ecological Quality Constraints: The Case of the Wuhan Urban Agglomeration Area, China, over 2020-2030. *Int. J. Environ. Res. Public Health* **2022**, *19*, 6095. [CrossRef] [PubMed]

74. Landis, J.R.; Koch, G.G. Measurement of Observer Agreement for Categorical Data. *Biometrics* **1977**, *33*, 159–174. [CrossRef]

75. Xu, L.F.; Liu, X.; Tong, D.; Liu, Z.X.; Yin, L.R.; Zheng, W.F. Forecasting Urban Land Use Change Based on Cellular Automata and the PLUS Model. *Land* **2022**, *11*, 652. [CrossRef]

76. Guo, R.; Wu, T.; Wu, X.C.; Luigi, S.; Wang, Y.Q. Simulation of Urban Land Expansion Under Ecological Constraints in Harbin-Changchun Urban Agglomeration, China. *Chin. Geogr. Sci.* **2022**, *32*, 438–455. [CrossRef]

77. Li, C.X.; Wu, J.Y. Land Use Transformation and Eco-Environmental Effects Based on Production-Living-Ecological Spatial Synergy: Evidence from Shaanxi Province, China. *Environ. Sci. Pollut. Res.* **2022**, *29*, 41492–41504. [CrossRef]
78. Sun, S.; Zhang, X. Analysis on the Spatiotemporal Evolution of Land Use Transformation and Its Ecological Environment Effect in Shaanxi Province. *Res. Soil Water Cons.* **2021**, *28*, 356.
79. Zhang, Y.; Li, Y. Ecosystem Service Value(Esv)and Spatial-Temporal Difference of Urban Agglomeration in Guanzhong Plain from 1995 to 2015. *J. Zhejiang Univ. Sci. Edit.* **2020**, *47*, 615.
80. Wu, C.Y.; Chen, B.W.; Huang, X.J.; Wei, Y.H.D. Effect of land-use change and optimization on the ecosystem service values of Jiangsu province, China. *Ecol. Indic.* **2020**, *117*, 14. [CrossRef]
81. Kubiszewski, I.; Muthee, K.; Rasheed, A.R.; Costanza, R.; Suzuki, M.; Noel, S.; Schauer, M. The costs of increasing precision for ecosystem services valuation studies. *Ecol. Indic.* **2022**, *135*, 108551. [CrossRef]
82. Ma, X.F.; Zhu, J.T.; Zhang, H.B.; Yan, W.; Zhao, C.Y. Trade-offs and synergies in ecosystem service values of inland lake wetlands in Central Asia under land use/cover change: A case study on Ebinur Lake, China. *Glob. Ecol. Conserv.* **2020**, *24*, e01253. [CrossRef]
83. Cui, X.; Liu, C.; Shan, L.; Lin, J.; Zhang, J.; Jiang, Y.; Zhang, G. Spatial-Temporal Responses of Ecosystem Services to Land Use Transformation Driven by Rapid Urbanization: A Case Study of Hubei Province, China. *Int. J. Environ. Res. Public Health* **2021**, *19*, 178. [CrossRef] [PubMed]
84. Zheng, D.F.; Wan, J.Y.; Bai, L.N.; lv, L.T. Multi-Scale Analysis of Ecosystem Service Trade-Offs/Synergies in Yanshan-Taihang Mountains Area. *J. Ecol. Rural Environ.* **2022**, *38*, 409–417.
85. Yang, Q.; Xu, G.; Li, A.; Liu, Y.; Hu, C. Evaluation and Trade-Off of Ecosystem Services in the Qingyijiang River Basin. *Acta Ecol. Sin.* **2021**, *41*, 9315–9327.
86. Zhang, J.; Qu, M.; Wang, C.; Zhao, J.; Cao, Y. Quantifying landscape pattern and ecosystem service value changes: A case study at the county level in the Chinese Loess Plateau. *Glob. Ecol. Conserv.* **2020**, *23*, e01110. [CrossRef]
87. Le Blanc, D. Towards Integration at Last? The Sustainable Development Goals as a Network of Targets. *Sustain. Dev.* **2015**, *23*, 176–187. [CrossRef]
88. Eisenmenger, N.; Pichler, M.; Krenmayr, N.; Noll, D.; Plank, B.; Schalmann, E.; Wandl, M.T.; Gingrich, S. The Sustainable Development Goals prioritize economic growth over sustainable resource use: A critical reflection on the SDGs from a socio-ecological perspective. *Sustain. Sci.* **2020**, *15*, 1101–1110. [CrossRef]
89. Han, R.; Feng, C.C.E.; Xu, N.Y.; Guo, L. Spatial heterogeneous relationship between ecosystem services and human disturbances: A case study in Chuandong, China. *Sci. Total Environ.* **2020**, *721*, 137818. [CrossRef]
90. Holden, M.; Roseland, M.; Ferguson, K.; Perl, A. Seeking urban sustainability on the world stage. *Habitat Int.* **2008**, *32*, 305–317. [CrossRef]
91. Jabareen, Y.R. Sustainable urban forms-Their typologies, models, and concepts. *J. Plan. Educ. Res.* **2006**, *26*, 38–52. [CrossRef]
92. Caragliu, A.; Del Bo, C.; Nijkamp, P. Smart Cities in Europe. *J. Urban Technol.* **2011**, *18*, 65–82. [CrossRef]
93. Mora, L.; Bolici, R.; Deakin, M. The First Two Decades of Smart-City Research: A Bibliometric Analysis. *J. Urban Technol.* **2017**, *24*, 3–27. [CrossRef]
94. Bibri, S.E.; Krogstie, J. Smart sustainable cities of the future: An extensive interdisciplinary literature review. *Sust. Cities Soc.* **2017**, *31*, 183–212. [CrossRef]
95. Huang, C.; Sun, Z.; Jiang, H.; Wang, J.; Wang, P.; Tao, J.; Liu, H.; Liu, N. Big Earth Data Supports Sustainable Cities and Communities: Progress and Challenges. *Bull. Chin. Acad. Sci.* **2021**, *36*, 914–922.

 *sustainability*

Article

Evaluation of ESV Change under Urban Expansion Based on Ecological Sensitivity: A Case Study of Three Gorges Reservoir Area in China

Hongjie Peng [1,2], Lei Hua [3], Xuesong Zhang [1,2,*], Xuying Yuan [1,2] and Jianhao Li [1,2]

[1] Department of Geographic Information Science, College of Urban and Environmental Sciences, Central China Normal University, Wuhan 430079, China; penghongjie@mails.ccnu.edu.cn (H.P.); Yuanxuying@mail.ccnu.edu.cn (X.Y.); ljh@mails.ccnu.edu.cn (J.L.)
[2] Key Laboratory for Geographical Process Analysis & Simulation, Wuhan 430079, China
[3] Department of Public Service Management, School of Public Affairs, Chongqing University, Chongqing 400044, China; hualei@mails.ccnu.edu.cn
[*] Correspondence: zhangxuesong@mail.ccnu.edu.cn

Abstract: In recent years, ecosystem service values (ESV) have attracted much attention. However, studies that use ecological sensitivity methods as a basis for predicting future urban expansion and thus analyzing spatial-temporal change of ESV are scarce in the region. In this study, we used the CA-Markov model to predict the 2030 urban expansion under ecological sensitivity in the Three Gorges Reservoir area based on multi-source data, estimations of ESV from 2000 to 2018 and predictions of ESV losses from 2018 to 2030. Research results: (i) In the concept of green development, the ecological sensitive zone has been identified in Three Gorges Reservoir area; it accounts for about 35.86% of the study area. (ii) It is predicted that the 2030 urban land will reach 211,412.51 ha by overlaying the ecological sensitive zone. (iii) The total ESV of Three Gorges Reservoir area showed an increasing trend from 2000 to 2018 with growth values of about USD 3644.26 million, but the ESVs of 16 districts were decreasing, with Dadukou and Jiangbei having the highest reductions. (iv) New urban land increases by 80,026.02 ha from 2018 to 2030. The overall ESV losses are about USD 268.75 million. Jiulongpo, Banan and Shapingba had the highest ESV losses.

Keywords: ecological sensitivity; ecosystem service values; CA-Markov model; urban expansion; Three Gorges Reservoir area

Citation: Peng, H.; Hua, L.; Zhang, X.; Yuan, X.; Li, J. Evaluation of ESV Change under Urban Expansion Based on Ecological Sensitivity: A Case Study of Three Gorges Reservoir Area in China. *Sustainability* **2021**, *13*, 8490. https://doi.org/10.3390/su13158490

Academic Editors: Kikuko Shoyama, Rajarshi Dasgupta and Ronald C. Estoque

Received: 18 June 2021
Accepted: 26 July 2021
Published: 29 July 2021

Publisher's Note: MDPI stays neutral with regard to jurisdictional claims in published maps and institutional affiliations.

1. Introduction

An ecosystem is a unity of the biota and the abiotic surroundings in a certain space of nature [1]. It is also an ecological functional unit formed by the continuous exchange of energy and materials between organisms and their abiotic environment [2–4]. Ecosystems can provide a range of services for human production and livelihoods [5–7]. Ecosystem services are the materials that humans need to obtain from the Earth's ecosystems and natural environment [8,9]. Ecosystem services are diverse, and the services are unique and irreplaceable for different ecosystem [10–12]. For example, forests are the most essential ecosystems for human life, providing wood, regulating air, promoting soil formation and supporting social benefits such as spiritual, landscape and educational values [13]. The evaluation of ESV can measure the status of a regional ecosystem.

The evaluation of ESV is a method based on ecology, economics and sociology [14–16]. It can quantitatively evaluate ecosystem services from the perspective of monetary value and better reflect the change in ESV and provide decision makers with knowledge to make policy adjustments [16,17]. In the early days, its evaluation methods made some progress in emphasizing the impact of socio-environmental change on ecosystem services (supply, regulation, support and entertainment services) [18–20]. Research scholars began to use

RS and GIS technology to assess ESV for the region by remote sensing technology, such as China at the national scale [9,21] and Leipzig at the municipal scale [22].

The delimitation of the ecological sensitive zone is important to improve ESV in the region [5]. It is believed that an ecological sensitive zone is characterized by the low stability of an ecosystem that is easily affected by external activities, ecological degradation and difficulty with self-restoration [23]. When humans develop them unreasonably, ecological sensitive zones are prone to environmental problems. Delineating the ecological sensitive zone plays an important role in maintaining the health of the ecosystem [24,25]. The ecological sensitivity evaluation method can quantitatively identify the ecological sensitive zone. The evaluation results provide the possibility to effectively control and protect the target zone [7]. It also provides a practical way for developing countries to realize its regional sustainable development strategy [26]. Starting from the perspective of ecological sensitivity, providing scientific guidance methods for realizing the sustainable development of regional ecosystem [27].

The assessment of ESV is closely linked to land use patterns. Land use simulations can be modelled using spatial tools such as cellular automata (CA), which are temporal, spatial and state-discrete models that can simulate complex dynamic systems with spatial characteristics [28–31]. CA model express geographic entity information to simulate and predict complex geographic processes by constructing a systemic spatial concept system [32]. Thus, CA is commonly used to predict future land use changes through endogenous and exogenous drivers, which have the roles of human disturbance as well as socioeconomic and institutional factors [33,34]. Estimations of future ESVs are obtained by addressing the interactions between drivers, land use changes and ESV changes. These interactions are usually considered as the dependent variable component of the CA model. For example, logistic regression is combined with a CA model, which is used by assuming that the development probability of a location is a function of a set of independent variables [35]. The ANN-CA model is a combination of artificial neural networks (ANN) and CA models, which combines the nonlinear processing capability of artificial neural networks and the spatial simulation capability of CA models [36]. In the CA-Markov model, the Markov chain process controls the temporal variation between land use types according to the transition matrix, which facilitates the use of a wider range of spatial variables and improves the accuracy of the model [37,38].

As urbanization progresses, it is necessary to delimit ecological sensitive zones and predict the future urban development direction of the region through the CA-Markov model to promote sustainable urbanization development. On this basis, estimating the dynamic change of ESV in the region will provide a full understanding of the services provided by ecosystems for human social development, providing relevant reference for ecological environmental protection and the comprehensive control of ecological reserves. Therefore, this paper takes the Three Gorges Reservoir area as an example to simulate the future development rate and direction of urban land in different districts and counties of the Three Gorges Reservoir area under the role of the ecological sensitive zone and urban land drivers. The quantitative estimation of change in ESV and the prediction of future ESV losses in the region are performed by ecosystem service value equivalence tables. This study can provide scientific reference to promote ecological civilization construction and sustainable development in developing countries.

2. Materials and Methods

2.1. Description of the Study Area

The Three Gorges Reservoir area is located at the end of the upper reaches of the Yangtze River Basin, with an area of approximately 68,772.93 km^2 (Figure 1). It is bordered by three provinces and a municipality, namely Hubei Province, Sichuan Province, Guizhou Province and Chongqing City. The Three Gorges Reservoir area includes 26 districts and counties (autonomous regions). The relative height shows a gradual increase from west to east. The central and western parts of the reservoir area are dominated by platforms and hills. The eastern part is close to the Daba Mountain and has many rolling hills [39].

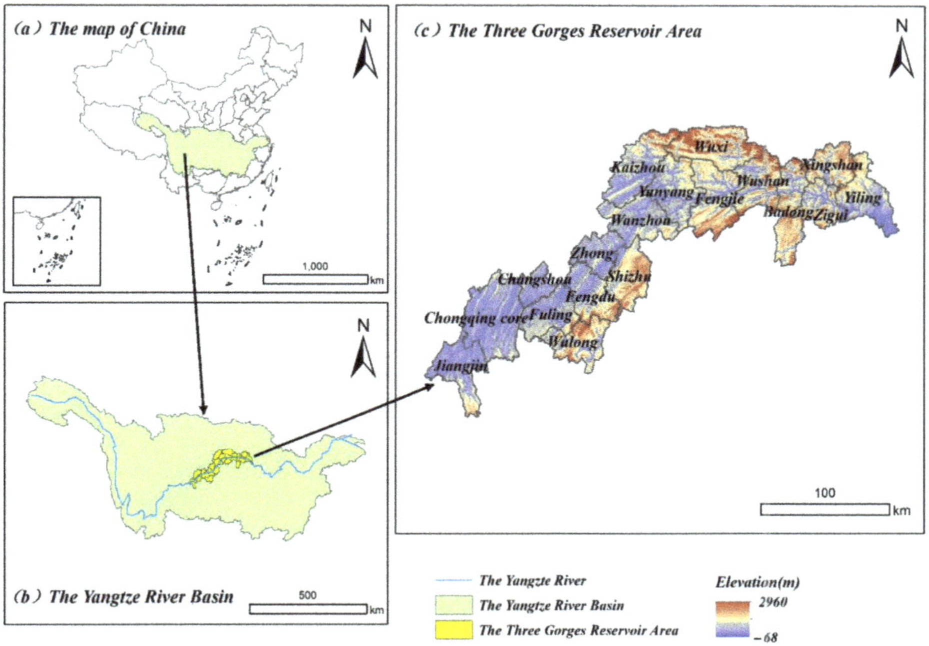

Figure 1. Location and elevation of the study area. (**a**) The map of China; (**b**) the Yangtze River basin; (**c**) the Three Gorges Reservoir area.

There are abundant vegetation types in the Three Gorges Reservoir area, and the overall spatial distribution of vegetation coverage shows the characteristics of high in the east and low in the west [40]. The east is mostly broad-leaved forests, bushes and grasslands; the central and western area are farming area with many cultivated plants and crops [41]. The Three Gorges Reservoir area is an important ecological barrier in the Yangtze River Basin, China's strategic water resources reserve. Therefore, the prediction of ESV change in the Three Gorges Reservoir area is important for China's sustainable development under the future urban expansion.

2.2. Materials

The two main types of data used in this study are spatial data and statistical yearbook data.

Spatial Data. (1) Administrative boundary vector data of Three Gorges Reservoir area (SHP format). (2) Soil dataset provided by Harmonized World Soil Database (HWSD) and Cold Arid Regions, Available online: http://www.westdc.westgis.ac.cn (accessed on 17 April 2019), which contains the spatial coordinates and properties of the soil (GRID format). (3) Digital elevation model (DEM), which was downloaded from the Geospatial Data Cloud Available online: http://www.gscloud.cn/ (accessed on 26 April 2020). In the above data, the DEM data, with a resolution of 30 m, can be extracted into slope and elevation. (4) The highway, the primary road and railroad data procured from OpenStreetMap Available online: http://www.openstreetmap.org (accessed on 13 July 2020). The population density data was provided by Landscan Available online: https://landscan.ornl.gov/ (accessed on 18 December 2020). (5) The 1000 m Normalized Difference Vegetation Index (NDVI) data and 30 m land use and land cover change (LUCC) data were obtained from the Data

Centre for Resources and Environmental Sciences, Chinese Academy of Sciences (RESDC) Available online: http://www.resdc.cn (accessed on 26 April 2020). NDVI and LUCC data were obtained by remote sensing image processing. NDVI data was based on inversion of SPOT/VEGETATION and MODIS satellite remote sensing and LUCC data was generated by manual visual interpretation based on Landsat 8 remote sensing images. (6) China Nature Reserve, the earthquake and landslide vector data acquired from RESDC. China Nature Reserve are surface vector data, the rest are point vector data.

Statistical data: The statistical data include average annual rainfall (2018), average annual temperature (2018), number of days with wind and sand wind speed $\geq$10 m/s (2018), annual evaporation (2018), groundwater mineralization and groundwater burial depth. Meteorological data were mainly obtained from 22 meteorological stations around the Three Gorges Reservoir area by the Chinese Meteorological Science Data Sharing Service Available online: http://data.cma.cn/site/index.html (accessed on 17 January 2020). According to each weather station, the inverse distance interpolation (IDW) method of ArcGIS was used to convert meteorological statistics into a raster image of meteorological data in the study area. Groundwater mineralization and groundwater depth of burial data were obtained by querying the statistical yearbooks of different districts and counties in the Three Gorges Reservoir area.

2.3. Methods

In this paper, we use the overlay analysis function of GIS, Markov chain and cellular automata (CA) to simulate the future urban expansion. From the scenario of ecological conservation, predicting future urban expansion, estimating ESV during 2018–2030 and forecasting ESV losses at 2030 in the Three Gorges Reservoir area were performed. The research framework of this study is presented in Figure 2.

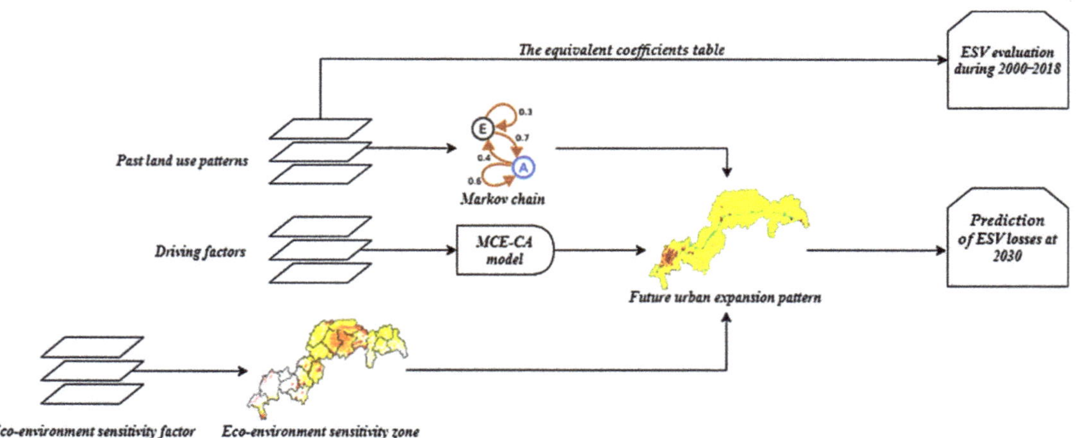

Figure 2. Research framework.

2.3.1. Ecological Comprehensive Sensitivity and Zone Identification

In this research, the comprehensive ecological sensitivity was calculated from three sensitivities of soil erosion, land desertification and soil salinization. The calculation formula is shown below:

$$EES = \omega a * SSi + \omega b * Zi + \omega c * Si \tag{1}$$

where EES denotes the comprehensive ecological sensitivity, and SS_i, Z_i and S_i are the weight values of soil erosion, land desertification and soil salinization, respectively. The coefficient of variation method was used to calculate ω_a, ω_b and ω_c as 49.63%, 32.00% and 18.37%, respectively [42].

R_i, SG_i, LS_i and C_i were selected for the evaluation of SS_i. The ecological sensitivity factors were classified into five classes named insensitive, mildly sensitive, moderately sensitive, highly sensitive and extremely sensitive, which were categorized into classes 1, 2, 3, 4 and 5, respectively (Table 1). The four indicators are calculated as follows:

$$SSi = \sqrt[4]{Ri * SGi * LSi * Ci} \tag{2}$$

where SS_i denotes the soil erosion sensitivity; R_i is the rainfall erosivity factor; SG_i is the soil type factor; LS_i is the relative height factor; and C_i is the normalized difference vegetation index factor. The ecological sensitivity factors were divided into five classes named insensitive, mildly sensitive, moderately sensitive, highly sensitive and extremely sensitive, which were categorized into classes 1, 2, 3, 4 and 5 (Table 1).

Table 1. Criteria for the soil erosion sensitivity.

Sensitivity Degree	R_i	SG_i	LS_i	C_i
Insensitive (1)	<25	Paddy soil, urban area, rock and river	<20	>0.49
Mildly sensitive (2)	25–100	Limestone soil, rock-soil and mountain meadow soil	20–50	0.39–0.49
Moderately sensitive (3)	100–400	Dark brown soil and yellow-cinnamon soil	50–100	0.28–0.39
Highly sensitive (4)	400–600	Yellow loam, yellow-brown soil and skeleton soil	100–300	0.16–0.28
Extremely sensitive (5)	>600	Purple soil	>300	<0.16

The evaluation of the Z_i required the I_i, W_i and SL_i. Its classification and assignment methods were the same as those of SS_i (Table 2). The formula of Z_i was as follows:

$$Zi = \sqrt[3]{Ii * Wi * SLi} \tag{3}$$

where Z_i denotes the land desertification sensitivity; I_i is the dryness index factor; W_i is the number of days on which wind-blown sand speeds are 6m/s; SL_i is the slope factor.

Table 2. Criteria for the land desertification sensitivity.

Sensitivity Degree	I_i	W_i	SL_i
Insensitive (1)	<0.96	<2	≤5
Mildly sensitive (2)	0.96–1.01	2–4	5–8
Moderately sensitive (3)	1.01–1.08	4–6	8–15
Highly sensitive (4)	1.08–1.17	6–8	15–25
Extremely sensitive (5)	>1.17	>8	>25

The evaluation of the S_i required the E_i, GS_i, GD_i and $LUCC_i$. Its classification and assignment methods were the same as those of SS_i (Table 3). The formula of S_i is as follows:

$$Si = \sqrt[4]{Ei * GSi * GDi * LUCCi} \tag{4}$$

where S_i is the salinization sensitivity; E_i is the evaporation index factor; GS_i is the degree of mineralization of ground water factor; GD_i is the groundwater depth factor; $LUCC_i$ is the land use and land cover change factor.

Table 3. Criteria for the salinization sensitivity.

Sensitivity Degree	E_i	GS_i	GD_i	$LUCC_i$
Insensitive (1)	<0.6	<0.1	>20	Water body
Mildly sensitive (2)	0.6–0.7	0.1–0.2	18–20	Impervious surface
Moderately sensitive (3)	0.7–0.8	0.2–0.3	16–18	Forest and grassland
Highly sensitive (4)	0.8–0.9	0.3–0.4	14–16	Farmland
Extremely sensitive (5)	>0.9	>0.4	<14	Unused land

The assessment results of ecological sensitivity were divided into 5 degrees, namely, insensitive, mildly sensitive, moderately sensitive, highly sensitive and extremely sensitive. The eleven factors required to calculate the spatial distributions of SS_i, Z_i and S_i were presented in Figure 3.

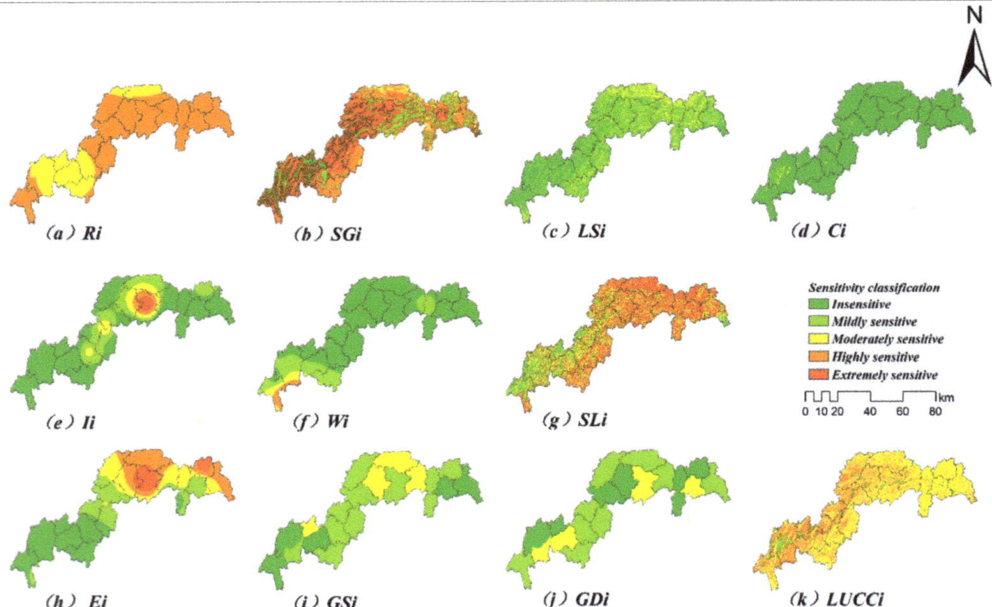

Figure 3. The spatial distribution of ecological sensitivity of 11 indicators in 2018: (**a**) rainfall erosion sensitivity; (**b**) soil type sensitivity; (**c**) relief of topography sensitivity; (**d**) vegetation cover sensitivity; (**e**) dryness sensitivity; (**f**) wind-blown sand sensitivity; (**g**) slope sensitivity; (**h**) evaporation sensitivity; (**i**) mineralization of ground water sensitivity; (**j**) groundwater depth sensitivity; (**k**) land use and land cover change sensitivity.

Based on the results of the SS_i, Z_i, S_i and EES (Figure 4), using the natural breakpoint method in ArcGIS, the high sensitivity zone in the EES is proposed to determine the location of ecological sensitive zone.

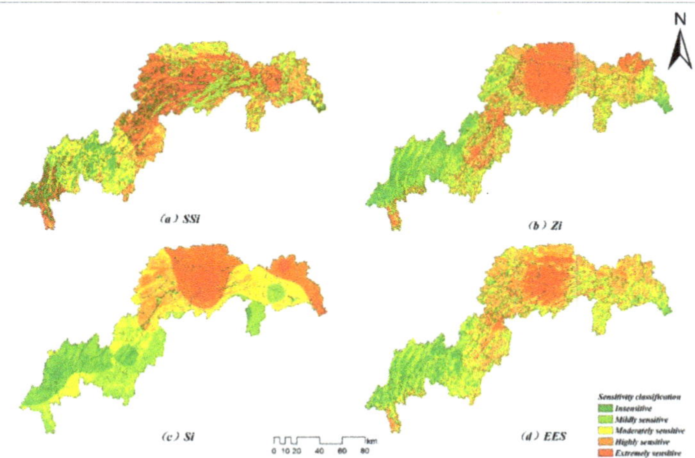

Figure 4. The spatial distribution of SS_i, Z_i, S_i and EES: (**a**) soil erosion sensitivity; (**b**) land desertification sensitivity; (**c**) salinization sensitivity; (**d**) comprehensive ecological sensitivity.

2.3.2. CA-Markov Model

The CA-Markov model is composed of CA model, Markov chain and multi-criteria evaluation (MCE) [29]. The CA model is a discrete, finite state composition of the meta-cell model. It can simulate complex dynamic systems with spatial-temporal characteristics according to certain local rules [43]. Markov chains create the transfer matrix and probability between land use types for multiple time periods in the past through spatial comparison analysis, which are the basic data for predicting future land use patterns. MCE refers to the selection of expansion factors to construct a land use transition suitability image collection. CA-Markov model can effectively predict future land use dynamics [44]. The prediction process equation of the CA-Markov model is shown below [45]:

$$Ctj + 1 = F[Ctj, N] \tag{5}$$

where $C(t_j)$ and $C(t_{j+1})$ are the states of the cell at time t_j and t_{j+1}, respectively; F is the transition rule; N is the domain of the cellular. In this study, elevation, slope, earthquake, landslide, highway, main road, railroad and population density were selected as the driving factors affecting urban expansion. Among them, elevation, slope, earthquake and landslide are negative indicators, and the rest are positive indicators. The suitability evaluation maps for earthquakes, landslides, highways, main roads and railroads were calculated by the kernel density tool, and finally, the normalized driver maps were obtained (Figure 5).

The Kappa coefficient was applied to check the accuracy of the CA-Markov model simulation results [46]. The formula for calculating the Kappa coefficient is shown below:

$$Kappa = \frac{P_a - P_c}{1 - P_c} \tag{6}$$

$$P_a = \frac{s}{n} \tag{7}$$

$$P_c = \frac{a_1 * b_1 + a_2 * b_2}{n * n} \tag{8}$$

where n is the total number of cell sizes in the raster; a_1 is the number of cell sizes in the real raster of the urban land; a_2 is the number of cell sizes in the non-urban land; b_1 is the number of cell sizes in the simulated raster of the urban land; b_2 is the number of cell sizes in the non-urban land; s is the number of cell sizes in the real raster and the simulated raster that correspond to each other.

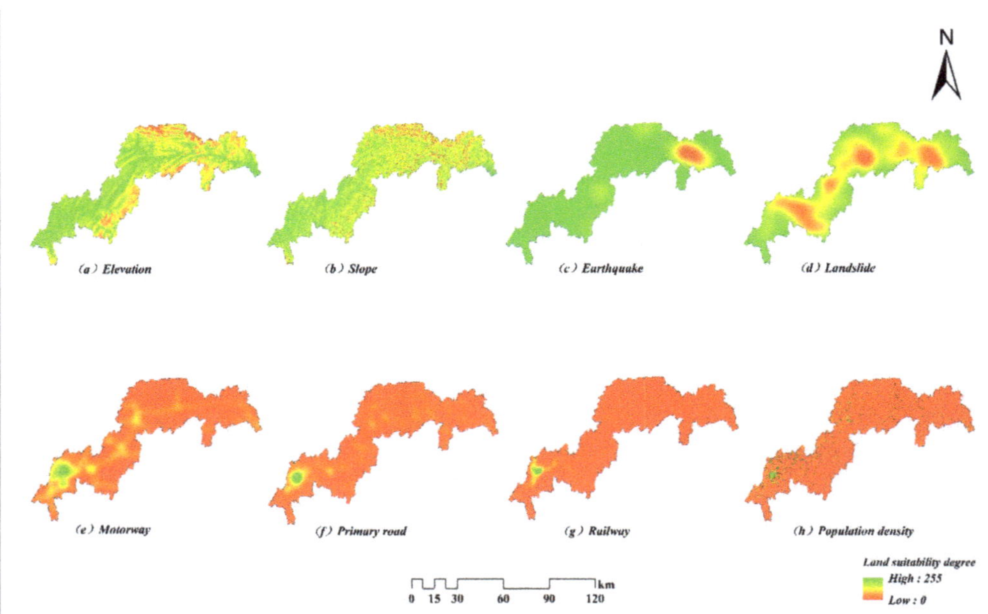

Figure 5. The spatial distribution of driving factors of urban expansion in Three Gorges Reservoir area: (**a**) elevation drive factor; (**b**) slope drive factor; (**c**) earthquake drive factor; (**d**) landslide drive factor; (**e**) motorway drive factor; (**f**) primary road drive factor; (**g**) railway drive factor; (**h**) population drive factor.

2.3.3. Evaluation of ESV

Ecosystems provide the ecological products and services that humans need [47]. In order to estimate the value of ecosystems, basic value transfer, expert modified value transfer and spatial explicit function modeling were used [5,14,47]. Costanza et al. proposed to divide ecosystem services into 17 services and established a global value equivalence factor system to calculate the global ESV. This method has been widely accepted and used, but it cannot be directly applied to evaluate ESV in China [48]. This paper will be based on the recent research results of Xie et al. in Table 4, classify 6 types of ecosystems into the 4 following types of ecosystem services: provisioning services, regulating services, habitat services and entertainment services. A VPUA-based work has published that in China, and unit equivalent coefficients were estimated as USD 503.2 of ESV per ha [19]. For each ESV category, the calculation formula of ESV_k is as follows:

$$ESV_k = \sum_{i=1}^{n} ESVi \tag{9}$$

$$ESV_k = \sum_{j=1}^{m} EC_{i,j} * \frac{Area_j}{Area_u} \tag{10}$$

where ESV_k is total value of ecosystem services; ESV_i is the ESV for a particular primary service class (i); $EC_{i,j}$ is the equivalent coefficients of the secondary service class (j) in a particular primary service class (i); n is the four kinds of primary service class that include provisioning services, regulating services, habitat services and entertainment services; m is the total number of the secondary service class; $Area_j$ is the area of class (j) in a land type of a hectare and $Area_u$ is a hectare. Table 4 shows the equivalent coefficients table for ESV per unit area in China.

Table 4. The equivalent coefficients table for ESV per unit area in China.

| Primary Service | Secondary Service | Farmland | | Forest | | | | Grassland | Water Body | | Unused Land |
		Dry Land	Paddy Field	Coniferous Forest	Mixed Forest	Broadleaved Forest	Bush	Meadow	Wetland	Lake and River	Barren
Provisioning services	Food	0.85	1.36	0.22	0.31	0.29	0.19	0.22	0.51	0.80	0.00
	Materials	0.40	0.09	0.52	0.71	0.66	0.43	0.33	0.50	0.23	0.00
	Water	0.02	−2.63	0.27	0.37	0.34	0.22	0.18	2.59	8.29	0.00
Regulating services	Air quality regulation	0.67	1.11	1.70	2.35	2.17	1.41	1.14	1.90	0.77	0.02
	Climate regulation	0.36	0.57	5.07	7.03	6.50	4.23	3.02	3.60	2.29	0.00
	Waste treatment	0.10	0.17	1.49	1.99	1.93	1.28	1.00	3.60	5.55	0.10
	Water flow regulation	0.27	2.72	3.34	3.51	4.74	3.35	2.21	24.23	102.24	0.03
	Erosion prevention	1.03	0.01	2.06	2.86	2.65	1.72	1.39	2.31	0.93	0.02
	Maintenance of soil fertility	0.12	0.19	0.16	0.22	0.20	0.13	0.11	0.18	0.07	0.00
Habitat services		0.13	0.21	1.88	2.60	2.41	1.57	1.27	7.87	2.55	0.02
Entertainment services		0.06	0.09	0.82	1.14	1.06	0.69	0.56	4.73	1.89	0.01
Total		4.01	3.89	17.53	23.09	22.95	15.22	11.43	52.02	125.61	0.2

3. Results

3.1. Ecological Sensitivity Zone Identification

3.1.1. The Evaluation of Soil Erosion Sensitivity

Soil erosion is the serious damage to natural resources of water, soil and land productivity [49].The soil erosion sensitivity result was calculated by formula (2), as shown in Table 5 and Figure 4a. It shows a lower sensitivity at both ends and a higher sensitivity in the middle. The percentages of mildly sensitive zones and moderately sensitive zones are 30.00% and 22.99%, respectively. Next, the highly sensitive and insensitive zones accounted for 19.88% and 15.85%, respectively. Finally, the least percentage of the extremely sensitive zone is 11.28%, and it is mainly concentrated in Yunyang, Zigui, Fengjie, Kaizhou and Wanzhou.

Table 5. The comprehensive assessment on ecological sensitivity.

Sensitivity Classification	Soil Erosion		Land Desertification		Soil Salinization		The Comprehensive Evaluation	
	Area (km^2)	Percentage (%)	Area (km^2)	Percentage (%)	Area (km^2)	Percentage (%)	Area (km^2)	Percentage (%)
Extremely sensitive	6494.08	11.28	10,587.74	18.40	5145.91	8.94	6279.62	10.91
Highly sensitive	11,442.09	19.88	10,246.66	17.80	16,129.51	28.02	14,360.09	24.95
Moderately sensitive	13,232.75	22.99	13,845.73	24.06	16,224.49	28.19	17,261.63	29.99
Mildly sensitive	17,265.48	30.00	13,213.63	22.96	14,444.79	25.10	15,234.81	26.47
Insensitive	9120.03	15.85	9660.65	16.79	5609.71	9.75	4418.26	7.68

3.1.2. The Evaluation of Land Desertification Sensitivity

Land desertification is generally defined as the loss of surface soil due to soil erosion, a reduction of agricultural land use value and ecological degradation [50]. In the spatial distribution of land desertification sensitivity (Figure 4b), extremely and highly sensitive zones occur mainly in Fengjie County, Wuxi County, Yunyang County and the upper part of Xingshan County. From the amount of land desertification sensitivity, as shown in Table 5, the moderately sensitive zone and the mildly sensitive zone accounted for the highest percentages of 24.06% and 22.96%, respectively. The extremely sensitive zone, highly sensitive zone and insensitive zone accounted for 18.40%, 17.80% and 16.79%, respectively.

3.1.3. The Evaluation of Soil Salinization Sensitivity

Soil salinization is a process in which salts from the soil substrate or groundwater rise to the surface, causing salts to accumulate in the surface soil after the water evaporates [51]. The spatial heterogeneity of the distribution of soil salinity sensitivity in the study area is shown in Figure 4c. The characteristics of the spatial distribution showed a gradual decrease from northeast to southwest. The higher sensitivity zones are mainly in Fengjie County, Wuxi County and Xingshan County. In the evaluation table of soil salinity sensitivity (Table 5), the highest percentages of the moderately and highly sensitive zones are 28.19% and 28.02%, respectively. The mildly sensitive and insensitive zones accounted for 25.10% and 9.75%, respectively. The least percentage was observed in the extremely sensitive zone at 8.94%.

3.1.4. The Comprehensive Evaluation and Zone Identification of Ecological Sensitivity

The comprehensive sensitivity of the ecological environment was calculated by formula (1), as shown in Table 5 and Figure 4d. In general, the highly sensitive zone is mainly concentrated in Wuxi, Yunyang and Fengjie. It is close to the Daba Mountain Nature Reserve and belongs to high forest cover areas with high terrain elevation [52]. The moderately sensitive zone is mainly concentrated in Zhong, Fengdu, Badong and Zigui.

The following zone of low sensitivity is mainly concentrated in the nine central urban areas of Chongqing, Fuling and Jiangjin. As shown in the comprehensive sensitivity evaluation table of the ecological environment (Table 5), the study area mainly showed moderate sensitivity (about 29.99%). The proportions of extremely sensitive, highly sensitive, low sensitive and insensitive zone are 10.91%, 24.95%, 26.47% and 7.68%, respectively.

In the identification of ecological sensitive zone, we use the spatial analyst tool in ArcGIS to extract the extremely and highly sensitive zone in the comprehensive ecological sensitivity, the vector data of Chinese nature reserves were converted into raster grid data, and then both were mosaicked. Finally, we obtained the spatial distribution map of the ecological key zone and the ecological core zone in the Three Gorges Reservoir area (Figure 6).

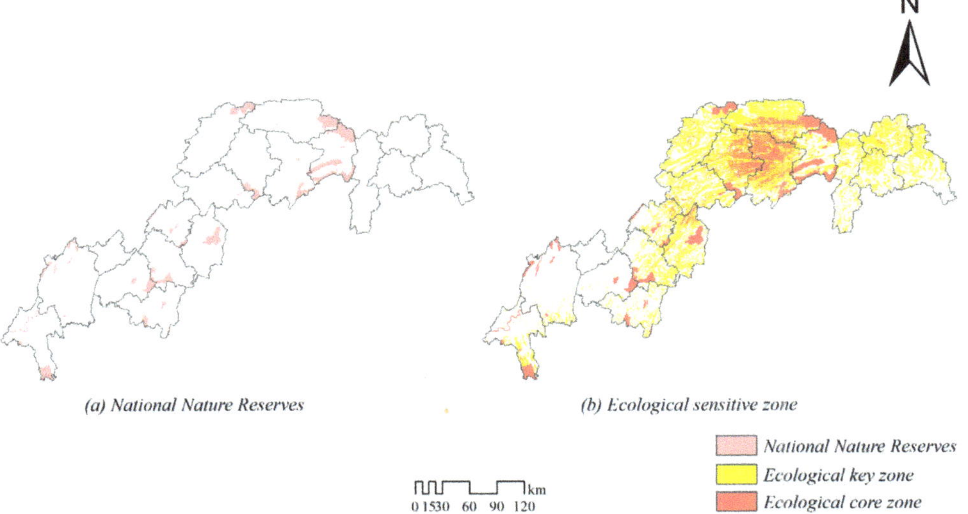

Figure 6. The spatial distribution of ecological key zone and ecological core zone.

3.2. The Urban Expansion Simulation

3.2.1. The Model Validation and Assessment

Using the 2000 and 2010 actual map as the base data and the 2010 land use data as the initial state and driving factors of urban expansion, we predicted the 2018 urban development pattern (Figure 7d). The accuracy of the model was verified by using the 2018 actual pattern with the 2018 simulated pattern to judge the accuracy of the model. By comparing with the actual 2018 urban pattern (Figure 7c), using the Kappa coefficient formula to verify the accuracy, the Kappa coefficients for the three land types and the whole were calculated in Figure 8. The Kappa coefficients for nonurban, urban and water bodies are 86.73%, 83.14% and 92.19%, respectively, while the accuracy of the overall pattern was 86.74%, which indicated that the overall simulation accuracy was better [53], it can provide a basis for further simulation.

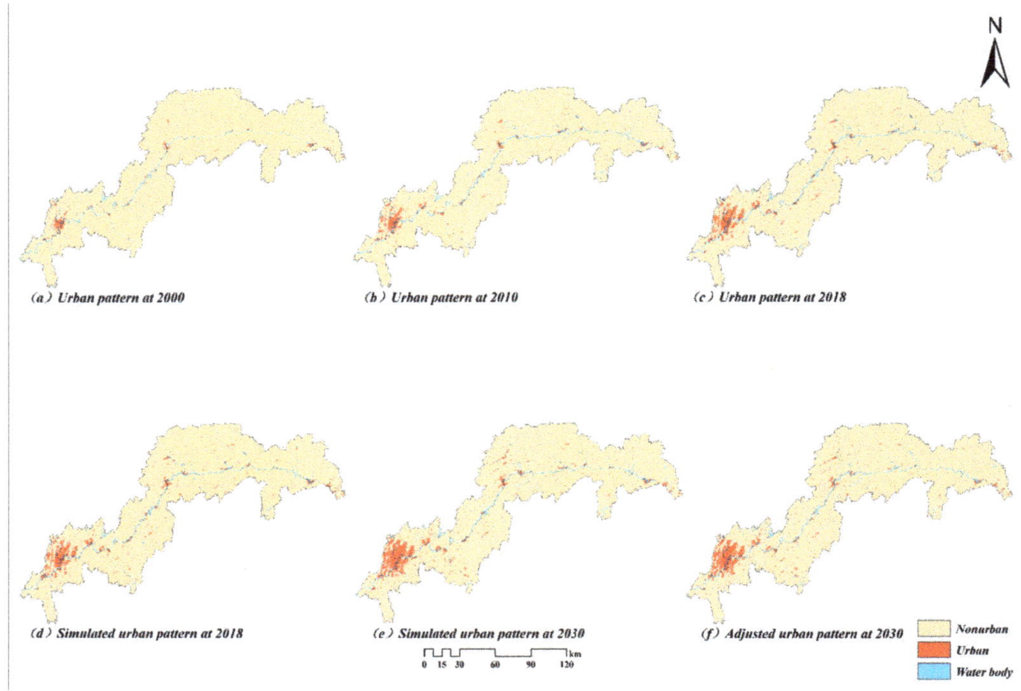

Figure 7. Urban patterns for 2000, 2010 and 2018, simulated scenarios for 2018, 2030 and adjusted scenario for 2030 in three classes: (**a**) urban pattern in 2000; (**b**) urban pattern in 2010; (**c**) urban pattern in 2018; (**d**) simulated urban pattern in 2018; (**e**) simulated urban pattern in 2030; (**f**) adjusted urban pattern in 2030.

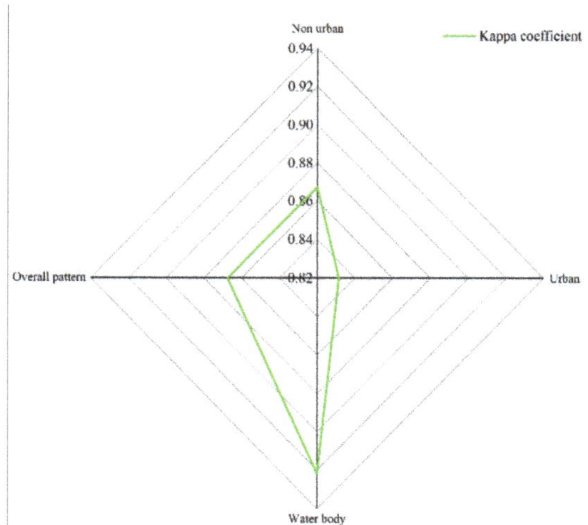

Figure 8. Kappa coefficient of three classes and overall pattern.

3.2.2. The Simulation of Urban Expansion Based on Ecological Sensitivity

This study will use the trained sample to predict the state of urban expansion in the Three Gorges Reservoir area in 2030, the urban expansion simulation map of the study area in 2030 (Figure 7e) and the area of urban land is about 226,594.43 ha. The focus is on the protection of ecological core zone and ecological key zone in the context of "Ecological

Priority and Green Development" so that the urban expansion can be effectively adjusted. Using spatial overlay and analysis tools, the adjusted land use is shown in Figure 7f. Excluding the part of urban land that overlaps with the ecological core zone and ecological key zone, the area of this part of land is about 15,181.92 ha, and finally, the area of urban land is adjusted to 211,412.51 ha in 2030.

3.2.3. The Analysis of the Urban Expansion Simulation

The results of the 2030 urban land expansion were simulated based on ecological sensitivity in the Three Gorges Reservoir area, and we identified the development direction and area change of urban expansion in different districts in the study area. A map of the urban expansion change in the study area from 2018–2030 is shown in Figure 9, the faster growth in urban land is mainly occurring around the main urban areas of Chongqing. The statistical analysis yields a statistical map of the urban expansion area as shown in Figure 10, the northwestern part of Banan shows the most growth in urban land, about 9903.69 ha; followed by the northwestern part of Yubei, with an increase of 9400.41 ha; the growth area of Jiulongpo, Shapingba and Beibei are about 7448.76 ha, 7308.09 ha and 6133.68 ha, respectively; Yuzhong growth area is only 100.80 ha in urban land, and it is the only city-wide urban functional core area in Chongqing and past large-scale urban expansion has resulted in less available land for development and construction. Next, the smallest areas of urban land growth are in Xingshan, Zigui and Wushan. These areas are constrained by natural conditions, economic development and policy regulation, making them unsuitable for future urban expansion.

3.3. Estimation of ESV Change

3.3.1. ESV Change from 2000 to 2018

The ESV was calculated by Formulas (9) and (10) and land use data as shown in Table 6, Figures 11 and 12.

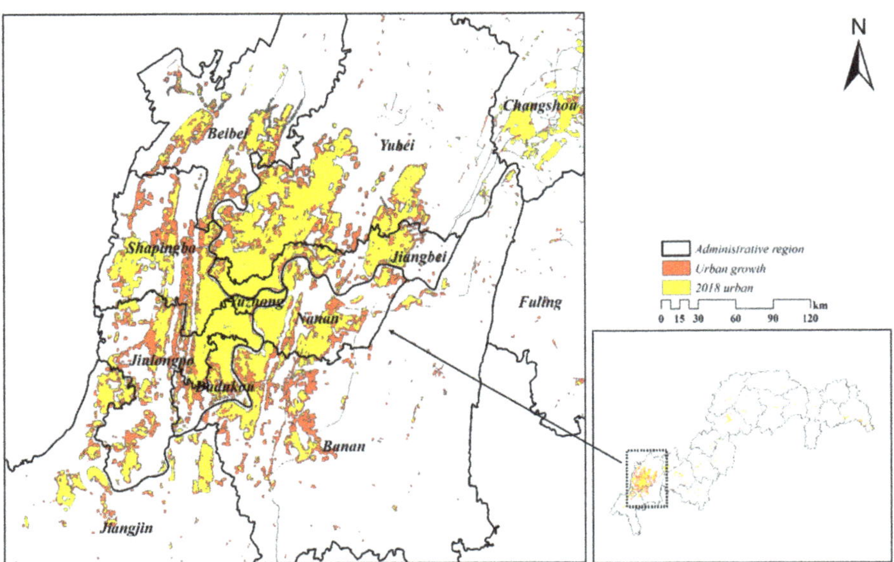

Figure 9. Modeled urban expansion from 2018 to 2030.

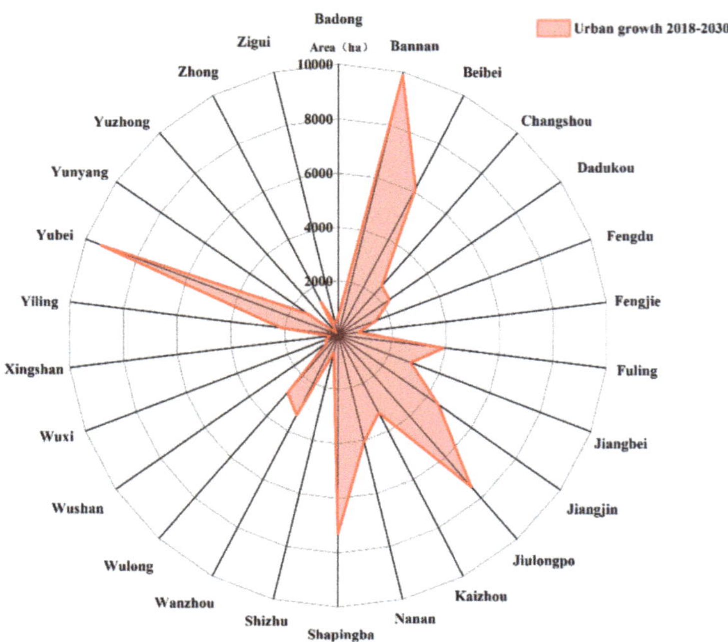

Figure 10. The urban growth area of various districts from 2018 to 2030 in the Three Gorges Reservoir area.

Table 6. The ESV of different land use types in the study area from 2000 to 2018.

Land Use Type	2000		2010		2018	
	ESV (Million USD)	Percentage (%)	ESV (Million USD)	Percentage (%)	ESV (Million USD)	Percentage (%)
Dry land	3188.74	8.32	3135.26	7.75	3105.17	7.40
Paddy field	1229.56	3.21	1196.58	2.96	1139.29	2.71
Coniferous forest	7960.03	20.77	7959.81	19.68	7171.30	17.09
Mixed forest	719.17	1.88	883.25	2.18	876.58	2.09
Broadleaved forest	10,893.48	28.43	11,314.77	27.97	16,575.17	39.50
Bush	5923.07	15.46	6150.63	15.20	3372.12	8.04
Meadow	4252.15	11.10	3578.59	8.85	3413.38	8.13
Wetland	537.39	1.40	459.82	1.14	515.56	1.23
Lake and river	3616.79	9.44	5775.83	14.28	5796.12	13.81
Barren	0.10	0.00	0.05	0.00	0.05	0.00
Total	38,320.49	—	40,454.61	—	41,964.75	—

From Table 6, it can be found that the total ESV of the study area is USD 38.32 billion (2000), USD 40.45 billion (2010) and USD 41.96 billion (2018), respectively. In 18 years, the total ESV of the study area increased by USD 3644.26 million. The ecosystem services of the broadleaved forests, the coniferous forests, the bush, the meadows, the lakes and the rivers provide the main systems in various ecosystem services, and their value accounted for 85.20% (2000), 85.98% (2010) and 86.57% (2018) of the total value. The broadleaved forests had the highest ESV, about 1 USD 0.89 billion (2000), USD 11.31 billion (2010) and USD 16.58 billion (2018); the coniferous forests have the second highest value, about USD

7.96 billion (2000), USD 7.96 billion (2010) and USD 7.17 billion (2018), respectively. From 2000 to 2018, the value of broadleaved forests has increased most significantly, while the value of the dry land and paddy fields has been decreasing. The results indicate that although the Three Gorges Reservoir area has been affected by urban expansion to some extent, the overall value has maintained an increasing trend, mainly due to the increasing area of broadleaved forests, which provide a large amount of ESV.

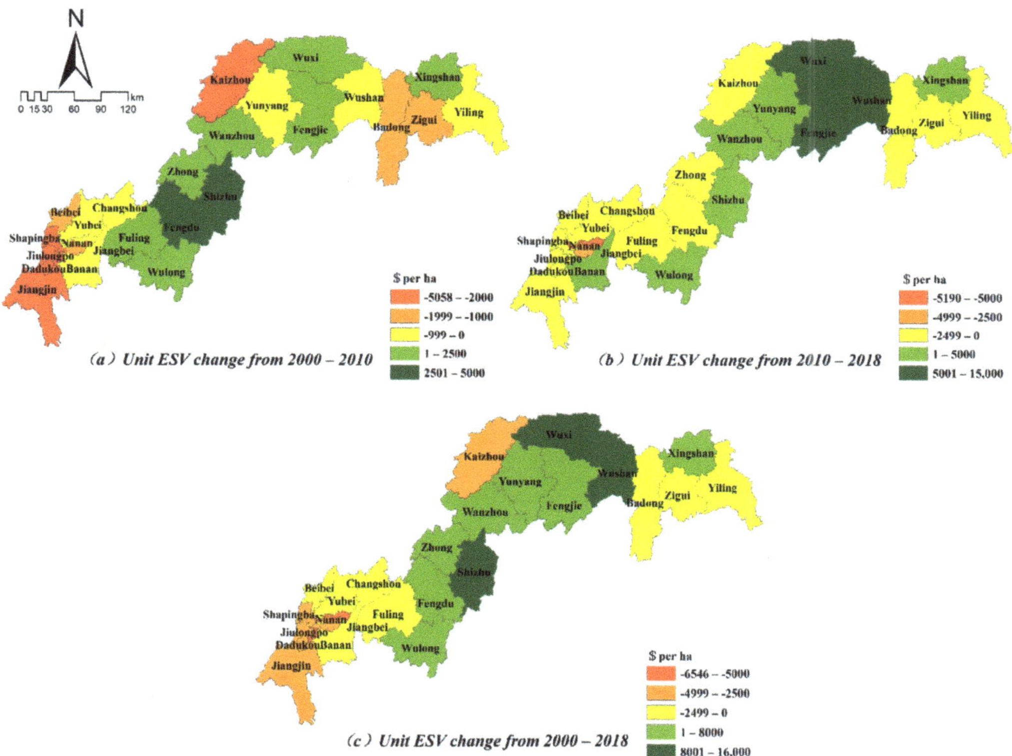

Figure 11. The change in unit ESV from 2000 to 2018.

At the district scale, nine districts and counties had increasing unit ESVs from 2000 to 2010 (Figure 11a), this growth is partially concentrated in the middle section of the Three Gorges Reservoir area, with the highest gain in Shizhu (4052.54 USD/ha). However, the decrease in unit ESV is mainly concentrated in the vicinity of the main city of Chongqing: Dadukou had the largest decrease, about 5058.49 USD/ha. The total value showed an increasing trend in the Three Gorges Reservoir area between 2010 and 2018, but there are still 17 districts that declined in unit ESV (Figure 11b), and the most significant decrease was found in Jiangbei (5190.16 USD/ha). The increase was concentrated in the key protective areas of the Three Gorges Reservoir area, such as Wushan and Wuxi, and the highest increase was found in Wuxi County (14,530.61 USD/ha). Overall, the increase in unit ESV during the 18 years was concentrated in the central part of the Three Gorges Reservoir area (Figure 11c), with Wuxi, Wushan and Shizhu growing faster than 8000 USD/ha, and Dadukou and Jiangbei showing the most significant decrease, both exceeding 5000 USD/ha.

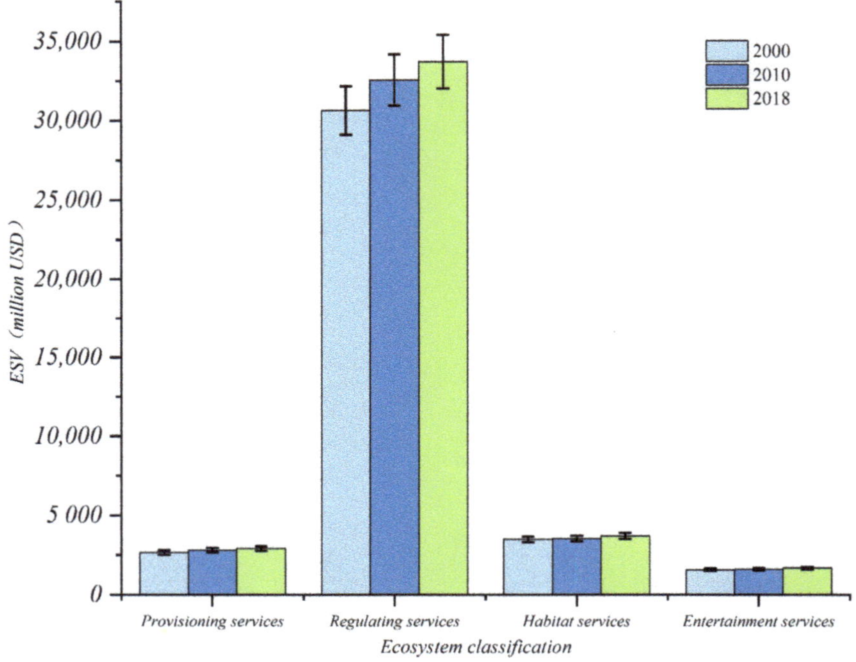

Figure 12. The ESV of four ecosystem classifications from 2000 to 2018.

As shown in Figure 12, the four types of value in the study area showed an increasing trend from 2000 to 2018. We found that regulating services accounted for the highest proportion of overall ESV, about 79.88% (2000), 80.40% (2010) and 80.29% (2018), respectively. The proportion of each type of ecosystem service function had the following order of magnitude: regulating services > habitat services > provisioning services > cultural and amenity services.

3.3.2. Prediction of ESV Losses from 2018 to 2030

The CA-Markov model predicts the urban expansion pattern of the study area in 2030, and the results show that there will be a certain degree of outward urban expansion, mainly in the vicinity of the main city of Chongqing. The urban expansion will lead to a decrease in the value of ecosystem services. Our model suggests that the newer urban area is 80,026.02 ha from 2018 to 2030. In the ten land use types (Figure 13a), The most transferred land to urban areas is dry land and paddy field, about 33,903.54 ha and 28,381.23 ha, respectively, while the least transferred area is wetland, lake and river, about 6.48 ha and 61.38 ha, respectively.

Table 7 shows the significant differences in the losses of other ecological land under urban expansion in different districts from 2018 to 2030. Dry land losses in Banan and Yubei comprised an area of 5083.38 ha and 4681.71 ha, respectively, the sum of which accounted for 28.80% of the total area. The area of paddy field losses in Yubei, Banan, Shapingba and Jiulongpo accounted for 45.69% of the total area. The largest coniferous and mixed forest area losses were in Wulong, Beibei and Kaizhou with areas of about 293.22 ha and 268.65 ha lost, respectively. The losses of broadleaved forest mainly occurred in Jiulongpo (1177.83 ha) and Shapingba (1130.31 ha). The losses of meadows mainly occurred in Fuling (509.67 ha). The largest losses occurred in Zhong (4.05 ha) and Beibei (45.90 ha), respectively. In barren land, Jiangbei and Banan lost the largest area of 376.65 ha and 285.48 ha, respectively.

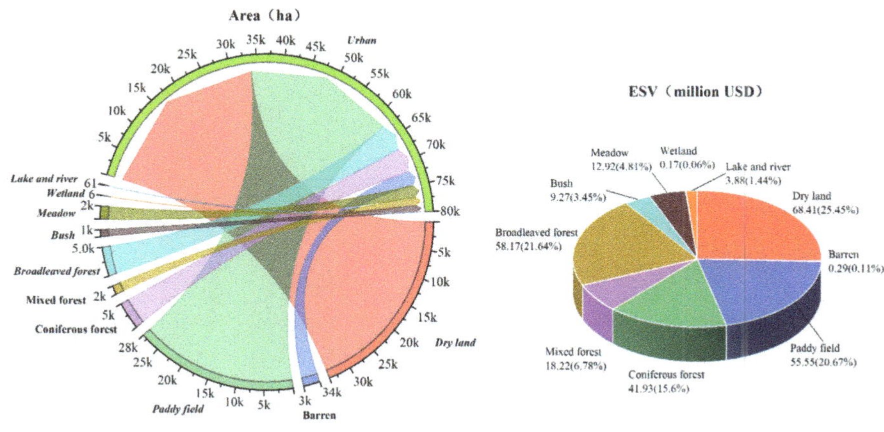

Figure 13. The losses area of land-use area and ESV from 2018 to 2030.

Table 7. The land-use losses in different districts from 2018–2030.

Region (ha)	Farmland		Forest			Grassland	Water Body		Unused Land
	Dry Land	Paddy Field	Coniferous Forest	Mixed Forest	Broadleaved Forest	Bush	Meadow	Wetland	Lake and River	Barren
Badong	138.6	143.19	216.81	76.05	54.36	1.89	6.21	0	0	43.2
Xingshan	47.79	61.29	42.93	73.89	15.66	13.95	1.26	0	0	51.12
Yiling	209.88	767.88	318.69	71.82	402.21	271.26	4.14	0	0	95.58
Zigui	22.41	108.09	201.87	13.23	45.72	21.15	2.79	0	0	24.75
Banan	5083.38	3623.40	302.04	83.79	434.88	29.25	55.44	0	6.03	285.48
Beibei	3111.93	2213.37	113.49	293.22	139.5	64.17	0.18	0	45.9	151.92
Changshou	679.77	1546.38	45	0	87.12	20.07	27.09	0.99	0	18.54
Dadukou	1135.71	679.86	175.32	26.64	232.47	0	0	0	0	48.87
Fuling	1281.42	1466.01	337.32	14.4	153.36	1.71	509.67	0	0	160.47
Jiangbei	1221.75	840.24	168.93	214.2	38.43	0	0	0	0	376.65
Jiangjin	2434.32	986.67	321.3	57.42	134.91	304.02	176.49	0	0	73.8
Jiulongpo	3026.34	2533.14	327.24	116.19	1177.83	152.01	49.95	0	0	66.06
Nanan	1906.20	1587.96	227.7	8.64	125.19	9.36	0	0	0	171
Shapingba	2778.84	2978.82	225.63	1.35	1130.31	8.01	6.3	0	0	178.83
Wanzhou	1035.72	1462.14	191.88	109.26	14.31	34.56	313.65	0	0	161.91
Yubei	4681.71	3830.94	282.87	49.05	347.04	11.25	64.8	1.44	9.45	121.86
Yuzhong	0	0	0	9.18	0	0	0	0	0	91.62
Fengdu	663.21	533.97	77.49	6.93	59.67	39.87	57.78	0	0	58.68
Fengjie	196.56	235.71	86.85	0.54	51.93	8.19	96.03	0	0	98.91
Kaizhou	1342.89	1261.80	78.3	268.65	10.44	1.98	203.85	0	0	62.28
Shizhu	234.36	266.4	40.05	0	20.43	3.24	73.35	0	0	30.87
Wushan	231.75	78.03	4.23	1.89	119.43	5.58	16.29	0	0	54.18
Wuxi	232.38	101.07	13.41	0	22.5	2.34	133.83	0	0	14.85
Wulong	1284.03	434.97	484.38	0	207.63	174.96	156.69	0	0	95.22
Yunyang	377.01	245.79	193.05	14.49	11.16	16.74	250.47	0	0	271.8
Zhong	545.58	394.11	276.21	57.42	0.18	15.48	39.87	4.05	0	49.86
Total	33,903.54	28,381.23	4752.99	1568.25	5036.67	1211.04	2246.13	6.48	61.38	2858.31

The increasing urbanization will encroach on other ecological lands and directly lead to the losses of ESV by the equivalent coefficients. We found that the total ESV losses is USD 268.81 million in the Three Gorges Reservoir area (Figure 13b). The highest values of dry land and broadleaved forest were lost with USD 68.41 million (25.45%) and USD 58.17 million (21.64%), respectively. The variation of the overall ESV losses was explored from the district scales (Figure 14a). Jiulongpo, Banan, Shapingba and Yubei have higher losses (more than USD 25 million). This result indicates that these areas are at risk of ecological degradation. Yuzhong, Xingshan and Wuxi have the lowest losses, mainly due to the highest level of urbanization in Yuzhong, resulting in limited land for urban development. In addition, Xingshan and Wuxi are as important ecological reserves in China, which limit the large-scale expansion of urban land use. On average, Dadukou, Jiulongpo, Shapingba and Nanan have the highest unit ESV losses (more than 400 USD/ha) (Figure 14b).

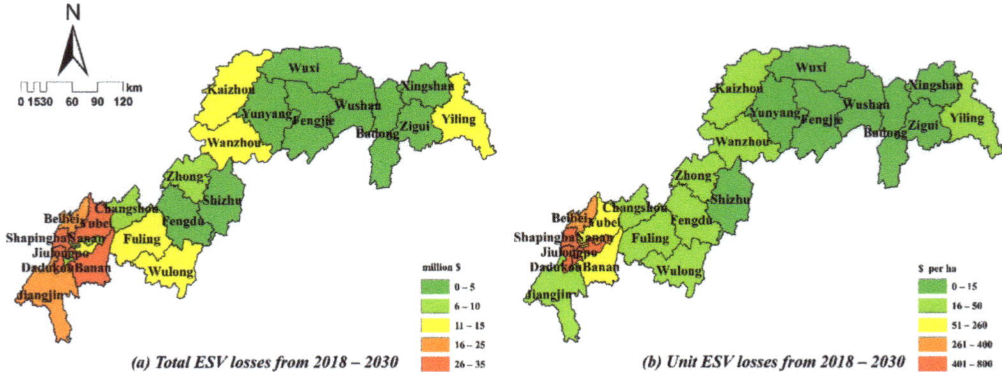

Figure 14. The losses in the total and unit ESV from 2018 to 2030.

Figure 15 summarizes the amount of loss of the four primary services in different districts. In Figure 15a, we find that Banan, Jiulongpo and Jiangjin have the largest loss of provisioning services values (more than USD 1.2 million). There are 13 districts and counties with a loss of less than USD 0.3 million, mainly concentrated in the upper half of the Three Gorges Reservoir area of Wushan, Badong and Zigui, etc. The losses of regulation services values are the largest among the four types of services. Four districts lost more than USD 20 million: Jiulongpo, Banan, Shapingba and Yubei (Figure 15b). These districts are important places for future urban expansion and their ecological regulating services functions will decline from 2018 to 2030. Among the habitat services values and entertainment services values (Figure 15c,d), Jiulongpo and Shapingba have the largest losses.

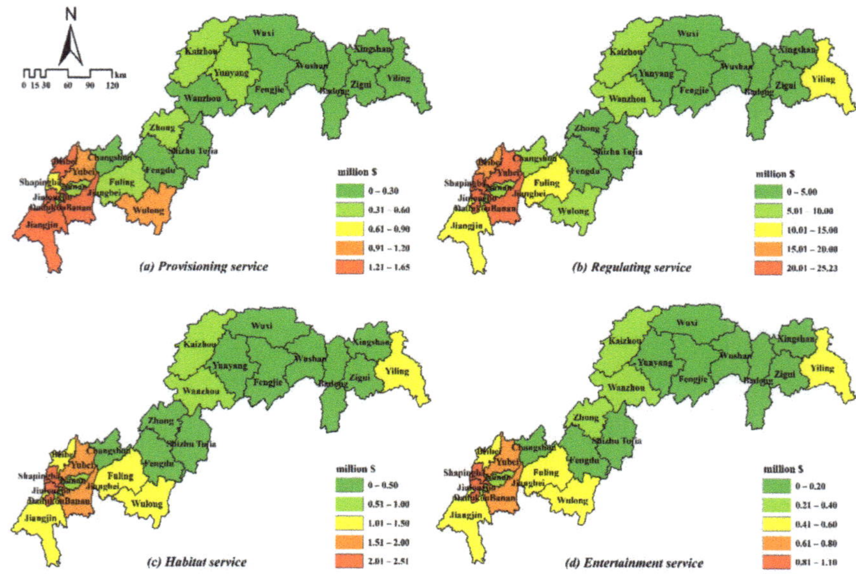

Figure 15. The ESV losses in primary services from 2018 to 2030.

4. Discussion

4.1. The Rationality of Delimiting Ecological Sensitive Areas

First, because of the limitation of collecting data and the need to delimit ecological sensitive areas, according to the actual situation of the Three Gorges Reservoir area, this study selects relevant indicators from 2018 to identify the ecological sensitive areas in the study area in terms of three sensitivities (i.e., soil erosion sensitivity, land desertification sensitivity and soil salinity sensitivity) so as to identify the scope and number of ecological sensitive areas and provide a basis for achieving green development in the region [54]. As of now, the indicators and data will be further updated, which is a hot topic for future research.

Second, the grid size used to identify ecological sensitive areas is 30 m × 30 m in this paper. However, some of the raw data (NDVI, soil type, groundwater mineralization and groundwater depth) are of higher resolution, but still can accurately identify the extent of ecological sensitive zone. We believe that higher resolution data will be used in the future to improve the accuracy of delimiting ecological sensitive zones.

Third, in the comprehensive ecological sensitivity evaluation, the areas with high sensitivity or above account for 35.86% of the study area, with an area of 20,639.71 km^2, mainly in Wushan, Fengjie and Yunyang. This part is a forest reserve with mainly mountainous terrain and high elevation, and its ecosystem types are diversified and spatially heterogeneous, which are vulnerable to some natural disasters and human factors. The harm to the ecological sensitive zone should be reduced. Meanwhile, a corresponding ecological compensation mechanism is established [55], the ecological advantages will be transformed into economic advantages, the ecological and economic benefits of the study area will be improved, and the green development concept of "making all-out efforts to protect it, and forbidding large-scale development" will be fully implemented.

4.2. The Accuracy of Simulated Urban Expansion

This study combines the ecological sensitivity, CA and Markov models in the context of green development to simulate and predict the future urban spatial patterns in the Three Gorges Reservoir area. The urban construction land area is predicted to increase by 80,026.02ha between 2018 and 2030, while the main expansion areas are in the northwestern part of Banan, the northwestern part of Yubei, Jiulongpo, Shapingba and Beibei. The analysis revealed that the future urban expansion in the study area is mainly occupying paddy fields and drylands, occupying an area of 33,903.54ha and 28,381.23ha, respectively.

The driving factors affecting urban expansion were mainly selected as elevation, slope, earthquake, landslide, highway, main road, railroad and population density. These factors performed relatively well in simulating urban expansion in 2018, with an overall kappa coefficient of 86.74% compared to the actual 2018 urban sites, but urban expansion is a complex dynamic system [56]. It is not enough to consider only the above eight driving factors. In the future, more socioeconomic and ecological data should be obtained to provide a basis for improving the accuracy of urban expansion simulation. At the same time, different prediction scenarios [57–59] and different grids sizes [33,60] play different roles in the simulation of urban expansion.

4.3. Estimation of ESV Dynamics and Prediction of Future ESV Losses as an Effective Tool for Ecological Safety Management

The estimation of ESV can provide a basis for the region to achieve sustainable development [61]. The ESV change in the study area were calculated by using the land use data from 2000–2018 and the ESV value equivalent table. The results show that the ESV in the Three Gorges Reservoir area increased from USD 38,320.49 million in 2000 to USD 41,964.75 million in 2018, and the overall ESV change showed an increasing trend. This may be due to the continuous increase in the area of forest land (especially broadleaf forest) due to the implementation of the reforestation project, which also brings an increase in the value of different ecosystem service functions [62,63]. However, it is noteworthy that between 2000 and 2018, there are 16 districts and counties with decreasing ESV per unit. The most significant decreases (more than 5000 USD/ha) were observed in Dadukou and Jiangbei. With increasing urbanization, we see that ESV will show a decreasing trend and ecosystems will face increasing pressure in the region. In the four different ecosystem services, regulating services provide the highest value, followed by habitat services, and finally provisioning services and entertainment services.

As urbanization continues, it may cause degradation of the ecosystem [64,65]. This study uses the CA-Markov model to predict the trend of urban expansion in 2030, and the study finds that the growth of urban land in the study area is mainly through the occupation of dry land and paddy fields. From 2018 to 2030, the overall ESV of the Three Gorges Reservoir area will be reduced by about USD 268.81 million. At the district and county scales, the ESV of Jiulongpo, Banan and Shapingba will be reduced most significantly, and the ecological pressure in the area will be further increased. In starting from the perspective of the four ESV, we found that regulating services values decreased the most. Therefore, it is necessary to combine the ecological sensitivity method and ESV evaluation results to achieve a more reasonable and targeted implementation of ecological spatial management and steadily realize the green development strategy.

This study provides insights into ecological conservation under sustainable urbanization by predicting change in urban land expansion and calculating ESV, using the Three Gorges Reservoir area as an example, which can be of guidance to other countries. Especially, developing countries are facing different problems caused by rapid urbanization, such as India [66], Sri Lanka [53] and Egypt [67]. Each country faces different ecological problems. We should consider geohazard sensitivity and select suitable indicators for the region where geohazards occur frequently. In ESV evaluation, there are differences in different countries' ecosystem value, and the equivalent coefficients table should be changed appropriately according to the regional characteristics. Although there are differences between the spatial scales of different countries, it is necessary to identify the ecological sensitive zones and to predict the future development direction of urban land and the dynamic change of ESV through scientific and rational identification. This quantitative evaluation provides a reference basis for ecological spatial control and effective macro-control.

5. Conclusions and Outlook

In the context of economic globalization, the urbanization process of human beings has become irreversible. Handling the relationship between environmental protection and economic development has become particularly critical. In this study, we simulate the future urban expansion and ESV change in the study area based on ecological sensitivity. Under the guidance of green development concept, the socioeconomic maximization is achieved, and the relationship between ecological protection and urban expansion is coordinated. This study shows that:

1. In the comprehensive ecological sensitivity assessment, we found that the ecological sensitive zone is about 20,639.71 km^2 in the Three Gorges Reservoir area, accounting for 35.86% of the total study area. This part of the area is in Wushan, Fengjie and Yunyang.
2. The results of the study show that the overall ESV in the Three Gorges Reservoir area showed an increasing trend from 2000 to 2018. The growth was about USD 3644.26 million. From the perspective of ESV change in districts and counties, we found that 16 districts and counties were decreasing in unit ESV, Dadukou and Jiangbei decreased most significantly. In four ecosystem services, regulating services provided the highest ESV.
3. In the context of ecological priority and green development, the 2030 urban land was predicted and simulated. In 2018–2030, about 80,026.02 ha of new construction land will be added to the Three Gorges Reservoir area, and the overall ESV will lose USD 268.75 million. The largest losses are in Jiulongpo, Banan and Shapingba.

Although, this method can better delimit ecologically sensitive zones and estimate the dynamic changes of ESV in the Three Gorges Reservoir area, there is still much room for improvement. We will use high-resolution land use data and remote sensing inversion data (FVC, WET, NDBSI, LST and NPP) in the future work. This allows us to better identify zones of ecological sensitivity. Meanwhile, we should also make different simulations of urban expansion for different scenarios, such as developmental orientation or ecological orientation. It can provide better prediction and estimation of ESV changes for future urban expansion in the Three Gorges Reservoir area.

Author Contributions: Data curation, H.P. and J.L.; Funding acquisition, X.Z.; Investigation, H.P. and J.L.; Methodology, H.P., L.H. and X.Z.; Supervision, X.Z. and X.Y.; Visualization, H.P.; Writing—original draft, H.P.; Writing—review & editing, L.H., X.Y. and J.L. All authors have read and agreed to the published version of the manuscript.

Funding: This research was supported by the special scientific research project of Hubei Provincial Land Consolidation Center in 2021 and the special scientific research project of Hubei Research Station for Integrated Land Remediation and Restoration in the Jianghan Plain in 2021.

Institutional Review Board Statement: Not applicable.

Informed Consent Statement: Not applicable.

Data Availability Statement: Not applicable.

Conflicts of Interest: The authors declare no conflict of interest.

References

1. Haeckel, E. *The History of Creation*; Taylor and Francis: London, UK, 1899.
2. Pennekamp, F.; Pontarp, M.; Tabi, A.; Altermatt, F.; Alther, R.; Choffat, Y.; Fronhofer, E.A.; Ganesanandamoorthy, P.; Garnier, A.; Griffiths, J.I.; et al. Biodiversity increases and decreases ecosystem stability. *Nature.* **2018**, *563*, 109–112. [CrossRef] [PubMed]
3. Klain, S.C.; Satterfield, T.A.; Chan, K.M.A. What matters and why? Ecosystem services and their bundled qualities. *Ecol. Econ.* **2014**, *107*, 310–320. [CrossRef]
4. Palomo, I.; Martín-López, B.; Alcorlo, P.; Montes, C. Limitations of Protected Areas Zoning in Mediterranean Cultural Landscapes under the Ecosystem Services Approach. *Ecosystems.* **2014**, *17*, 1202–1215. [CrossRef]
5. Li, R.Q.; Dong, M.; Cui, J.Y.; Zhang, L.L.; Cui, Q.G.; He, W.M. Quantification of the impact of land-use changes on ecosystem services: A case study in Pingbian County, China. *Env. Monit. Assess.* **2007**, *128*, 503–510. [CrossRef]
6. Costanza, R.; de Groot, R.; Sutton, P.; van der Ploeg, S.; Anderson, S.J.; Kubiszewski, I.; Farber, S.; Turner, R.K. Changes in the global value of ecosystem services. *Glob. Environ. Chang.* **2014**, *26*, 152–158. [CrossRef]
7. Yang, Y.; Song, G.; Lu, S. Study on the ecological protection redline (EPR) demarcation process and the ecosystem service value (ESV) of the EPR zone: A case study on the city of Qiqihaer in China. *Ecol. Indic.* **2020**, *109*, 105754. [CrossRef]
8. Bommarco, R.; Vico, G.; Hallin, S. Exploiting ecosystem services in agriculture for increased food security. *Glob. Food Secur.* **2018**, *17*, 57–63. [CrossRef]
9. Xie, G.; Zhang, G.; Zhang, L.; Chen, W.; Li, M. Improvement of the Evaluation Method for Ecosystem Service Value Based on Per Unit Area. *J. Nat. Resour.* **2015**, *30*, 1243–1254.
10. Bastian, O.; Haase, D.; Grunewald, K. Ecosystem properties, potentials and services—The EPPS conceptual framework and an urban application example. *Ecol. Indic.* **2012**, *21*, 7–16. [CrossRef]
11. MEA (Middle East Airline). *Ecosystems and Human Wellbeing: Biodiversity Synthesis*; Island Press: Washington, DC, USA, 2005.
12. Anaya-Romero, M.; Muñoz-Rojas, M.; Ibáñez, B.; Marañón, T. Evaluation of forest ecosystem services in Mediterranean areas. A regional case study in South Spain. *Ecosyst. Serv.* **2016**, *20*, 82–90. [CrossRef]
13. Groot, R.S.D. A typology for the classification, description and valuation of ecosystem functions, goods and services. *Ecol. Econ.* **2002**, *41*, 393–408. [CrossRef]
14. Costanza, R.; D'Arge, R.; DeGroot, R.; Farberk, S.; Grasso, M.; Hannon, B.; Limburg, K.; Naeem, S.; O'Neill, R.V.; Paruelo, J.; et al. The value of the world's ecosystem services and natural capital. *Nature.* **1997**, *387*, 253–260. [CrossRef]
15. Qiu, J.; Zipper, S.C.; Motew, M.; Booth, E.G.; Kucharik, C.J.; Loheide, S.P. Nonlinear groundwater influence on biophysical indicators of ecosystem services. *Nat. Sustain.* **2019**, *2*, 475–483. [CrossRef]
16. Zhou, D.; Tian, Y.; Jiang, G. Spatio-temporal investigation of the interactive relationship between urbanization and ecosystem services: Case study of the Jingjinji urban agglomeration, China. *Ecol. Indic.* **2018**, *95*, 152–164. [CrossRef]
17. Daily, G.C.; Polasky, S.; Goldstein, J.; Kareiva, P.M.; Mooney, H.A.; Pejchar, L.; Ricketts, T.H.; Salzman, J.; Shallenberger, R. Ecosystem services in decision making: Time to deliver. *Front. Ecol. Environ.* **2009**, *7*, 21–28. [CrossRef]

18. La Notte, A.; Liquete, C.; Grizzetti, B.; Maes, J.; Egoh, B.N.; Paracchini, M.L. An ecological-economic approach to the valuation of ecosystem services to support biodiversity policy. A case study for nitrogen retention by Mediterranean rivers and lakes. *Ecol. Indic.* **2015**, *48*, 292–302. [CrossRef]

19. Xie, G.; Zhang, C.; Zhen, L.; Zhang, L. Dynamic changes in the value of China's ecosystem services. *Ecosyst. Serv.* **2017**, *26*, 146–154. [CrossRef]

20. Wainger, L.A.; King, D.M.; Mack, R.N.; Price, E.W.; Maslin, T. Can the concept of ecosystem services be practically applied to improve natural resource management decisions? *Ecol. Econ.* **2010**, *69*, 978–987. [CrossRef]

21. Ouyang, Z.; Zheng, H.; Xiao, Y.; Polasky, S.; Liu, J.; Xu, W.; Wang, Q.; Zhang, L.; Xiao, Y.; Rao, E.; et al. Improvements in ecosystem services from investments in natural capital. *Science.* **2016**, *352*, 1455–1459. [CrossRef]

22. Lautenbach, S.; Kugel, C.; Lausch, A.; Seppelt, R. Analysis of historic changes in regional ecosystem service provisioning using land use data. *Ecol. Indic.* **2011**, *11*, 676–687. [CrossRef]

23. MEP; NDRC. *Guidelines for the Delimitation of Red Lines for Ecological Protection*; Ministry of Environmental Protection of the People's Republic of China (MEP): Beijing, China; National Development and Reform Commission of the People's Republic of China (NDRC): Beijing, China, 2017. (In Chinese)

24. Ding, Q.; Shi, X.; Zhuang, D.; Wang, Y. Temporal and Spatial Distributions of Ecological Vulnerability under the Influence of Natural and Anthropogenic Factors in an Eco-Province under Construction in China. *Sustainability* **2018**, *10*, 3087. [CrossRef]

25. Chi, Y.; Zhang, Z.; Gao, J.; Xie, Z.; Zhao, M.; Wang, E. Evaluating landscape ecological sensitivity of an estuarine island based on landscape pattern across temporal and spatial scales. *Ecol. Indic.* **2019**, *101*, 221–237. [CrossRef]

26. Yuan, W.; Li, J.; Meng, L.; Qin, X.; Qi, X. Measuring the area green efficiency and the influencing factors in urban agglomeration. *J. Clean. Prod.* **2019**, *241*, 118092. [CrossRef]

27. Sun, Y.; Tong, L.; Liu, D. An Empirical Study of the Measurement of Spatial-Temporal Patterns and Obstacles in the Green Development of Northeast China. *Sustainability.* **2020**, *12*, 10190. [CrossRef]

28. García, A.M.; Santé, I.; Boullón, M.; Crecente, R. Calibration of an urban cellular automaton model by using statistical techniques and a genetic algorithm. Application to a small urban settlement of NW Spain. *Int. J. Geogr. Inf. Sci.* **2013**, *27*, 1593–1611. [CrossRef]

29. Ghosh, P.; Mukhopadhyay, A.; Chanda, A.; Mondal, P.; Akhand, A.; Mukherjee, S.; Nayak, S.K.; Ghosh, S.; Mitra, D.; Ghosh, T.; et al. Application of Cellular automata and Markov-chain model in geospatial environmental modeling—A review. *Remote Sens. Appl. Soc. Environ.* **2017**, *5*, 64–77. [CrossRef]

30. Kim, I.; Arnhold, S.; Ahn, S.; Le, Q.B.; Kim, S.J.; Park, S.J.; Koellner, T. Land use change and ecosystem services in mountainous watersheds: Predicting the consequences of environmental policies with cellular automata and hydrological modeling. *Environ. Model. Softw.* **2019**, *122*, 103982. [CrossRef]

31. Gounaridis, D.; Chorianopoulos, I.; Symeonakis, E.; Koukoulas, S. A Random Forest-Cellular Automata modelling approach to explore future land use/cover change in Attica (Greece), under different socio-economic realities and scales. *Sci. Total Environ.* **2019**, *646*, 320–335. [CrossRef]

32. Ren, W.; Zhang, X.; Shi, Y. Evaluation of Ecological Environment Effect of Villages Land Use and Cover Change: A Case Study of Some Villages in Yudian Town, Guangshui City, Hubei Province. *Land.* **2021**, *10*, 251. [CrossRef]

33. Chen, S.; Feng, Y.; Tong, X.; Liu, S.; Xie, H.; Gao, C.; Lei, Z. Modeling ESV losses caused by urban expansion using cellular automata and geographically weighted regression. *Sci. Total Environ.* **2020**, *712*, 136509. [CrossRef] [PubMed]

34. Thapa, R.B.; Murayama, Y. Drivers of urban growth in the Kathmandu valley, Nepal: Examining the efficacy of the analytic hierarchy process. *Appl. Geogr.* **2010**, *30*, 70–83. [CrossRef]

35. Shu, B.; Zhu, S.; Qu, Y.; Zhang, H.; Li, X.; Carsjens, G.J. Modelling multi-regional urban growth with multilevel logistic cellular automata. *Comput. Environ. Urban Syst.* **2020**, *80*, 101457. [CrossRef]

36. Yang, X.; Chen, R.; Zheng, X.Q. Simulating land use change by integrating ANN-CA model and landscape pattern indices. *Geomat. Nat. Hazards Risk.* **2015**, *7*, 918–932. [CrossRef]

37. Xu, X.; Du, Z.; Zhang, H. Integrating the system dynamic and cellular automata models to predict land use and land cover change. *Int. J. Appl. Earth Obs. Geoinf.* **2016**, *52*, 568–579. [CrossRef]

38. Adhikari, S.; Southworth, J. Simulating Forest Cover Changes of Bannerghatta National Park Based on a CA-Markov Model: A Remote Sensing Approach. *Remote Sens.* **2012**, *4*, 3215–3243. [CrossRef]

39. Ma, X.; Li, Y.; Li, B.; Han, W.; Liu, D.; Gan, X. Nitrogen and phosphorus losses by runoff erosion: Field data monitored under natural rainfall in Three Gorges Reservoir Area, China. *Catena.* **2016**, *147*, 797–808. [CrossRef]

40. Lu, C.-J.; Luo, D.; Junaid, M.; Duan, J.-J.; Ding, S.-M.; Dai, A.-X.; Cao, T.-W.; Pei, D.-S. The Status of Pollutants in the Three Gorges Reservoir Area, China and its Ecological Health Assessment. *Am. J. Environ. Sci.* **2016**, *12*, 308–316. [CrossRef]

41. Chu, S.; Chen, L. Evaluation of energy conservation and emission reduction in Anhui Province based on coefficient of variation method. *China Popul. Resour. Environ.* **2011**, *21*, 512–516. (In Chinese)

42. Guan, Q.; Wang, L.; Clarke, K.C. An Artificial-Neural-Network-based, Constrained CA Model for Simulating Urban Growth. *Cartogr. Geogr. Inf. Sci.* **2005**, *32*, 369–380. [CrossRef]

43. Wang, Z.-G.; Qiao, F.-W.; Zhou, L.; Che, L.; Bai, Y.-P.; Yang, X.-D. Delimitation of urban growth boundary in ecologically vulnerable areas in the Upper Yellow River: Take Linxia Hui Autonomous Prefecture as an example. *J. Nat. Resour.* **2021**, *36*, 162–175. (In Chinese) [CrossRef]

44. Guan, D.; Li, H.; Inohae, T.; Su, W.; Nagaie, T.; Hokao, K. Modeling urban land use change by the integration of cellular automaton and Markov model. *Ecol. Model.* **2011**, *222*, 3761–3772. [CrossRef]
45. Cohen, J. A Coefficient of Agreement for Nominal Scales. *Educ. Psychol. Meas.* **1960**, *20*, 37–46. [CrossRef]
46. Shiferaw, H.; Bewket, W.; Alamirew, T.; Zeleke, G.; Teketay, D.; Bekele, K.; Schaffner, U.; Eckert, S. Implications of land use/land cover dynamics and Prosopis invasion on ecosystem service values in Afar Region, Ethiopia. *Sci. Total Environ.* **2019**, *675*, 354–366. [CrossRef]
47. Howarth, R.B.; Farber, S. Special Issue: The Dynamics and Value of Ecosystem Services: Integrating Economic and Ecological Perspectives Accounting for the value of ecosystem services. *Ecol. Econ.* **2002**, *41*, 421–429. [CrossRef]
48. Zhang, B.; Li, W.; Xie, G. Ecosystem services research in China: Progress and perspective. *Ecol. Econ.* **2010**, *69*, 1389–1395. [CrossRef]
49. Wang, X.; Ouyang, Z.; Xiao, H.; Miao, H.; Fu, B. Distribution and division of sensitivity to water-caused soil loss in China. *Acta Ecol. Sin.* **2001**, *21*, 14–19. (In Chinese)
50. Tang, J.; Tang, X.; Qin, Y.; He, Q.; Yi, Y.; Ji, Z. Karst rocky desertification progress: Soil calcium as a possible driving force. *Sci. Total Environ.* **2019**, *649*, 1250–1259. [CrossRef] [PubMed]
51. Chen, W.; Lu, S.; Pan, N.; Jiao, W. Impacts of long-term reclaimed water irrigation on soil salinity accumulation in urban green land in Beijing. *Water Resour. Res.* **2013**, *49*, 7401–7410. [CrossRef]
52. Xie, J.; Xu, W.; Zhang, M.; Qiu, C.; Liu, J.; Wisniewski, M.; Ou, T.; Zhou, Z.; Xiang, Z. The impact of the endophytic bacterial community on mulberry tree growth in the Three Gorges Reservoir Ecosystem, China. *Environ. Microbiol.* **2020**, *23*, 1858–1875. [CrossRef] [PubMed]
53. Landis, J.R.; Koch, G.G. The Measurement of Observer Agreement for Categorical Data. *Biometrics.* **1977**, *33*, 159–174. [CrossRef]
54. Pecci, A.; Patil, G.; Rossi, O.; Rossi, P. Biodiversity protection funding preference: A case study of hotspot geoinformatics and digital governance for the Map of Italian Nature in the presence of multiple indicators of ecological value, ecological sensitivity and anthropic pressure for the Oltrepò Pavese and Ligurian-Emilian Apennine study area in Italy. *Environ. Ecol. Stat.* **2010**, *17*, 473–502. [CrossRef]
55. Li, S.; Nie, X.; Zhang, A. Research progress on farmland ecological compensation mechanism based on ecosystem service evaluation. *Resour. Sci.* **2020**, *42*, 2251–2260. (In Chinese) [CrossRef]
56. Long, H.; Liu, Y.; Hou, X.; Li, T.; Li, Y. Effects of land use transitions due to rapid urbanization on ecosystem services: Implications for urban planning in the new developing area of China. *Habitat Int.* **2014**, *44*, 536–544. [CrossRef]
57. Ku, C.-A. Incorporating spatial regression model into cellular automata for simulating land use change. *Appl. Geogr.* **2016**, *69*, 1–9. [CrossRef]
58. Liao, W.; Liu, X.; Xu, X.; Chen, G.; Liang, X.; Zhang, H.; Li, X. Projections of land use changes under the plant functional type classification in different SSP-RCP scenarios in China. *Sci. Bull.* **2020**, *65*, 1935–1947. [CrossRef]
59. Liang, X.; Liu, X.; Li, X.; Chen, Y.; Tian, H.; Yao, Y. Delineating multi-scenario urban growth boundaries with a CA-based FLUS model and morphological method. *Landsc. Urban Plan.* **2018**, *177*, 47–63. [CrossRef]
60. Feng, Y.; Wang, J.; Tong, X.; Liu, Y.; Lei, Z.; Gao, C.; Chen, S. The Effect of Observation Scale on Urban Growth Simulation Using Particle Swarm Optimization-Based CA Models. *Sustainability.* **2018**, *10*, 4002. [CrossRef]
61. Liu, C.; Li, Y.; Yang, H.; Min, J.; Wang, C.; Zhang, H. RS and GIS-based Assessment for Eco-environmental Sensitivity of the Three Gorges Reservoir Area of Chongqing. *Acta Geogr. Sin.* **2011**, *66*, 631–642. (In Chinese)
62. Wu, D.; Huang, Z.; Xiao, W.; Zeng, L. Control of soil nutrient loss in a typical conversion model of farmland to forest in the Three Gorges Reservoir area. *Environ. Sci.* **2015**, *36*, 3825–3831. (In Chinese)
63. Yang, B.; Zhiyun, O.; Hua, Z.; Weihua, X.; Bo, J.; Yu, F. Evaluation of the forest ecosystem services in Haihe River Basin, China. *Acta Ecol. Sin.* **2011**, *31*, 2029–2039. (In Chinese)
64. Lambin, E.F.; Meyfroidt, P. Global land use change, economic globalization, and the looming land scarcity. *Proc. Natl. Acad. Sci. USA* **2011**, *108*, 3465–3472. [CrossRef] [PubMed]
65. Su, S.; Xiao, R.; Jiang, Z.; Zhang, Y. Characterizing landscape pattern and ecosystem service value changes for urbanization impacts at an eco-regional scale. *Appl. Geogr.* **2012**, *34*, 295–305. [CrossRef]
66. Ramachandra, T.V.; Bharath, A.H.; Sowmyashree, M.V. Monitoring urbanization and its implications in a mega city from space: Spatiotemporal patterns and its indicators. *J. Environ. Manag.* **2015**, *148*, 67–81. [CrossRef] [PubMed]
67. Abd-Elmabod, S.K.; Fitch, A.C.; Zhang, Z.; Ali, R.R.; Jones, L. Rapid urbanisation threatens fertile agricultural land and soil carbon in the Nile delta. *J. Environ. Manag.* **2019**, *252*, 109668. [CrossRef] [PubMed]

 *sustainability*

Article

Ecosystem Services Monitoring in the Muthurajawela Marsh and Negombo Lagoon, Sri Lanka, for Sustainable Landscape Planning

Darshana Athukorala [1,*], Ronald C. Estoque [2], Yuji Murayama [3] and Bunkei Matsushita [3]

1 Graduate School of Life and Environmental Sciences, University of Tsukuba, 1-1-1 Tennodai, Tsukuba 305-8572, Ibaraki, Japan
2 Center for Biodiversity and Climate Change, Forestry and Forest Products Research Institute, Tsukuba 305-8687, Ibaraki, Japan; estoquerc21@affrc.go.jp
3 Faculty of Life and Environmental Sciences, University of Tsukuba, 1-1-1 Tennodai, Tsukuba 305-8572, Ibaraki, Japan; mura@geoenv.tsukuba.ac.jp (Y.M.); matsushita.bunkei.gn@u.tsukuba.ac.jp (B.M.)
* Correspondence: s1830207@s.tsukuba.ac.jp or darshana12594@gmail.com

Abstract: In this study, we examined the impacts of urbanization on the natural landscape and ecosystem services of the Muthurajawela Marsh and Negombo Lagoon (MMNL) located in the Colombo Metropolitan Region, Sri Lanka, with the goal to help inform sustainable landscape and urban planning. The MMNL is an important urban wetland ecosystem in the country but has been under the immense pressure of urbanization where the natural cover (e.g., marshland and mangrove areas) is continuously being converted to urban use (e.g., residential and commercial). Here, we estimated and assessed the changes in the ecosystem service value (ESV) of the MMNL based on land use/cover (LUC) changes over the past two decades (1997–2017). Considering two plausible scenarios, namely a business-as-usual (BAU) scenario and ecological protection (EP) scenario, and using a spatially explicit land change model, we simulated the future (2030) LUC changes in the area and estimated the potential consequent future changes in the ESV of the MMNL. The results revealed that from 1997 to 2017, the ESV of the MMNL decreased by USD 8.96 million/year (LKR 1642 million/year), or about 33%, primarily due to the loss of mangrove and marshland from urban expansion. Under a BAU scenario, by 2030, it would continue to decrease by USD 6.01 million/year (LKR 1101 million/year), or about 34%. Under an EP scenario, the projected decrease would be lower at USD 4.79 million/year (LKR 878 million/year), or about 27%. Among the ecosystem services of the MMNL that have been, and would be, affected the most are flood attenuation, industrial wastewater treatment, agriculture production, and support to downstream fisheries (fish breeding and nursery). Overall, between the two scenarios, the EP scenario is the more desirable for the sustainability of the MMNL. It can help flatten its curve of continuous ecological degradation; hence, it should be considered by local government planners and decision-makers. In general, the approach employed is adaptable and applicable to other urban wetland ecosystems in the country and the rest of the world.

Keywords: wetland ecosystem; urban wetland; wetland ecosystem services; Muthurajawela Marsh; Negombo Lagoon; sustainability; land change modeling; scenario modeling

Citation: Athukorala, D.; Estoque, R.C.; Murayama, Y.; Matsushita, B. Ecosystem Services Monitoring in the Muthurajawela Marsh and Negombo Lagoon, Sri Lanka, for Sustainable Landscape Planning. *Sustainability* **2021**, *13*, 11463. https://doi.org/10.3390/su132011463

Academic Editor: Elena Cristina Rada

Received: 4 August 2021
Accepted: 12 October 2021
Published: 17 October 2021

Publisher's Note: MDPI stays neutral with regard to jurisdictional claims in published maps and institutional affiliations.

1. Introduction

Wetland ecosystems provide a wide range of valuable ecosystem services, such as flood control, pollution control, water conservation, and climate regulation [1,2]. They feature prominently in the United Nations' Sustainable Development Goals (SDGs) and targets [3]. In urban areas, urban wetland ecosystems play important roles, providing ecosystem services that contribute to the maintenance and sustainability of urban ecological environment and the overall safety and livability of urban regions [4–6]. However, anthropogenic activities such as industrialization, agricultural expansion, and urbanization have changed, diminished, or destroyed most wetlands in recent decades [7,8], including urban wetland ecosystems [9].

In this study, we examined the impacts of urbanization on the natural landscape and ecosystem services of the Muthurajawela Marsh and Negombo Lagoon (MMNL), an important urban wetland ecosystem and one of the top priority wetlands in Sri Lanka. Here, we used settlement expansion as a proxy indicator of urbanization, where settlement is a land use/cover (LUC) class that includes low-intensity and high-intensity urban areas, industrial zones, transportation hubs, airports, home gardens, asphalt areas, and residential areas (more on Section 2.2). Owing to its geographical and biophysical characteristics, the MMNL is a source of valuable ecosystem services, such as flood attenuation, water purification, carbon sequestration, and fish breeding and nursery [10,11].

However, because of rapid and uncontrolled urbanization that has led to the loss of natural cover [9,12], we hypothesize that these ecosystem services, including the ecosystem service value (ESV) of the MMNL have been affected. Among the major challenges that local government planners and decision-makers and other concerned groups and individuals are facing today is how the MMNL's curve of continuous ecological degradation can be flattened. The goal of our study is to help inform landscape and urban planning towards this context and for the sustainability of the MMNL in general.

In previous studies, the concept of ecosystem services has been included in spatiotemporal monitoring and assessments of the impacts of urbanization in many parts of the world, both in non-wetland urban regions [13–17] and in urban wetland regions [4,6]. Advances in geospatial technology, including the increasing availability of Earth observation (remote sensing) data at various spatial and temporal resolutions, have helped to improve social-ecological monitoring and assessments. Furthermore, the development of land change models has helped researchers to project future LUC changes and explore different scenarios [13,14,18–20].

Several studies have been conducted to investigate various aspects of the MMNL. For example, Athukorala et al. [9] have studied the impacts of urbanization on the MMNL, emphasizing implications for landscape planning towards a sustainable urban ecosystem. Jayathilake and Chandrasekara [21] have investigated the variations of avifaunal diversity concerning land use modifications in the Negombo estuary. Emerton and Kekulandala [11] have assessed the economic value of the Muthurajawela Marsh. Bambaradeniya et al. [22] have studied the biodiversity status in the Muthurajawela wetland sanctuary. However, we are not aware of any study that monitored the past-present changes and/or projected the future changes in the ecosystem services and ESV of the MMNL based on Earth observation data and geospatial techniques.

Hence, in this study, we assessed changes in the ESV of the MMNL based on LUC changes over the past two decades (1997–2017) using Earth observation data. Considering two plausible scenarios, namely a business-as-usual (BAU) scenario and ecological protection (EP) scenario, and using a spatially explicit land change model, we simulated future (2030) LUC changes in the area and estimated potential consequent future changes in the ESV of the MMNL. We discussed the implications of the results in the context of the MMNL's sustainability.

2. Methods

2.1. Study Area

Figure 1 shows a cross-section of the MMNL, classified into five ecological zones, each with a description of the level of anthropogenic activities, water conditions, soil groups, vegetation types, and some ecosystem services. The MMNL is situated on the western coastal plain of Sri Lanka, within the Colombo Metropolitan Region (Figure 2). It has a total area of approximately 134 km^2 [9,10]. In 1996, the Government of Sri Lanka designated the northern section of the MMNL as a wetland sanctuary (Muthurajawela Sanctuary) owing to its high ecological and biological significance [10] (Figure 2, PA1). In 2006, another protected area was designated by the government (Muthurajawela Environmental Protection Area) for ecosystem services, including flood control [23] (Figure 2, PA2).

The Negombo Lagoon is linked to the Indian Ocean by a single narrow opening at the northern end of the channel segment (Figure 2). The Muthurajawela Marsh stretches southward from the lagoon, forming the largest coastal peat bog in the country [9]. The elevation range is approximately from −13 to 44 m above sea level. This urban wetland ecosystem receives plenty of rainwater from the southwest monsoon. Annual average rainfall ranges from 2000 to 2500 mm, and annual average temperatures are from 22.5 °C to 25.0 °C [24].

The marsh plant vegetations are in their final stages of succession, leading to dry land formation [10]. In the MMNL, 194 species of flora have been recorded under seven major plant communities—marsh, reed swamp, short grassland, shrubland, lentic, stream bank, and mangrove swamp. For species of fauna, 40 fishes (4 of which are endemic and nationally endangered), 14 amphibians, 31 reptiles, 102 birds (1 endemic and 19 winter migratory birds), and 22 mammals have also been recorded [10]. The aquatic resources are abundant in phytoplankton, phosphors, and algae, all of which are essential components in the food web of various organisms [25].

Today, the sustainability of this valuable urban wetland ecosystem is under threat from the growing pressure of urbanization. Flattening the MMNL's curve of continuous ecological degradation is important, not only as a research endeavor but also as a landscape and urban planning priority.

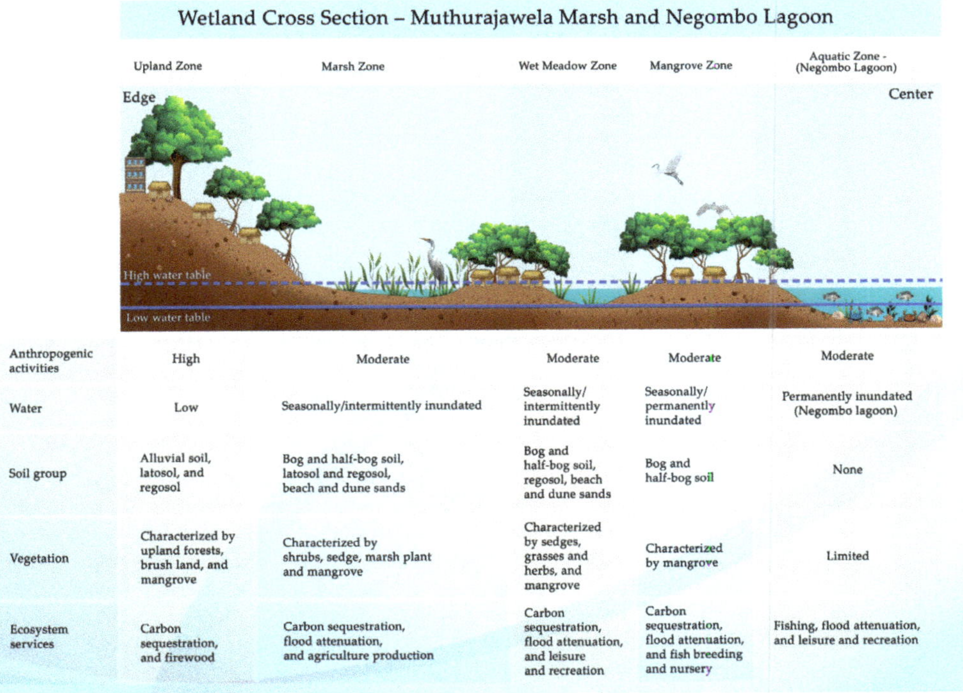

	Upland Zone	Marsh Zone	Wet Meadow Zone	Mangrove Zone	Aquatic Zone - (Negombo Lagoon)
Anthropogenic activities	High	Moderate	Moderate	Moderate	Moderate
Water	Low	Seasonally/intermittently inundated	Seasonally/ intermittently inundated	Seasonally/ permanently inundated	Permanently inundated (Negombo lagoon)
Soil group	Alluvial soil, latosol, and regosol	Bog and half-bog soil, latosol and regosol, beach and dune sands	Bog and half-bog soil, regosol, beach and dune sands	Bog and half-bog soil	None
Vegetation	Characterized by upland forests, brush land, and mangrove	Characterized by shrubs, sedge, marsh plant and mangrove	Characterized by sedges, grasses and herbs, and mangrove	Characterized by mangrove	Limited
Ecosystem services	Carbon sequestration, and firewood	Carbon sequestration, flood attenuation, and agriculture production	Carbon sequestration, flood attenuation, and leisure and recreation	Carbon sequestration, flood attenuation, and fish breeding and nursery	Fishing, flood attenuation, and leisure and recreation

Figure 1. Graphical illustration of the cross-section of the MMNL divided into zones (aquatic, mangrove, wet meadow, marsh, and upland). These zones are further characterized based on anthropogenic activities, water level, soil group, vegetation, and ecosystem services. Source: National wetland directory of Sri Lanka [10], Environmental Profile of Muthurajawela and Negombo Lagoon [25], Athukorala et al. 2021 [9].

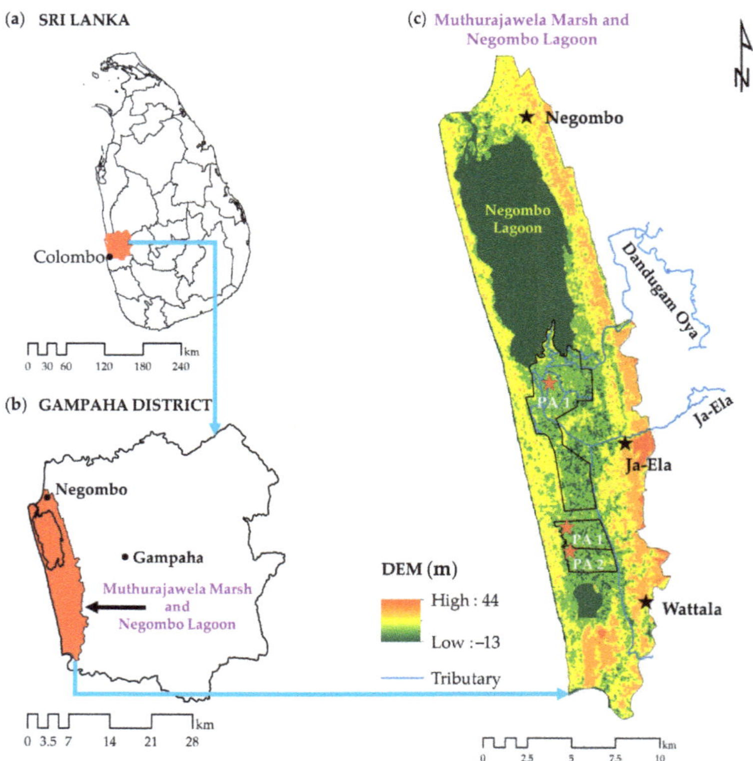

Figure 2. Location of the study area: (**a**) map of Sri Lanka [26]; (**b**) Gampaha District; and (**c**) Muthurajawela Marsh and Negombo Lagoon. PA 1 = Protected area 1 (Muthurajawela Sanctuary). PA 2 = Protected area 2 (Muthurajawela Environmental Protection Area).

2.2. LUC Change Analysis

We used three LUC maps in this study (1997, 2007, and 2017) (Figure 3). We classified these LUC maps from cloud-free Landsat images (https://earthexplorer.usgs.gov/, accessed on 1 October 2021) captured on 7 February 1997, 2 January 2007, and 13 January 2017. We used a supervised classification method, employing the maximum likelihood algorithm [27–29].

Our LUC classification system included four classes, namely marshland, mangrove, settlement, and water. The marshland class included seasonally and intermittently flooded areas, abandoned paddy lands, agricultural lands, marsh plant vegetation, trees, grassland and scrub, peat and bog soil areas, and other cropland areas. The mangrove class included seasonally and intermittently flooded areas with mangroves. The settlement class included low-intensity and high-intensity urban areas, industrial zones, transportation hubs, airports, home gardens, asphalt areas, and residential areas. The water class included the lagoon and other bodies of water such as canals, streams, and ponds.

The accuracy of each LUC map was assessed using 400 reference points generated using a random sampling technique. Google Earth images were used as sources of reference data for 2007 and 2017, while topographic maps of Sri Lanka were used as sources of reference data for 1997. The overall accuracy was 86.50%, 84.25%, and 84.50% for the 1997, 2007, and 2017 LUC maps, respectively.

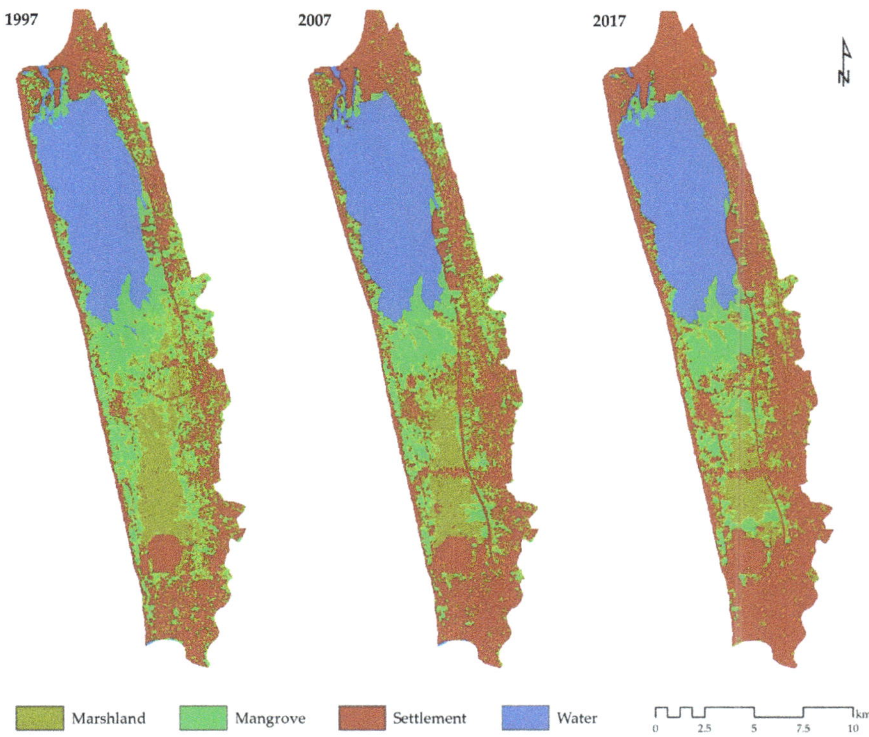

Figure 3. LUC maps of the MMNL in 1997, 2007 and 2017.

2.3. LUC Change Modeling

2.3.1. Model Calibration and Validation

We used the Land Change Modeler (LCM) [30–32], which is available in geospatial monitoring and modeling software called TerrSet (https://clarklabs.org/terrset/, accessed on 1 October 2021), to simulate future LUC changes in the area and examine potential future impacts of urbanization on the natural landscape and ecosystem services of the MMNL (2017–2030). To do this, we first calibrated the model by simulating the observed LUC changes between 2007 and 2017. We considered two LUC transitions: (i) marshland to settlement and (ii) mangrove to settlement. We used the Markov chain algorithm [30,32] to derive a transition matrix that contained the rate or proportion of the area of a particular LUC class that would persist (non-change) or transition to another class (change) from 2007 to 2017 (in our case, these were from marshland to settlement and from mangrove to settlement).

To spatialize the projected quantities of LUC changes from the two transitions considered, we used six spatial variables (variables that we hypothesized to have influenced LUC change patterns in the area) and the multi-layer perception neural network (MLP NN) algorithm [30,32] to model two transition potential maps (one for each transition). These variables included distance to road, distance to growth nodes, distance to the lagoon, distance to the protected areas, elevation, and slope (Figure 4). They were identified and selected based on the literature [33–36], our knowledge of the study area, and the availability of data. The same set of spatial variables was used for both transitions.

To simulate the LUC changes between 2007 and 2017, we ran the model (LCM) with the following inputs: 1997 and 2007 LUC maps, the six spatial variables for each transition (for the modeling of transition potential maps using the MLP NN algorithm), and a transition matrix for the 2007–2017 period (derived using Markov chain based on the 1997 and

2007 LUC transitions). The output was a simulated LUC map in 2017 that depicted the projected LUC changes from 2007 based on the two LUC transitions considered (marshland to settlement and mangrove to settlement).

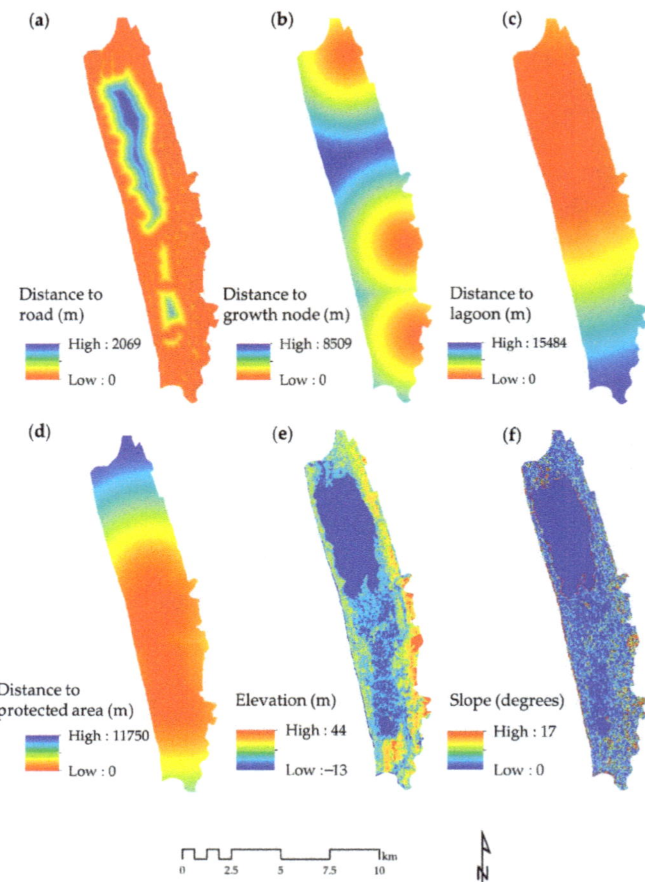

Figure 4. Spatial variables used in the modeling of LUC transition potential maps: (**a**) distance to road, (**b**) distance to growth node, (**c**) distance to lagoon, (**d**) distance to protected area, (**e**) elevation, and (**f**) slope.

We validated the simulation result by calculating the figure of merit (FoM) statistic [18,37,38] for each transition. The FoM was derived based on a three-map comparison technique: LUC 2007 (observed), LUC 2017 (observed), and LUC 2017 (simulated). More specifically, it was derived by taking the ratio of the intersection (H) of the observed change between 2007 and 2017 (H and M) and simulated change between 2007 and 2017 (H and F) to the union of the observed change and simulated change (Equation (1)).

$$\text{FoM} = \frac{H}{(H + M + F)} \times 100 \tag{1}$$

In Equation (1), H (hits) refers to the quantity of observed change pixels that were simulated as change. M (misses) refers to the quantity of observed change pixels that were simulated as non-change. F (false alarms) refers to the quantity of observed non-change pixels that were simulated as change.

2.3.2. Scenario-Based LUC Change Simulation

The trajectory (quantity and spatial pattern) of future LUC changes generally depends on various factors, such as future changes in various socioeconomic indicators (including development policy-related) and biophysical conditions in the area. In this context, a scenario analysis might be useful as scenarios are aimed at forward-looking adaptive development planning and decision making [39,40]. In fact, scenario analysis has become a useful technique in land change and sustainability research [13,41–44]. Scenario analysis is a structured process of exploring and evaluating plausible alternative futures [39,45].

In this study, we projected the future impacts of urbanization (2017–2030) on the natural landscape and ecosystem services of the MMNL. We considered two plausible development scenarios: a business as usual (BAU) scenario and an environment protection (EP) scenario.

In the BAU scenario, we allowed the model (LCM) to project and simulate future LUC changes in the MMNL based on the past rates (Markov transition matrix based on the 2007 and 2017 LUC maps) and spatial pattern (transition potential maps) of LUC changes as per the two transitions considered (marshland to settlement and mangrove to settlement). In a recent study, it has been shown that settlements have also been expanding and encroaching into the protected areas (PAs) [9]. In this scenario, we did not introduce any spatial constraints, allowing the observed LUC change pattern to continue. To run the scenario, we used the 2007 and 2017 LUC maps (Figure 3) and the six spatial variables (Figure 4) as inputs and considered 2030 as the end time (year) of the simulation.

In the EP scenario, we used the same data inputs as in the BAU scenario, but we also introduced some plausible policy and development-related assumptions. More specifically, we were interested in the potential impacts of urbanization on the ecosystem services of the MNNL under a scenario in which (i) the urbanization rate would slow down by 20%, and (ii) the two protected areas (PAs) in the area would be completely protected. To do this, first, we revised the Markov transition matrix by withholding (deducting) 20% of the proportion of the area of marshland and mangrove that would transition to settlement by 2030. Second, we introduced a spatial constraint disallowing LUC change to occur in the two PAs. The 20% rate is based on a previous study [37], and our assumption was that the rate (20%) is not that stringent, meaning plausible at given circumstances (e.g., protection of the protected areas, implementation of land use zoning, no illegal settlements, etc.).

2.4. Monitoring ESV Changes

We estimated the past changes in the ESV of the MMNL (1997–2017), as well as the potential future changes based on the BAU and EP scenarios (2017–2030). We considered 10 ecosystem services, namely flood attenuation, industrial wastewater treatment, agriculture production, support to downstream fisheries (fish breeding and nursery), firewood, fishing (fisheries production), leisure and recreation, domestic sewage treatment, freshwater supplies for the local population, and carbon sequestration (Table 1).

We sourced the needed ESV coefficients from an earlier study in the MMNL [11]. We converted the ESV coefficients, which were originally expressed in Sri Lankan Rupee (LKR)/year (2003 price level), into 2020 USD/ha/year equivalents (Equation (2)). To do this, we first expressed the coefficients into 2003 LKR/ha/year. We then converted these 2003 values to the 2020 price level, taking into account inflation. We used a deflator based on the average consumer price index (CPI, a measure of inflation) in 2003 (44.838) and 2020 (135.367) from https://www.imf.org/en/Home (accessed on 1 October 2021). Subsequently, we took the average USD–LKR conversion equivalent (CE) in 2020 (I USD = 183.23 LKR) from https://www.cbsl.gov.lk (accessed on 1 October 2021) and converted the derived 2020 LKR values to 2020 USD equivalents.

$$\text{ESV (2020 USD)} = \frac{\left(\text{ESV LKR 2003} \times \frac{\text{2020 CPI}}{\text{2003 CPI}}\right)}{\text{2020 CE}} \tag{2}$$

Using the ESV coefficients in Table 1, we estimated the ESV of the MMNL in 1997, 2007, 2017, and by 2030 (under two scenarios) following Equations (3) and (4):

$$ESV_f = \sum_{k=1}^{n} A_k \times VC_f \tag{3}$$

$$ESV = \sum_{k=1}^{n} \sum_{f=1}^{m} A_k \times VC_f \tag{4}$$

where ESV_f and ESV refer to the value of ecosystem service f and the ecosystem service value of the MMNL, respectively. A_k refers to the area (ha) of LUC class k, VC_f refers to the ESV coefficient of ecosystem service f (USD/ha/year) for LUC class k, and n and m refer to the number of LUC classes and ecosystem services considered, respectively. We considered two LUC classes (marshland and mangrove) and 10 ecosystem services (Table 1).

We also mapped the spatial distribution of the 99 Grama Niladhari (GN) divisions that cover the entire MMNL with their respective ESVs. GN divisions are the smallest administrative divisions in Sri Lanka. To do this, first, we conducted a zonal analysis (tabulate area) to determine the LUC composition and extent in each GN division in 1997, 2007, 2017, 2030 BAU, and 2030 EP using the polygon boundaries of the GN divisions as zones and the LUC maps as inputs. Second, we estimated the ESV of each GN division using Equations (3) and (4).

Table 1. Values of the ecosystem services considered for the MMNL's marshland and mangrove biomes. Source of original values: Emerton and Kekulandala [11].

Ecosystem Services	ESV Coefficients (2020 USD/ha/Year)
Flood attenuation	2607.43
Industrial wastewater treatment	871.69
Agriculture production	162.67
Support to downstream fisheries (fish breeding and nursery)	107.41
Firewood	42.75
Fishing (fisheries production)	33.62
Leisure and recreation	28.36
Domestic sewage treatment	23.20
Freshwater supplies for local population	20.30
Carbon sequestration	4.19

3. Results

3.1. Changes in LUC and ESV (1997–2017)

Over the past 20 years, the MMNL's landscape has undergone considerable changes. In 1997, the MMNL had a marshland and mangrove area of 4242 ha and 2637 ha, respectively (Figure 5). However, in 2017, their extent decreased to 3058 ha and 1523 ha, equivalent to a 28% and 42% decrease, respectively. By contrast, the area of the settlement has expanded rapidly over the past two decades at the expense of the MMNL's marshlands and mangroves, with 3368 ha in 1997 and 5741 ha in 2017, i.e., equivalent to a 70% increase.

As a consequence of the significant loss of marshland and mangrove due to urbanization (settlement expansion), the ESV of the MMNL decreased by USD 8.96 million/year, from USD 26.84 million/year in 1997 to 17.88 million/year in 2017, i.e., equivalent to a 33% decrease (Table 2). Among the ecosystem services considered, flood attenuation, industrial wastewater treatment, agriculture production, and support to downstream fisheries (fish breeding and nursery) were the top services that were affected the most. Altogether, they accounted for over 95% of the total decrease. The ESV loss of flood attenuation accounted for 67% (USD 6.0 million/year).

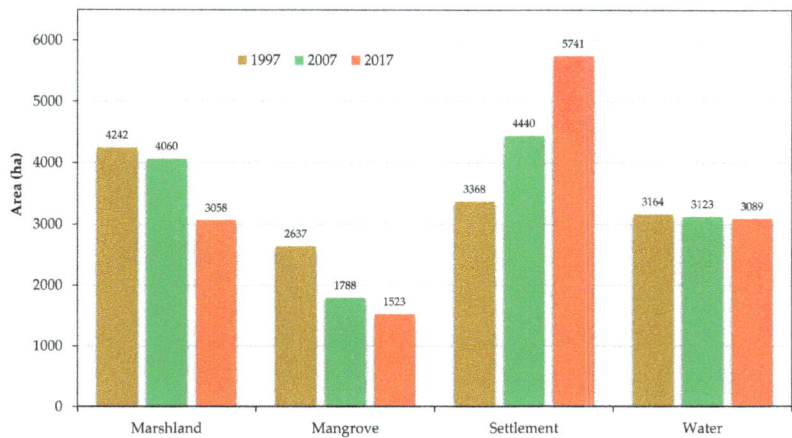

Figure 5. LUC changes in the MMNL (1997–2017).

Table 2. ESV changes in the MMNL.

Ecosystem Services	USD Million/Year						
	1997	2007	2017	Changes			
				1997–2007	% of 1997	2007–2017	% of 2007
Flood attenuation	17.94	15.25	11.94	−2.69	−14.99	−3.31	−21.70
Industrial wastewater treatment	6.00	5.10	4.00	−0.90	−15.00	−1.10	−21.57
Agriculture production	1.12	0.95	0.75	−0.17	−15.18	−0.20	−21.05
Support to downstream fisheries (fish breeding and nursery)	0.74	0.63	0.49	−0.11	−14.86	−0.14	−22.22
Firewood	0.29	0.25	0.20	−0.04	−13.79	−0.05	−20.00
Fishing (fisheries production)	0.23	0.20	0.15	−0.03	−13.04	−0.05	−25.00
Leisure and recreation	0.19	0.17	0.13	−0.02	−10.53	−0.04	−23.53
Domestic sewage treatment	0.16	0.13	0.11	−0.03	−18.75	−0.02	−15.38
Freshwater supplies for local population	0.14	0.12	0.09	−0.02	−14.29	−0.03	−25.00
Carbon sequestration	0.03	0.03	0.02	0.00	0.00	−0.01	−33.33
Total	**26.84**	**22.83**	**17.88**	**−4.01**		**−4.95**	

3.2. Projected Changes in LUC and ESV (2017–2030)

Under the BAU scenario, by 2030, the area of marshland and mangrove in the MMNL would decrease by 1329 ha and 213 ha, respectively, whereas the area of settlement would increase by 1542 ha (Figure 6, Table 3). By contrast, under the EP scenario in which urban expansion rate (settlement expansion) would slow down by 20% (Section 2.3.2), the decrease in the area of marshland and mangrove would be much lower at 1063 ha and 171 ha, respectively. In this scenario, the area of settlement would increase by 1234 ha.

As a consequence of the projected loss of marshland and mangrove by 2030, the ESV of the MMNL would also decrease (Table 4). Under the BAU scenario, the MMNL's ESV would decrease by USD 6.01 million/year, i.e., equivalent to a 34% decrease relative to 2017 (Table 2). Under the EP scenario, the decrease would be much less at USD 4.79 million/year, i.e., equivalent to a 27% decrease relative to 2017.

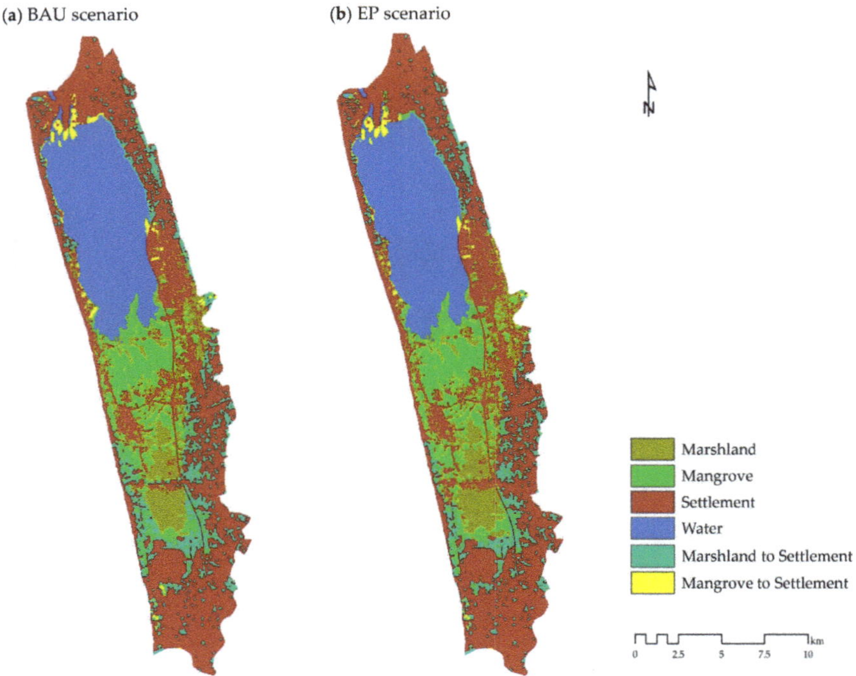

Figure 6. Projected LUC changes (loss of marshland and mangrove and gain of settlement) in the MMNL under the (**a**) BAU and (**b**) EP scenarios (2017–2030).

Table 3. Projected LUC changes in the MMNL under the BAU and EP scenarios (ha).

LUC Class	2017	2030 BAU	2030 EP	Changes			
				2017–2030 BAU	% of 2017	2017–2030 EP	% of 2017
Marshland	3058.47	1729.89	1995.66	−1328.58	−43.44	−1062.81	−34.75
Mangrove	1522.98	1309.5	1352.07	−213.48	−14.02	−170.91	−11.22
Settlement	5741.10	7283.16	6974.82	1542.06	26.86	1233.72	21.49

Table 4. Projected changes in the ESV of the MMNL under the BAU and EP scenarios.

Ecosystem Services	2017–2030 (BAU)			2017–2030 (EP)		
	Million USD/Year	% of 2017	% of Total Decrease	Million USD/Year	% of 2017	% of Total Decrease
Flood attenuation	−4.02	−33.67	66.89	−3.21	−26.88	67.01
Industrial wastewater treatment	−1.35	−33.75	22.46	−1.08	−27.00	22.55
Agriculture production	−0.26	−34.67	4.33	−0.21	−28.00	4.38
Support to downstream fisheries (fish breeding and nursery)	−0.16	−32.65	2.66	−0.13	−26.53	2.71
Firewood	−0.07	−35.00	1.16	−0.05	−25.00	1.04
Fishing (fisheries production)	−0.05	−33.33	0.83	−0.03	−20.00	0.63
Leisure and recreation	−0.04	−30.77	0.67	−0.03	−23.08	0.63
Domestic sewage treatment	−0.04	−36.36	0.67	−0.03	−27.27	0.63
Freshwater supplies for local population	−0.02	−22.22	0.33	−0.02	−22.22	0.42
Carbon sequestration	0.00	0.00	0.00	0.00	0.00	0.00
Total	**−6.01**		**100.00**	**−4.79**		**100.00**

3.3. ESV and Its Changes across the GN Divisions

Figure 7 shows the spatial distribution of the GN divisions in the MMNL with their respective ESVs in three time points. Of the 99 divisions, only three had a positive change between 1997 and 2017, and these are Katunayaka North (143), Munnakkarai North (156A), and Siriwardana Pedesa (156C) (Figure 7, Table A1). The top five ESV-losing divisions over the past 20 years were Kerawalapitiya (171), Pattiyawala (167B), Ambalammulla (146), Bolawalana (157), and Mahabage (178). In both scenarios (BAU and EP), the projected top five ESV-losing divisions were Pattiyawala (167B), Balagala (171B), Kunjawatta (166A), Siriwardana Pedesa (156C), and Mahabage (178). Overall, this GN division-level ESV monitoring can help in landscape and urban planning. For example, the projected top ESV-losing divisions should be given particular attention.

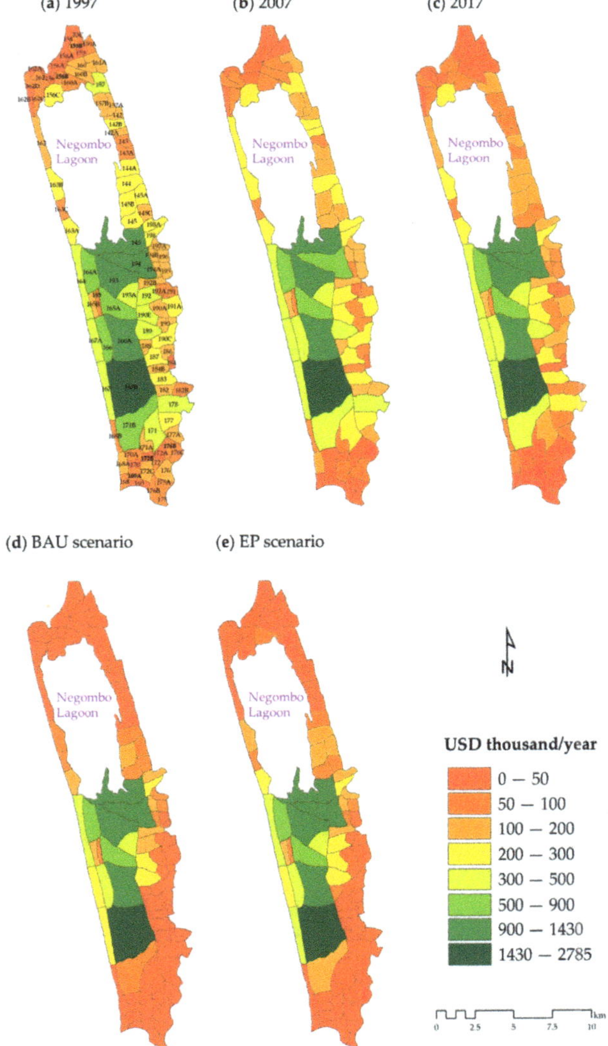

Figure 7. GN divisions in the MMNL with their respective ESVs on (**a**) 1997, (**b**) 2007 and (**c**) 2017 under the (**d**) BAU and (**e**) EP scenarios. The numbers and letters on the 1997 map refer to the GN codes as presented in Table A1.

3.4. LUC Change Model Validation

Our LUC change modeling focused on assessing the impacts of urbanization as proxied by settlement expansion on the natural landscape and ecosystem services of the MMNL. Two transitions were considered, namely marshland to settlement and mangrove to settlement, with the FoM being used to validate the LUC change modeling results (Section 2.3.1). The validation results revealed that the marshland to settlement transition had an FoM of 45.4%, whereas the mangrove to settlement transition had 29.5%. These FoM values are within the range of FoM values reported in other LUC change modeling studies. For example, in their LUC change modeling in connection with ecosystem services, Estoque and Murayama [18] recorded an FoM of 43%. In their LUC change modeling in the context of flooding, Johnson et al. [44] had an FoM of 20%. In an earlier seminal review of FoM applications in the validation of LUC change modeling studies, Pontius et al. [38] reported an FoM value range of 1–59%.

4. Discussion

The MMNL is among the 12 priority wetlands in Sri Lanka. The presence of two protected areas within the MMNL (Figure 2) is a direct manifestation of its ecological significance. However, our findings indicate that the sustainability of the MMNL is now in jeopardy; hence, urgent action has to be taken, landscape and urban planning wise.

In the early 2000s, in a seminal study that identified and quantified the ecosystem services of the MMNL, it was reported that the area had been experiencing intense and growing pressure from urbanization [11]. It had been observed that (i) wetland resources had been harvested at high and often unsustainable levels; (ii) lands were being rapidly reclaimed and modified for agricultural, commercial, and residential purposes; and (iii) heavy loads of industrial and domestic wastes were being discharged untreated into the MMNL. With all of these happening, the said study concluded that the MMNL has seriously degraded over time.

Nearly two decades have passed since the conduct of the said study, but the MMNL's curve of continuous ecological degradation has not been flattened out; instead, the degradation of this valuable urban wetland ecosystem has continued as indicated by our findings. For example, between 2007 and 2017, the MMNL lost another 1002 ha of marshland and 265 ha of mangrove (Figure 5). Between 1997 and 2007, these values were 182 ha and 849 ha, respectively. By contrast, another 1301 ha of natural cover were converted into settlement between 2007 and 2017. This value was even higher than during the 1997–2007 period (1072 ha). Consequently, the ESV of the MMNL has decreased by USD 8.96 million/year over the past 20 years (USD 4.01 million/year between 1997 and 2007, and USD 4.95 million/year between 2007 and 2017) (Table 2). Flood attenuation and industrial wastewater treatments were among the ecosystems that were greatly affected.

The MMNL has long been seen as having prime potential for industrial and urban development, but at the same time, it is considered as a coastal wetland ecosystem of high biodiversity and ecological significance [9–11]. Our findings indicate that while urbanization has been continuing at an unprecedented rate, the conservation of this critical urban wetland ecosystem has been neglected. One important earlier observation that remains valid until today is that there seems to be little appreciation of either the economic value attached to the conservation of the MMNL or the high and far-reaching economic costs arising from its degradation [11]. Decisions regarding how land and resources should be used have been based on development initiatives that favor the modification of the wetland for short-term economic gain over long-term benefits and the conservation and sustainability of the MMNL [11]. In fact, the loss of natural cover due to settlement expansion has been observed even within the boundaries of the PAs (Figures 2 and 3).

Our results have also shown that the future (2030) condition of the MMNL can be expected to be worse if the recent (2007–2017) rate and spatial pattern of urbanization (settlement expansion) continues (BAU scenario). It is because the projected ESV loss under this scenario between 2017 and 2030 (USD 6.01 million/year) (Table 4) would be greater than the ESV loss in the past decade (2007–2017, USD 4.95 million/year) (Table 2). This also means that the future of the MMNL will be much worse if the threat of marshland and mangrove loss due to urbanization grows and intensifies. Nonetheless, our study also demonstrates that under the EP scenario, while the continuous decline of the MMNL's ESV cannot be fully stopped, the rate of loss could be slowed down (USD 4.79 million/year) (Table 4). Hence, between the two scenarios, the EP scenario is the more desirable one for the MMNL.

Our study considered only two basic scenarios, and thus, the exploration of other more complex plausible scenarios can be considered for future research. Examples of more complex scenarios include those that incorporate future trajectories of relevant socioeconomic indicators, such as population growth, changes in economic, land use, and environmental conditions, as well as future development priorities and policy targets. The shared socioeconomic pathways (SSPs) are examples of such scenarios, though they are designed for a global scale analysis [46–48]. The adaption of these pathways to local-level analysis could be a future research direction. Other scenarios could focus more on conservation storylines [14] or other more complex and stringent versions of our EP scenario.

Nonetheless, despite their simplicity, the inclusion of our two basic scenarios in the analysis helped us demonstrate that, for the sustainability of the MMNL, it is still possible to flatten its curve of continuous ecological degradation. In fact, the simple full protection of the PAs inside the MMNL (EP scenario) could make a significant positive contribution. Furthermore, with the use of a monitoring scheme built on a state-of-the-art geospatial technique (including GIS, remote sensing, and scenario-based land change modeling) and the concept of ecosystem services, our study also makes important methodological and empirical contributions.

In fact, the economic value of wetland goods and services is rarely factored into LUC change decisions in the MMNL [11]. Our study offers a basic template that can be adopted and improved in future studies and/or considered in landscape and urban planning for the MMNL. In general, the valuation and monitoring of ecosystem services across space and time have many potential uses, including raising of awareness and interest, national income and well-being accounts, specific policy analyses, urban and regional planning, payment for ecosystem services, full cost accounting, and common asset trusts [13,49,50]. We argue that the MMNL can benefit from landscape and urban planning that considers the concept of ecosystem services.

5. Summary and Conclusions

The MMNL is an important urban wetland ecosystem in Sri Lanka, but its sustainability is now in jeopardy due to rapid and uncontrolled urbanization. Swift action must be taken in order to save this valuable urban wetland ecosystem. In this study, to help inform sustainable landscape and urban planning, we examined the impacts of urbanization on the natural landscape and ecosystem services of the MMNL over the past 20 years (1997–2017). We also projected landscape and ESV changes by 2030 under two plausible scenarios. We found that, due to rapid urbanization (settlement expansion equivalent to 70% from 1997 to 2017), the area of the MMNL's marshland and mangrove had decreased by 1184 ha and 1114 ha, respectively. Consequently, its ESV had decreased by USD 8.96 million/year (33%). If the current rate and spatial pattern of urbanization (2007–2017) continued in the future (BAU scenario), another 1329 ha of marshland and 213 ha of mangrove would be lost by 2030. The projected loss in ESV would be USD 6.01 million/year (34%). However, if the urbanization rate slowed down by 20% and the PAs were completely protected (EP scenario), the future loss of marshland and mangrove would only be around 1063 ha and 171 ha, respectively. The projected loss in ESV would be lower at USD 4.79 million/year (27%). Between the two scenarios, the EP scenario would be the more desirable one that should be considered by local government planners and decision-makers. The past, present, and future ESV maps of the GN divisions produced in this study can be used to identify hotspots. For future research, other more complex and stringent plausible scenarios need to be explored to help flatten the MMNL's curve of continuous ecological degradation. Overall, the results of this study can help provide landscape and urban planners with information useful to the sustainability of the MMNL. The approach employed is also adaptable and applicable to other urban wetland ecosystems in the country and the rest of the world.

Author Contributions: D.A. conceptualized the study. D.A. and R.C.E. conducted the research, performed the analysis, and wrote the paper. Y.M. and B.M. provided research supervisions. All authors have read and agreed to the published version of the manuscript.

Funding: This work was supported by the Japan Society for the Promotion of Science (JSPS) through a Grant-in-Aid for Early-Career Scientists (20K13262) (PI: Ronald C. Estoque).

Institutional Review Board Statement: Not applicable.

Informed Consent Statement: Not applicable.

Data Availability Statement: All data generated and analyzed in this study are included in this published article.

Acknowledgments: We thank Lucy Emerton for providing some guidance regarding ecosystem service valuation. Likewise, we are thankful to the four anonymous reviewers for their constructive and insightful comments and suggestions.

Conflicts of Interest: The authors declare no conflict of interest.

Appendix A

Table A1. ESV and its changes at the GN division level. Note: The GN code is linked with Figure 7.

GN Code	GN Name	ESV (USD Thousand/Year)					Change	
		1997	2007	2017	2030 BAU	2030 EP	2017–2030 BAU	2017–2030 EP
190A	Weligampitiya North	167.40	78.90	52.60	1.00	1.00	51.60	51.60
191A	Ja-Ela	279.00	241.20	131.40	0.60	0.60	130.80	130.80
165B	Pulluhena	171.30	168.20	133.80	133.80	133.80	0.00	0.00
175	Telangapatha	62.20	39.40	12.40	0.00	0.00	12.40	12.40
169	Hekitta	57.70	17.30	4.80	0.00	0.00	4.80	4.80
175A	Evariwatta	90.80	47.70	10.90	0.01	0.09	10.89	10.81
176B	Galwetiya	70.50	25.20	3.90	0.02	0.06	3.88	3.84
168	Palliyawatta South	64.10	32.70	34.10	0.01	0.08	34.08	34.02
169A	Kurunduhena	72.70	42.00	8.10	0.01	0.00	8.09	8.09
172C	Nayakakanda South	136.60	75.80	29.20	0.00	0.00	29.20	29.20
170	Thimbirigasyaya	49.20	26.50	1.30	0.00	0.00	1.30	1.30
176	Wattala	140.00	71.20	13.10	0.00	0.00	13.10	13.10
172	Hendala South	60.90	11.60	7.10	0.00	0.00	7.10	7.10
168A	Palliyawatta North	154.70	127.60	86.60	0.00	0.00	86.60	86.60
172B	Nayakakanda North	41.70	29.50	9.00	0.00	0.00	9.00	9.00
170A	Elakanda	76.80	35.00	6.40	0.00	0.00	6.40	6.40
176C	Welikadamulla	103.40	43.90	7.80	0.00	0.00	7.80	7.80
172A	Hendala North	50.00	19.60	9.10	0.00	0.00	9.10	9.10
176A	Mabola	60.90	35.10	18.20	0.00	0.00	18.20	18.20
171A	Matagoda	134.30	64.50	35.10	0.02	0.00	35.08	35.09
177A	Kerangapokuna	151.20	129.90	84.10	0.00	0.90	84.10	83.20
168B	Dikovita	115.60	100.00	83.20	24.50	25.60	58.70	57.60
177	Mattumagala	244.20	214.90	60.20	3.00	3.00	57.20	57.20

Table A1. *Cont.*

GN Code	GN Name	ESV (USD Thousand/Year)					Change	
		1997	2007	2017	2030 BAU	2030 EP	2017–2030 BAU	2017–2030 EP
171	Kerawalapitiya	500.10	243.80	136.50	16.20	16.20	120.30	120.30
178	Mahabage	445.60	391.90	224.60	29.50	29.50	195.10	195.10
171B	Balagala	607.50	468.60	412.10	54.60	104.20	357.50	307.90
182	Welisara	194.60	146.70	83.60	29.30	29.30	54.30	54.30
182B	Elehiwatta	118.50	131.70	58.80	0.00	0.00	58.80	58.80
183	Nagoda	321.70	238.30	166.80	27.70	27.70	139.10	139.10
184B	Uswatta	153.00	86.40	39.50	1.20	1.90	38.30	37.60
167	Uswetakeiyawa	391.80	397.40	314.90	245.80	280.50	69.10	34.40
167B	Pattiyawala	2785.40	2548.70	2493.10	1916.50	2278.10	576.60	215.00
184	Kandana West	28.40	20.90	4.30	0.00	0.00	4.30	4.30
187	Nedurupitiya	316.70	245.40	166.60	17.10	17.10	149.50	149.50
186	Rilavulla	127.10	111.40	42.70	3.50	3.50	39.20	39.20
188	Kalaeliya	171.00	142.80	94.70	19.00	38.40	75.70	56.30
190C	Kapuwatta	302.70	233.90	118.00	4.10	4.10	113.90	113.90
189	Wewala	428.10	368.10	297.90	183.30	224.90	114.60	73.00
167A	Paranambalama	407.20	371.30	324.40	174.40	234.40	150.00	90.00
190	Weligampitiya South	153.90	80.70	55.20	1.10	1.10	54.10	54.10
166	Nugape	594.10	541.90	448.80	439.30	448.80	9.50	0.00
166A	Kunjawatta	1658.50	1529.00	1458.80	1139.90	1198.00	318.90	260.80
190E	Indivitiya	383.40	263.40	255.90	170.70	211.20	85.20	44.70
165	Bopitiya	102.70	78.00	47.50	47.50	47.50	0.00	0.00
165A	Bopitiyathuduwa	887.70	780.10	796.20	703.00	708.30	93.20	87.90
192	Thudella West	379.30	288.10	209.60	116.30	176.90	93.30	32.70
192A	Thudella South	65.40	39.00	14.90	0.00	0.00	14.90	14.90
191	Kanuwana	88.60	76.70	14.30	0.00	0.00	14.30	14.30
193A	Delathura East	462.40	334.80	326.60	326.60	326.60	0.00	0.00
192B	Thudella North	194.70	178.60	127.20	105.40	111.20	21.80	16.00
194A	Dehiyagatha South	123.30	105.90	93.60	55.80	85.90	37.80	7.70
195	Kudahakapola South	163.40	119.40	63.80	4.10	18.20	59.70	45.60
193	Delathura West	1453.00	1415.20	1368.50	1358.50	1368.50	10.00	0.00
196	Kudahakapola North	167.50	137.00	97.80	54.80	74.40	43.00	23.40
164A	Maha Pamunugama	837.30	839.30	752.60	736.10	736.10	16.50	16.50
194	Dandugama	1090.30	952.30	1076.70	896.30	893.40	180.40	183.30
194B	Dehiyagatha North	158.60	147.90	120.10	120.10	120.10	0.00	0.00
164	Pamunugama	532.00	521.40	464.90	460.00	464.10	4.90	0.80
197A	Udammita South	159.50	156.10	109.00	94.60	107.90	14.40	1.10
198	Alawathupitiya	227.40	224.20	205.20	196.80	205.20	8.40	0.00
163A	Kepungoda	218.00	289.90	188.00	140.30	181.90	47.70	6.10
198A	Dambaduraya	285.30	259.10	160.50	160.50	160.50	0.00	0.00
146	Ambalammulla	1425.80	1263.20	1159.50	1079.40	1159.50	80.10	0.00
145	Bandarawatta West	308.80	169.50	92.90	91.90	91.80	1.00	1.10
145C	Bandarawatta East	167.50	175.20	89.90	85.10	89.90	4.80	0.00

Table A1. *Cont.*

GN Code	GN Name	ESV (USD Thousand/Year)					Change	
		1997	2007	2017	2030 BAU	2030 EP	2017–2030 BAU	2017–2030 EP
163C	Settappaduwa	102.90	109.80	73.10	48.40	48.40	24.70	24.70
145B	Mookalangamuwa West	305.80	177.70	122.00	120.10	122.00	1.90	0.00
145A	Mookalangamuwa East	273.90	212.90	137.40	72.80	131.50	64.60	5.90
144	Liyanagemulla South	244.80	200.60	134.10	70.50	108.50	63.60	25.60
163B	Dungalpitiya	258.10	258.70	172.10	58.60	104.90	113.50	67.20
144A	Liyanagemulla North	293.60	189.30	127.70	34.60	50.40	93.10	77.30
143A	Katunayaka South	81.20	140.40	63.20	0.00	0.90	63.20	62.30
143	Katunayaka North	60.80	82.50	132.10	0.00	0.00	132.09	132.09
142A	Kurana Katunayaka South	213.20	215.50	58.40	0.00	0.00	58.40	58.40
142B	Kurana Katunayaka Central	265.70	256.50	80.20	0.00	0.00	80.19	80.19
163	Thalahena	179.30	218.50	111.60	0.01	0.02	111.58	111.57
142	Kurana Katunayaka North	133.60	145.80	90.30	0.00	0.00	90.30	90.30
157B	Kurana West	196.90	150.70	45.30	0.01	0.02	45.28	45.27
157A	Kurana East	133.10	135.50	75.50	0.09	0.02	75.41	75.48
162C	Pitipana Southeast	107.60	110.00	49.90	0.00	0.00	49.90	49.90
162B	Pitipana South -West	54.40	54.00	8.90	0.00	0.00	8.90	8.90
156C	Siriwardana Pedesa	233.60	221.10	240.50	0.00	0.00	240.50	240.50
156	Munnakkarai	18.30	2.00	10.50	0.10	0.00	10.49	10.49
160A	Thaladoowa	82.80	85.30	70.30	0.00	42.80	70.29	27.50
162D	Pitipana Central	99.70	94.30	77.30	0.00	0.00	77.30	77.30
157	Bolawalana	376.00	299.30	147.00	0.00	0.30	147.00	146.70
156B	Munnakkarai East	61.00	48.10	58.90	0.00	26.40	58.90	32.50
162	Pitipana North	108.00	102.80	31.50	0.00	0.00	31.50	31.50
162A	Doowa	33.70	39.10	29.10	0.05	0.01	29.05	29.09
160B	Udayarthoppuwa South	161.40	89.00	12.40	0.00	0.30	12.40	12.10
160	Udayarthoppuwa	122.50	71.90	1.20	0.00	0.30	1.20	0.90
156A	Munnakkarai North	32.30	54.60	71.20	0.00	11.20	71.20	60.00
161A	Angurukaramulla	209.30	124.60	3.90	0.00	0.00	3.90	3.90
158A	Wella Weediya South	10.70	9.60	1.20	0.00	0.00	1.20	1.20
159	Periyamulla	47.30	29.20	6.70	0.00	0.00	6.70	6.70
158	Wella Weediya	12.50	12.30	5.00	0.00	0.00	5.00	5.00
73C	Kudapaduwa South	45.20	38.00	6.20	0.00	0.00	6.20	6.20
158B	Wella Weediya East	63.70	29.90	0.20	0.00	0.00	0.20	0.20
159A	Hunupitiya	74.70	28.70	2.00	0.00	0.00	2.00	2.00

References

1. Sinclair, M.; Vishnu Sagar, M.K.; Knudsen, C.; Sabu, J.; Ghermandi, A. Economic appraisal of ecosystem services and restoration scenarios in a tropical coastal Ramsar wetland in India. *Ecosyst. Serv.* **2021**, *47*, 101236. [CrossRef]
2. Ghermandi, A.; Sheela, A.M.; Justus, J. Integrating similarity analysis and ecosystem service value transfer: Results from a tropical coastal wetland in India. *Ecosyst. Serv.* **2016**, *22*, 73–82. [CrossRef]
3. Sustainable Development Goals. Available online: https://sdgs.un.org/goals (accessed on 23 September 2021).
4. Assefa, W.W.; Eneyew, B.G.; Wondie, A. The impacts of land-use and land-cover change on wetland ecosystem service values in peri-urban and urban area of Bahir Dar City, Upper Blue Nile Basin, Northwestern Ethiopia. *Ecol. Process.* **2021**, *10*, 39. [CrossRef]

5. Huang, Q.; Zhao, X.; He, C.; Yin, D.; Meng, S. Impacts of urban expansion on wetland ecosystem services in the context of hosting the Winter Olympics: A scenario simulation in the Guanting Reservoir Basin, China. *Reg. Environ. Chang.* **2019**, *19*, 2365–2379. [CrossRef]
6. Tong, C.; Feagin, R.A.; Lu, J.; Zhang, X.; Zhu, X.; Wang, W.; He, W. Ecosystem service values and restoration in the urban Sanyang wetland of Wenzhou, China. *Ecol. Eng.* **2007**, *29*, 249–258. [CrossRef]
7. Kharazmi, R.; Tavili, A.; Rahdari, M.R.; Chaban, L.; Panidi, E.; Rodrigo-Comino, J. Monitoring and assessment of seasonal land cover changes using remote sensing: A 30-year (1987–2016) case study of Hamoun Wetland, Iran. *Environ. Monit. Assess.* **2018**, *190*, 1–23. [CrossRef]
8. Wu, W.; Zhou, Y.; Tian, B. Coastal wetlands facing climate change and anthropogenic activities: A remote sensing analysis and modelling application. *Ocean Coast. Manag.* **2017**, *138*, 1–10. [CrossRef]
9. Athukorala, D.; Estoque, R.C.; Murayama, Y.; Matsushita, B. Impacts of Urbanization on the Muthurajawela Marsh and Negombo Lagoon, Sri Lanka: Implications for landscape planning towards a sustainable urban wetland ecosystem. *Remote Sens.* **2021**, *13*, 316. [CrossRef]
10. CEA; IUCN; IWMI. *National Wetland Directory of Sri Lanka*; The Central Environmental Authority (CEA); The World Conservation Union (IUCN); The International Water Management Institute (IWMI): Colombo, Sri Lanka, 2006; ISBN 9558177547.
11. Emerton, L.; Kekulandala, L.D.C.B. *Assessment of the Economic Value of Muthurajawela Wetland*; IUCN-Sri Lanka: Colombo, Sri Lanka, 2003; ISBN 9558177199.
12. Subasinghe, S.; Estoque, R.; Murayama, Y. Spatiotemporal Analysis of Urban Growth Using GIS and Remote Sensing: A Case Study of the Colombo Metropolitan Area, Sri Lanka. *ISPRS Int. J. Geo-Inform.* **2016**, *5*, 197. [CrossRef]
13. Estoque, R.C.; Murayama, Y. Quantifying landscape pattern and ecosystem service value changes in four rapidly urbanizing hill stations of Southeast Asia. *Landsc. Ecol.* **2016**, *31*, 1481–1507. [CrossRef]
14. Sun, X.; Crittenden, J.C.; Li, F.; Lu, Z.; Dou, X. Urban expansion simulation and the spatio-temporal changes of ecosystem services, a case study in Atlanta Metropolitan area, USA. *Sci. Total Environ.* **2018**, *622–623*, 974–987. [CrossRef]
15. Zhou, D.; Tian, Y.; Jiang, G. Spatio-temporal investigation of the interactive relationship between urbanization and ecosystem services: Case study of the Jingjinji urban agglomeration, China. *Ecol. Indic.* **2018**, *95*, 152–164. [CrossRef]
16. Dai, X.; Wang, L.; Huang, C.; Fang, L.; Wang, S.; Wang, L. Spatio-temporal variations of ecosystem services in the urban agglomerations in the middle reaches of the Yangtze River, China. *Ecol. Indic.* **2020**, *115*, 106394. [CrossRef]
17. He, Y.; Wang, W.; Chen, Y.; Yan, H. Assessing spatio-temporal patterns and driving force of ecosystem service value in the main urban area of Guangzhou. *Sci. Rep.* **2021**, *11*, 1–18. [CrossRef]
18. Estoque, R.C.; Murayama, Y. Examining the potential impact of land use/cover changes on the ecosystem services of Baguio city, the Philippines: A scenario-based analysis. *Appl. Geogr.* **2012**, *35*, 316–326. [CrossRef]
19. Yirsaw, E.; Wu, W.; Shi, X.; Temesgen, H.; Bekele, B. Land Use/Land Cover change modeling and the prediction of subsequent changes in ecosystem service values in a coastal area of China, the Su-Xi-Chang region. *Sustainability* **2017**, *9*, 1204. [CrossRef]
20. Swetnam, R.D.; Fisher, B.; Mbilinyi, B.P.; Munishi, P.K.T.; Willcock, S.; Ricketts, T.; Mwakalila, S.; Balmford, A.; Burgess, N.D.; Marshall, A.R.; et al. Mapping socio-economic scenarios of land cover change: A GIS method to enable ecosystem service modelling. *J. Environ. Manage.* **2011**, *92*, 563–574. [CrossRef]
21. Jayathilake, M.B.; Chandrasekara, W.U. Variation of avifaunal diversity in relation to land-use modifications around a tropical estuary, the Negombo estuary in Sri Lanka. *J. Asia-Pacific Biodivers.* **2015**, *8*, 72–82. [CrossRef]
22. Bambaradeniya, C.N.B.; Ekanayake, S.P.; Kekulandala, L.D.C.B.; Samarawickrama, V.A.P.; Ratnayake, N.D.; Fernando, R.H.S.S. *An Assessment of the Status of Biodiversity in the Muthurajawela Wetland Sanctuary*; IUCN—The World Conservation Union, Sri Lanka Country Office: Colombo, Sri Lanka, 2002; ISBN 9558177172.
23. Central Environment Authority, Sri Lanka. Available online: http://www.cea.lk/web/?option=com_content&view=article&layout=edit&id=1155 (accessed on 14 June 2021).
24. Department of Meteorology, Sri Lanka. Available online: https://www.meteo.gov.lk/index.php?lang=en (accessed on 15 June 2021).
25. Greater Colombo Economic Commision. *Environmental Profile of Muthurajawela and Negombo Lagoon*; Greater Colombo Economic Commision: Euroconsult, The Netherlands, 1991.
26. DIVA-GIS, Free, Simple & Effective. Available online: https://www.diva-gis.org/ (accessed on 15 June 2021).
27. Estoque, R.C.; Murayama, Y. Landscape pattern and ecosystem service value changes: Implications for environmental sustainability planning for the rapidly urbanizing summer capital of the Philippines. *Landsc. Urban Plan.* **2013**, *116*, 60–72. [CrossRef]
28. Zhang, X.; Estoque, R.C.; Murayama, Y. An urban heat island study in Nanchang City, China based on land surface temperature and social-ecological variables. *Sustain. Cities Soc.* **2017**, *32*, 557–568. [CrossRef]
29. Hou, H.; Estoque, R.C.; Murayama, Y. Spatiotemporal analysis of urban growth in three African capital cities: A grid-cell-based analysis using remote sensing data. *J. African Earth Sci.* **2016**, *123*, 381–391. [CrossRef]
30. Eastman, J.R. *TerrSet, Geospatial Monitoring and Modeling System, Manual*; Clark Labs, Clark University: Worcester, MA, USA, 2016.
31. Eastman, J.R.; Toledano, J.A. A Short Presentation of the Land Change Modeler (LCM). In *Geomatic Approaches for Modeling Land Change Scenarios*; Lecture Notes in Geoinformation and Cartography; Springer International Publishing AG: Cham, Switzerland, 2018; Volume 6, pp. 499–506, ISBN 9781441980731.

32. Mas, J.F.; Kolb, M.; Paegelow, M.; Camacho Olmedo, M.T.; Houet, T. Inductive pattern-based land use/cover change models: A comparison of four software packages. *Environ. Model. Softw.* **2014**, *51*, 94–111. [CrossRef]
33. Peng, K.; Jiang, W.; Deng, Y.; Liu, Y.; Wu, Z.; Chen, Z. Simulating wetland changes under different scenarios based on integrating the random forest and CLUE-S models: A case study of Wuhan Urban Agglomeration. *Ecol. Indic.* **2020**, *117*, 106671. [CrossRef]
34. Zhang, Z.; Hu, B.; Jiang, W.; Qiu, H. Identification and scenario prediction of degree of wetland damage in Guangxi based on the CA-Markov model. *Ecol. Indic.* **2021**, *127*, 107764. [CrossRef]
35. Ghosh, S.; Das, A. Wetland conversion risk assessment of East Kolkata Wetland: A Ramsar site using random forest and support vector machine model. *J. Clean. Prod.* **2020**, *275*, 123475. [CrossRef]
36. Ansari, A.; Golabi, M.H. Prediction of spatial land use changes based on LCM in a GIS environment for Desert Wetlands—A case study: Meighan Wetland, Iran. *Int. Soil Water Conserv. Res.* **2019**, *7*, 64–70. [CrossRef]
37. Estoque, R.C.; Murayama, Y. Measuring Sustainability Based Upon Various Perspectives: A Case Study of a Hill Station in Southeast Asia. *Ambio* **2014**, *43*, 943–956. [CrossRef]
38. Pontius, R.G.; Boersma, W.; Castella, J.C.; Clarke, K.; Nijs, T.; Dietzel, C.; Duan, Z.; Fotsing, E.; Goldstein, N.; Kok, K.; et al. Comparing the input, output, and validation maps for several models of land change. *Ann. Reg. Sci.* **2008**, *42*, 11–37. [CrossRef]
39. Costanza, R.; Kubiszewski, I.; Cork, S.; Atkins, P.W.B.W.B.; Bean, A.; Diamond, A.; Grigg, N.; Korb, W.; Logg-Scarvell, J.; Navis, R.; et al. Scenarios for Australia in 2050. *J. Futur. Stud.* **2015**, *19*, 49–76.
40. Estoque, R.C.; Gomi, K.; Togawa, T.; Ooba, M.; Hijioka, Y.; Akiyama, C.M.; Nakamura, S.; Yoshioka, A.; Kuroda, K. Scenario-based land abandonment projections: Method, application and implications. *Sci. Total Environ.* **2019**, *692*, 903–916. [CrossRef]
41. Kubiszewski, I.; Costanza, R.; Anderson, S.; Sutton, P. The future value of ecosystem services: Global scenarios and national implications. *Ecosyst. Serv.* **2017**, *26*, 289–301. [CrossRef]
42. DasGupta, R.; Hashimoto, S.; Okuro, T.; Basu, M. Scenario-based land change modelling in the Indian Sundarban delta: An exploratory analysis of plausible alternative regional futures. *Sustain. Sci.* **2019**, *14*, 221–240. [CrossRef]
43. Chen, G.; Li, X.; Liu, X.; Chen, Y.; Liang, X.; Leng, J.; Xu, X.; Liao, W.; Qiu, Y.; Wu, Q.; et al. Global projections of future urban land expansion under shared socioeconomic pathways. *Nat. Commun.* **2020**, *11*, 1–12. [CrossRef] [PubMed]
44. Johnson, B.A.; Estoque, R.C.; Li, X.; Kumar, P.; Dasgupta, R.; Avtar, R.; Magcale-Macandog, D.B. High-resolution urban change modeling and flood exposure estimation at a national scale using open geospatial data: A case study of the Philippines. *Comput. Environ. Urban Syst.* **2021**, *90*, 101704. [CrossRef]
45. Estoque, R.C.; Ooba, M.; Avitabile, V.; Hijioka, Y.; DasGupta, R.; Togawa, T.; Murayama, Y. The future of Southeast Asia's forests. *Nat. Commun.* **2019**, *10*, 1829. [CrossRef]
46. O'Neill, B.C.; Kriegler, E.; Ebi, K.L.; Kemp-Benedict, E.; Riahi, K.; Rothman, D.S.; van Ruijven, B.J.; van Vuuren, D.P.; Birkmann, J.; Kok, K.; et al. The roads ahead: Narratives for shared socioeconomic pathways describing world futures in the 21st century. *Glob. Environ. Chang.* **2017**, *42*, 169–180. [CrossRef]
47. Riahi, K.; van Vuuren, D.P.; Kriegler, E.; Edmonds, J.; O'Neill, B.C.; Fujimori, S.; Bauer, N.; Calvin, K.; Dellink, R.; Fricko, O.; et al. The Shared Socioeconomic Pathways and their energy, land use, and greenhouse gas emissions implications: An overview. *Glob. Environ. Chang.* **2017**, *42*, 153–168. [CrossRef]
48. Popp, A.; Calvin, K.; Fujimori, S.; Havlik, P.; Humpenöder, F.; Stehfest, E.; Bodirsky, B.L.; Dietrich, J.P.; Doelmann, J.C.; Gusti, M.; et al. Land-use futures in the shared socio-economic pathways. *Glob. Environ. Chang.* **2017**, *42*, 331–345. [CrossRef]
49. Costanza, R.; de Groot, R.; Sutton, P.; van der Ploeg, S.; Anderson, S.J.; Kubiszewski, I.; Farber, S.; Turner, R.K. Changes in the global value of ecosystem services. *Glob. Environ. Chang.* **2014**, *26*, 152–158. [CrossRef]
50. Costanza, R.; de Groot, R.; Braat, L.; Kubiszewski, I.; Fioramonti, L.; Sutton, P.; Farber, S.; Grasso, M. Twenty years of ecosystem services: How far have we come and how far do we still need to go? *Ecosyst. Serv.* **2017**, *28*, 1–16. [CrossRef]

 *sustainability*

Article

The Impact of Impervious Surface Expansion on Soil Organic Carbon: A Case Study of 0–300 cm Soil Layer in Guangzhou City

Jifeng Du [1,2,3], Mengxiao Yu [1] and Junhua Yan [1,*]

[1] Key Laboratory of Vegetation Restoration and Management of Degraded Ecosystems, South China Botanical Garden, Chinese Academy of Sciences, Guangzhou 510650, China; dujifeng@scau.edu.cn (J.D.); yumx@scib.ac.cn (M.Y.)
[2] College of Resources and Environment, University of Chinese Academy of Sciences, Beijing 100049, China
[3] School of Public Administration, South China Agricultural University, Guangzhou 510642, China
* Correspondence: jhyan@scib.ac.cn; Tel.: +86-20-37252720

Abstract: Empirical evidence shows that the expansion of impervious surface threatens soil organic carbon (SOC) sequestration in urbanized areas. However, the understanding of deep soil excavation due to the vertical expansion of impervious surface remains limited. According to the average soil excavation depth, we divided impervious surface into pavement (IS_{20}), low-rise building (IS_{100}) and high-rise building (IS_{300}). Based on remote-sensing images and published SOC density data, we estimated the SOC storage and its response to the impervious surface expansion in the 0–300 cm soil depth in Guangzhou city, China. The results showed that the total SOC storage of the study area was 8.31 Tg, of which the top 100 cm layer contributed 44%. The impervious surface expansion to date (539.87 km^2) resulted in 4.16 Tg SOC loss, of which the IS_{20}, IS_{100} and IS_{300} contributed 26%, 58% and 16%, respectively. The excavation-induced SOC loss (kg/m^2) of IS_{300} was 1.8 times that of IS_{100}. However, at the residential scale, renovating an IS_{100} plot into an IS_{300} plot can substantially reduce SOC loss compared with farmland urbanization. The gains of organic carbon accumulation in more greenspace coverage may be offset by the loss in deep soil excavation for the construction of underground parking lots, suggesting a need to control the exploitation intensity of underground space and promote residential greening.

Keywords: expansion of impervious surface; underground space development; deep soil excavation; SOC loss in deep soil; urban renovation; Guangzhou city

Citation: Du, J.; Yu, M.; Yan, J. The Impact of Impervious Surface Expansion on Soil Organic Carbon: A Case Study of 0–300 cm Soil Layer in Guangzhou City. *Sustainability* **2021**, *13*, 7901. https://doi.org/10.3390/su13147901

Academic Editors: Kikuko Shoyama, Rajarshi Dasgupta and Ronald C. Estoque

Received: 9 April 2021
Accepted: 6 July 2021
Published: 15 July 2021

Publisher's Note: MDPI stays neutral with regard to jurisdictional claims in published maps and institutional affiliations.

1. Introduction

The share of urban people in the world has increased from 30% in 1950 to 55% in 2018 and this proportion is projected to reach 68% by 2050 [1]. Urbanized areas in 2030 will nearly triple from those in 2000 [2]. With the rapid expansion of urban areas, many agricultural or natural ecosystem areas have been converted into greenspace or impervious surface, which greatly changes the structure and function of the original ecosystems and has a far-reaching impact on soil organic carbon (SOC) storage [2–5]. Previous studies have shown that urban vegetation has high carbon sequestration capacity [6–11], and the SOC density observed in greenspace is higher than that of grassland or farmland in some cities [8,12,13]. In New York City, the SOC density of the greenspace is comparable to that of suburban and rural forest soil [14]. However, the vegetation and soil removal and soil sealing during the installation of impervious surface have seriously disturbed soil functions [15], which can counterbalance the positive effects from urban greenspace ecosystems [14]. Impervious surface has become the dominant land cover in urbanized areas. Impervious surface covers approximately 31% [13] and 65.91% [16] of the land within urban areas in the United States of America and China, respectively, and more than 50% in many European cities [17]. Therefore, it is necessary to study the influence of impervious surface construction on SOC storage to comprehensively quantify the impact of urbanization on regional carbon budgets [18,19].

The effect of impervious surface expansion on soil carbon storage has been studied from two perspectives. Some studies revealed the differences in SOC storage between sealed soils and surrounding greenspace soil at the sample site scale [14,19–23], and others have used spatially explicit models driven by land cover change and/or other multiple environmental covariates to analyze the total SOC loss and its spatial and/or temporal variability at regional, national and even global scales [24–28]. These models provide a powerful and effective tool for presenting carbon changes in space and time. However, they usually treat urban area or impervious surface as a homogeneous space unit, and interpret land cover types from medium or low resolution remote sensing images, which cannot reflect the differences of soil carbon disturbance among impervious surface types. For urban carbon management, research based on high-resolution images and refined types of impervious surface could be more beneficial to policy making.

Topsoil removal is the main cause of depletion in SOC content under impervious surface [29]. Although the depth of soil removal varies in different types of impervious surface, few studies have explored in detail the difference of soil carbon disturbance between different impervious surfaces. In China, the depth of soil extraction for pavement is approximately 15–30 cm, and that for low-rise buildings without underground parking lots is approximately 80–100 cm. High-rise buildings usually include underground garages. According to the Design Code for Residential Buildings (GB 50096-2011) [30], the net height of underground garage driveways should not be less than 220 cm. Adding spaces for the construction of building foundations and for the installation of facilities, such as pipes, wires and ventilating ducts, the depth of soil extraction for buildings with underground parking requires a deeper depth than 300 cm. Large utility lines, subways, car park facilities and malls have been constructed under the urban surface in many big cities around the world [31,32], especially in China where high-rise buildings with underground parking lots are widely distributed in many medium-sized or big cities. Although the depth of soil excavation in many underground space utilizations is deeper than the top 100 cm soil layer, most published studies on urban SOC storage ignored the carbon below a depth of 100 cm. Soil at deeper depths (e.g., 100–300 cm) still has large carbon storage [33]; therefore, consideration of 0–20 cm or 0–100 cm soil depth cannot reveal the influence of the vertical expansion of impervious surface on SOC pools.

Soil is the largest contributor to the urban ecosystem carbon pool [13,34]. Although the SOC density beneath impervious surface is lower than that in adjacent greenspace areas [14,19,20,28,29], the contribution of soil under impervious surface to urban SOC pool is still large due to the large coverage area. For instance, in the Chicago and Boston metropolitan areas, the impervious surfaces contribute 28.76% and 22.46% to the total soil and plant carbon pools with aerial coverages of 60% and 53.9%, respectively [18]. In the China Urumqi urban area, the impervious surface covers 63% of the urban area and it contributes 57% of the SOC pool. Buildings account for a large proportion of impervious surface in cities; however, previous studies usually sample soils under pavement, such as roads, sidewalks, parking lots, paved backyards and paved squares [14,19,20,22,23,29], because it is hard to sample soil under buildings. The SOC stock in building-covered soil was assumed to be 0 kg/m^2 [19,23], or was designated a proxy such as the SOC density of clean fill soil [13,18] or the provincial average SOC density [34], which greatly increased the uncertainty of urban SOC pool estimation and its response to the expansion of urban impervious surface. Thus, it is necessary to try other methods to estimate SOC storage in building-covered areas.

Due to the rapid land use/cover changes in China, ecosystem services intensity experienced a continuously decreasing trend from 1995 to 2015, especially in large megacities [35], and ecosystem service values for provision, regulation, support and culture also decreased during the period of 1988–2008 [36]. Guangzhou is a representative megacity with rapid urbanization and increasing impervious surface in China. Urban area in Guangzhou has increased from 187.40 km^2 in 1990 to 1324.17 km^2 in 2019 with an annual expansion area of 39 km^2 [37]. Urban underground space use has become an essential part of the urban master plan in China. According to the urban subsurface development target of "The master plan of Guangzhou city (2011–2020)", the city-wide area of developed underground space will reach 90 million square meters, of which 8 million square meters will be used as commercial space. This large-scale soil excavation will cause large disturbances to SOC storage, which should not be ignored in the environmental assessment of underground space exploitation. The objectives of this study were: (1) to estimate the SOC stocks of pavement, low-rise buildings and high-rise buildings based on their average soil extraction depth and the SOC density of pre-urbanization pervious surfaces; (2) to estimate the SOC storage in the 0–300 cm soil layer and then quantify the impacts of urban impervious surface expansion on the SOC stock at the residential and regional scale. To our knowledge, this is the first study in which the urban SOC and its response to the expansion of impervious surface in the 0–300 cm soil layer have been quantified. This research provides implications for policy making on how to reduce disturbances to the deep soil carbon pool during urban underground space development.

2. Materials and Methods

2.1. Study Area

Guangzhou city (112°57′~114°03′ E, 22°26′~23°56′ N) is located in the lower reaches of the Pearl River Delta near the South China Sea. This city is the economic and transportation center of South China, with a built-up area of 1263.34 km^2 and a large population of 14.49 million in 2018 [37], making it a key area to study the effects of impervious surface expansion on soil carbon sequestration. Guangzhou city, which is dominated by lateritic red soil and paddy soil, has a subtropical monsoon climate with an average annual temperature of 21.4–22.6 °C and an average annual precipitation of 1600–2300 mm [37]. The northeastern part of this city is dominated by middle and low mountains, the central part contains a hilly basin and the southern part contains a coastal alluvial plain. We delineated the urban core area (735.66 km^2) as our study area on Google Earth images (0.27 m resolution) using geographic information system software (ArcGIS version 10.4, ESRI) (Figure 1). This area is separated from the surrounding regions by rivers, woodlands, farmland and rural settlements.

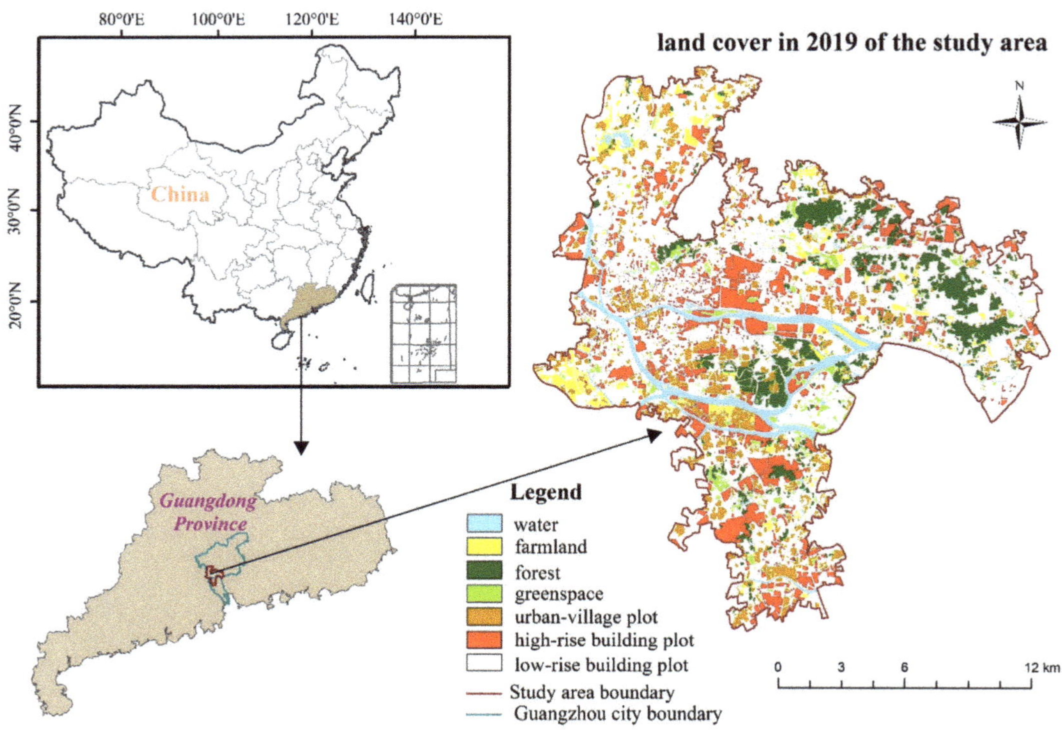

Figure 1. Location of Guangzhou city and its land use/cover in the urban core area in 2019.

2.2. Mapping Land Use/Cover in 2019

We first classified land use/cover into nine types: urban village plot, high-rise building plot (>28 m), farmland, conservatory, residual forest, tree orchard, water, greenspace with area larger than 1500 m^2 and low-rise building (<28 m) plot. Then, using manual digitization techniques, we digitized the first eight land use/cover types from Google Earth images (0.27 m resolution) using the software of ArcGIS (version 10.4, ESRI). The area of the low-rise building plot was the total size of this study area minus the summed area of the other eight land cover types. Second, we grouped the above nine land use/cover types into six categories: greenspace, woodland, farmland, pavement (excavated soil thickness $\leq$ 20 cm, IS$_{20}$), low-rise building (excavated soil thickness $\leq$ 100 cm, IS$_{100}$) and high-rise building (excavated soil thickness $\geq$ 300 cm, IS$_{300}$). According to codes for road and building foundation construction, consultation with engineers and our field investigation, we assigned the evacuation thickness of IS$_{20}$, IS$_{100}$ and IS$_{300}$ as 0–20, 0–100 and 0–300 cm, respectively, and detailed information was presented in the Section 1 of the Supplementary Materials [30,38]. The area of green space was the difference between the total area of this study and the summed area of the other five land cover types. Tree orchard land was grouped with forestland. Conservatory land was classified as farmland. Buildings located in urban village plots and low-rise building plots were grouped as IS$_{100}$. IS$_{300}$ included buildings located in high-rise building plots and low-rise building plots or villas equipped with underground parking or other kinds of underground space development. To calculate the area covered by buildings, we multiplied the area of each type of building plot by its building density. The building densities of low-rise building plots, high–rise building plots and urban village plots were randomly sampled based on Google Earth images (0.27 m resolution) using the software of ArcGIS. To calculate the area of pavement, we first calculated the sum area of the impervious surface and bare soil and then subtracted the area of the buildings and conservatory areas (according to Google Earth images, most of the bare soil plots are land parcels under construction, and they were digitized as high-rise building plots; impervious surface interpreted from Landsat images included conservatories that were grouped with farmland). We used the method developed by Fan et al. [39] to estimate the area of imperious surface (Figure S1) and bare soil from Landsat8 OLI images (30 m resolution) at the subpixel scale. The detailed framework for the land use/cover type interpretation was presented in Figure S1 and Table S1. The accuracy assessment of impervious surface map and land use/cover data was presented in the Section 2 of the Supplementary Materials [39–42].

2.3. Compiled SOC Densities for Pervious Surfaces

By assuming that the 100–200 cm soil layer had a constant SOC density, we calculated the SOC density of the 100–200 cm layer according to the difference in SOC density between the 0–180 cm and 0–100 cm soil layers. The SOC density of woodland was an area-weighted value of the SOC density in residual forest and tree orchard soil. The SOC density of the 0–20 cm, 0–100 cm and 0–180 cm soil layers in farmland, woodland, orchard and greenspace were derived from Zhu [39,40].

For the SOC density of the 200–300 cm layers in woodland and farmland soil, we used the globally averaged SOC density of 200–300 cm soil in tropical evergreen forest and farmland [33] as a proxy, respectively. Urban land in this study area was mainly converted from farmland; therefore, the SOC density for greenspace in the 200–300 cm layer was assigned the value of the corresponding layer in the farmland soil.

2.4. Estimates of SOC Density for Impervious Surfaces

Previous studies have suggested that there is no significant difference between SOC density at the equivalent depth and soil sampled under impervious surface and greenspace [21,23]. The pattern of vertical decline of the SOC density in impervious-covered soil is similar to the trend of rural soils, and the disturbance of pavement installation to the subsoil layers may not be serious; therefore, the vertical distribution of the SOC density before soil sealing is largely maintained [19]. Due to the inaccessibility of soil sampling beneath impervious surface, especially soil under buildings, we estimated the SOC density of sealed soil according to the SOC density of the residual soil layer of the land use/cover before urbanization and the average depth of soil excavation for an impervious surface installation. We hypothesize that: (1) the mineralization rate of residual original soil layers was zero and that all excavated soil was removed from this study area; (2) for regional SOC loss estimation, soil excavation for the construction of buildings and underground facilities was limited to the areas where the buildings would be located. The specific assignment method was as follows.

The average excavation depth was included in the profile of impervious-covered soil; however, its SOC density was assigned as 0 kg/m^2. The SOC density of the soil layer beneath impervious surface was equal to the SOC density of the residual soil layer of the land use/cover before urbanization. The average soil excavation depths were 0–20 cm, 0–100 cm and 0–300 cm for IS_{20}, IS_{100} and IS_{300}, respectively. By subtracting the SOC_{0-20} density from the SOC_{0-100} density of the pre-urbanization land use/cover, we obtained the SOC_{0-100} density of IS_{20}. By assuming that the SOC density between 100 cm and 200 cm soils was a constant value, we evaluated the $SOC_{100-200}$ density for IS_{20} or IS_{100} according to the averaged SOC density of the 0–180 cm layer in the original land use/cover soil before urbanization. We assigned the $SOC_{200-300}$ density of farmland soil from Jobbágy and Jackson [33] as the $SOC_{200-300}$ density of IS_{20} and IS_{0-100}.

Field investigations [19,21,23] and estimation based on the measured SOC density of interval soil layers [14,22] are the main methods by which to obtain the SOC density of deep soil under pavement. SOC density of building-capped soil was assigned as 0 kg/m^2 [14,19] or the density of clean-fill soil [13,18]. For comparison with the other methods and their values (Tables S2 and S3 within which references [19,21,23,42–46] were cited). we sampled soil (0–20 cm) beneath impervious surface from 11 new road reconstruction projects and clean-fill soil (a kind of backfill soil excavated from natural sedimentary soil layer, and into which organic waste, stone or white lime should not be mixed) from three road greening projects. We also collected SOC content data from granitic residual soil and oceanic-continental sedimentary clay soil (the widely distributed parent material in Guangzhou), as well as SOC content data from the bottom soil layers of paddy soil and lateritic red soil (the widely distributed soils in Guangzhou). Based on these literature-derived SOC data and the code of soil compaction used for road construction (to evaluate the bulk density of soil under impervious surface), we calculated the SOC density of the deep-soil layer beneath buildings and pavement (Tables S2 and S3). Methods of soil sample and SOC density calculation for the sealed soil and clean-fill soil were present in the Section 3 of the Supplementary Materials [38,46–61]. The SOC density estimated according to the SOC density of the remaining soil layer in the original pervious soil profile was used to assess the total SOC pool and its loss due to the expansion of impervious surface, while the SOC densities calculated by the other methods were mainly used for comparison.

2.5. Estimate of SOC Storage Loss Caused by Impervious Surface Expansion

Topsoil removal largely contributed to the depletion in SOC content under impervious surface [21,23,29]. We used the SOC storage in the removal soil layer as a representation of the carbon loss caused by impervious surface expansion since it is almost impossible to survey soil beneath buildings. SOC_{0-300} density changes and the total SOC_{0-300} pool loss were estimated to reveal the impact of impervious surface expansion at the residential and regional scale, respectively. For farmland urbanization, the loss of SOC_{0-300} storage at the residential scale was estimated by the area-weighted average SOC_{0-300} loss in pavement, building and greenspace. For impervious surface and green space, we assumed that the SOC density loss for each impervious surface was the difference between the SOC density of the original pervious surface before urbanization and the estimated SOC density of each surface type. It is very difficult to conduct a spatially explicit study on the conversion between each impervious surface and original pervious surfaces; therefore, we took the average SOC_{0-300} density of woodland (including tree orchard) and farmland as a proxy for the SOC density of original pervious surfaces (farmland, tree orchard and woodland were the main land use/cover types before urbanization in this region, and the differences in SOC_{0-180} density between these land covers were small [40]). For urban renewal plots, the loss of SOC storage was the difference of SOC_{0-300} density before and after urban renewal. The detailed method of this section is present in the Table S5 [62–64].

The rapid urbanization of Guangzhou city began around 1990; however, the earliest fine resolution images that we could get were from 2000. Therefore, we did not consider the specific time period of impervious surface expansion, but rather assessed the total SOC loss according to the current distributions and areas of these impervious surface types. To estimate the SOC pool loss caused by the expansion of each impervious surface, we multiplied the area of each impervious surface by its loss of SOC_{0-300} density. Summing the SOC pool losses caused by each impervious surface, we obtained the total SOC pool loss due to the expansion of impervious surface to date. A summary of the method and characteristics of the study area are shown in Figure 2.

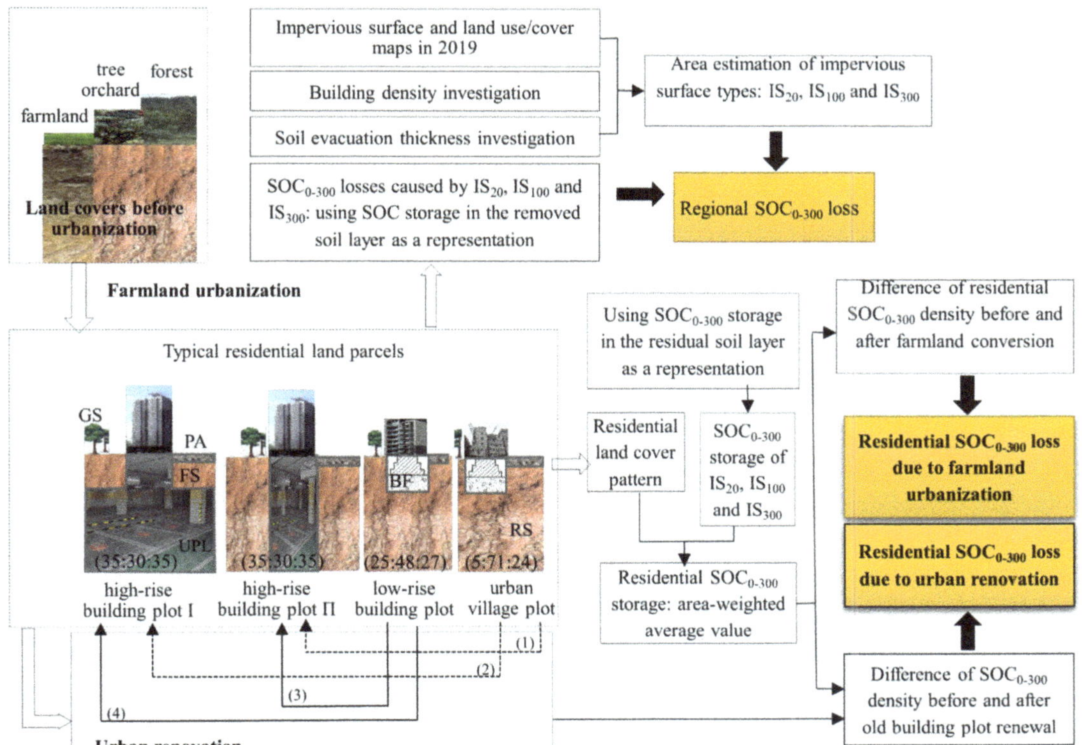

Figure 2. The summary of the methodology and land use/cover characteristics of the study area. Numbers in the round brackets represent the coverage proportion of green space, building and pavement in each typical residential land parcel. The GS, PA, FS, UPL, BF and RS represent green space, pavement, fill soils, underground parking lots, building foundations and residual soils, respectively. IS$_{20}$, IS$_{100}$ and IS$_{300}$ represent pavement with a soil evacuation thickness of around 20 cm, buildings with soil evacuation thickness $\leq$ 100 cm and buildings with evacuation thickness $\geq$ 300 cm, respectively. (1) and (2) represent renovating urban village plots into high-rise building plots I and Ⅱ, respectively; (3) and (4) represent renovating low-rise building plots into high-rise building plots I and Ⅱ, respectively. Areas of high-rise building land parcels, low-rise building land parcels and urban village plots in 2019 were 121.41, 364.97 and 89.53 km^2, respectively, and IS$_{20}$, IS$_{100}$ and IS$_{300}$ totally covered 73% of the study area (Table S1).

3. Results

3.1. Area Estimation by Land Cover in 2019

IS$_{20}$ had the largest area (35.54%), followed by IS$_{100}$ (32.88%), greenspace (9.13%), woodland (8.02%), IS$_{300}$ (4.96%) and farmland (3.39%) (Figure 1). The areas of urban village plots, low-rise building plots and high-rise building plots were 89.53, 364.97 and 121.41 km^2, respectively (Table S1). Large areas of urban roads were subsumed in the low-rise building plots. However, these roads were not considered in the building density sample, which may result in an overestimation of 16% of the low-rise building's area (Section 2 of the Supplementary Materials). We used the root mean square error (RMSE) to assess the accuracy of the impervious surface map, and the RMSE citywide was 0.20 (Section 2, Figure S1). The total area of IS$_{20}$, IS$_{100}$ and IS$_{300}$ (539.87 km^2) was much higher than the impervious surface area interpreted from Landsat 8 OLI images (Figures S1 and S2), because the bare soil areas were regarded as high-rise building plots under construction.

3.2. SOC Stock and Its Profile Distribution by Land Cover

When the thickness of the excavated soil layer was included in the profile of impervious-covered soil, the SOC density at the 0–300 cm layer of farmland, woodland, greenspace, IS_{20}, IS_{100} and IS_{300} was 19.04 kg/m^2, 17.32 kg/m^2, 17.15 kg/m^2, 4.04 kg/m^2, 8.22 kg/m^2 and 0 kg/m^2, respectively (Table 1). Relative to the top 100 cm, the percentage of SOC density in the 0–20 cm layer was 26%, 28% and 26% for farmland, woodland and greenspace, respectively; the proportion of SOC density in the 100–200 cm layer was 0.73%, 0.67%, and 0.75%, respectively (Figure 3). The SOC_{0-100} density and $SOC_{100-200}$ density for IS_{20} accounted for 41% and 39% of SOC_{0-300} density, respectively. The SOC density of IS_{100} in the 100–200 cm layer was 67% of that in the 0–300 cm layer (Table 1).

Table 1. Soil organic carbon density by land cover.

Land Use/Cover	Area (km^2)	SOC Density (kg/m^2)				
		0–20 cm	20–100 cm	100–200 cm	200–300 cm	0–300 cm
Farmland	24.97	2.49	6.97	6.88	2.70	19.04
Woodland	59.02	2.22	5.82	5.38	3.90	17.32
Greenspace	67.16	2.13	6.14	6.18	2.70	17.15
IS_{20}	261.46	0	5.82	5.52	2.70	14.04
IS_{100}	241.90	0	0	5.52	2.70	8.22
IS_{300}	36.51	0	0	0.00	0.00	0.00

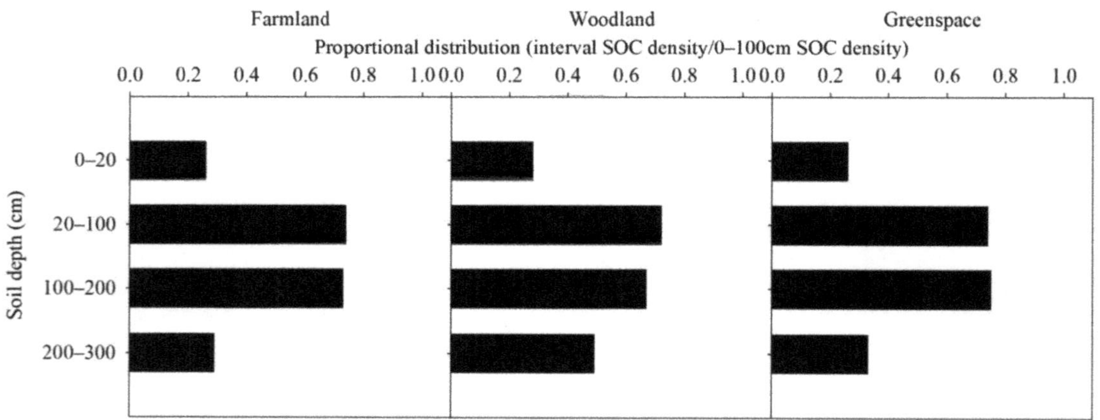

Figure 3. Proportional distributions of SOC density by land cover.

The total SOC pool of 300 cm was 8.31 Tg (the thickness of the excavated soil layer was included in the profile of impervious-covered soil), and the largest pool by land cover was IS_{20} (3.67 Tg), followed by IS_{100} (1.99 Tg), greenspace (1.15 Tg), woodland (1.02 Tg), farmland (0.48 Tg) and IS_{300} (0 Tg) (Figure 4). By soil horizon, the contributions of the 0–20 cm, 20–100 cm, 100–200 cm and 200–300 cm layers to the total SOC pool were 4%, 30%, 44% and 22%, respectively (Figure 4, Table S4). Pervious surface carbon pools contributed 32% of the total SOC pool, of which 48% was stored in the 0–100 cm soil layer; impervious surface carbon pools contributed 69% of the total SOC pool, of which 49% was stored in the 100–200 cm layer (Table S4).

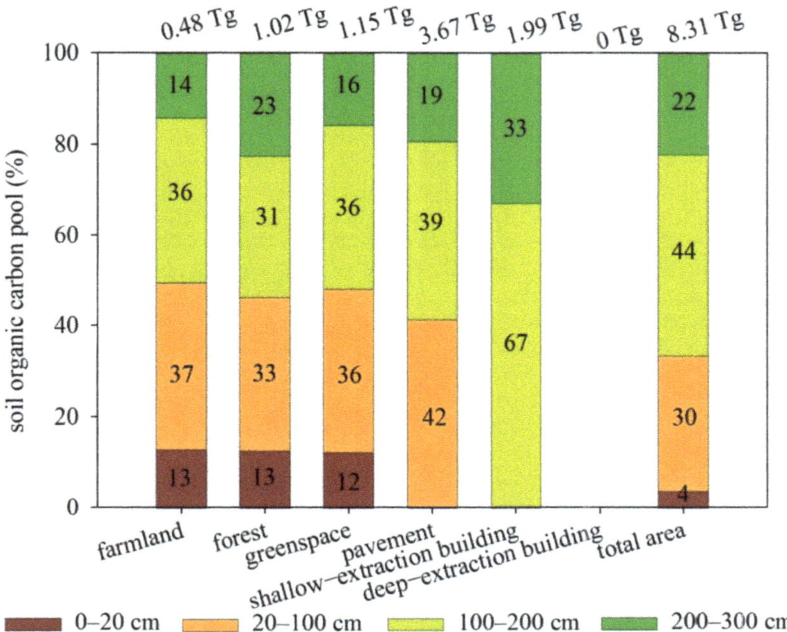

Figure 4. Profile distributions of SOC pools by land cover.

3.3. SOC Loss Caused by Soil Removal for the Installation of Impervious Surface

All residential districts converted from farmland had high SOC_{0-300} loss (6.16–14.18 kg/m^2), particularly the high-rise building plots for which underground parking lots covered the whole residential land (14.18 kg/m^2) (Figure 5, Table S5). The SOC_{0-300} losses of renovating urban village into high-rise building plot were lower than that of renovating low-rise building plot. If underground parking lot covers 30% of the residential area, urban village renovation may even slightly increase SOC storage (0.85 kg/m^2) (Figure 5, Table S5).

The total SOC loss caused by the expansion of impervious surface (539.87 km^2) was 4.16 Tg (0–300 cm), of which the IS_{100}, IS_{20} and IS_{300} contributed 58%, 26% and 16%, respectively (Table 2). The SOC density loss by impervious surface type was 18.18 kg/m^2, 9.96 kg/m^2 and 4.14 kg/m^2 for IS_{300}, IS_{100} and IS_{20}, respectively (Table 2).

Table 2. The SOC loss at a depth of 0–300 cm caused by the expansion of impervious surface.

Impervious Surface Type	SOC Density Loss kg/m^2	SOC Pool Loss Tg	Fraction of SOC Pool Loss %	Impervious Surface Coverage %
IS_{300}	18.18	0.66	16	5
IS_{100}	9.96	2.41	58	33
IS_{20}	4.14	1.08	26	35
Total	7.67 *	4.16	100	73

Note: * The area-weighted loss of SOC density due to the construction of the IS_{20}, IS_{100} and IS_{300}.

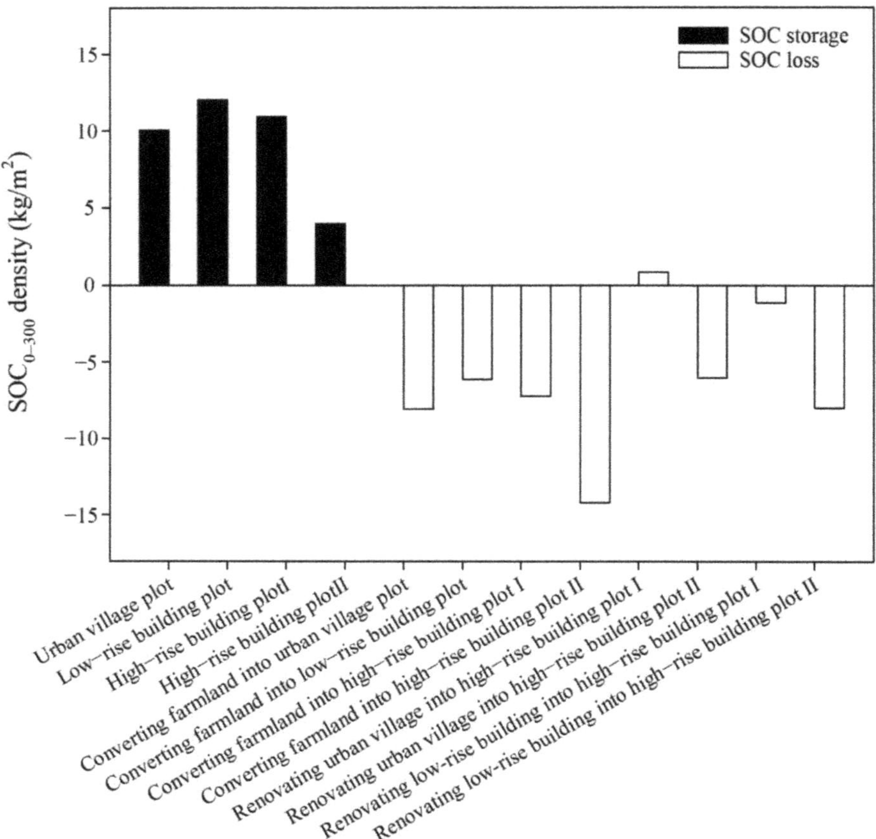

Figure 5. SOC storage (black columns) and SOC loss of farmland urbanization and urban renovation (white columns) at the residential scale; I and II represent the coverage fraction of underground parking lots (30% and 100%, respectively). The detailed estimation method is presented in Table S5.

4. Discussion

4.1. Comparison Analysis of SOC Density for Pervious Surfaces

Our results showed that the SOC_{0-20} density and SOC_{0-100} density in farmland were much lower than the global mean values of the corresponding soil layers [33]. The average proportion of SOC_{0-20} density to SOC_{0-100} density in farmland was 26% (Figure 3), which was close to the sampled value of 29% in the Ten-Thousand-Mu Orchard in the Haizhu District of Guangzhou city [65], but slightly lower than the citywide average value of paddy soils (33%) in Guangzhou [66]. All of the above-mentioned ratios in the farmland of Guangzhou were much lower than the global mean value of 40% [33], and farmland management practices may be the main reason for this difference. Compared with the utilization of chemical fertilizer, less organic fertilizer was applied in the farmland of Guangzhou, which was disadvantageous to the SOC accumulation, especially in the topsoil. Sulman et al. [67] reported that tilled soil had low organic carbon concentrations in the top mineral soil. In Guangzhou city, farmers rarely let their land lie fallow, instead tilling their farmland more frequently because of the farming system of the rice-rice-vegetable rotation or the rotation of various vegetables. Farmers plow paddy fields into deep furrows (to a depth of 20–30 cm covered with water) and then heap up soil into rows of broad ridges to grow vegetables on them. Therefore, soils layers deeper than 20 cm are often flooded, which is beneficial to the accumulation of organic carbon in the deep soil. The SOC_{0-200} density in farmland (Table 1) was slightly higher than the global averaged value of 15.0 kg/m^2 [33]. This may contribute to landform and soil parent material in this region. Buried soil layers in the alluvial plain of a river can increase the organic carbon storage of the deep soil layer [68,69]. Our study area is located in the lower reaches of the Pearl River Delta and near the South China Sea. Organic-rich, oceanic-continental sedimentary clay soils are widely distributed in this area. Furthermore, large areas of this clay soil are buried no deeper than 300 cm [70], and its average organic carbon concentration is as high as 2% [46].

The average SOC_{0-100} density of woodland was only 43% of the global average value [33]; the low stand age and anthropogenic disturbance may be the main reasons for these lower carbon stocks. Guo and Gifford [71] noted that the conversion of natural forests into plantations reduced soil carbon storage. Many forests in the study area are secondary forests or young forests planted in the middle of the 1980s. The average SOC density of the young, middle-aged and mature forests in the Pearl River Delta (where Guangzhou city is located) was only 6.67 kg/m^2, 8.41 kg/m^2 and 10.26 kg/m^2, respectively [72], all of which are much lower than the values reported by Jobbágy and Jackson [33]. The $SOC_{100-200}$ density in woodland (Table 1) was comparable to the average value in tropical evergreen forest soil [33], which may be due to the deeply buried organic-rich clay soil under the orchard land (tree orchards were grouped with woodland).

We assigned the $SOC_{200-300}$ density of farmland reported by Jobbágy and Jackson [33] as a proxy to the $SOC_{200-300}$ density of farmland soil in our study area. Turnover profiles, which were removed from the dataset of Jobbágy and Jackson, were widely distributed in our study area because of the organic-rich sedimentary clay soil that developed in the deep layer of paddy soil [43]; therefore, this proxy maybe an underestimated value.

The proportion of SOC_{0-20} density to SOC_{0-100} density in greenspace (Figure 3) was equivalent to the value of 0.25 in a greenspace in the Ten-Thousand-Mu Orchard in the Haizhu District of Guangzhou city [65]. Many topsoil layers in the greenspace originated from clean fill soil, plus the younger age of the vegetation and the carbon storage in the upper 20 cm soil was low. The proportion of $SOC_{100-200}$ density to SOC_{0-100} density in the greenspace was as high as 0.75, which was slightly higher than the proportion in the woodland. Most of the greenspaces were converted from farmland. Perhaps because the high carbon pool of 100–200 cm in farmland had not been greatly disturbed during the conversion, the greenspace soil inherited the high carbon density at a depth of 100–200 cm (Table 1 and Figure 3). In addition, areas with thick organic-rich clay soil are usually developed as greenspaces rather than buildings because of their low bearing capacity.

4.2. Comparison Analysis of SOC Density of Impervious-Covered Soil

For soils under pavement, the SOC_{0-100} density estimated from the residual original pervious surface was similar to the value calculated by the measured SOC density of 0–20 cm soil beneath pavement (Table S2 and Section 3). Our estimated SOC_{0-100} density in soils sealed by pavement was comparable to that in Urumqi, China [19] and in New York City, USA as reported by Raciti et al. [14]; however, it was much lower than that in Leicester, UK [23] and in New York City, USA as reported by Cambou et al. [21] (Table S2). Clay content and climatic differences may account for the carbon density differences among cities [23]. For clean-fill soil or constructional layers (mainly carbon-free materials) under pavement, SOC_{0-100} density estimation based on the SOC density of residual pervious surface may lead to a large overestimation. For instance, the SOC density estimated based on the measured organic carbon content in clean-fill soil (100 cm thick) in this study was 3.94 kg/m^2 for artificially tampered soil and 3.30 kg/m^2 for untampered soil (Table S2 and Section 3), which was similar to the estimation by Pouyat et al. in USA [18]. The sampled SOC density of filled soil under pavement of the Seoul residential area in South Korea was as low as 2.7 kg/m^2 [42], and the sampled value for constructional layers (mainly gravel, crushed rock and sand) below the road surface in Lahti, Finland, was as low as 1.20 kg/m^2 [22].

For the SOC density of the 0–100 cm soil layer under buildings, Pouyat et al. [18] took the estimated SOC density of clean fill soil as a proxy, which was equivalent to the value calculated in this study (Table S2). If we are using the SOC density of the residual original soil layer before urbanization as a proxy, assuming that the decomposition rate of the residual soil was zero, the estimated values of 5.52 kg/m^2 (an estimation based on the SOC density of the residual original soil layer) and 5.43 kg/m^2 (an estimation based on the averaged SOC content of the bottom soil in original pervious soil layer) were both higher than the SOC density of the clean fill soil (Table S2 and Section 3). Fill soil mainly comes from excavated deep soil, which has low organic matter content; therefore, the SOC density of clean fill soil may be used as an underestimated proxy for the 0–100 cm soil under IS_{100}, especially for buildings without a clean fill soil layer.

For the $SOC_{200-300}$ density of pavement and IS_{100} covered soil, we took the $SOC_{200-300}$ density (Table 1) in farmland as a proxy, which was close to the estimated SOC density of the granitic residual soil (100 cm thickness) (Table S2). The proxy coincided with the measured SOC density of the 0–100 cm fill soil under impervious surface in a Seoul residential area in South Korea [42] (Table S2). Weathered granitic residual soils excavated during the construction of building foundations is commonly used as a kind of high-quality fill soil. This coincidence may support the rationality of the above-mentioned proxy, since most fill soil came from excavated deep soil layers. However, if the organic-rich, oceanic-continental sedimentary clay soil (Table S2) developed beneath pavement and IS_{100}, this proxy may underestimate the $SOC_{200-300}$ density of pavement and IS_{100} covered soil.

Topsoil removal, decomposition and leaching of the residual soil layers beneath impervious surface may attribute to the carbon loss from sealed soil [14]. A controlled field study reported that the removal of 0–10 cm soil contributed 57% to the total carbon loss (0–30 cm) [29]. We assumed that the SOC mineralization of the residual soil was zero, which would underestimate the carbon loss. However, SOC decomposition in impervious-covered areas will be slowed because of anoxic conditions [20,29]. The SOC turnover rate varies with the type, degree and extent of sealing [20,23,29]. Evidence has shown that the SOC density decreases from the edge to center in imperviously covered areas [19]. For small patches of impervious surface surrounded by greenspace, soil could still be colonized by vegetation roots and retain the organic carbon turnover rate to some extent, while organic carbon could be more stable in areas with impervious surface largely distributed [20,23]. Large areas of impervious surface continuously distributed in the study area, and the soil was excavated deeper than the organic-rich topsoil for the construction of building foundations. Therefore, the carbon mineralization in residual soil may contribute less to the carbon loss. Thus, the underestimation of carbon loss due to the above-mentioned assumption may be small.

4.3. Impacts of Impervious Surface Expansion on SOC Storage

Our result showed that there was a large amount of organic carbon in impervious-covered areas. SOC stored in sealed soil in the 0–300 cm layer and 100–300 cm layer in our study area was equivalent to 9% and 6% of the 0–100 cm SOC pool in Guangzhou city [66], respectively. If SOC density in building-covered soil were assigned 0 kg/m^2 (Table S6, scenario 3), the total SOC pool in the 0–300 cm soil layer would decrease 34% (Table S6, scenario 1) or 42% (Table S6, scenario 2), and the loss of the SOC pool due to impervious surface expansion would increase by 48% or 82%, respectively (Table S6). Therefore, ignoring the SOC pool under IS_{100} will greatly underestimate the urban carbon pool and then greatly overestimate the negative impact of impervious surface expansion on urban soil storage.

At the residential scale, carbon loss per unit area varied with the coverage of greenspace and pavement, building density and development intensity of underground space. Foundation-excavation-induced SOC loss of IS_{300} was 1.8 times that of IS_{100} (Table 2);

however, at the residential scale, if the intensity of underground space development is limited within a reasonable range, the SOC loss of IS_{300} plots can by roughly equivalent to that of IS_{100} plots (Figure 5, Table S5). Underground parking lots in many high-rise residential districts had extended to the underground spaces covered by greenspace and pavement. According to the urban construction code of the Planting Soil for Greening (CJ/T 340-2011) [64], the thickness of an effective soil layer for tree planting should not be less than 100 cm. Appropriately raising this thickness criterion may increase SOC stock in greenspace. Compared with farmland urbanization, renovating urban village or low-rise building plots into high-rise building plots has potential to substantially reduce SOC loss (Figure 5, Table S5). Adding the increment of vegetation carbon due to the increased greenspace coverage, urban renewal will be more profitable to urban carbon management. Analysis of the total organic carbon loss per unit area in vegetation and soil at the residential scale can further our understanding.

In areas with deep soil layers and intensive utilization of underground space, studies at conventional depths of 0–20 cm or 0–100 cm cannot fully reveal the negative impact of the vertical expansion of impervious surface on SOC storage. The SOC pool estimated in the 0–100 cm soil layer was 73% (scenario 1) or 54% (scenario 2) lower than that in the 0–300 cm soil layer, and the SOC loss of the 0–100 cm layer was 23% (scenario 1) or 47% (scenario 2) lower than that in the 0–300 cm soil layer (Table S6). We used SOC storage in the excavated soil layer as a representation of carbon loss caused by the expansion of impervious surface. However, these removed SOC cannot simply be considered as a carbon source because they were not necessarily mineralized rapidly [29].

The difficulty of sealed soil sampling, the higher heterogeneity and the lack of standardized methods for urban soil sampling are the main barriers to a robust estimation of the urban soil pool and its response to urbanization [21]. If the thickness of the excavated soil layer is included in the soil profile of the building-covered soil (0 kg/m^2 is assigned as the SOC density of the excavated soil layer), and the SOC density of the residual pervious soil layer is taken as a proxy for soil under buildings, the great difficulty of deep soil sampling conducted beneath buildings could be avoided. Indeed, soil surveys are the basic method to accurately assess the urban soil carbon pool; however, the method presented here may be used as a last resort.

4.4. Uncertainties of SOC Pool Evaluation

Buried cultural layers may greatly contribute to the urban deep soil carbon pool [42,73]. We neglected carbon pools stored in the buried A horizon and buried sediment from ponds, which underestimated the regional carbon pool. The possibility of buried A horizons in greenspaces and paved ground is relatively high. If there are buried A horizons under all the greenspaces and pavement in the study area, and by assigning the SOC density of the 0–20 cm layer in farmland (the main land use before urbanization) as a proxy for these horizons, then the total SOC pool of these buried layers is 0.82 Tg. Rapid urbanization in Guangzhou city began in the 1990s. From 1990 to 2018, a total of 7.74 km^2 of water area (mainly fish ponds and streams) was converted into construction land, according to the dataset provided by the Data Center for Resources and Environmental Sciences, Chinese Academy of Sciences [74]. The SOC density of this sludge in the 0–100 cm layer can reach 19.64 kg/m^2 [65]. If we assume that all these converted water areas contain a buried sludge layer (100 cm depth with a SOC density of 19.64 kg/m^2), then the SOC pool of this buried sediment is 0.15 Tg. Taken together, the underestimation of SOC stored in buried soil layers may not be larger than 12% of the total carbon pool.

We ignored the SOC losses in deep soil under pavement and greenspace. If we assume that all of the 100–300 cm soil layers under greenspaces and pavement in high-rise building plots have also been removed during the construction of underground parking lots, then the SOC pool loss of IS_{300} will increase by 0.75 Tg (the SOC density of greenspace and pavement at the 0–100 cm layer is derived from Table 1), which is 9% and 18% of the total SOC pool and SOC pool loss, respectively. In addition, a small number of low-rise building plots also have small underground parking lots. The construction of subways and underground shopping malls also results in a larger amount of soil excavation. However, carbon loss due to these kinds of underground space developments is difficult to estimate due to the inaccessibility of data on the volume of the soil cuts and fills.

5. Conclusions

We estimated the total SOC pool and its response to impervious surface expansion in the 0–300 cm layer in the core area of Guangzhou city. Our results suggest that there was a large amount of SOC in low-rise building- and pavement-covered soils. Deep soil excavation during the vertical expansion of impervious surface can seriously disturb soil carbon sequestration, and the conventional 0–100 cm soil profile cannot fully reveal the negative effect of this vertical expansion on the regional soil carbon storage. For high-rise residential areas, the gains of organic carbon accumulation in more greenspace coverage may be offset by the loss in deep soil excavation for the construction of underground parking lots. However, compared with farmland urbanization, renovating urban village or low-greenspace-coverage low-rise building plots into high-rise building plots can generate lower SOC_{0-300} loss. Large areas of deep soil excavation and sealing could greatly disturb the capacity of soil to provide supporting, provisioning, regulatory and cultural services. Possible disposals of these excavated soils include discarding, dumping into designated landfill sites and using as backfill soil to foundation ditch, greenspace, ponds, etc. For regional sustainable development, it is necessary for policy-makers to manage the intensity of underground space development and to promote urban greening. It is also very necessary to investigate the disposal method of excavated soil and its corresponding effects on SOC so as to find the proper solution to meet both the requirements of urban construction and soil protection.

There are some uncertainties in SOC estimations because they are based on data compiled from the literature and a series of assumptions. For the inaccessibility of fine resolution images circa 1990 and the impossibility of deep soil sampling in impervious surface area, we used the statistical summary method to evaluate the possible excavation-induced SOC losses caused by existing impervious surface. Soil profiles formed during the construction of building foundation or underground parking lots can be used to investigate the SOC content in deep soil. Models driven by space-time explicit land cover map (fine resolution) and other natural or human-mediated covariates (such as sediment characteristics, disposal method of excavated soil and its corresponding SOC mineralization characteristics, evacuation thickness, etc.) can provide more useful information for urban construction and soil protection at site at the regional scale.

Supplementary Materials: The following are available online at https://www.mdpi.com/article/10.3390/su13147901/s1, Section 1 The basis of assuming evacuation thickness as IS20, IS100 and IS300, Section 2 Accuracy assessment of impervious surface map and land use/cover data, Section 3 Calculation method of SOC density in clean fill soil and sealed soil, Figure S1: Framework for land use/cover type interpretation based on Google Earth and Landsat 8 OLI images, Figure S2: Impervious surface fraction of the urban core area of Guangzhou city in 2018, Table S1: Land use/cover classification and its area, Table S2: Comparison of SOC densities and their calculation methods by impervious surface type, Table S3: Parameters for SOC density calculation for impervious-covered soil, Table S4: Profiles of SOC pool distribution by land cover in the urban core area of Guangzhou city, Table S5: SOC storage and SOC losses due to farmland urbanization & urban renovation at the residential scale (0–300 cm), Table S6: Losses of SOC stock due to the imperious surface expansion under different scenarios.

Author Contributions: Conceptualization, J.D. and J.Y.; methodology, J.D. and M.Y.; software, J.D. and M.Y.; validation, J.Y. and M.Y.; investigation, J.D. and M.Y.; writing—original draft preparation, J.D.; writing—review and editing, J.Y. and M.Y.; funding acquisition, J.Y. and M.Y. All authors have read and agreed to the published version of the manuscript.

Funding: This study was supported by the National Science Fund for Distinguished Young Scholars (41825020), the China Postdoctoral Science Foundation (2020M672863) and Guangdong Basic and Applied Research Foundation (2021A1515012147).

Institutional Review Board Statement: Not applicable.

Informed Consent Statement: Not applicable.

Data Availability Statement: The data that support the findings of this study are available from the corresponding author upon request.

Acknowledgments: We thank H. Li for her help with laboratory soil carbon measurements, Z. Wu for his help in the impervious surface extraction and S. Kong, W. Liang and Z. Liang for their help with soil sample.

Conflicts of Interest: The authors declare no conflict of interest.

References

1. United Nations. *World Urbanization Prospects: The 2018 Revision*; United Nations: New York, NY, USA, 2019.
2. Seto, K.C.; Guneralp, B.; Hutyra, L.R. Global forecasts of urban expansion to 2030 and direct impacts on biodiversity and carbon pools. *Proc. Natl. Acad. Sci. USA* **2012**, *109*, 16083–16088. [CrossRef]
3. Hutyra, L.R.; Duren, R.; Gurney, K.R.; Grimm, N.; Kort, E.A.; Larson, E.; Shrestha, G. Urbanization and the carbon cycle: Current capabilities and research outlook from the natural sciences perspective. *Earth's Future* **2014**, *2*, 473–495. [CrossRef]
4. Pickett, S.T.; Cadenasso, M.L.; Grove, J.M.; Groffman, P.M.; Band, L.E.; Boone, C.G.; Burch, W.R.; Grimmond, C.S.B.; Hom, J.; Jenkins, J.C.; et al. Beyond urban legends: An emerging framework of urban ecology, as illustrated by the Baltimore Ecosystem Study. *Bioscience* **2008**, *58*, 139–150. [CrossRef]
5. Churkina, G. The role of urbanization in the global carbon cycle. *Front. Ecol. Evol.* **2016**, *3*, 1–9. [CrossRef]
6. Gregg, J.W.; Jones, C.G.; Dawson, T.E. Urbanization effects on tree growth in the vicinity of New York City. *Nature* **2003**, *424*, 183–187. [CrossRef] [PubMed]
7. Kaye, J.P.; Mcculley, R.L.; Burke, I.C. Carbon fluxes, nitrogen cycling, and soil microbial communities in adjacent urban, native and agricultural ecosystems. *Glob. Chang. Biol.* **2005**, *11*, 575–587. [CrossRef]
8. Golubiewski, N.E. Urbanization increases grassland carbon pools: Effects of landscaping in Colorado's front range. *Ecol. Appl.* **2006**, *16*, 555–571. [CrossRef]
9. Zhao, S.; Liu, S.; Zhou, D. Prevalent vegetation growth enhancement in urban environment. *Proc. Natl. Acad. Sci. USA* **2016**, *113*, 6313–6318. [CrossRef]
10. Pretzsch, H.; Biber, P.; Uhl, E.; Dahlhausen, J.; Schütze, G.; Perkins, D.; Rötzer, T.; Caldentey, J.; Koike, T.; van Con, T.; et al. Climate change accelerates growth of urban trees in metropolises worldwide. *Sci. Rep.* **2017**, *7*, 1–10. [CrossRef]
11. Jia, W.; Zhao, S.; Liu, S. Vegetation growth enhancement in urban environments of the Conterminous United States. *Glob. Chang. Biol.* **2018**, *24*, 4084–4094. [CrossRef]
12. Pouyat, R.V.; Yesilonis, I.D.; Golubiewski, N.E. A comparison of soil organic carbon stocks between residential turf grass and native soil. *Urban Ecosyst.* **2009**, *12*, 45–62. [CrossRef]
13. Churkina, G.; Brown, D.G.; Keoleian, G. Carbon stored in human settlements: The conterminous United States. *Glob. Chang. Biol.* **2010**, *16*, 135–143. [CrossRef]
14. Raciti, S.M.; Hutyra, L.R.; Finzi, A.C. Depleted soil carbon and nitrogen pools beneath impervious surfaces. *Environ. Pollut.* **2012**, *164*, 248–251. [CrossRef]
15. Scalenghe, R.; Marsan, F.A. The anthropogenic sealing of soils in urban areas. *Landsc. Urban Plan.* **2009**, *90*, 1–10. [CrossRef]
16. Kuang, W.; Liu, J.; Zhang, Z.; Lu, D.; Xiang, B. Spatiotemporal dynamics of impervious surface areas across China during the early 21st century. *Chin. Sci. Bull.* **2013**, *58*, 1691–1701. [CrossRef]
17. Fuller, R.A.; Gaston, K.J. The scaling of green space coverage in European cities. *Biol. Lett.* **2009**, *5*, 352–355. [CrossRef] [PubMed]
18. Pouyat, R.V.; Yesilonis, I.D.; Nowak, D.J. Carbon storage by urban soils in the United States. *J. Environ. Qual.* **2006**, *35*, 1566–1575. [CrossRef]
19. Yan, Y.; Kuang, W.; Zhang, C.; Chen, C. Impacts of impervious surface expansion on soil organic carbon—A spatially explicit study. *Sci. Rep.* **2015**, *5*, 1–9. [CrossRef] [PubMed]
20. Wei, Z.; Wu, S.; Yan, X.; Zhou, S. Density and stability of soil organic carbon beneath impervious surfaces in urban areas. *PLoS ONE* **2014**, *9*, 1–7. [CrossRef]
21. Cambou, A.; Shaw, R.K.; Huot, H.; Vidal-Beaudet, L.; Hunault, G.; Cannavo, P.; Nold, F.; Schwartz, C. Estimation of soil organic carbon stocks of two cities, New York City and Paris. *Sci. Total Environ.* **2018**, *644*, 452–464. [CrossRef]
22. Lu, C.; Kotze, D.J.; Setl, H.M. Soil sealing causes substantial losses in C and N storage in urban soils under cool climate. *Sci. Total Environ.* **2020**, *725*, 138369. [CrossRef]
23. Edmondson, J.L.; Davies, Z.G.; Mchugh, N.; Gaston, K.J.; Leake, J.R. Organic carbon hidden in urban ecosystems. *Sci. Rep.* **2012**, *2*, 1–7. [CrossRef]

24. Tian, H.; Melillo, J.; Lu, C.; Kicklighter, D.; Liu, M.; Ren, W.; Xu, X.; Chen, G.; Zhang, C.; Pan, S.; et al. China's terrestrial carbon balance: Contributions from multiple global change factors. *Glob. Biogeochem. Cycles* **2011**, *25*, 1–16. [CrossRef]
25. Zhang, C.; Tian, H.; Chen, G.; Chappelka, A.; Xu, X.; Ren, W.; Hui, D.; Liu, M.; Lu, C.; Pan, S.; et al. Impacts of urbanization on carbon balance in terrestrial ecosystems of the Southern United States. *Environ. Pollut.* **2012**, *164*, 89–101. [CrossRef]
26. Jiang, W.; Deng, Y.; Tang, Z.; Lei, X.; Chen, Z. Modelling the potential impacts of urban ecosystem changes on carbon storage under different scenarios by linking the CLUE-S and the InVEST models. *Ecol. Model.* **2016**, *345*, 30–40. [CrossRef]
27. Lyu, R.; Mi, L.; Zhang, J.; Xu, M. Modeling the effects of urban expansion on regional carbon storage by coupling SLEUTH model and InVEST model. *Ecol. Res.* **2019**, *34*, 380–393. [CrossRef]
28. Heuvelink, G.B.; Angelini, M.E.; Poggio, L.; Bai, Z.; Batjes, N.H.; Bosch, R.; Bossio, D.; Estella, S.; Lehmann, J.; Olmedo, G.; et al. Machine learning in space and time for modelling soil organic carbon change. *Eur. J. Soil Sci.* **2021**, *72*, 1607–1623. [CrossRef]
29. Majidzadeh, H.; Graeme, L.B.; Robert, P.; Robin, G. Soil Carbon and Nitrogen Dynamics beneath Impervious Surfaces. *Soil Sci. Soc. Am. J.* **2018**, *82*, 1–8. [CrossRef]
30. Ministry of Housing and Urban-Rural Development of the People's Republic of China. *Design Code for Residential Buildings (GB 50096-2011)*; China Building Industry Press: Beijing, China, 2011; pp. 18–19.
31. Bobylev, N. Mainstreaming sustainable development into a city's Master plan: A case of Urban Underground Space use. *Land Use Policy* **2009**, *26*, 1128–1137. [CrossRef]
32. Cui, J.; Broere, W.; Lin, D. Underground space utilization for urban renewal. *Tunn. Undergr. Space Technol.* **2021**, *108*, 1–10. [CrossRef]
33. Jobbágy, E.G.; Jackson, R.B. The vertical distribution of soil organic carbon and its relation to climate and vegetation. *Ecol. Appl.* **2000**, *10*, 423–436. [CrossRef]
34. Zhao, S.; Zhu, C.; Zhou, D.; Huang, D.; Jeremy, W.; Ben, B.L. Organic carbon storage in China's urban areas. *PLoS ONE* **2013**, *8*, 1–8. [CrossRef]
35. Chen, W.; Chi, G.; Li, J. The spatial association of ecosystem services with land use and land cover change at the county level in China, 1995–2015. *Sci. Total Environ.* **2019**, *669*, 459–470. [CrossRef] [PubMed]
36. Song, W.; Deng, X. Land-use/land-cover change and ecosystem service provision in China. *Sci. Total Environ.* **2017**, *576*, 705–719. [CrossRef] [PubMed]
37. Statistics Bureau of Guangzhou City. *Guangzhou Statistical Yearbook of 2020 and 2000*; China Statistics Press: Beijing, China, 2021; Available online: http://112.94.72.17/portal/queryInfo/statisticsYearbook/index (accessed on 7 February 2021).
38. Ministry of Housing and Urban-Rural Development of the People's Republic of China. *Code for Design of Building Foundation GB 50007-2002*; China Building Industry Press: Beijing, China, 2002; pp. 17–35.
39. Fan, F.; Fan, W.; Weng, Q. Improving urban impervious surface mapping by linear spectral mixture analysis and using spectral indices. *Can. J Remote Sens.* **2015**, *41*, 577–586. [CrossRef]
40. Zhu, X. Temporal and spatial variation of organic carbon and soil carbon storage in the Pearl River Delta Economic Zone. *Chin. J. Geol. Miner. Resour. South China* **2014**, *30*, 176–185.
41. Transport Engineering Center of Ministry of Housing and Urban-Rural Development of the People's Republic of China; China Academy of Urban Planning and design; Beijing Siwei Tuxin Science and Technology Co., Ltd. *Annual Report on Road Network Density in Major Chinese Cities (2020)*; China Academy of Urban Planning and Design: Beijing, China, 2020.
42. Bae, J.; Ryu, Y. High soil organic carbon stocks under impervious surfaces contributed by urban deep cultural layers. *Landsc. Urban Plan.* **2020**, *204*, 103953. [CrossRef]
43. Soil Census Office of Guangdong Province. *Soil Species Records of Guangdong Province*; Science Press: Beijing, China, 1996; pp. 85–88, 91–94, 117–123, 393–396, 409–410, 413–414.
44. Soil Census Office of Guangdong Province. *Soils of Guangdong Province*; Science Press: Beijing, China, 1996; pp. 142–144, 147–152.
45. Wang, Q. Study on the structure and composition of granite residual soil in the Eastern China. *Chin. J. Changchun Univ. Earth Sci.* **1991**, *21*, 70–81.
46. Ding, L.; Jiang, Y.; Chen, D.; Zhang, X. Analysis of mechanical properties and correlation of soft soil developed in Guangzhou City. *Chin. J. Railw. Eng.* **2011**, 75–78. [CrossRef]
47. Nelson, D.W.; Sommers, L.E. Total Carbon, Organic Carbon, and Organic Matter. In *Methods of Soil Analysis*; Sparks, D., Page, A., Helmke, P., Loeppert, R., Soltanpour, P.N., Tabatabai, M.A., Johnston, C.T., Sumner, M.E., Eds.; Wiley: Hoboken, NJ, USA, 1996. [CrossRef]
48. Ministry of Housing and Urban-Rural Development of the People's Republic of China. *Code for Design of Urban Road Engineering CJJ37-2012*; China Building Industry Press: Beijing, China, 2012; pp. 45–46.
49. Pan, X.; Huan, Q. Experimental study on cement modified red clay compaction. *Chin. J. Sci. Techn. Vis.* **2018**, 4–85. [CrossRef]
50. Zhou, J. Modification and modification mechanism of red clay. Master's Thesis, Chongqing University, Chongqing, China, May 2014.
51. Li, H.; Yu, G.; Liang, X. Study on variation of maximum dry density with time for modified mild clay. *Chin. J. Subgrade Eng.* **2007**, 95–97. [CrossRef]
52. Liu, X. Research on the Application of Construction Waste Regeneration in Highway Subgrade. *Chin. J. Highw. Eng.* **2019**, *44*, 208–212.

53. Zhang, Y.; Xu, Q.; Peng, D.; Zhao, K.; Guo, C. An experimental study of the permeability of the catastrophic landslide at the Shenzhen Landfill. *Chin. J. Hydrogeol. Eng. Geol.* **2017**, *44*, 149.
54. Ministry of Housing and Urban-Rural Development of the People's Republic of China. *Technical Code for Ground Treatment of Buildings JGJ79-2012*; China Building Industry Press: Beijing, China, 2012; p. 13.
55. Liang, S.; Zhou, S.; Zhang, L.; Wang, M. Statistical analysis of physical and mechanical indexes of granite residual soil in eastern Guangzhou. *Chin. J. Guangdong Uni. Tech.* **2015**, *32*, 49.
56. USDA–NRCS. *Soil Survey Laboratory Information Manual, Version 2.0*; Soil Survey Investigations Report No. 45; U.S. Government Printer: Washington, DC, USA, 2011; pp. 235–253.
57. Cao, M. Comparative analysis of granite residual soil developed in Yan Mountains period and sinian system. *Chin. J. Sichuan Cem.* **2016**, 53–55. [CrossRef]
58. Xiang, Y. Statistical Analysis of physical and mechanical properties of granite residual soil and weathered layer in Luogang station of Guangzhou Metro Line 6. *Chin. J. Constr. Sci. Technol.* **2012**, 92–93. [CrossRef]
59. Huang, H. Engineering geological characteristics of granite residual soil and application in Panyu District, Guangdong Province. *Chin. J. Hydrogeol. Eng. Geol.* **2000**, 38–39. [CrossRef]
60. Zeng, F. The engineering geological feature of granite residual soil in Guangdong Shiqiao Region. *Chin. J. Soil Eng. Found.* **1993**, *7*, 29–34.
61. Wu, F. Study of engineering geological characteristics of granitic rocks weathered in Guangzhou. Master's Thesis, Guangzhou University, Guangzhou, China, 30 May 2013.
62. Technical Regulations on Urban and Rural Planning of Guangzhou City (Government Decree of the People's Government of Guangzhou Municipality [2015] No. 133). Available online: http://www.gz.gov.cn/zwgk/fggw/zfgz/content/mpost_4756949.html (accessed on 4 November 2015).
63. Ten Years of Urban Renewal in Guangzhou: From Single Goal to Multiple Goals. 2020. Available online: https://www.xkb.com.cn/article_634086 (accessed on 27 November 2020).
64. Ministry of Housing and Urban-Rural Development of the People's Republic of China. *Planting Soil for Greening CJ/T 340-2011*; Standards Press of China: Beijing, China, 2011; p. 4.
65. Zhu, Y.; Luo, Q.; Chen, Y.; Huang, Y.; Guan, D. Carbon density and distribution of main ecosystems in Ten-Thousand-Mu Orchard at Haizhu District, Guangzhou. *Chin. J. Ecol.* **2016**, *35*, 164–169. [CrossRef]
66. Wu, Z.; Huang, Y.; Jiang, C. Soil and vegetation carbon storage and its spatial pattern analysis of Guangzhou City, China. *J. Guangzhou Univ.* **2014**, *13*, 73–79.
67. Sulman, B.N.; Harden, J.; He, Y.; Treat, C.; Nave, L.E. Land use and land cover affect the depth distribution of soil carbon: Insights from a large database of soil profiles. *Front. Environ. Sci.* **2020**, *8*, 1–12. [CrossRef]
68. D'Elia, A.H.; Liles, G.C.; Viers, J.H.; Smart, D.R. Deep carbon storage potential of buried floodplain soils. *Sci. Rep.* **2017**, *7*, 1–7. [CrossRef]
69. Steger, K.; Fiener, P.; Marvin-DiPasquale, M.; Viers, J.H.; Smart, D.R. Human-induced and natural carbon storage in floodplains of the Central Valley of California. *Sci. Total Environ.* **2019**, *651*, 851–858. [CrossRef]
70. Bi, L.; Chen, X.; Lu, B. Distribution features of soft soil and the prediction of seismic subsidence in Guangzhou. *Chin. J. China Earthq. Eng. J.* **2018**, *40*, 1026–1033.
71. Guo, L.B.; Gifford, R.M. Soil carbon stocks and land use change: A meta analysis. *Glob. Chang. Biol.* **2010**, *8*, 345–360. [CrossRef]
72. Zhang, X.; Xu, Z.; Zeng, F.; Hu, X.; Han, Q. Carbon density distribution and storage dynamics of forest ecosystem in Pearl River Delta of low subtropical China. *China Environ. Sci.* **2011**, *31*, 69–77. [CrossRef]
73. Vasenev, V.; Yakov, K. Urban soils as hot spots of anthropogenic carbon accumulation: Review of stocks, mechanisms and driving factors. *Land Degrad. Dev.* **2018**, *29*, 1607–1622. [CrossRef]
74. Xu, X.L.; Liu, J.Y.; Zhang, S.W.; Li, R.D.; Yan, C.Z.; Wu, S.X. *China's Multi-Period Land Use/Cover Remote Sensing Monitoring Dataset (CNLUCC)*; Data Registration and Publishing System of the Resource and Environmental Science Data Center of the Chinese Academy of Sciences: Beijing, China, 2018. [CrossRef]

sustainability

Article

Spatial Variation and Terrain Gradient Effect of Ecosystem Services in Heihe River Basin over the Past 20 Years

Lingge Zhang and Ningke Hu *

School of Geography and Tourism, Shaanxi Normal University, Xi'an 710119, China; zlg98@snnu.edu.cn
* Correspondence: hunk2014@snnu.edu.cn

Abstract: With the advent of large-scale development, extreme imbalance in the ecology of the Heihe River Basin (HRB) has caused a series of ecological problems. In order to explore the spatiotemporal variation of ecosystem services (ESs) and to assess the characteristics of ESs under the terrain gradient effect (TGE), the three key ESs were quantified based on the InVEST model using five series of land-use data obtained from remote sensing images from 2000 to 2020 in this study. The terrain index was used to analyze the influence of terrain on ESs. The results show that most of the ESs were in high numbers in the south and low numbers in the north, as well as high numbers in the middle and upper reaches and low numbers at downstream locations. It was found that high-quality habitats degrade to general-quality habitats, and poor-quality habitats evolve into general-quality habitats. It was also found that the water production volume continues to decline and soil conservation becomes relatively stable with little change. This study illustrates different ESs showing obvious TGE with changes in elevation and slope. These results indicate that the effect of land-use change is remarkable and TGE is highly important to ESs in inland watersheds. This research study can provide a scientific basis for the optimization of regional ecosystem patterns. The results are of great significance in terms of rational planning land use, constructing ecological civilizations, and maintaining the physical conditions of land cover at inland river basins.

Keywords: land use; ecosystem services; InVEST; topographic index; ecosystem pattern

Citation: Zhang, L.; Hu, N. Spatial Variation and Terrain Gradient Effect of Ecosystem Services in Heihe River Basin over the Past 20 Years. *Sustainability* **2021**, *13*, 11271. https://doi.org/10.3390/su132011271

Academic Editors: Kikuko Shoyama, Rajarshi Dasgupta and Ronald C. Estoque

Received: 1 September 2021
Accepted: 1 October 2021
Published: 13 October 2021

Publisher's Note: MDPI stays neutral with regard to jurisdictional claims in published maps and institutional affiliations.

1. Introduction

Ecosystem services (ESs), which are the benefits that humans receive directly or indirectly from ecosystems, have played an important role in the development of human civilization [1]. Generally, ESs include food supply, climate regulation, and cultural and supporting services. In recent years, the conflicts between the rapid socio-economic development and the consumption of natural resources and environmental change have resulted in the degradation of ESs and frequent natural disasters in some areas [2–4]. ESs have been considered as indispensable features in land-use planning and resource adjustment, becoming one of the research hotspots over recent decades. Currently, ES assessments mostly focus on different land-use scenarios, such as croplands, forests, oceans, lakes, wetlands, and others [5–8]. In order to ensure the sustainable provision of ESs for human beings, it is necessary to evaluate ESs in order to support human wellbeing. ES evaluations consist of a value assessment and material quality assessment [9–11]. The former measures the economic value of ESs by using ecological economics methods [12–14] that are based on the value equivalent factor of a unit area or the price of a unit service function [15–21]. The value measurement method is primarily used to evaluate supply services, and the results are of high economic significance [22–24]. With the refinement of the ES classification, material quality methods are gradually used as the mainstream assessment methods of ESs. The material quality of various ESs can be quantitatively evaluated based on ecological principles in this method. By contrast, the credibility and accuracy of the quality assessment method are higher than that of the value assessment method, which is influenced subjectively to some extent [25]. It is necessary to conduct research on the sustainability of an ecosystem and to provide a more scientific and reliable basis for decision making [11]. Material quality assessment methods can be roughly divided into emergy methods and model methods [26]. The emergy method is based on the emergy theory for evaluating different types of emergy and material flows in the system [27]. It combines the total amount of effective energy invested directly and indirectly in the ecosystem with the energy conversion rate to calculate the final energy value. The energy flow and utilization rate of ESs can be expressed better by emergy evaluation. This method is often used to describe the regional differences with the utilization of ESs in large-scale areas and to evaluate urban ecosystems [28]. However, with increasing studies on the coupling of multiple ESs and the relationship between ESs and human wellbeing during these years, the emergence and development of evaluation models represent a major breakthrough in the field of ES assessment. The results can be presented in the form of a map incorporating more spatial and intuitive information by using model assessments. It can simulate and predict the changing scenario of future ESs.

According to the formation mechanism of ESs, studies on ES evaluation have increased due to the quality assessments of ecosystems generated through comprehensive models. The commonly used models include Integrated Valuation Ecosystem Services and Tradeoffs (InVEST), Artificial Intelligence for Ecosystem Services (ARIES), Social Value for Ecosystem Services (SolVES), and Multi-scale Integrated Models of Ecosystem Services (MIMES) [29,30]. Some models, such as SoLVES, ARIES, and MIMES, output good evaluations for specific regions and have promising application prospects, but they have not been fully developed yet. The InVEST model is the most widely used tool for the quality assessment of ESs. The advantages of the InVEST model have been highlighted by integrating various service production functions or simulating service dynamic changes. Previous assessment studies using the InVEST model are mostly concentrated in small-scale areas, such as administrative regions, urban economic zones, or lakes and rivers [31–36]. ES evaluations in medium and large watershed scales are relatively rare. In addition, most of the studies analyzed spatiotemporal pattern changes or showed cold and hot spots in the service space by spatial mapping for individual or several services, such as biodiversity conservation, water conservation, soil conservation, and carbon storage [33–38]. Relatively few studies discuss the connection and divergence between different ES supplies and specific natural environmental factors [39,40]. As an important factor in the natural environment, the terrain factor is of great significance for fully understanding the spatiotemporal difference of ESs [41,42]. It is well known that ESs are closely related to their terrain [43,44]. Some terrain-based studies analyzed the spatial distribution of land-use patterns [45,46] and some were focused on mountain areas [47,48]. However, there are only a few studies assessing the impact of topographic factors on the spatial heterogeneity of ESs. The evaluation of the characteristics of ESs with terrain gradients has not been explored in depth. Understanding the influence of topographic elements on the spatial difference of ESs comprehensively is of great significance for deepening the understanding of ES differentiation and for ensuring effective ecosystem management.

The Heihe River Basin (HRB) is not only a key area with respect to the ecological safety barrier, but it is also an important node in the layout of the national "Belt and Road" initiative [49,50]. Typical natural conditions and long-term, frequent human activities have resulted in ecological environmental deterioration in the basin, resulting in the degradation of biodiversity, water production, soil, and water conservation. Therefore, the problem of how to maintain the stability of regional ecosystems as an economy develops is an important issue for all inland river basins in arid zones. To summarize, combined with the geographical conditions and characteristics of the HRB, we discuss the relationship between ESs and land-use changes in typical arid inland river basins from a topographical perspective on the basis of previous research studies. In this paper, we selected three key ESs (i.e., habitat quality, water yield, and soil conservation) of the HRB in Northwestern China to study. Based on the proposed framework (Figure 1), our analysis focuses on addressing three questions at fine scales: (1) How have ESs changed over the past 20 years and where are the areas of high and low distribution of individual ESs? (2) How are different ESs affected by terrain factors with respect to inland watershed scales? (3) How can variations of ESs guide land-use management in a river basin?

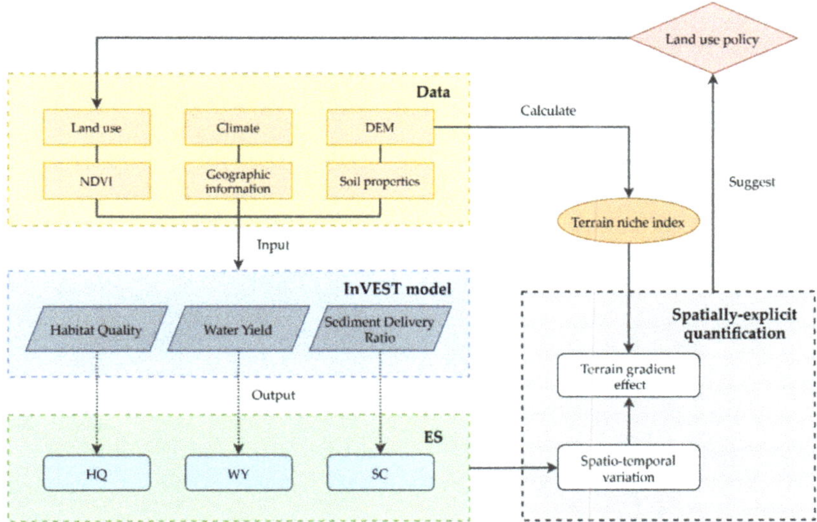

Figure 1. Framework of the methodological process in this study.

2. Materials and Methods

2.1. Study Area

The HRB is the second largest inland river basin in arid region of Northwestern China, within the range of 96°42′–102°00′ E, 34°41′–42°42′ N (Figure 2). The area of core drainage is approximately 128,900 km^2, with a mainstream length of 821 km. Different reaches of the river are distributed in the three provinces or autonomous region, belonging to Qinghai, Gansu, and Inner Mongolia, respectively. Climatic characteristics of the basin are a typical continental arid climate, little but concentrated precipitation, ample sunshine, and greater diurnal temperature range. With an average altitude over 1200 m, the topography varies significantly from south to north [51]. The regional climate is significantly different in the basin. Therefore, it is an important for ecological function in terms of water production, soil conservation, biodiversity protection, windbreak, and sand fixation in Northwestern China [52].

This basin can be divided into three parts according to the locations of the gorge stations. The upstream region is located in the Qilian Mountain, with low population density, high vegetation coverage, and good ecological environment. This segment, the main contributing area, is bounded by the Yingluo Gorge. The midstream area is a main field for developing human activities, extending from the Yingluo Gorge to the Zhengyi Gorge. This region is rich in light and heat resources. The downstream region, below the Zhengyi Gorge, constitutes areas where runoff disappears. It is mostly made up of desert and bare land [53]. More than two thousand years of different human activities in HRB have caused collision and exchange of various cultures, making natural and human processes meet together. This is an ideal region for studying the variation of different ESs.

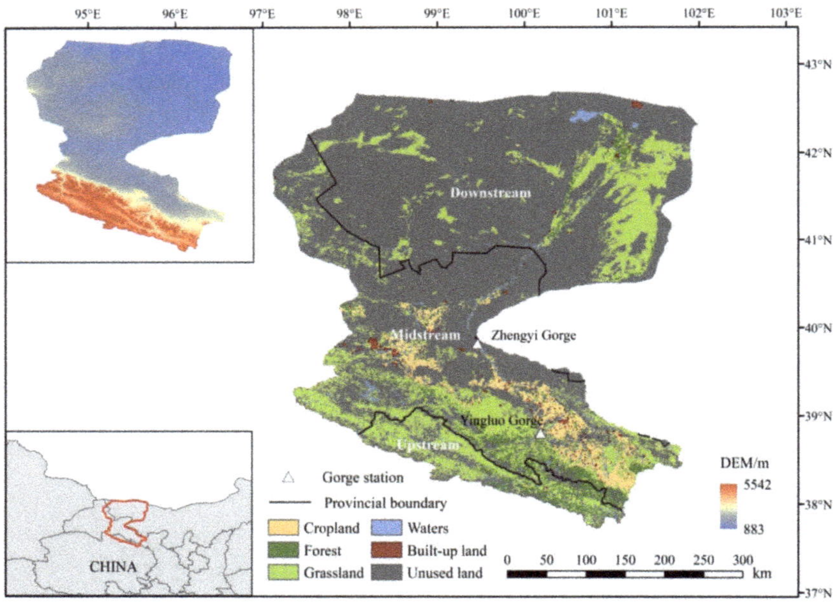

Figure 2. Location of Heihe River Basin and land use and land cover in 2015.

2.2. Data Sources

The datasets used in this study included: (1) land-use types in 2000, 2005, 2010, 2015 and 2020 (with a spatial resolution of 1km) and NDVI (Normalized Difference Vegetation Index). These were both obtained from the Resource and Environment Science and Data Center (http://www.resdc.cn accessed on 21 March 2021). According to the model input acquirements and the reference of land-use classification system of the Resources and Environment Database, the land-use data were reclassified into six categories: cropland, forest, grassland, waters, built-up land, and unused land; (2) DEM (Digital Elevation Model) and potential evapotranspiration data (with a spatial resolution of 250m) were derived from the National Tibetan Plateau/Third Pole Environment Data Center (https://data.tpdc.ac. cn/zh-hans/accessed on 24 March 2021); (3) basic geographic information data (including regional boundaries, roads, settlements, etc.) were acquired from the "Digital Heihe River" project of the Cold and Arid Region Scientific Data Center; (4) soil attribute data were derived from Chinese soil datasets (1995) in the Food and Agriculture Organization's Harmonized World Soil Database; and (5) meteorological data were retrieved from the China Meteorological Data Service Center (http://www.nmic.cn/ accessed on 26 March 2021) and the National Earth System Science Data Center (http://www.geodata.cn accessed on 06 April 2021), including annual, monthly, and daily value data from 2000 to 2020.

2.3. Methodology

In this study, terrain analysis and integrated maps were carried out by using the ArcGIS (Ver.10.2) software. Regression analyses to detect trend significance were performed in SPSS (Ver. 21.0), and visualized charts were generated by using Microsoft Excel (Ver. 2017). All the spatial assessments of ESs were processed by using InVEST (Ver. 3.8.0). As a strong technical support in ES research, the InVEST model is used for a quantitative assessment of ESs on the basis of ecosystem processes, and a full demonstration of their spatially distributed characteristics [54,55]. The three sub-models, habitat quality, water yield, sediment delivery and retention were selected to quantitatively evaluate and analyze the changes of three ESs (including habitat quality, water production and soil conservation) in this work.

2.3.1. Habitat Quality (HQ)

The HQ in this model can be described as the connection between different types of land use and threat sources. The results and scarcity of HQ can be obtained in accordance with the response degree of different habitats to various threat sources.

The formula for calculating habitat quality (HQ_{ij}) is:

$$HQ_{ij} = H_j \left(1 - \frac{D_{xj}^2}{D_{xj}^2 + k^2} \right) \tag{1}$$

where H_j is the habitat suitability of land-use type j; D_{xj} is the total threat level; k is the half-saturation value.

The input data of this sub-module include current land-use and threat data. ArcGIS was used to process the original data (i.e., mosaic, reclassify, merge, etc.) in order to obtain the land-use and threat factor layer data. The HQ correlation function in the model corresponds to the four variables: the relative impact of each threat, the relative sensitivity of each habitat to each threat, the distance between habitat and threat and impact from the threat, and the level at which a grid cell of the habitat is legally protected. In order to reduce the interference of human factor on results, the impacts of the first three variables were discussed in this study. We selected cropland, town, residential area, and road as habitat threats. The corresponding parameters of the model (Tables 1 and 2) were set by referring to the user guide and previous works with similar eco-environments in arid region [56,57], as well as the suggestions from experts and professors in the ecological field. The half-saturation constant defaults to 0.5 according to the user guide.

Table 1. Threat factor properties.

Threat	Max-Dist/km	Weight
Cropland	1	0.2
Town	8	0.9
Residential area	3	0.8
Road	10	0.6

Table 2. Sensitivity of habitat types to each threat.

Habitat Type	Habitat Suitability	Cropland	Town	Residential Area	Road
Cropland	0.4	0.5	0.4	0.5	0.45
Forest	0.8	0.4	0.8	0.6	0.5
Grassland	0.6	0.4	0.5	0.6	0.6
Waters	0.9	0.5	0.6	0.6	0.55
Built-up land	0	0	0	0	0
Unused land	0	0	0	0	0

2.3.2. Water Yield (WY)

The WY estimation is based on the principle of water balance, which simplifies the flow process. It does not distinguish between surface runoff, runoff in the soil, and base flow. WY defines the amount of water produced within each grid range by the difference between precipitation and evapotranspiration (including plant transpiration and surface evaporation). Its calculation is based on the acquired data, such as topography, land use, meteorological factors, etc.

The water yield (WY_{xj}) is calculated using the following formulas:

$$WY_{xj} = \left(1 - \frac{AET_{xj}}{P_x}\right) \times P_x \tag{2}$$

$$\frac{AET_{xj}}{P_x} = \frac{1 + \omega_x R_{xj}}{1 + \omega_x R_{xj} + \frac{1}{R_{xj}}} \tag{3}$$

$$\omega_x = Z\frac{AWC_x}{P_x} \tag{4}$$

$$R_{xj} = \frac{K_{xj}ET_{0_x}}{P_x} \tag{5}$$

where AET_{xj} is the annual actual evapotranspiration for land-use type in category j on grid x (*mm*); P_x is the average annual value of precipitation on grid x (*mm*); ω_x is the ratio of the year's available water for modified vegetation to the expected amount of water; R_{xj} is the ratio of potential evaporation to rainfall; Z is the Zhang coefficient, which represents the parameters of seasonal rainfall distribution and precipitation depth, ranging from 1 to 30; AWC_x is the average annual value of available water capacity on grid x (*mm*); ET_{0_x} is the potential evapotranspiration on grid x (*mm*); K_{xj} is the evapotranspiration coefficient of vegetation. The input data of the model include precipitation, reference evapotranspiration, depth to root restricting layer, plant available water fraction, land use, and watersheds. The parameters include Z parameter and evapotranspiration coefficient of vegetation. The precipitation data were interpolated by using ArcGIS. The reference evapotranspiration was calculated by the modified formula. The soil depth data were derived by rasterizing the soil spatial attribute data. The plant available water fraction was set according to the reference results [58]. The watershed boundary was extracted by hydrological analysis with DEM in ArcGIS. The Zhang coefficient was set as 2.2 referring to the instruction manual and previous reference [59]. The plant evapotranspiration coefficient was defined according to the calculation method proposed by FAO and the reference value was given by the model.

2.3.3. Soil Conservation (SC)

The SC can be calculated with soil erosion reduction and sediment retention using the sediment delivery and retention sub-module. The decrease in soil erosion was expressed by the difference between potential erosion and actual erosion. The amount of sediment retention refers to the amount of uphill sand retained by plot. On the basis of cell scale USLE (Universal Soil Loss Equation) calculation method, the calculation on the grid unit scale was simulated using the following data, including land-use data, soil attribute, DEM, rainfall data, vegetation cover factor, and soil and water protection measure factor. The rainfall erosivity was calculated using observation data from meteorological stations in the basin and then interpolated with Kriging method in ArcGIS. The soil erodibility factor was calculated from the soil attribute data by formula-10. The vegetation cover factor and soil and water protection measure factor can be defined by consulting the user guide.

The formulas for calculating SC are defined as the following:

$$SC = USLE - RKLS + SEDR \tag{6}$$

$$USLE = R \times K \times LS \times C \times P \tag{7}$$

$$RKLS = R \times K \times LS \tag{8}$$

$$R = \sum_{i=1}^{12}\left[1.735 \times 10^{\left(1.5lg\frac{P_i^2}{P_y} - 0.8188\right)}\right] \times 17.02 \tag{9}$$

$$K = 0.1317 \times \left\{ 0.2 + 0.3 \times exp\left[-0.0265 SAN\left(1 - \tfrac{SIL}{100} \right) \right] \right\} \times \left[\tfrac{SIL}{CLA+SIL} \right]^{0.3}$$
$$\times \left\{ 1 - 0.25 \times \tfrac{C_0}{C_0 + exp(3.72 - 2.95C_0)} \right\} \tag{10}$$
$$\times \left\{ 1 - 0.7 \times \tfrac{SN_1}{SN_1 + exp(22.9SN_1 - 5.51)} \right\}$$

$$SN_1 = 1 - \frac{SAN}{100} \tag{11}$$

$$c = \frac{NDVI - NDVI_{min}}{NDVI_{max} - NDVI_{min}} \tag{12}$$

$$\begin{cases} C = 1 & c = 0 \\ C = 0.6508 - 0.3436 logc & 0 < c < 78.3\% \\ C = 0 & c > 78.3\% \end{cases} \tag{13}$$

where SC is the amount of soil conservation ($t \cdot hm^{-2} \cdot a^{-1}$); $USLE$ is the amount of potential soil loss in the original land-use cover ($t \cdot hm^{-2} \cdot a^{-1}$); $RKLS$ is the amount of potential soil loss for bare soil ($t \cdot hm^{-2} \cdot a^{-1}$); $SEDR$ is the retention from the upstream sediment ($t \cdot hm^{-2} \cdot a^{-1}$); R is the rainfall erosivity ($MJ \cdot mm \cdot hm^{-2} \cdot h^{-1} \cdot a^{-1}$); K is the soil erodibility factor ($t \cdot hm^2 \cdot h \cdot hm^{-2} \cdot MJ^{-1} \cdot mm^{-1}$); LS is the slope length gradient factor; P is the soil and water protection measure factor; P_i is the average monthly rainfall (mm); P_y is the average annual precipitation (mm); SAN, SIL and CLA are the content values of sand grains, powder grains and sticky grains (%); C_0 indicates the organic carbon content value (%); c is the vegetation cover; $NDVI$ is the normalized difference vegetation index; C is the crop management factor (value range is between 0 and 1). If C is 0, the vegetation cover of the land surface is good and almost not eroded; if C is 1, there is almost no vegetation cover on the surface.

2.4. Terrain Niche Index

Terrain is one of the important factors affecting spatial distribution and changes in land use. TGE is also manifested as the variations in ES supply that are caused by land-use changes to some extent. The terrain differences are outstanding in HRB, which has a variety of mountain, plain, basin, and hilly geomorphological types. In order to mirror the comprehensive relationship between terrain condition and ESs, the terrain niche index was selected as a geographical factor to analyze TGE between land-use patterns and terrain gradient in this study. The terrain niche index (T) can be calculated using the following formula:

$$T = lg\left[\left(\frac{E}{\overline{E}} + 1 \right) \times \left(\frac{S}{\overline{S}} + 1 \right) \right] \tag{14}$$

where E and $\overline{E}$, respectively, indicate the elevation value of any point in the region and that of the areas where the point is located; S and $\overline{S}$, respectively, represent the slope value of any point in the region and the average slope value of the areas where the point is located. If terrain level index of a point is low, the elevation and the slope of this point are low values; if that of a point is centered, the values are high and low. Both are high values if that of a point is large.

3. Results

3.1. Spatiotemporal Variations of ESs

3.1.1. Spatial Distribution and Variation of HQ

HQ score is a comprehensive indication of the influence of each threat factor on itself and the impact of habitat on threat resistance. The higher the number is, the better the HQ is. The HQ in this study was classified as higher (0.8–1), high (0.5–0.8), general (0.3–0.5) and poor (0–0.3). Figure 3 shows the proportion of different HQ grades in the basin. More than half of the HRB was in poor and average HQ over the past 20 years. Higher and high HQ account for less than 30% of the total area, and the overall proportion composition is relatively stable from 2000 to 2020. In order to further clearly explore the changes of HQ spatial distribution over time, we calculated the difference based on the HQ data and obtained the interannual changes in four five-year periods (Figure 4). Spatial differences and variations of HQ are obvious throughout the basin. Most of the upstream region has high- and higher-grade HQ. There have been many changes over the past two decades, especially in the two phases of 2005–2010 and 2015–2020. The area of higher-grade HQ increased and the HQ showed an upward trend. The overall ecological environment became better, which might be the benefit of a sparse population and dense vegetation in the upper mountains. A different layout was presented in the middle stream. The south-central region was dominated by general- and high-grade HQ. The area of general HQ level increased, which was due to the human reclamation of wasteland, grasslands, and forest land, and the expansion of cultivated land and construction land in the active areas. At the same time, the area of high-level HQ also reduced with the reclamation and the degradation of grassland and forest land. By incorporating the changes of four stages, it can be found that mostly the HQ in the northern part of the midstream area was at the poor level and it increased over the past 20 years, due to large distribution of unused land. Downstream areas were clearly affected by land use. The HQ was higher in most of the east, with grasslands, woodlands and waters. Obvious increases can be seen through the four periods in this region, while the central and western regions were dominated by unused land with low coverage and the HQ levels therefore dropped. There were few changes in the areas during 2000–2005 and 2010–2015, while the HQ changed significantly and mainly increased in the other two periods. With the transformation of unused land into grasslands and woodlands, as well as the increase in water area over the past 20 years, the area of high-grade HQ in the downstream region increased and that of poor-grade HQ decreased gradually. This indicates that the overall HQ of the downstream area improved, which may be related to the implementation of the Heihe Ecological Water-divide Project (2000).

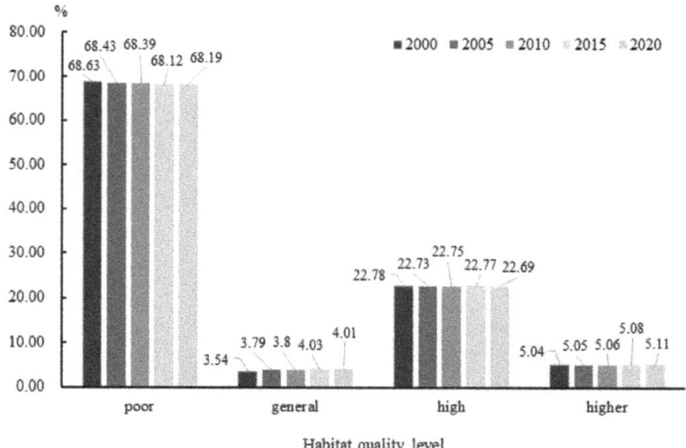

Figure 3. HQ grade ratio of different habitats in HRB from 2000 to 2020.

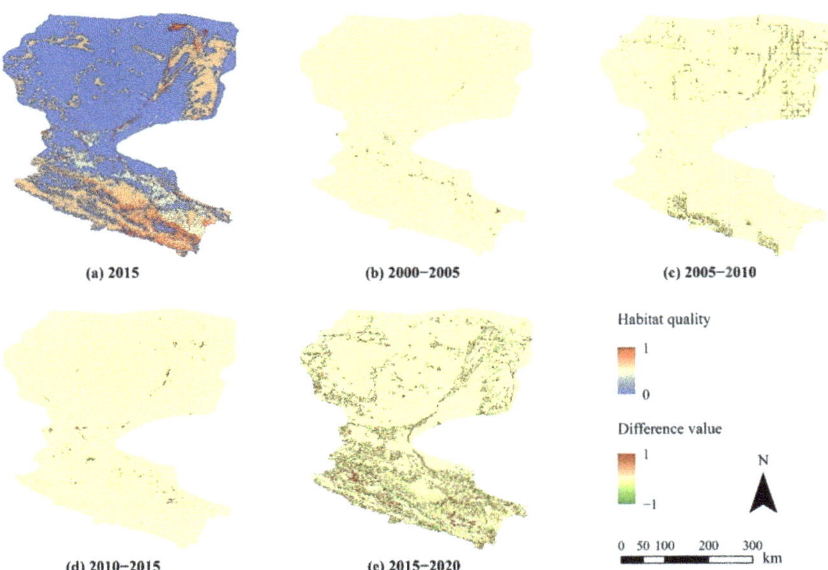

Figure 4. Variation of HQ in HRB from 2000 to 2020. (**a**) Spatial distribution of HQ in 2015; (**b**)–(**e**) four difference maps of the five-year changes.

Totally, the quality of habitats at all levels changed by a small margin, with a slight decrease in poor and high HQ areas and a slight increase at general and high levels. The overall HQ in the basin became better and the improvement was mostly concentrated in the upstream and downstream areas in the northeast, while the quality levels deteriorated in the midstream and lower-west regions.

3.1.2. Spatial Distribution and Variation of WY

The annual WY in this study was classified into one, two, three and four levels from low to higher weight (Figure 5). In 2000–2015, the proportion of low-grade WY gradually increased, while that of high-grade WY decreased bit by bit. However, during the five-year period from 2015 to 2020, the proportion of low-value WY decreased, and that of high-value WY increased. However, the WY in most areas was at a lower level and the depth of water production was not large. As can be seen in Figure 6, the spatial distribution of WY shows great difference. The upstream region was the principal origin for high-value water production, and the area of high-value WY gradually decreased. In particular, some of them degraded to medium-yielding areas. However, with the exception of large-scale decline in the five years from 2010 to 2015, the changes of WY still increased in the other years. The overall trend also caused an increase in the upstream water production. The mid-range showed a decreasing trend from south to north. The high-value water-producing region in the southern gradually decreased, yet some low-value water-producing areas increased to medium-yielding areas. The central and northern regions were dominated by low-value water production and continued to decrease in most areas over the past 20 years. Compared with the upper and middle reaches, the downstream area is a region with a low-value WY. Although the WY partially increased between 2005 and 2015, there has still been a slight decline in the past 20 years. The overall change was not significant.

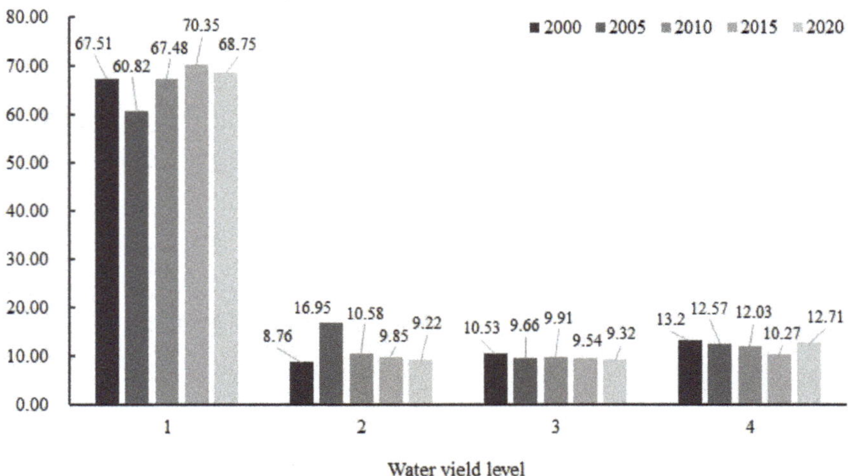

Figure 5. WY ratio of different grades in HRB from 2000 to 2020.

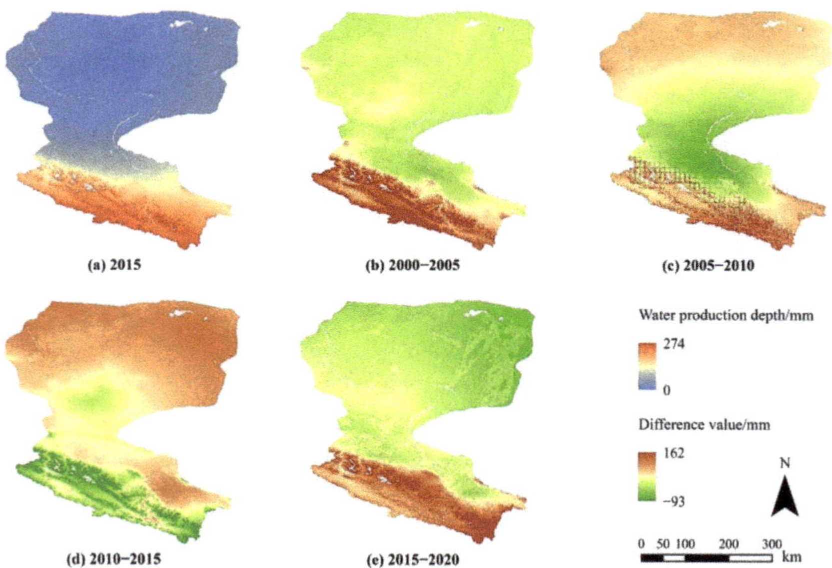

Figure 6. Variation of WY in HRB from 2000 to 2020. (**a**) Spatial distribution of WY in 2015; (**b**)–(**e**) four difference maps of the five-year changes.

Combined with precipitation and land use, it can be found that the spatial distribution shows consistency for WY, precipitation and vegetation in the basin. The region with strong WY capacity has high average annual precipitation and high vegetation cover and low steaming emission. On the other hand, the areas with low annual precipitation and low vegetation cover and strong steaming had a low-value WY. In terms of land-use changes from 2000 to 2020, except rainfall, the variation of WY capacity in HRB was dominated by the area decrease in grassland and forest and the area increase in unused land. Considering the effects of grassland and forest degradation through the whole basin, the overall WY finally showed continuous reduction. Compared to land use in recent years, it can be found that grassland provided the largest proportion among the overall WY services in the basin, reaching more than half of the total WY. This was followed by forest land and cropland.

3.1.3. Spatial Distribution and Variation of SC

The annual SC was classified into one, two, three and four levels, from low to higher values (Figure 7). The soil-conservation composition strongly varied through the whole basin. The low-grade SC accounts for about 70% of the total area, but there was a downward trend year by year. The proportion of SC changed little at middle and high levels. As shown in Figure 8, the spatial distribution of SC in HRB showed that high-value SC had more in the east and less in the north in the southern area. From a regional perspective, the soil holds in the upstream area was relatively small. The SC decreased in 2000–2005 and 2010–2015 and increased significantly with a wide range in the five years from 2005 to 2010. The mesh distribution of high-value SC increased over the past 20 years. The high-value SC focused on the southern part of the mid-range areas. It expanded year by year and the amount of hold increased. Conversely, soil erosion in the north region was more serious and the changes mainly occurred in the mutual transformation of low SC into medium-grade SC. As for the four stages, the changes slightly increased or decreased but were not obvious. So, the soil-keeping ability in mid-range areas has been improved since 2000. High soil holds in the downstream area were few. SC in the first decade showed an upward trend and declined in the next decade. Some soil-hold areas previously of a general level have degraded to that of low level.

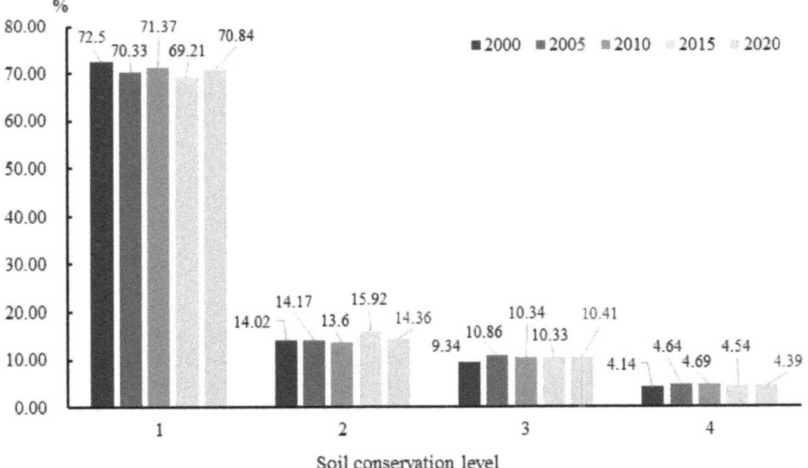

Figure 7. Proportion of SC grade in HRB from 2000 to 2020.

Incorporating the analysis of land use, the expanded utilization of cropland and forest land has brought an increase in SC in the upper and the middle regions. However, the main body to provide SC supply was grassland throughout the whole basin. The areas decrease in grassland caused the decline of SC. At the same time, the growth of cropland and the reduction in unused land compensated the loss of SC caused by the area drop of grassland to some extent. So, in terms of total volume, the soil holds in the study area showed a slight upward trend from 2000 to 2015, followed by a downward trend from 2015 to 2020.

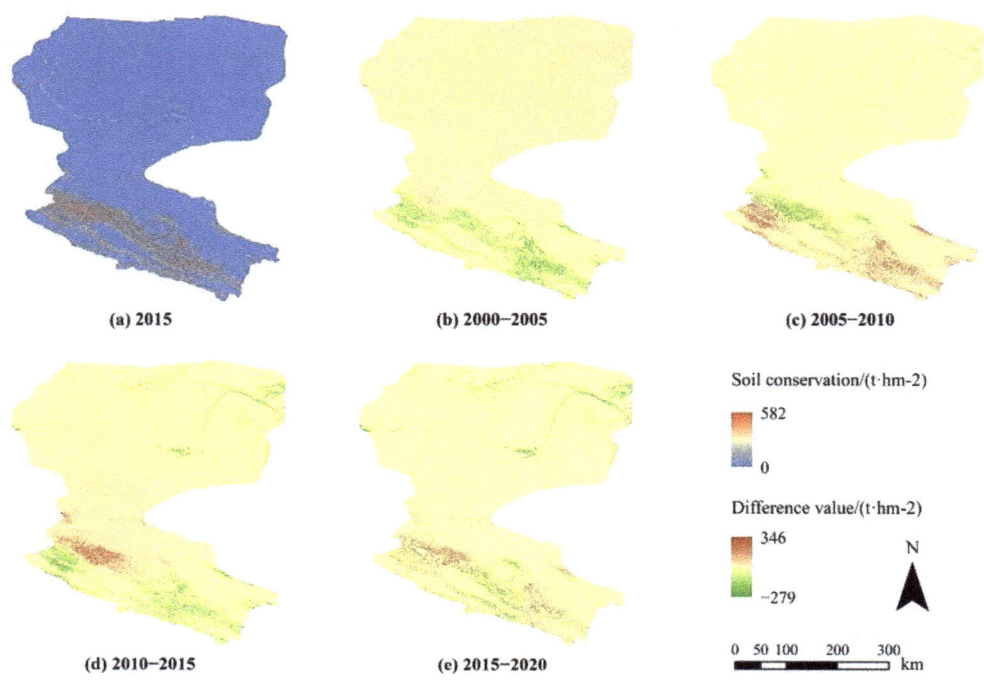

Figure 8. Variation of SC in HRB from 2000 to 2020. (**a**) Spatial distribution of SC in 2015; (**b**)–(**e**) four difference maps of the five-year changes.

3.2. TGE of Different ESs

In order to analyze the TGE of different ESs, 2015-HRB was selected to conduct this practice.

3.2.1. Terrain Niche Index of HRB

The slope and the elevation data were extracted from DEM, and the topographic index of HRB was calculated by the terrain niche index formula-14 (Figure 9). It was divided into six levels: one (0.179–0.305), two (0.305–0.463), three (0.463–0.709), four (0.709–1.005), five (1.005 –1.282), and six (1.282–1.786). The topography was dominated by one and two levels, the area of which accounts for 70.96% of the whole basin. As seen from the spatial distribution (Figure 10), the topographic index was low in the south (east) and high in the north (west). In other words, the high-grade topographic level was concentrated in the upper and middle reaches of the basin, while low- and medium-level terrain areas were heavily distributed in the north of midstream region and throughout the downstream region.

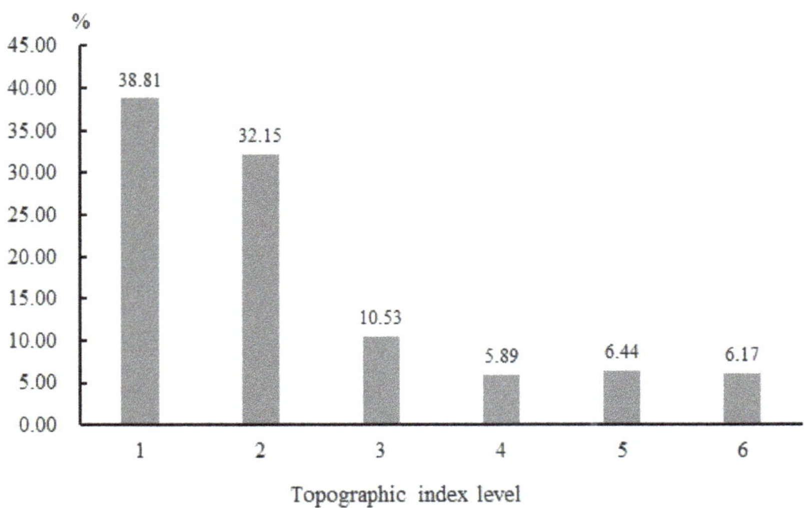

Figure 9. Proportion of different topographic positions index in HRB.

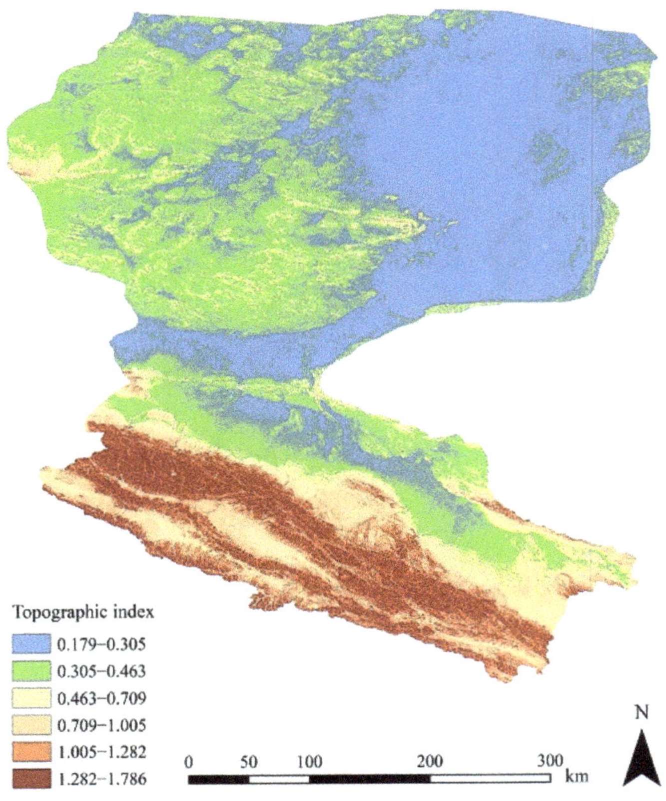

Figure 10. Spatial distribution of topographic position index in HRB in 2015.

3.2.2. TGE of HQ

The HQ was reclassified into four levels: one (0–0.3), two (0.3–0.5), three (0.5–0.8), and four (0.8–1), which correspond to poor, general, high, and higher grades (see Section 3.1.1), respectively. The scatter plot was obtained by the statistical re-regression analysis on the HQ and topographic index (Figure 11). The HQ in the HRB has an upward trend with the growth of the topographic index, and there is a significant positive correlation between the two. The curve fitting (R^2) is 0.441, indicating that the logarithmic function can effectively describe their relationship.

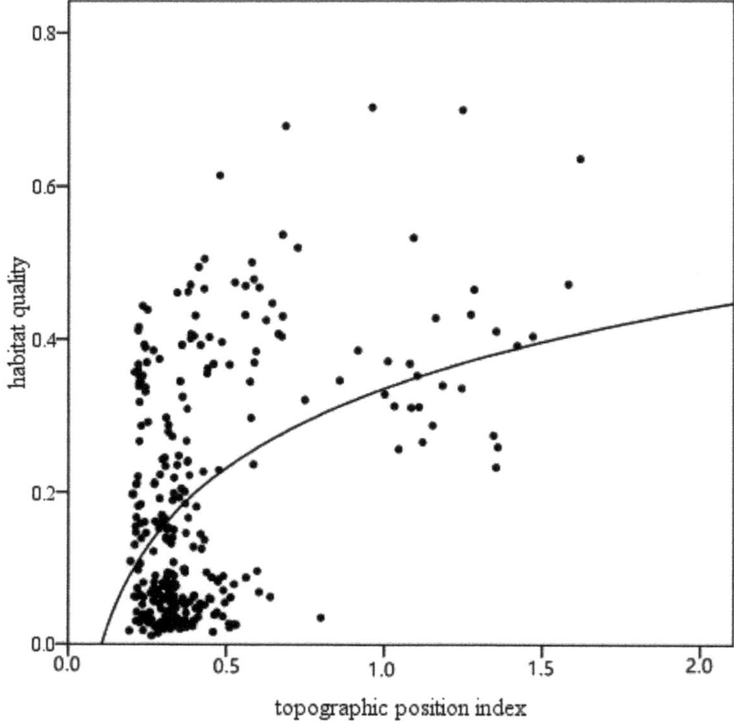

Figure 11. Correlation between terrain index and HQ in HRB in 2015.

As can be seen from Figure 12, the majority of low-grade habitat mass was distributed in the low-grade topographic index in HRB in 2015. With the growth of the topographic level index, the low-quality habitat area gradually reduced with the area increase in high-quality habitats. Additionally, it can be clearly seen that the proportion of high-grade HQ significantly increased with the growth of the terrain niche index from the proportion of change matrix during 2000–2015 (Table 3). The dominant interval was the third HQ level for the fourth and fifth levels of the terrain index, while the other levels of terrain were the first level of HQ. Over the past 15 years, the HQ of second- and fourth-grade topography positions declined. By incorporating land-use changes, this appearance was caused by the decrease in grassland and the increase in built-up land and unused land. Conversely, the increase in HQ at other topographical levels was caused by the increase in cropland and the decrease in unused land. It is sufficient to show that the transformation of nature has a considerable impact on HQ. Therefore, under the impact of rapid urbanization on ecological environment in this basin, relevant departments should take some ecological measures to deal with this.

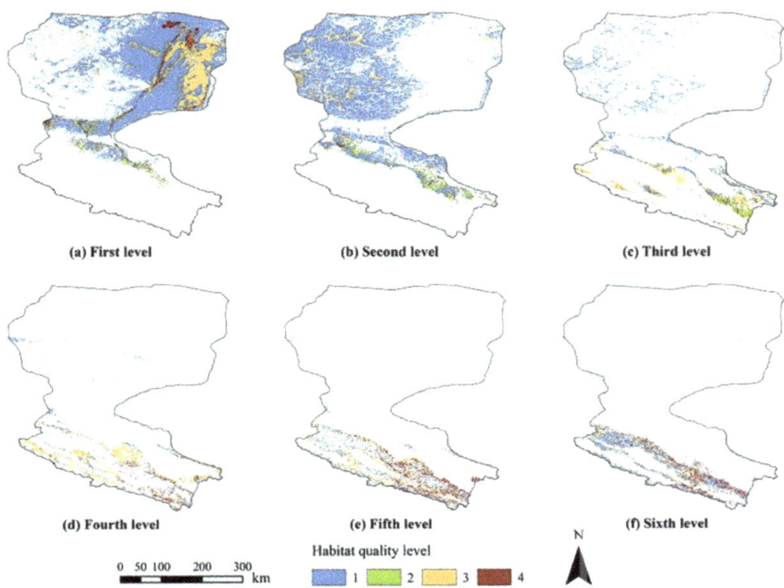

Figure 12. Spatial distribution of HQ at various topographic levels in HRB in 2015.

Table 3. Proportion of HQ distribution at different topographic locations (2000/change value from 2000 to 2015) (%).

Terrain Niche Index Level	HQ Level							
	1		**2**		**3**		**4**	
1	75.05	−0.99	2.58	0.64	19.20	0.26	3.17	0.05
2	84.33	−0.36	4.37	0.59	10.64	−0.32	0.65	0.03
3	58.65	−0.11	9.34	0.35	28.95	0.03	3.07	−0.16
4	22.93	0.20	2.18	0.13	62.90	−1.13	11.99	0.33
5	27.61	0.34	0.35	0.02	49.02	0.15	23.02	−0.56
6	49.51	−0.82	0.04	0.01	32.77	0.60	17.68	0.68

3.2.3. TGE of WY

The water depth in the basin was divided into four levels based on the raster data obtained from model (see Section 3.1.2). The scatter plot was obtained by the statistical re-regression analysis of WY and the topographic index (Figure 13). The WY in the HRB has an upward trend with the growth of the topographic index. There is a significant positive correlation between the two. The curve fitting (R^2) is 0.566, which indicates that the logarithmic function can successfully describe their relationship. The overlaid spatial distribution of the terrain niche index and WY shows the distinct discrepancy in the distribution of water depth levels on the local shape level index in HRB in 2015 (Figure 14). Similar to the distribution of HQ with the local level index, the WY distributed in the low-terrain-level index areas significantly decreased. With the growth of the topographic level index, the depth of water production also gradually increased. There was little low-grade water depth in the areas of the sixth topographic level index.

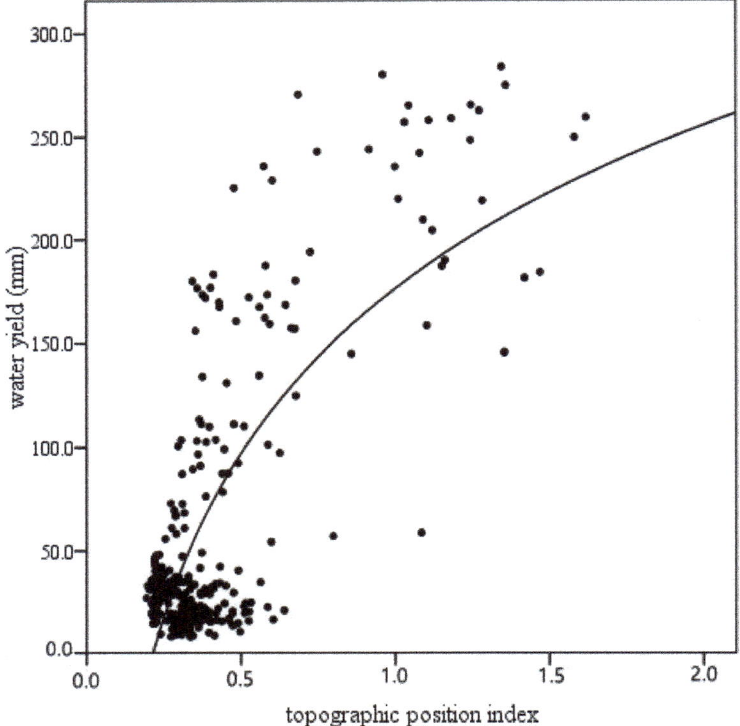

Figure 13. Correlation between terrain index and WY in 2015.

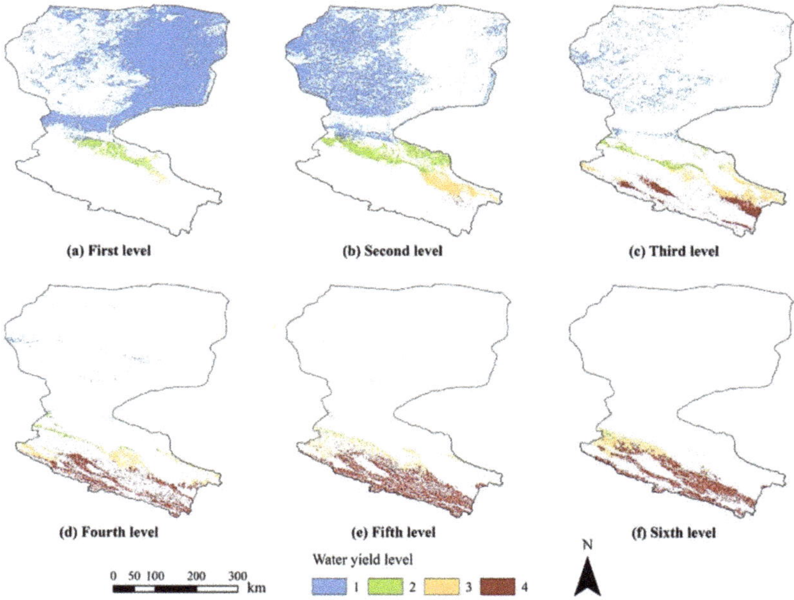

Figure 14. Spatial distribution of WY depth at various topographic levels in HRB in 2015.

The regional statistics of topographic level can be obtained based on the spatial distribution of WY in each year (Table 4). WY in the low terrain level was dominated by the low depth of first and second levels. The water production depth increased to three or four levels in this basin when it raised to fifth and sixth topographic indexes. This is enough to show that WY is closely related to the topographic gradient. The reason for this is that land use in low-terrain areas is majorly unused land with low grassland cover and low water conservation. High-terrain areas mostly have with medium- and high-density grassland cover, and the climate is humid at higher altitudes, resulting in higher water production. Comparing the water depth of 2015 with that of 2000, a decline in the water depth of the low-grade topographic index and the increase in water-producing depth of medium- and high-level topographic index areas was found. In addition to the differential rainfall, this difference may be due to the increasing area of grassland and forest land cover in high-altitude and high-terrain areas. The ecological environment in the unused land of the desert-dominated region continues to deteriorate in the north due to the continuous and extensive human activities. This needs to arouse people's vigilance with regard to, e.g., reducing human extrusion of original animal and plant living space, improving bad ecological environments, and maintaining good surroundings together.

Table 4. Distribution proportion of WY in different topographic levels (2000/change value from 2000 to 2015) (%).

Terrain Niche Index Level	WY Level							
	1		2		3		4	
1	97.75	−3.57	2.25	0.01	0.00	0.50	0.00	0.00
2	84.29	−4.23	14.52	−0.04	1.19	5.38	0.00	0.13
3	42.26	1.58	25.08	0.19	25.95	−1.23	6.71	10.40
4	4.69	1.43	16.72	−0.03	43.63	−7.87	34.96	13.42
5	0.21	0.61	4.98	−0.03	35.41	−8.89	59.41	8.84
6	0.01	0.01	1.61	0.00	28.41	3.60	69.97	−4.43

3.2.4. TGE on SC

The scatter plot was obtained by the statistical re-regression analysis of SC and the topographic index (Figure 15). The SC in the HRB has an upward trend with the increase in the topographic index, and there is a significant positive correlation between the two. The curve fitting (R^2) is 0.687, indicating that the logarithmic function can properly describe their relationship.

As seen in Figure 16, it was in a poor condition for the whole basin in 2015. However, it can also be seen that the TGE of SC is similar to that of HQ and WY, although the SC in different locations is primarily of a low grade. The SC gradually became better with the increase in the terrain niche index. Furthermore, in order to analyze the proportions of each part, the SC was divided into four levels using the natural break point method (Table 5) and was then overlaid with the terrain niche index. Most of the low-grade areas transformed into high-grade areas and the proportion decreased gradually with the increase in the terrain niche index. This is mainly due to the concentration of high-terrain zones in the upper reaches of the basin, which have large areas of forest land, grassland and cropland, as well as dense vegetation and thick and fertile soil. The downstream area was dominated by desert ecosystems, with sparse vegetation and poor soil. The area proportion of low soil levels in various shapes decreased slightly, while that of high-grade soil levels increased. However, soil conditions in the basin remained poor and large amounts of soil were at risk of erosion, particularly in the southern mountainous areas and unused land in the north. Administrators should develop reasonable management measures. Only by strengthening the protection of woodland and grassland, planting trees and expanding vegetation cover can we reduce the threat of soil erosion and decrease the harm.

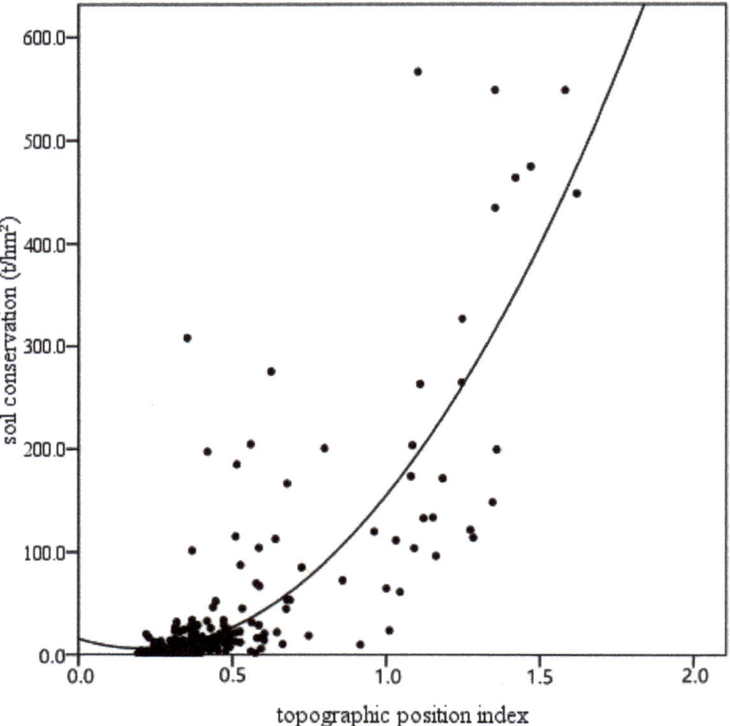

Figure 15. Correlation between terrain index and SC in 2015.

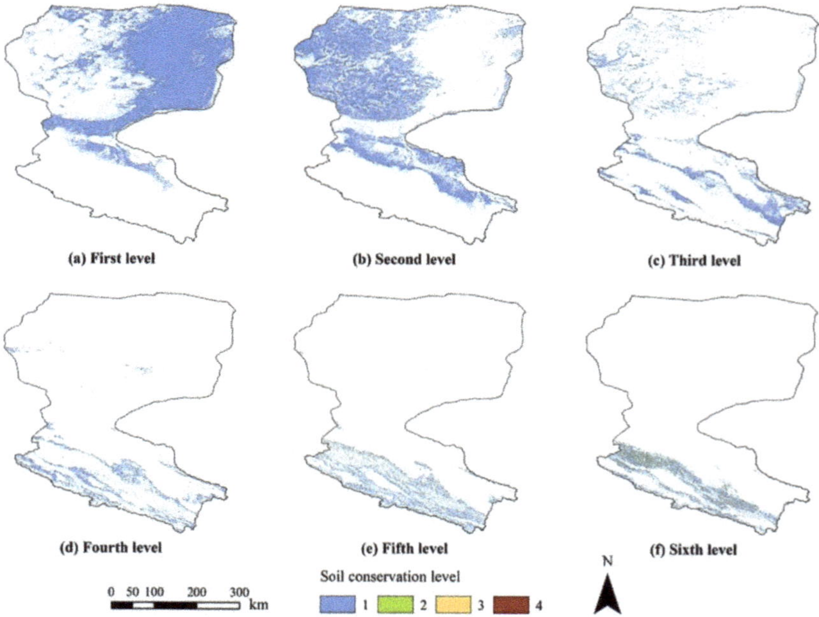

Figure 16. Spatial distribution of SC at various topographic levels in HRB in 2015.

Table 5. Distribution proportion of SC at different topographic levels (2000/change value from 2000 to 2015) (%).

Terrain Niche Index Level	SC Level							
	1		2		3		4	
1	99.83	−2.70	0.14	0.07	0.03	0.10	0.01	2.53
2	99.75	−1.73	0.20	0.96	0.04	0.02	0.01	0.75
3	98.45	0.39	1.38	−0.58	0.12	−0.04	0.05	0.23
4	92.42	−2.33	7.24	−0.03	0.23	1.13	0.11	1.23
5	90.12	1.62	8.68	−2.65	1.07	0.29	0.13	0.74
6	81.21	−4.91	12.70	3.24	0.83	−0.37	0.18	2.04

4. Discussion

In this study, we evaluated the changes in ES patterns in the HRB from 2000 to 2020 by using the three sub-modules in the InVEST model. The spatial variations of HQ, WY and SC were revealed for the past 20 years. We also analyzed the distribution characteristics of three ESs from the perspective of topography. This can provide the scientific basis for optimizing regional ecosystem patterns and land-use planning and management. The results are of great significance for rational planning of land use and constructing ecological civilizations.

4.1. Validation of Results with Previous Works

In general, the overall trends of the spatiotemporal changes of ESs are the same compared with the similar studies in the region [52,57]. Therefore, the reliable results can objectively mirror the spatial differentiation characteristics of ESs in the HRB. Otherwise, elevation and slope were used to analyze the changes in ESs on different terrain gradients and the results illustrate different ESs showing obvious TGE in the HRB. The distribution pattern of ESs is generally high in the mountainous areas and low in the plain, which is consistent with the relevant research [60,61]. In order to further verify the accuracy and reliability of the results, some specific cross-validations were conducted between this paper and previous studies [62,63], including the values of different ESs and spatial distributions (Table 6). The range of HQ in this result is the same with the referenced work. It shows complete consistency with the distribution of high values and low values. The WY value in our study is slightly higher than that in the referenced work, but the difference is within a reasonable range. The major spatial differentiation is that the high value in our work is continuously distributed in the south of middle reaches, while it is dispersed with fragments in the referenced study. This may be due to different methods used to calculate the potential evapotranspiration in the two studies. Additionally, owing to lacking SC in previous studies in HRB, research from 2011 was selected to compare with our results in 2010. The result of this study is numerically low. The spatial difference lies in the shortage of high-value areas for SC in the northern basin and the distribution is not obvious along the riverside. This may be the factor parameter difference in the biophysical coefficient table. In addition, different formulas used to calculate the K factor in the two studies may also cause a discrepancy in the final outcomes.

Table 6. Comparison between the results in this paper and those in the referenced work.

ESs Supply	Results in This Paper			
	Year	**Area**	**Range**	**Spatial Distribution**
HQ	2015	Midstream (by provincial boundaries)	0.0–1.0	General and higher in the central region while poor in the northern and southern regions
WY	2000	Midstream (by provincial boundaries)	0.0–314.7 mm	More southeast, less northwest; obvious difference between the north–south and continuous distribution of high-value areas
SC	2010	HRB	0–644 t/hm^2	More west and less east, more south and less north; high-value concentrated area in the southern part of middle reaches
ESs Supply	**Results in Referenced work**			
	Year	**Area**	**Range**	**Spatial Distribution**
HQ	2015	Midstream (by Yingluo gorge and Zhengyi gorge)	0.0–1.0	General and higher-grade in the central and southern regions while poor in the northern regions
WY	2000	Midstream (by Yingluo gorge and Zhengyi gorge)	0.0–306.0 mm	More southeast, less northwest; Dispersed distribution of high value areas
SC	2011	HRB	0–646 t/hm^2	More west and less east; Sparse distribution of high-value areas along the riverside in the north and south basin

4.2. Driving Forces of ESs Variation

The changes of ESs are affected by natural and human factors. Climate change is the main driving issue for natural factors. Furthermore, human factors, such as activity intensity on different topographic gradient will also have an important impact on ESs [64]. We qualitatively analyze the natural and human factors of ES changes in the HRB in terms of temperature, precipitation and human activities.

Research indicated that the temperature in the HRB shows a significant upward trend in the recent decades [65]. The increase rate of temperature in the study area is much higher than that of the average in the northern hemisphere during the same period. In particular, a large increase in temperature around 2010 was rare [66]. Rising temperature will promote the conversion of some unused land covered by glaciers and snow into grassland and grassland into woodland in the basin. The general trend of precipitation is increasing but the magnitude is negligible. The change is not obvious. The precipitation in this basin is not enough to balance the increase in evapotranspiration caused by the rising temperature. This gave rise to a further reduce in surface water, while the glacier melt water alleviated the decrease in runoff [66]. So, runoff changes little as temperature rises. The water supply gradually decreased, and some rivers and lakes dried up. The water area shrunk and desertification increased significantly [67].

Human activity is another essential driving force for land-use change in the watersheds. As be seen from Table 7, urban and rural industrial and other residential land continue to expand with the gradual deepening of human influence over the past 20 years. A large amount of unused land and grassland were reclaimed as arable land due to more intensive human activities in the middle reaches, which causes the area growth of cropland significantly. On the other hand, the manual allocation of water resources has given rise to a large amount of water diversion to the downstream region of the HRB. As a result, some unused land has been restored to grasslands. The ecological environment has been remarkably improved. A series of ecological management measures, such as natural forest and grass fence enclosures, have provoked the area increase in forest land in the upstream and the midstream areas. However, the growth of the population has brought about an increase in water demand. Then, adding up the decrease in rainfall, forest land and grassland shows a degradation trend in some areas. Therefore, the land-use change in the watersheds is a result of the joint power of human activities and climate change [37]. Furthermore, the land-use change will have an impact on the variation of different ESs.

Table 7. Area of land use in different reaches of HRB from 2000 to 2020 (km^2).

Land-Use Type	2000			2005		
	Upstream	Midstream	Downstream	Upstream	Midstream	Downstream
Cropland	11.66	4617.33	60.41	11.66	4939.70	69.22
Forest	1487.75	2950.35	878.84	1487.75	2941.53	870.45
Grassland	5457.08	13,654.56	11,069.79	5457.08	13,598.84	11,053.02
Waters	228.56	805.92	332.60	228.56	803.08	356.77
Built-up land	10.24	414.33	135.16	10.24	449.15	135.19
Unused land	2756.17	33684.40	54,012.44	2757.61	33,394.58	54,003.63

Land-Use Type	2010			2015		
	Upstream	Midstream	Downstream	Upstream	Midstream	Downstream
Cropland	10.24	4959.17	66.95	10.24	5255.53	65.53
Forest	1513.77	2933.15	845.86	1511.92	2932.72	823.55
Grassland	5459.92	13,631.82	11,046.05	5459.92	13,535.16	11,170.57
Waters	217.90	799.38	388.32	216.48	870.45	377.66
Built-up land	10.24	452.43	135.12	10.24	572.25	133.33
Unused land	2755.76	33,327.20	54,016.13	2758.23	32,937.04	53,806.20

Land-Use Type	2020		
	Upstream	Midstream	Downstream
Cropland	16.45	6217.98	76.42
Forest	1401.17	3292.52	876.20
Grassland	5732.36	14,570.27	11,568.41
Waters	188.73	921.64	386.24
Built-up land	12.24	641.19	116.89
Unused land	2616.08	30,459.55	53,352.68

4.3. Suggestions for Sustainable Improvement of ESs

The implementation of ecological protection policies and effective management in the Qilian Mountain reserve have enhanced the grassland coverage and water conservation capacity in the upstream area over the years. The Forestry Bureau should continue to strengthen the protection and surveillance of forest and grassland [68]. In the upper mountain region, climate change is always a dominant element for water resource allocation. Furthermore, except for water conservation, the increase in forest and grassland will also play an important role in adjusting runoff in the form of transpiration interception due to the warming and drying trend in recent years [69,70]. Grass planting and forestation measures should also be taken under the guidance of the Forestry Bureau within a scientific and reasonable range to stabilize runoff according to local conditions. In addition, thoughtful reservoir construction by the Water Resources Bureau is also useful in the regulation of seasonal runoff and in the efficient implementation of the ecological water diversion project.

The optimization and adjustment of the agricultural planting structure have been implemented in the midstream area in recent years, which yielded preliminary results for the construction of a water-saving society. However, the rapid development of the agricultural economy has brought the expansion of the irrigation area, which has caused the increase in water consumption, offsetting the consequences of saving water [70]. According to the analysis of long-term climate change in the basin, runoff may turn into the dry season in the future [71]. The implementation of the water transfer project may not be enough to support the rapid growth of population and fast development of the agricultural economy in the middle reaches [69–72]. In order to adapt to the changes in water resources in various stages, it is necessary to control the cropland expansion in the midstream region as soon as possible and the Bureau of Land Management needs to adjust the industrial structure. Moreover, during the implementation of the Heihe River "97" water diversion scheme, the continuous expansion of the oasis has begun to consume a large amount of groundwater and the ecological environment has a tendency of degradation [70]. It is essential to strictly limit the exploitation of groundwater, make zoning planning and continue to implement ecological protection projects in fragile ecological environment areas by the Natural Resources Planning Bureau. This basin needs to develop urbanization without destroying natural environment and follow the principles of sustainable development.

Since the implementation of the Heihe River "97" water diversion scheme in 2000, the amount of water flowing into the downstream region has increased significantly. The water area gradually recovered and steadily expanded, and the growth of vegetation improved. This alleviated the deterioration of the ecological environment in the downstream area, remarkably restored the ecological environment and initially realized the control goal [69,73]. However, the continuous expansion of cropland and construction land still brings great pressure on the environment. So, to improve ecological conditions, the Environmental Protection Agency should strictly control overgrazing and deforestation to allow the self-healing function of the ecosystem to take place. According to the local conditions, planting trees and grass can alleviate the ecological deterioration of desert. The government should further optimize and adjust the amount of water allocation and conduct a more reasonable allocation of resources for the sustainable development of the basin.

4.4. Limitations and Future Study

Although some reliable results have been obtained in this study, we recognize some limitations of this research. First of all, more attention should be paid in future studies. For example, only three key services provided by the ecosystems in the HRB were considered in our study due to the urgency to solve eco-environmental problems, such as the deterioration of the ecological environment, shortage of water resources and soil erosion. At the same time, the eco-environmental characteristics and data availability of the research area were also taken into consideration. However, the temporal and spatial changes of other ESs in the basin cannot be ignored, and they can be further discussed in future research. Then, the terrain niche index was used to comprehensively mirror the effect of topography on ESs. However, the respective weights of elevation and slope in the impact is not clear, and this needs to be further explored so as to formulate more practical ecological protection policies. Additionally, distinct operation mechanisms of different models may result in dissimilar outputs. It is necessary to focus on the comparative verification of the evaluated results between various models in future work.

5. Conclusions

In this study, three key ESs (e.g., HQ, WY and SC) were evaluated using the InVEST model in HRB from 2000 to 2020, and the TGE of different ESs was analyzed based on the terrain niche index. First of all, HQ and SC in the basin improved overall, while WY was in continuous decline. The values of various ESs changed little and the spatial distribution of them always presented similar discrepancy laws. The distribution of ESs in the basin shows lower values in the north and higher values in the south, high values at upstream areas and low values at downstream area. Furthermore, the correlations between different ESs and topography are strong, showing obvious TGE. HQ, WY and SC increased with the growth of the topographic level index. A high-quality habitat has high rates of water production. A low-quality habitat has a low yield of water. High SC is mostly distributed in the mountainous hilly areas with high grassland coverage, while a low SC is related to the accumulation of built-up land and unused land.

We analyze the variation of different ESs on the two scales: watersheds and sub-stream. Relevant suggestions should be adopted for the distinct changes of ESs on different scales. In addition, topography and land use were combined to analyze the ES variation; in particular, the terrain niche index was specifically used to analyze the influence of

TGE on different ESs. The results provide useful information for policy proposals to comprehensively consider the effect of terrain factors on the spatial differences of Ess and its drivers. The method provides a new perspective for studying the spatial differentiation of various ecological problems in the inland watersheds and can be applied to other similar areas, and then be incorporated into the formulation of watershed ecological zoning management policies.

Author Contributions: Conceptualization, L.Z. and N.H.; methodology, L.Z. and N.H.; software, L.Z.; validation, L.Z. and N.H.; formal analysis, L.Z.; data curation, L.Z.; writing—original draft preparation, L.Z.; writing—review and editing, L.Z. and N.H.; supervision, N.H. All authors have read and agreed to the published version of the manuscript.

Funding: This research was funded by the National Natural Science Foundation of China (No. 41701478) and by the Fundamental Research Funds for the Central Universities (No. GK201903069).

Institutional Review Board Statement: Not applicable.

Informed Consent Statement: Informed consent was obtained from all subjects involved in the study.

Data Availability Statement: Data sharing not applicable.

Acknowledgments: We thank the useful suggestions from the editors to improve this work. We also appreciate the constructive comments from the two anonymous reviewers that greatly improved this manuscript.

Conflicts of Interest: The authors declare no conflict of interest.

References

1. Fu, B.; Zhou, G.; Bai, Y.; Song, C.; Liu, J.; Zhang, H.; Lü, Y.; Zheng, H.; Xie, G. The main terrestrial ecosystem services and ecological security in China. *Adv. Earth Sci.* **2009**, *24*, 571–576. [CrossRef]
2. Chen, Y.-F.; Zhou, J.-X.; Xiao, J.; Li, Y.-P. Forecasting loss of ecosystem service value using a BP network: A case study of the impact of the South-to-north Water Transfer Project on the ecological environmental in Xiangfan, Hubei Province, China. *Biomed. Environ. Sci.* **2003**, *16*, 379–391. [CrossRef] [PubMed]
3. Zhang, H.; Wang, Q.; Li, G.; Zhang, H.; Zhang, J. Losses of ecosystem service values in the Taihu Lake Basin from 1979 to 2010. *Front. Earth Sci.* **2017**, *11*, 310–320. [CrossRef]
4. Wang, C.; Li, X.; Yu, H.; Wang, Y. Tracing the spatial variation and value change of ecosystem services in Yellow River Delta, China. *Ecol. Indic.* **2019**, *96*, 270–277. [CrossRef]
5. Gao, Y.; Wang, W.; Liu, X.; Zhang, L.; Dend, F.; Yang, C.; Sun, Q. Evaluation of carbon sequestration of forest ecosystem in Xiamen City. *Res. Environ. Sci.* **2019**, *32*, 2001–2007. [CrossRef]
6. Zhao, Q.; Wen, Z.; Chen, S.; Ding, S.; Zhang, M. Quantifying land use/land cover and landscape pattern changes and impacts on ecosystem services. *Int. J. Environ. Res. Public Health* **2019**, *17*, 126. [CrossRef] [PubMed]
7. Zhang, X.; He, S.; Yang, Y. Evaluation of wetland ecosystem services value of the yellow river delta. *Environ. Monit. Assess.* **2021**, *193*, 353. [CrossRef] [PubMed]
8. Ye, Y.; Zhang, J.; Wang, T.; Bai, H.; Wang, X.; Zhao, W. Changes in land-use and ecosystem service value in Guangdong Province, Southern China, from 1990 to 2018. *Land* **2021**, *10*, 426. [CrossRef]
9. Xie, G.; Cao, S.; Lu, C.; Xiao, Y.; Zhang, Y. Human's consumption of ecosystem services and ecological debt in China. *J. Nat. Resour.* **2010**, *25*, 43–51. [CrossRef]
10. Ouyang, Z.; Wang, X.; Miao, H. Preliminary study on the service function and eco economic value of terrestrial ecosystem in China. *Acta Ecol. Sin.* **1999**, *19*, 19–25. [CrossRef]
11. Zhao, J.; Xiao, H.; Wu, G. Comparison analysis on physical and value assessment methods for ecosystems services. *Chin. J. Appl. Ecol.* **2000**, *11*, 290–292. [CrossRef]
12. Groot, R.D.; Brander, L.; Ploeg, S.V.D.; Costanza, R.; Bernard, F.; Braat, L.; Christie, M.; Crossman, N.; Ghermandi, A.; Hein, L.; et al. Global estimates of the value of ecosystems and their services in monetary units. *Ecosyst. Serv.* **2012**, *1*, 50–61. [CrossRef]
13. Costanza, R.; Groot, R.D.; Sutton, P.; Ploeg, S.V.D.; Anderson, S.J.; Kubiszewski, I.; Farber, S.; Turner, R.K. Changes in the global value of ecosystem services. *Glob. Environ. Chang.* **2014**, *26*, 152–158. [CrossRef]
14. Chen, Z.; Zhang, X. The value of ecosystem benefits in China. *Chin. Sci. Bull.* **2000**, *45*, 17–22. [CrossRef]
15. Xie, G.; Zhang, C.; Zhang, L.; Chen, W.; Li, S. Improvement of the evaluation method for ecosystem service value based on per unit area. *J. Nat. Resour.* **2015**, *30*, 1243–1254. [CrossRef]
16. Costanza, R.; d'Arge, R.; Groot, R.D.; Farber, S.; Grasso, M.; Hannon, B.; Limburg, K.; Naeem, S.; O'Neill, R.V.; Paruelo, J.; et al. The value of the world's ecosystem services and natural capital. *Nature* **1997**, *387*, 253–260. [CrossRef]

17. Xie, G.; Lu, C.; Leng, Y.; Zheng, D.; Li, S. Ecological assets valuation of the Tibetan Plateau. *J. Nat. Resour.* **2003**, *18*, 189–196. [CrossRef]
18. Li, H.; Li, Z.; Li, Z.; Yu, J.; Liu, B. Evaluation of ecosystem services: A case study in the middle reach of the Heihe River Basin, Northwest China. *Phys. Chem. Earth* **2015**, *89–90*, 40–45. [CrossRef]
19. Sutton, P.C.; Anderson, S.J.; Costanza, R.; Kubiszewski, I. The ecological economics of land degradation: Impacts on ecosystem service values. *Ecol. Econ.* **2016**, *129*, 182–192. [CrossRef]
20. Jiang, W.; Fu, B.; Lü, Y. Assessing impacts of land use/land cover conversion on changes in ecosystem services value on the Loess Plateau, China. *Sustainability* **2020**, *12*, 7128. [CrossRef]
21. Wang, Y.; Zhang, S.; Zhen, H.; Chang, X.; Shataer, R.; Li, Z. Spatiotemporal evolution characteristics in ecosystem service values based on land use/cover change in the Tarim River Basin, China. *Sustainability* **2020**, *12*, 7759. [CrossRef]
22. Yin, N.; Wang, S.; Liu, Y. Ecosystem service value assessment: Research progress and prospects. *Chin. J. Ecol.* **2021**, *40*, 233–244. [CrossRef]
23. Wu, C.; Ma, G.; Yang, W.; Zhou, Y.; Peng, F.; Wang, J.; Yu, F. Assessment of ecosystem service value and its differences in the Yellow River Basin and Yangtze River Basin. *Sustainability* **2021**, *13*, 3822. [CrossRef]
24. Su, K.; Wei, D.-Z.; Lin, W.-X. Evaluation of ecosystem services value and its implications for policy making in China—A case study of Fujian province. *Ecol. Indic.* **2020**, *108*, 105752. [CrossRef]
25. Ouyang, Z.; Zheng, H.; Xiao, Y.; Polasky, S.; Liu, J.; Xu, W.; Wang, Q.; Zhang, L.; Xiao, Y.; Rao, E.; et al. Improvements in ecosystem services from investments in natural capital. *Science* **2016**, *352*, 1455–1459. [CrossRef]
26. Yuan, Z.; Wan, R. A review on the methods of ecosystem service assessment. *Ecol. Sci.* **2019**, *38*, 210–219. [CrossRef]
27. Odum, H.-T.; Odum, E.-P. The energetic basis for valuation of ecosystem services. *Ecosystems* **2000**, *3*, 21–23. [CrossRef]
28. Sherrouse, B.-C.; Clement, J.-M.; Semmens, D.-J. A GIS application for assessing, mapping, and quantifying the social values of ecosystem services. *Appl. Geogr.* **2011**, *31*, 748–760. [CrossRef]
29. Sharps, K.; Masante, D.; Thomas, A.; Jackson, B.; Redhead, J.; May, L.; Prosser, H.; Cosby, B.; Emmett, B.; Jones, L. Comparing strengths and weaknesses of three ecosystem services modelling tools in a diverse UK river catchment. *Sci. Total Environ.* **2017**, *584*, 118–130. [CrossRef]
30. Villa, F.; Bagstad, K.-J.; Voigt, B.; Johnson, G.-W.; Portela, R.; Honzak, M.; Batker, D. A methodology for adaptable and robust ecosystem services assessment. *PLoS ONE* **2014**, *9*, e91001. [CrossRef]
31. Wang, X.; Liu, X.; Long, Y.; Liang, W.; Zhou, J.; Zhang, Y. Analysis of Soil retention service function in the North Area of Guangdong based on the InVEST model. In *IOP Conference Series: Earth and Environmental Science, Proceedings of the Fourth International Workshop on Renewable Energy and Development, Sanya, China, 24–26 April 2020*; IOP: Bristol, UK, 2020; Volume 510, p. 510. [CrossRef]
32. Gao, J.; Zuo, L. Revealing ecosystem services relationships and their driving factors for five basins of Beijing. *J. Geogr. Sci.* **2021**, *31*, 111–129. [CrossRef]
33. Hu, W.; Li, G.; Gao, Z.; Jia, G.; Wang, Z.; Li, Y. Assessment of the impact of the Poplar Ecological Retreat Project on water conservation in the Dongting Lake wetland region using the InVEST model. *Sci. Total. Environ.* **2020**, *733*, 139423. [CrossRef]
34. Zhang, X.; Zhou, J.; Li, M. Analysis on spatial and temporal changes of regional habitat quality based on the spatial pattern reconstruction of land use. *Acta Geogr. Sin.* **2020**, *75*, 160–178. [CrossRef]
35. Bai, Y.; Chen, Y.; Alatalod, J.; Yang, Z.; Jiang, B. Scale effects on the relationships between land characteristics and ecosystem services—A case study in Taihu Lake Basin, China. *Sci. Total. Environ.* **2020**, *716*, 137083. [CrossRef]
36. Aneseyee, A.B.; Elias, E.; Soromessa, T.; Feyisa, G.L. Land use/land cover change effect on soil erosion and sediment delivery in the Winike watershed, Omo Gibe Basin, Ethiopia. *Sci. Total. Environ.* **2020**, *728*, 138776. [CrossRef] [PubMed]
37. Geng, X.; Wang, X.; Yan, H.; Zhang, Q.; Jin, G. Land use/land cover change induced impacts on water supply service in the upper reach of Heihe River Basin. *Sustainability* **2014**, *7*, 366–383. [CrossRef]
38. Wang, X.; Dai, E.; Zhu, J. Spatial patterns of forest ecosystem services and influencing factors in the Ganjiang River Basin. *J. Resour. Ecol.* **2016**, *7*, 439–452. [CrossRef]
39. Fang, Z.; Bai, Y.; Jiang, B.; Alatalo, J.-M.; Liu, G.; Wang, H. Quantifying variations in ecosystem services in altitude-associated vegetation types in a tropical region of China. *Sci. Total. Environ.* **2020**, *726*, 138565. [CrossRef]
40. Aneseyee, A.B.; Noszczyk, T.; Soromessa, T.; Elias, E. The InVEST habitat quality model associated with land use/cover changes: A qualitative case study of the Winike Watershed in the Omo-Gibe Basin, Southwest Ethiopia. *Remote Sens.* **2020**, *12*, 1103. [CrossRef]
41. Zhang, J.; Ren, Z. Spatiotemporal pattern and terrain gradient effect of land use change in Qinling-Bashan mountains. *Trans. Chin. Soc. Agric. Eng.* **2016**, *32*, 250–257. [CrossRef]
42. Gong, W.; Yuan, L.; Fan, W. Analysis on land use pattern changes in Harbin based on terrain gradient. *Trans. Chin. Soc. Agric. Eng.* **2013**, *29*, 250–259. [CrossRef]
43. Chen, Y.; Li, J.; Xu, J. The impact of socio- economic factors on ecological service value in Hubei Province: A geographically weighted regression approach. *Chin. Land Sci.* **2015**, *29*, 89–96. [CrossRef]
44. Zeng, J.; Li, J.; Yao, X. Spatio-temporal dynamics of ecosystem service value in Wuhan Urban Agglomeration. *Chin. Appl. Ecol.* **2014**, *25*, 883–891. [CrossRef]

45. Wang, Q.; Li, Y.; Liu, Y.; Hu, X. Analysis of spatial pattern of land use based on terrain gradient in karst trough valley area. *Acta Ecol. Sin.* **2019**, *39*, 7866–7880. [CrossRef]
46. Yang, B.; Wang, Z.; Yao, X.; Zhang, L. Terrain gradient effect and spatial structure characteristics of land use in mountain areas of northwestern Hubei province. *Resour. Environ. Yangtze Basin* **2019**, *28*, 313–321. [CrossRef]
47. Ma, S.; Qiao, Y.; Wang, L.; Zhang, J. Terrain gradient variations in ecosystem services of different vegetation types in mountainous regions: Vegetation resource conservation and sustainable development. *For. Ecol. Manag.* **2021**, *482*, 118856. [CrossRef]
48. Wang, L.; Ma, S.; Jiang, J.; Zhao, Y.; Zhang, J. Spatiotemporal variation in ecosystem services and their drivers among different landscape heterogeneity units and terrain gradients in the southern hill and mountain belt, China. *Remote Sens.* **2021**, *13*, 1375. [CrossRef]
49. Liu, W. Scientific understanding of the Belt and Road Initiative of China and related research themes. *Prog. Geogr.* **2015**, *34*, 538–544. [CrossRef]
50. Liu, H.; Yeerken, W.; Wang, C. Impacts of the Belt and Road Initiative on the spatial pattern of territory development in *China*. *Prog. Geogr.* **2015**, *34*, 545–553. [CrossRef]
51. Wang, Y.; Li, B.; Liu, J.; Feng, Q.; Liu, W.; Wang, X.; He, Y. Effects of cascade reservoir systems on the longitudinal distribution of sediment characteristics: A case study of the Heihe River Basin. *Environ. Sci. Pollut. Res.* **2021**, *28*, 1–13. [CrossRef]
52. Wang, B.; Zhao, J.; Hu, X. Spatial pattern analysis of ecosystem services based on InVEST in Heihe River Basin. *Chin. J. Ecol.* **2016**, *35*, 2783–2792. [CrossRef]
53. Ning, B.; He, Y.; He, X.; Li, Z. Research progress of water resources in Heihe River Basin. *J. Desert Res.* **2008**, *28*, 1180–1185.
54. Tallis, H. *Natural Capital: Theory and Practice of Mapping Ecosystem Services*; Oxford University Press: Oxford, UK, 2011.
55. Lin, Y.-P.; Lin, W.-C.; Wang, Y.-C.; Wang, Y.-C.; Lien, W.-Y.; Huang, T.; Hsu, C.-C.; Schmeller, D.S.; Crossman, N.D. Systematically designating conservation areas for protecting habitat quality and multiple ecosystem services. *Environ. Model Softw.* **2017**, *90*, 126–146. [CrossRef]
56. Bao, Y.; Liu, K.; Li, T.; Hu, S. Effects of land use change on habitat based on InVEST model-taking Yellow River Wetland nature reserve in Shaanxi Province as an example. *Arid Zone Res.* **2015**, *32*, 622–629. [CrossRef]
57. Wang, Y.; Meng, J. Effects of land use change on ecosystem services in the middle reaches of the Heihe River Basin. *Arid Zone Res.* **2017**, *34*, 200–207. [CrossRef]
58. Zhou, W.; Liu, G.; Pan, J.; Feng, X. Distribution of available soil water capacity in China. *J. Geogr. Sci.* **2005**, *15*, 3–12. [CrossRef]
59. Zhao, Y. Evaluation of Watershed Ecosystem Service Function Based on Invest Model. Master's Thesis, Lanzhou University, Lanzhou, China, 2017. [CrossRef]
60. Gao, H.; Han, H.; Yu, H.; Han, M. Distribution characteristic of important ecosystem services in terrain gradient in Wujiang River Basin. *Ecol. Sci.* **2016**, *35*, 154–159. [CrossRef]
61. Xu, C.; Gong, J.; Li, Y.; Yan, L.; Gao, B. Spatial distribution characteristics of typical ecosystem services based on terrain gradients of Bailongjiang Watershed in Gansu. *Acta Ecol. Sin.* **2020**, *40*, 4291–4301. [CrossRef]
62. Jiang, S.; Meng, J.; Chen, Y. Analysis of hydrological effects of land use and landscape pattern in the middle reaches of Heihe River. *Sci. Soil Water Conser.* **2019**, *17*, 64–73. [CrossRef]
63. Li, Q.; Wang, L.; Yan, C.; Liu, H. Simulation of urban spatial expansion in oasis towns based on habitat quality: A case study of the middle reaches of Heihe River. *Acta Ecol. Sin.* **2020**, *40*, 2920–2931. [CrossRef]
64. Dun, Y.; Yang, C.; Yuan, X.; Yang, Y.; Xiao, F.; Liang, S.; Lu, X. Researches on watershed ecosystem services. *Ecol. Econ.* **2019**, *35*, 179–183.
65. Wang, P.; Li, Z.; Gao, W.; Yan, D.; Bai, J.; Li, K.; Wang, L. Glacier changes in the Heihe River Basin over the past 50 years in the context of climate change. *Res. Sci.* **2011**, *33*, 399–407. Available online: http://ir.casnw.net/handle/362004/8103 (accessed on 17 June 2021).
66. Wang, L.; Zhang, X. Effect of the recent climate change on water resource in Heihe River Basin. *J. Arid Land Res. Environ.* **2010**, *24*, 60–65. [CrossRef]
67. Zhang, K.; Wan, R.; Han, H.; Wang, X.; Si, J. Hydrological and water resources effects under climate change in Heihe River Basin. *Res. Sci.* **2007**, *29*, 77–83. [CrossRef]
68. Tian, F.; Lü, Y.; Fu, B.; Yang, Y.; Qiu, G.; Zang, C.; Zhang, L. Effects of ecological engineering on water balance under two different vegetation scenarios in the Qilian Mountain, northwestern China. *J. Hydrol.* **2016**, *5*, 324–335. [CrossRef]
69. Cheng, G.; Li, X.; Zhao, W.; Xu, Z.; Xiao, H. Integrated study of the water–ecosystem–economy in the Heihe River Basin. *Natl. Sci. Rev.* **2014**, *1*, 413–428. [CrossRef]
70. Xiao, S.; Xiao, H.; Mi, L.; Li, L.; Lu, Z.; Peng, X. Scientific evaluation on ecological effect of national comprehensive harnessing project of Heihe River Basin. *Bull. Chin. Acad. Sci.* **2017**, *32*, 45–54. [CrossRef]
71. Guo, K.; Xie, Y.; Wang, X.; Qiu, T. Characteristics of spatiotemporal distribution of temperature in the Heihe River Basin during the period 1960–2015. *Res. Soil Water Conserv.* **2020**, *27*, 253–260. [CrossRef]

72. Su, L. Response of land use changes on ecological water diversion in midstream of the Heihe River Basin. In *MATEC Web Conference, Proceedings of the First International Symposium on Water System Operations (ISWSO 2018), Beijing, China, 22–26 January 2018*; EDP Science: Les Ulis, France, 2018; Volume 246, p. 246. [CrossRef]
73. Zhou, Y.; Li, X.; Yang, K.; Zhou, J. Assessing the impacts of an ecological water diversion project on water consumption through high-resolution estimations of actual evapotranspiration in the downstream regions of the Heihe River Basin, China. *Agric. For. Meteorol.* **2018**, *249*, 210–227. [CrossRef]

sustainability

Article

Investigation of Spatio–Temporal Changes in Land Use and Heat Stress Indices over Jaipur City Using Geospatial Techniques

Suresh Chandra [1], Swatantra Kumar Dubey [2], Devesh Sharma [3,*], Bijon Kumer Mitra [4] and Rajarshi Dasgupta [4] 

[1] Centre of Excellence for Climate Change & Vector-Borne Diseases, ICMR-National Institute of Malaria Research, Department of Health Research, Government of India, New Delhi 110077, India; sureshchandra1987@hotmail.com

[2] Department of Environmental Engineering, Seoul National University of Science & Technology (SeoulTech), Gongneung-ro, Nowon-gu, Seoul 01811, Korea; swatantratech1@gmail.com

[3] Department of Atmospheric Science, School of Earth Sciences, Central University of Rajasthan, Ajmer 305817, India

[4] Integrated Sustainability Center, Institute for Global Environmental Strategies (IGES), 2108-11 Kamiyamaguchi, Hayama 240-0115, Kanagawa, Japan; b-mitra@iges.or.jp (B.K.M.); dasgupta@iges.or.jp (R.D.)

* Correspondence: deveshsharma@curaj.ac.in

Abstract: Heat waves are expected to intensify around the globe in the future, with a potential increase in heat stress and heat-induced mortality in the absence of adaptation measures. India has high current exposure to heat waves, and with limited adaptive capacity, impacts of increased heat waves might be quite severe. This paper presents a comparative analysis of urban heat stress/heatwaves by combining temperature and vapour pressure through two heat stress indices, i.e., Wet Bulb Globe Temperature (WBGT) and humidex index. For the years 1970–2000 (historical) and 2041–2060 (future), these two indicators were estimated in Jaipur. Another goal of this research is to better understand Jaipur land use changes and urban growth. For the land use study, Landsat 5 TM and Landsat 8 OLI satellite data from the years 1993, 2010, and 2015 were examined. During the research period, urban settlement increased and the majority of open land is converted to urban settlements. In the coming term, all months except three, namely July to September, have seen an increase in the WBGT index values; however, these months are classified as dangerous. Humidex's historical value has been 21.4, but in RCP4.5 and RCP8.5 scenarios, it will rise to 25.5 and 27.3, respectively, and slip into the danger and extreme danger categories. The NDVI and SAVI indices are also used to assess the city's condition during various periods of heat stress. The findings suggest that people's discomfort levels will rise in the future, making it difficult for them to work outside and engage in their usual activities.

Keywords: heat stress; WBGT index; climate change; land use; humidex index

Citation: Chandra, S.; Dubey, S.K.; Sharma, D.; Mitra, B.K.; Dasgupta, R. Investigation of Spatio–Temporal Changes in Land Use and Heat Stress Indices over Jaipur City Using Geospatial Techniques. *Sustainability* **2022**, *14*, 9095. https://doi.org/10.3390/su14159095

Academic Editor: Hossein Azadi

Received: 20 May 2022
Accepted: 20 July 2022
Published: 25 July 2022

Publisher's Note: MDPI stays neutral with regard to jurisdictional claims in published maps and institutional affiliations.

1. Introduction

The occurrence of more extreme climate events has been becoming more frequent and severe as global warming, and causes a distressing effect on human lives [1]. These changes can have both positive and negative impacts on urbanization and human health. Climate change will have a significant impact on metropolitan areas, and it may result in chronic health concerns [2]. Different climate change pathways affect human health between different time periods [3]. India has generated only 2% of total carbon emissions from fossil fuel combustion over the last 100 years [4], which is likely owing to the effects of extreme weather events (NIOO-KNAW, 2017). Human health risks related to climate change can, directly and indirectly, affect older people [5]. An urban heat island (UHI) is a metropolitan area which is significantly warmer than its surrounding rural areas due to human activities. In metropolitan regions, UHIs tend to amplify the impact of heat waves, and rising temperatures in the area contribute to the likelihood of heat-related deaths [6,7]. In 2003, 3500 deaths were estimated across Europe due to extreme heatwave [8]. People's conditions are worse at night due to the high temperature during heatwaves compared to the high daytime temperature, and it also increases the mortality rate at night-time [9]. The high temperature causes an increase in mortality in metropolitan areas, as well as various health conditions such as heat cramps, weariness, non-fatal heat stroke, and overall discomfort [10]. Climate change models anticipate that a gradual increase in summer temperatures and heat waves will exacerbate the situation [8].

India is most vulnerable to the increased temperature associated with climate change. It is estimated that from 1992, about 25,000 Indian people died because of heat waves [11]. In 2003, heatwaves hit parts of India (Uttar Pradesh, Haryana, Punjab, Rajasthan, Gujarat, Bihar, and Orissa), resulting in a higher fatality rate [12]. As a result of the increased number and frequency of heat waves, the death rate will rise in the future [13]. The climatic approaches such as El Niño-Southern Oscillation (ENSO) and fluctuations in the sea surface temperatures in the Bay of Bengal have been related to the heatwaves over India. Heatwaves may occur as a result of changes in wind direction and a lack of moisture in inland areas, resulting in heat waves. Despite the significant societal impact, no systematic attempt has been made to investigate the primary mechanism of heatwaves in India.

In different parts of the world, some authors employed the WBGT and humidex for heat stress assessments [14,15]. WBGT is an experimental index that was developed by Yaglue and Minard in 1957 and published as an ISO 7243 standard in 1989. It is used in both indoor and outdoor environments. It was recommended to eliminate the time-consuming process of calculating the effective temperature index (ET), which was developed from a series of laboratory investigations about 1920 and quickly became the standard approach for assessing heat stress [16]. Temperature, humidity, radiation, and wind were merged into a single figure that could be utilized for assessment (ISO, 1989). The natural wet bulb temperature, globe temperature, and air temperature are the key determinants of WBGT. The WBGT index's most important strength is its sensitivity to radiant heat and air movement, which are two important factors in estimating the ambient air temperature [15,17]. In tropical and subtropical areas of the world, climate change has resulted in temporal and spatial changes in workplace heat exposure, resulting in occupational health issues. In this regard, the results of prior studies show that WBGT values have been rising in recent years. Wet bulb globe temperature (WBGT) is used as a heat stress indicator for assessment of thermal comfort in environments [15,18,19]. Ref. [20] examined WBGT in the Coimbra region of Portugal and found a strong association between globe temperature of 2.8 percent and natural wet bulb temperature of 2.6 percent and WBGT. Ref. [21] assessed the thermal comfort in 15 regions with the help of WBGT by evaluating the past and future threshold exceedance rates concerning moderate (28 °C), high (32 °C) and extreme (35 °C) temperatures. They are using the WBGT for the 2020s and 2050s with A1B scenarios and in the HadCM3 model, and observed that heat events might become aggravated in regions of tropical humidity and mid-latitude even though the temperature there would be less than the global average, but the absolute humidity is on the rise. The authors of [22] projected the future heat waves in India using the WBGT index using the CMIP5 scenarios data. They used the three representative concentration pathways (RCPs) RCP2.6, RCP4.5, and RCP8.5 for the historical and future period and projected the severe heatwave in the future period. The study aims to calculate the heat stress in the study area and its effects on human health in the past and future scenarios. The state-of-the-art of the research in the study is presented in Section 2. Section 3 data and methodology describes the SWBGT, humidex, and NDVI procedures and defines the simulation flow. Section 4: Results and Discussion presents the simulation's results as well as a discussion on them. The study's findings and the most important outcomes are summarized in Section 5 Conclusions.

2. Study Area Description

Jaipur is Rajasthan state's capital, India, also called the Pink city, for its characteristics of the buildings' colour. Jaipur has a population of around 3.15 million people (Census of India, 2011). The city is mostly flat and is flanked on three sides by the Aravalli hill ranges: north, northeast, and east. The rest of the city is made up of a combination of barren ground, low to medium height vegetation, and built-up areas like as highways, buildings, and industries [23]. According to the Köppen climate classification, the Jaipur come under the hot semi-arid climate. It is located at an elevation of 431 m above mean sea level and at 26.92° N latitude and 75.82° E longitude. Jaipur covers approximately 1464 km^2 (JDA) area and this study cover the 472 km^2 area (Figure 1). Jaipur city has mostly as-associated a flat plain and hills encircle it in the northern, northeast, and east directions. The area around Jaipur city experiences three seasons each year: winter from November to February (cold nights with average air temperatures as low as 3 °C), summer from March to June (very hot during the day with maximum air temperatures as high as 48 °C), and monsoon from July to October (with extensive variations in daily average air temperature due to atmospheric conditions) [22]. The rainfall mainly occurs in the July and August months due to the monsoon. According to Chandra et al. 2018, the percentage change of the urban area of the Jaipur city was 13.54 (1993) to 57.32 (2015) and open land has been decreased by 45.84 (1993) to 19.4 (2015) [24]. They also explained the urban city expansion in the north, west, and south direction.

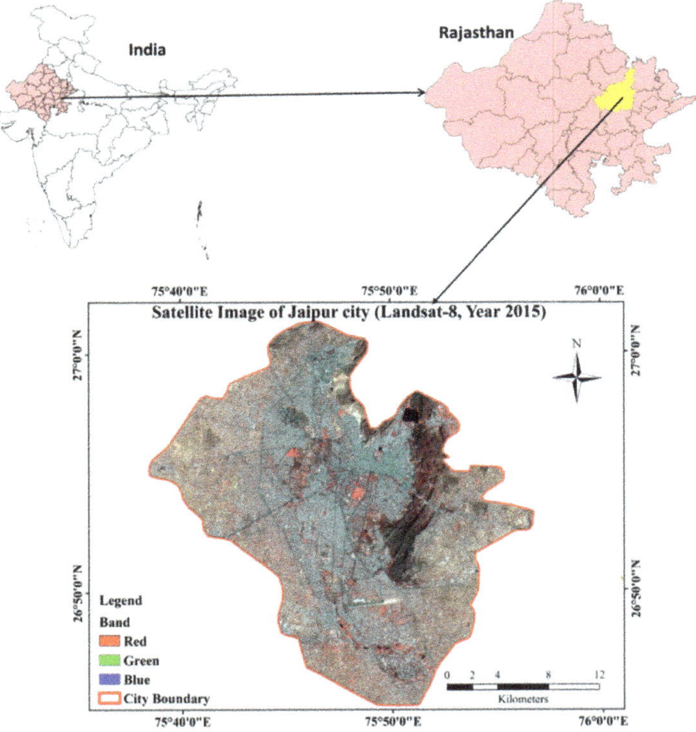

Figure 1. Study area map (Jaipur city).

3. Materials and Methods

The heat stress indicators were calculated using WorldClim's historical and future datasets. The WorldClim portal (http://worldclim.org (accessed on 2 May 2016) provides free access to WorldClim datasets for many climate indicators. Long-term average monthly climate data of maximum temperature and vapour pressure were acquired from the World-Clim data portal for the historical period (1970–2000) and future period 2050s (2041–2060) RCP4.5 and RCP8.5 scenarios. Table 1 lists all of the GCMs that were employed in the heat stress analysis. For the past and future eras, this study calculates two heat stress indicators for Jaipur. Monthly ensemble 17 GCMs are used to forecast the research area's future heat stress indices for the future timeframe.

Table 1. Detailed information of the GCMs (CMIP5) data of RCP4.5 and RCP8.5.

GCMs Information	Data Information
ACCESS1-0(AC), BCC-CSM1-1(BC), CCSM4(CC), CNRM-CM5(CN), GFDL-CM3(GF), GISS-E2-R(GS), HadGEM2-AO(HD), HadGEM2-CC(HG), HadGEM2-ES(HE), INMCM4(IN), IPSL-CM5A-LR(IP), MIROC-ESM-CHEM(MI), MIROC-ESM(MR), MIROC5(MC), MPI-ESM-LR(MP), MRI-CGCM3(MG), NorESM1-M(NO)	Monthly average maximum temperature (°C*10) GHG Scenarios: RCP4.5; RCP8.5

This analysis was conducted by combining temperature and vapour pressure through two heat stress indices, namely Simplified Wet Bulb Globe Temperature (SWBGT) and humidex. Many researchers used the SWBGT indicators to estimate the general heat stress index at various spatial and temporal scales [25,26].

The Australian Bureau of Meteorology [21] suggested the SWBGT indicator for spatial analysis. Equation (1) is used to calculate the *SWBGT* of Jaipur city.

$$SWBGT = 0.567Ta + 0.393e + 3.94 \tag{1}$$

where, *Ta* and *e* represent the air temperature (°C) and water vapour pressure (hPa) near the surface.

The humidex index was developed in Canada to estimate the humidity and consequence of high temperature on human health. The humidex indicator is assessed by using Equation (2) [27]:

$$Humidex = Ta + \left(\frac{5}{9}\right)(e - 10) \tag{2}$$

where, *Ta* is air temperature (°C) and *e* is the water vapour pressure (hPa) near the surface.

After an assessment of these indices, different categories are allocated based on these values. Each group represents a particular kind of condition and is linked with the heat stress situation for their effect on human health. Table 2 provides the classes of heat stress along with their consequence on human health.

Table 2. Categories of the heat stress, WBGT and humidex index with human effects.

Heat Stress Category	WBGT Index	Humidex Index	Inferences
Extreme danger	Greater and equal to 40	Greater and equal to 46	Dangerous and the risk of heat stroke
Danger	34–39	38–45	Very uncomfortable and avoid physical exertion
Extreme caution	28–33	30–37	Little uncomfortable
Caution	22–27	20–29	Comfortable

Source: http://www.crh.noaa.gov, http://www.ec.gc.ca/meteo-weather/ (accessed on 6 August 2016).

3.1. Image Classification and Accuracy Assessment

The study area is divided into five key groups using a supervised technique with the maximum likelihood classification method: water body, vegetation, urban settlement, open land, and hilly area/rocky area. The Kappa technique was used to examine the categorization accuracy [28,29].

Kappa coefficient (k) for the image classification is as follows:

$$k = \frac{N \sum_{i=1}^{r} xii - \sum_{i=1}^{r} xi + *xi + 1}{N^2 - \sum_{i=1}^{r} xi + *xi + 1} \tag{3}$$

$$k = \frac{(Total\ sum\ of\ correct) - Sum\ of\ the\ all\ the\ (row\ and\ column\ total)}{Total\ squared - Sum\ of\ the\ all\ the\ (row\ and\ column\ total)} \tag{4}$$

The Kappa coefficient should never be greater than or equal to one. The high Kappa value indicates accurate land use class information. According to [30] Monserud and Leemans (1992), Kappa coefficients ranging from 0.55 to 0.7 indicate good agreement, 0.7 to 0.85 indicate very good agreement, and values more than 0.85 indicate excellent agreement between image and ground.

3.2. Normalized Difference Vegetation Index (NDVI)

Vegetation cover plays a vital role in diminishing the conservation issues in urban areas. As indicated by Batista et al. 1997, the NDVI esteems went from -1 for the non-vegetated area to $+1$ for vegetation [31]. For the NDVI estimation red band and visible range band and the NIR band are utilized. The NDVI calculation is as follows:

$$NDVI = \frac{(Band\ 4 - Band\ 3)}{(Band\ 4 + Band\ 3)} \tag{5}$$

3.3. Soil-Adjusted Vegetation Index Calculate (SAVI)

The *SAVI* index also plays a role in the vegetation cover, but it adds the area's background soil conditions. *SAVI* calculation is as follow:

$$SAVI = (1 + L) * (band4 - band3) / (band4 + band3 + L) \tag{6}$$

where the TOA reflectance is used for each band and L is a soil brightness correction factor. From Huete (1988), $L = 0.5$ is used in most conditions. Figure 2 shows the methodology and the climatic data used in the study.

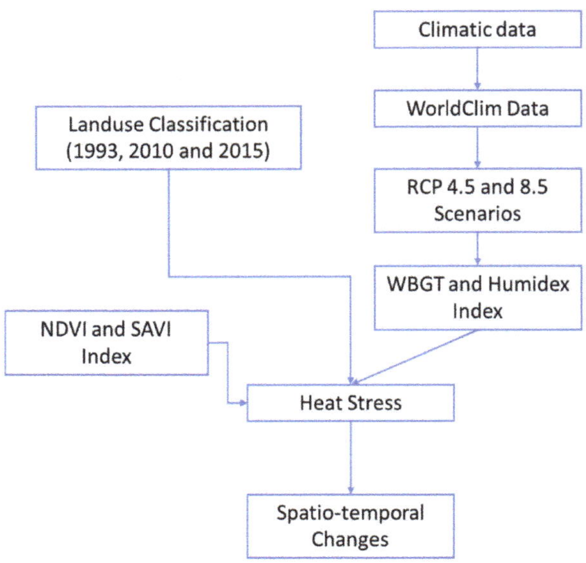

Figure 2. Flowchart of methodology and data used.

4. Results and Discussions

Heat stress is on the rise in various countries of the world, including India, and is to blame for the rising level of human misery. Heat stress is becoming more severe in cities as a result of urbanization and greenhouse gas emissions.

4.1. Land Used Classification

Water body, vegetation, urban settlement, open land, and hilly terrain/rocky area are the five primary land use types evaluated in this study. Land use classifications are carried out for 3 years: 1993, 2010, and 2015. The accuracy of the classified map was determined by a random selection of 330 points for each year. The overall accuracy of the classified maps was found to be 0.92, 0.97, and 0.95 for selected years. According to Table 3, the Kappa coefficients for the indicated years are 0.88, 0.95, and 0.93. In comparison to ground reality, the classified land use accuracy is shown to be good.

Table 3. Accuracy assessment of the land cover types.

Users Accuracy %							
Year	Water	Vegetation	Urban Settlement	Open Land	Hilly/Rocky Area	Overall Accuracy	Kappa Coefficient
1993	100.0	95.4	96.7	97.9	69.8	0.92	0.88
2010	100.0	94.7	100.0	96.6	91.2	0.97	0.95
2015	100.0	95.7	97.2	92.7	88.6	0.95	0.93
Producer Accuracy %							
Year	Water	Vegetation	Urban Settlement	Open Land	Hilly/Rocky Area		
1993	100.0	98.41	87.88	90.73	91.67		
2010	100.0	97.83	98.21	95.45	100.00		
2015	100.0	94.74	99.28	86.44	93.94		

Table 4 shows the total area covered by various categories and their percent coverage. It has been observed that the urban settlement of Jaipur city has grown over time. It was 63.9 km^2 in 1993, but by 2015, it expanded to 270.47 km^2. This indicates that during the course of 22 years, the area has changed nearly four times. In 2015, over 43.78% change of the studied area was under settlement, compared to the entire area. These trends suggest that the city is rapidly expanding, and it accelerated significantly after 2010.

Table 4. Land use area and percent change of different years.

Class Name	Area 1993	Area 2010	Area 2015	% Change (2010–1993)	% Change (2015–2010)	% Change (2015–1993)
Water.	0.4	0.9	0.8	0.10	−0.01	0.09
Vegetation	84.4	88.6	45.7	0.87	−9.09	−8.21
Urban Settlement	63.9	166.5	270.5	21.75	22.03	43.78
Open Land	216.3	159.8	91.5	−11.96	−14.47	−26.44
Hilly/Rocky Area	106.9	56.1	63.4	−10.76	1.55	−9.21

Figure 3 depicts the spatial distribution and patterns of land cover change during the three years. The image clearly shows the evolution of urban settlement in Jaipur city. In comparison to the 1993 map, there is a significant rise in of urban area in the Jaipur and found the maximum land use was converted into the urban settlement.

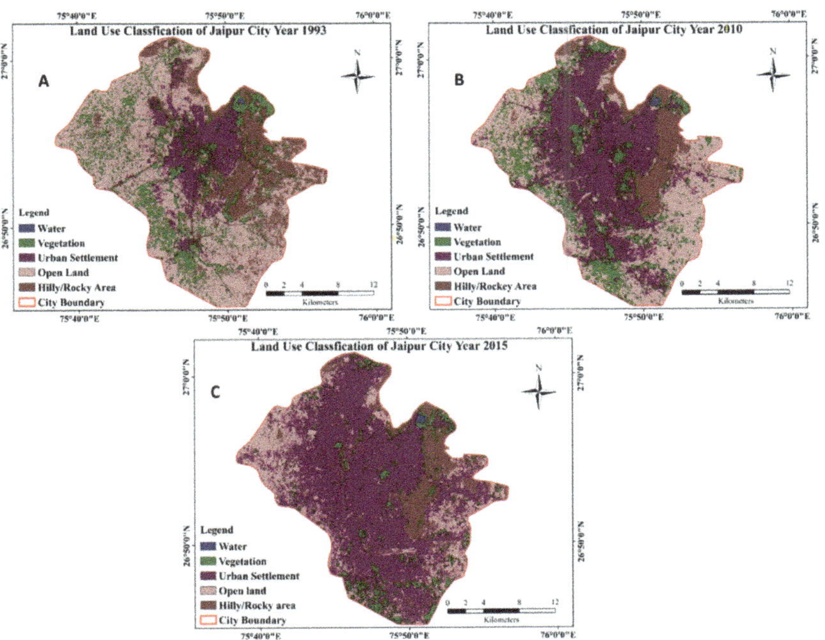

Figure 3. LULC maps for different years (**A**) 1993 (**B**) 2010 (**C**) 2015.

4.2. Humidex Index

For the historical and future RCP4.5 and RCP8.5 scenarios, the humidex index was calculated on a monthly and seasonal basis. All of the monthly and seasonal data were shown in Table 5. The lowest humidex was recorded in the month of January. The historical minimum humidex value has been 21.4, and in the RCP4.5 and RCP8.5 scenarios, it will rise to 25.5 and 27.3, respectively. The highest humidex values are observed to be 39.5, 43.2, and 46.4 for the historical and two future RCPs in the May month. In Table 5, the May and June months show the danger conditions in the humidex index for all three cases, but the RCP4.5 and RCP8.5 show the danger and extreme danger conditions in most of the months.

Table 5. Average monthly variation in humidex for historical and future periods.

	Historical	RCP4.5	RCP8.5	Historical	RCP4.5	RCP8.5
January	21.4	25.5	27.3	C	C	C
February	24.2	28.2	30.4	C	C	EC
March	29.9	34.1	36.7	EC	EC	EC
April	35.8	39.9	42.7	EC	D	D
May	**39.5**	**43.2**	**46.4**	D	D	ED
June	**39.5**	41.5	44.3	D	D	D
July	39.1	36.0	38.2	D	EC	D
August	37.7	33.4	35.4	D	EC	EC
September	37.7	35.0	37.3	D	EC	D
October	35.3	35.6	38.3	EC	EC	D
November	30.2	31.5	33.7	EC	EC	EC
December	25.5	26.9	25.5	C	C	EC
Winter	23.7	26.9	27.7	C	C	EC
Monsoon	**38.5**	36.5	38.8	D	EC	D
Summer	35.1	**39.1**	**41.9**	EC	D	D
Autumn	32.8	33.6	36.0	EC	EC	EC

C—caution; EC—extreme caution; D—danger; ED—extreme danger.

Figure 4 depicts the spatial distribution of Humidex for all of the months in the past. The months of May and June are classified as Danger and Extreme Danger. From January to May, the Humidex values rise, then begin to decrease until the month of December. There is a slight rise in value in September and October months compared to the decreasing trend. In majority of the months over the historical period, the humidex is high in the southeast and west.

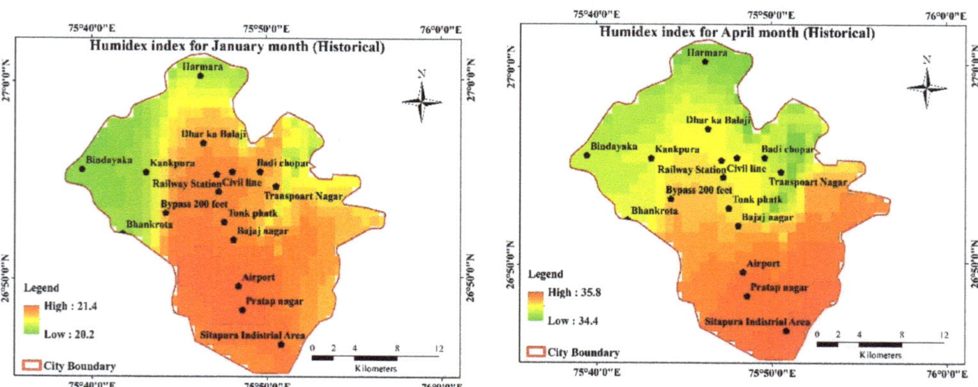

Figure 4. *Cont.*

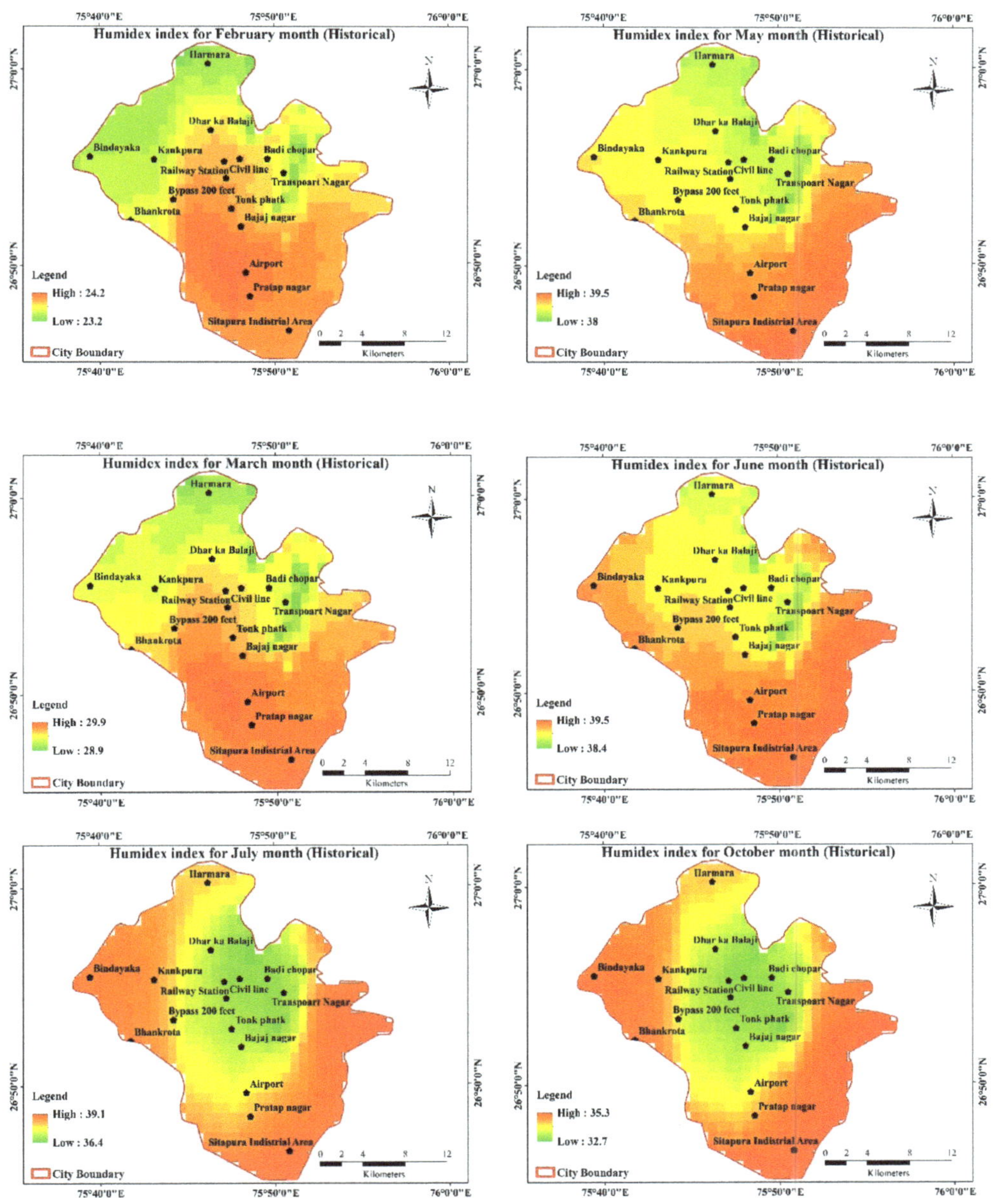

Figure 4. *Cont.*

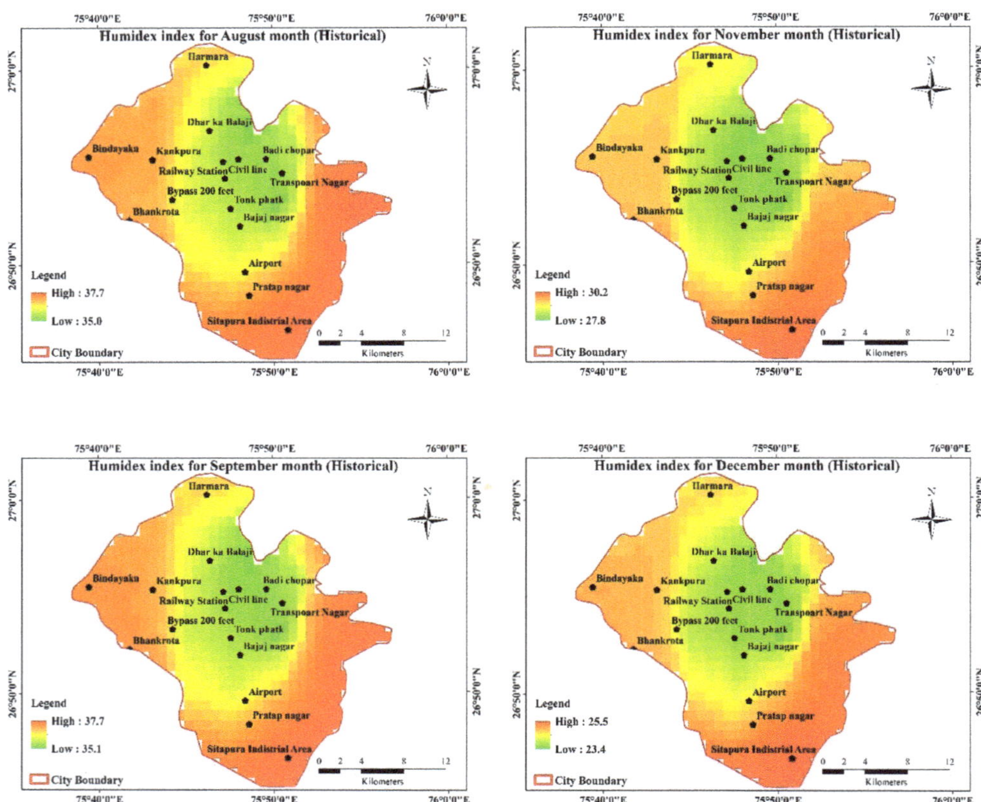

Figure 4. Spatial variation of humidex for historical in January to December months.

The spatial maps of humidex variations for January to December for the future RCP4.5 and RCP8.5 scenarios are shown in Figures 5 and 6. The majority of the month in these statistics depicts danger and extreme danger conditions in hypothetical futures. The months of May and June exhibit a danger situation, and the majority of the months fall into the danger and extreme dangerous categories. In the figure, the area with low values is represented by the colour green, while the area with high values is represented by the colour red. The humidex is elevated in the east and south as well as in a small portion of the west side between RCP4.5 and RCP8.5.

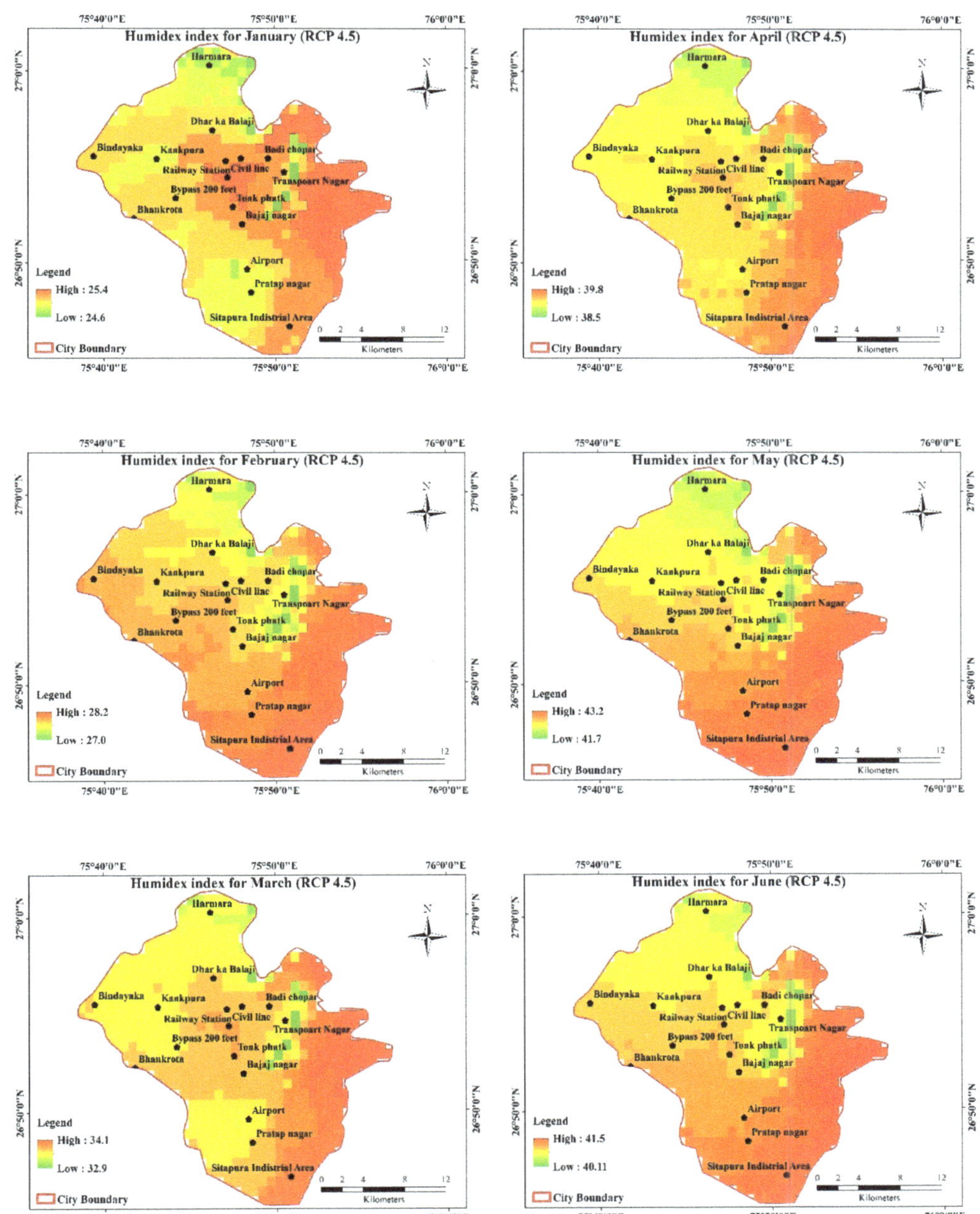

Figure 5. *Cont.*

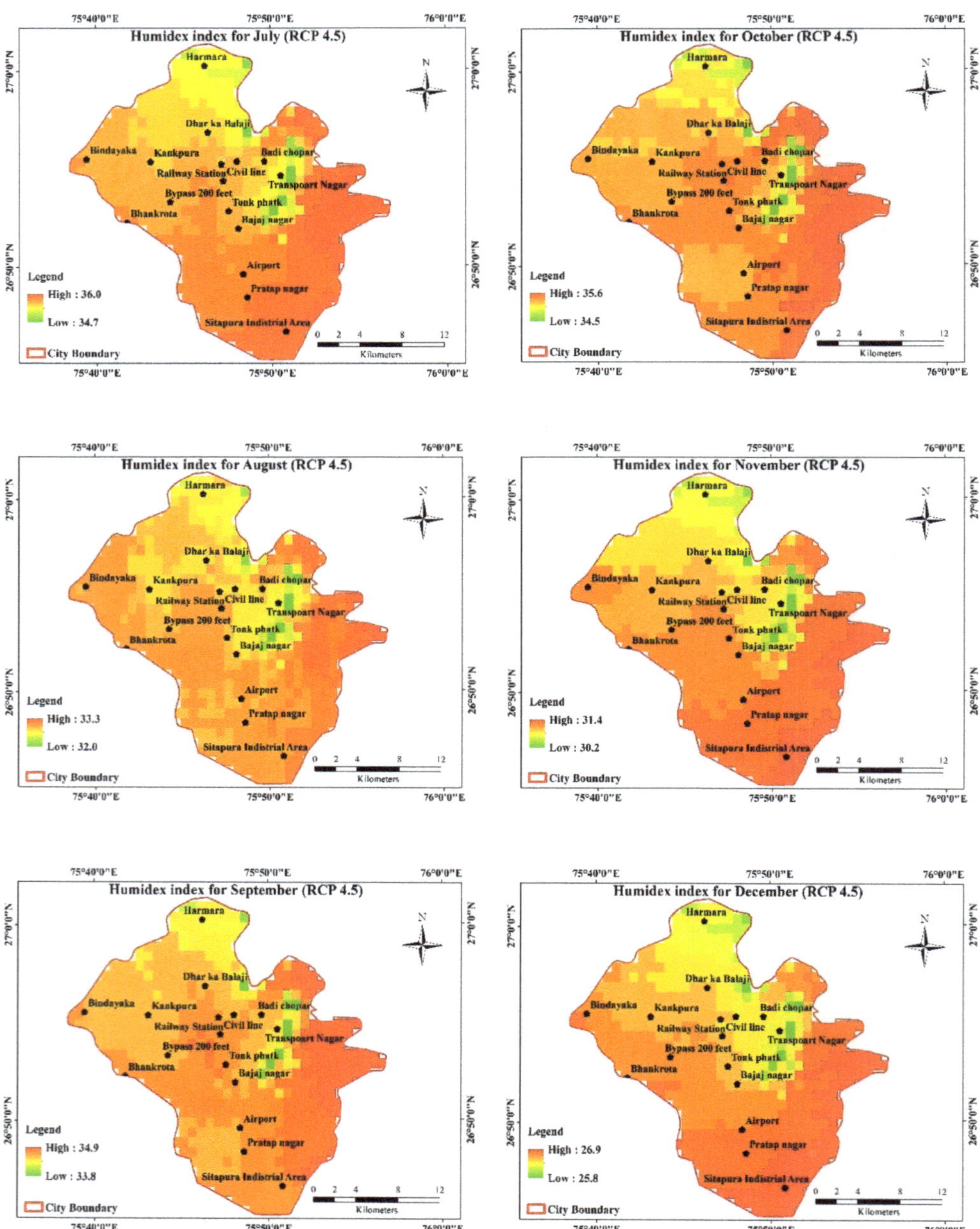

Figure 5. Spatial variation of the humidex for future (RCP4.5) in January to December months.

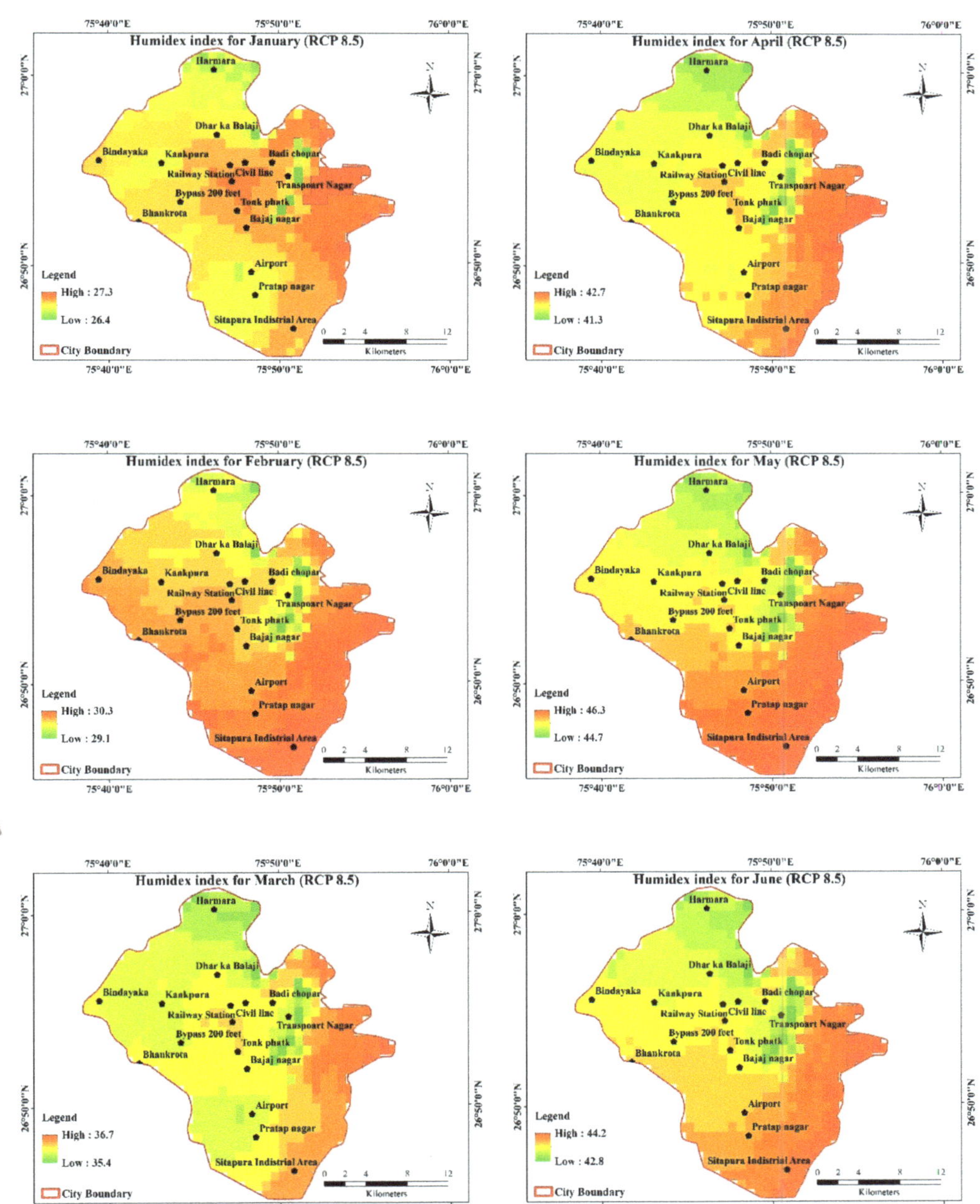

Figure 6. *Cont.*

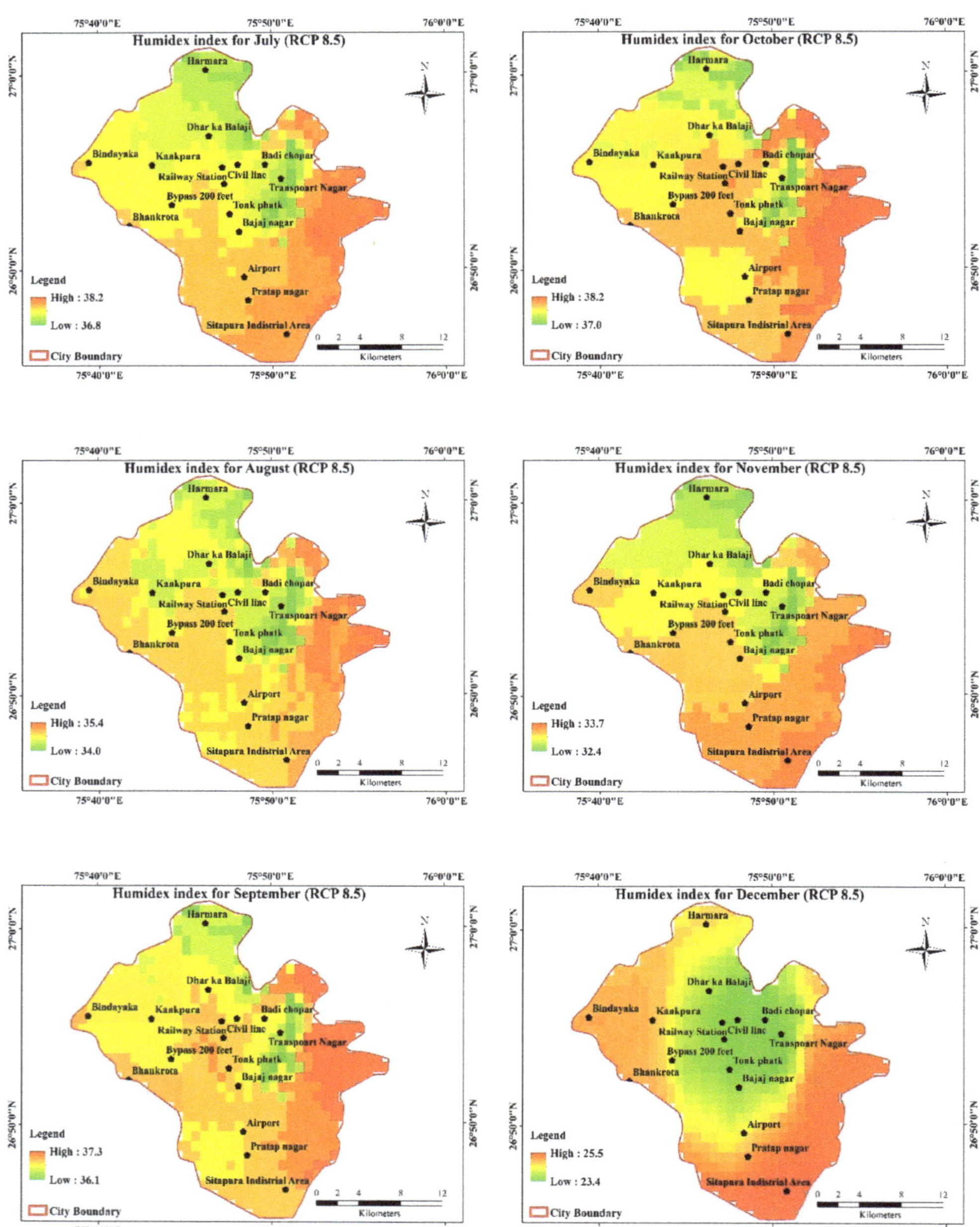

Figure 6. Spatial variation of the humidex for future (RCP8.5) in January to December months.

The monthly difference in humidex for historical and projected RCP4.5 and RCP8.5 is shown in Figure 7. In the three months, July to September, as well as throughout the monsoon season, humidex displays a drop. Because RCP8.5 represented the high emission scenario, there is always a significant disparity between RCP8.5 and RCP4.5. It has been demonstrated that the seasonal analysis helps to explain how the severe category shifts.

By the year 2050, the summer season displays a shift from extreme caution to danger and a rise in temperature in the urban region. Figure 8 displays the seasonal humidex variations for the past and future of the city border. The monsoon and autumn seasons show the maximum humidex value in all scenarios and cover the city's east, west, and north direction.

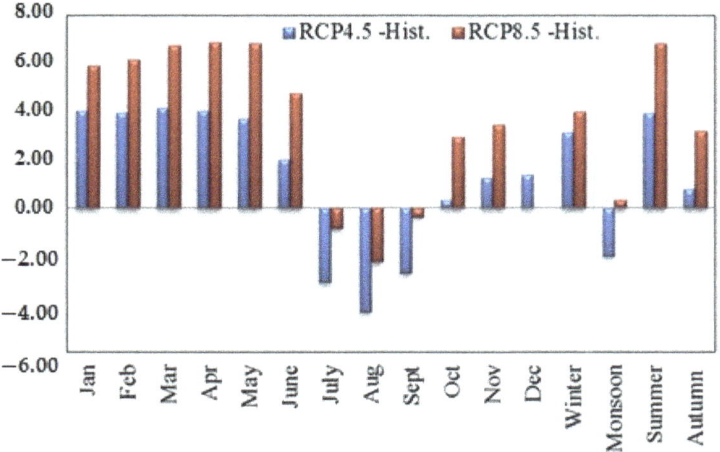

Figure 7. Difference in humidex values for future scenarios (2050s).

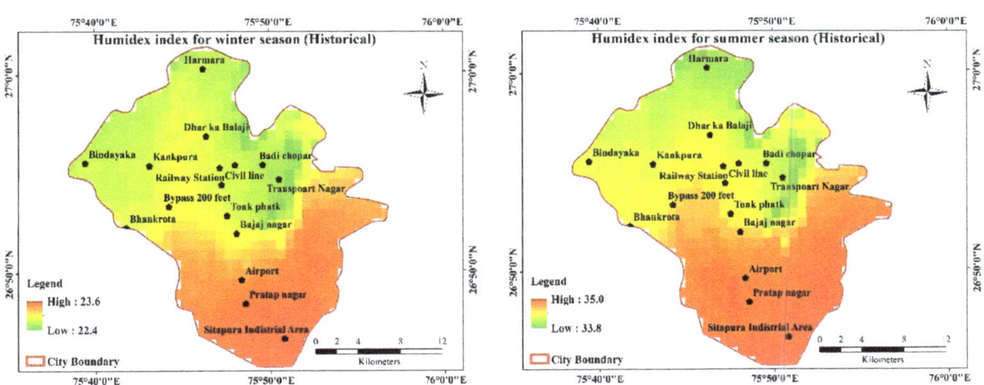

Figure 8. *Cont.*

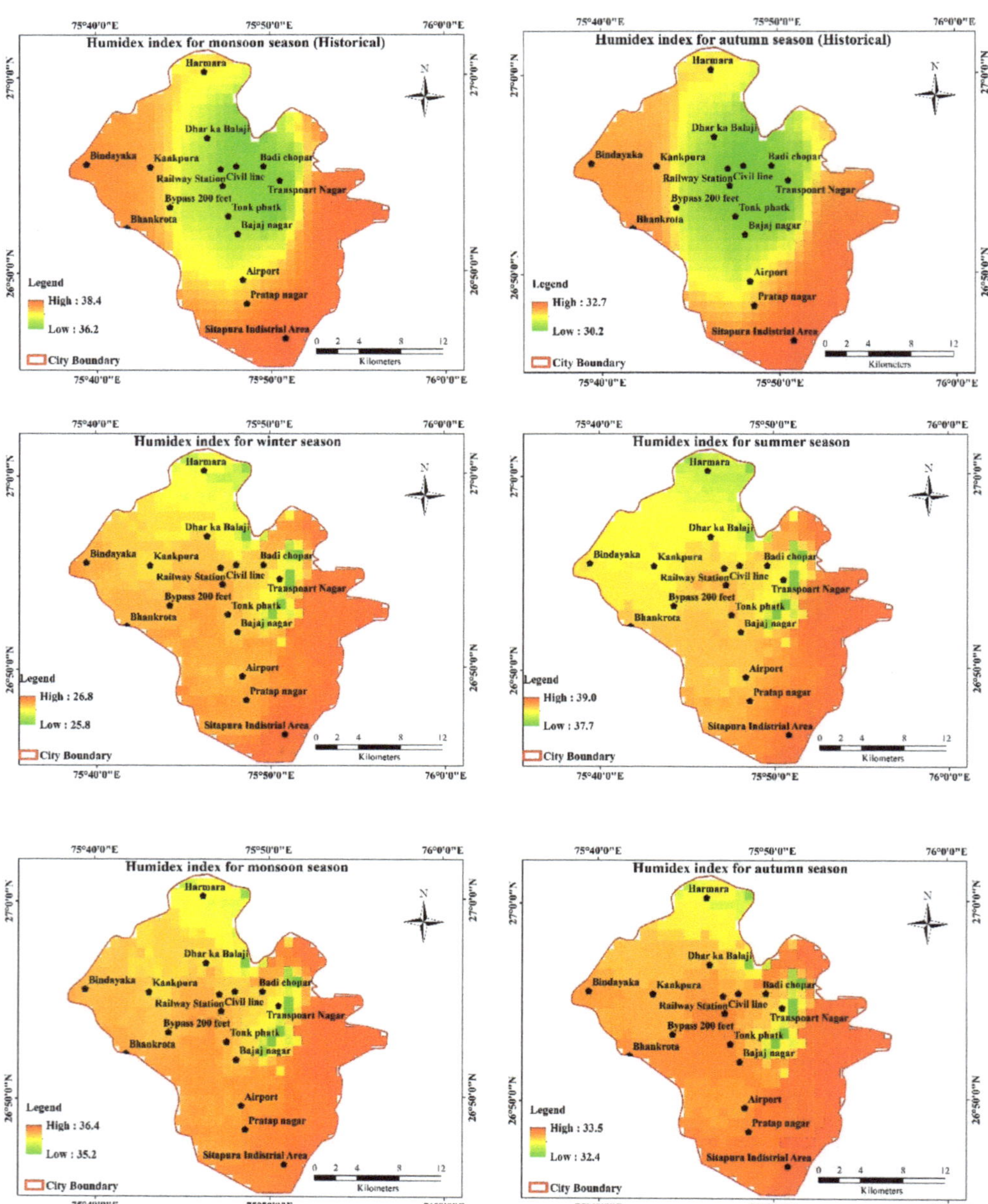

Figure 8. *Cont.*

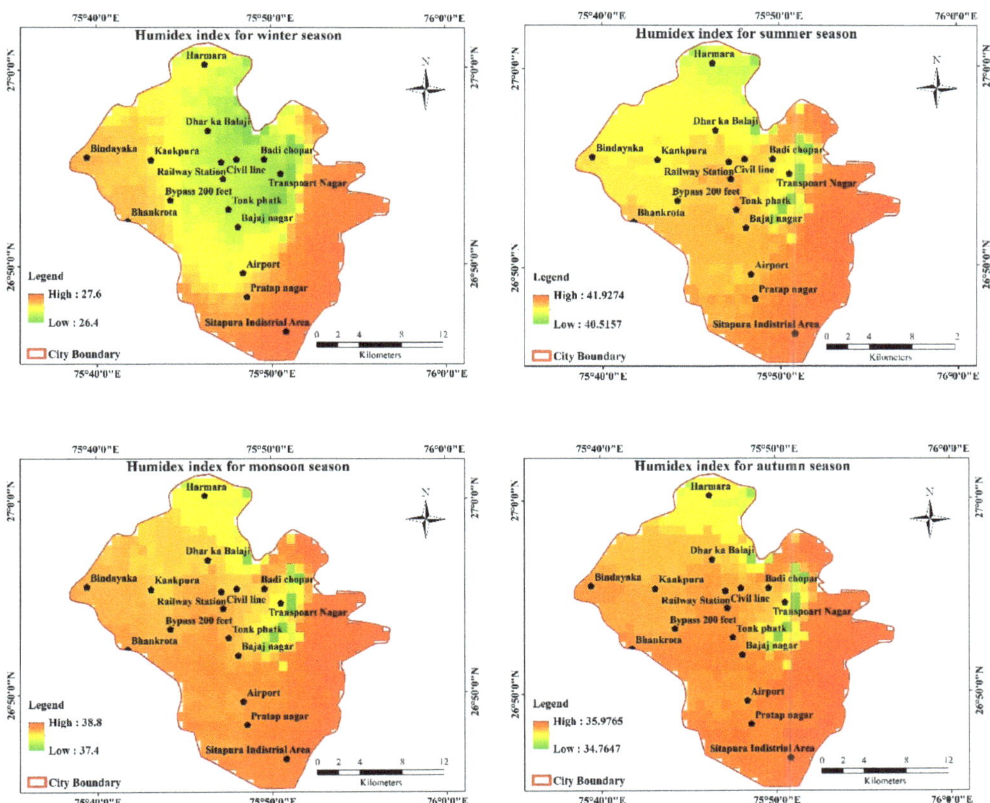

Figure 8. Seasonal map of the humidex index of Jaipur city (Historical, RCP4.5, RCP8.5).

4.3. WBGT Index

The WBGT index is computed in this study on a monthly and seasonal basis for both past and future periods. The average monthly seasonal fluctuations in the WBGT indicator, together with its stress category, are shown in Table 6. In all three scenarios—historical, RCP4.5, and RCP8.5—the danger categories are visible from June through September. In January, the value is at its lowest, and in June and July, it is at its highest. However, the WBGT, high in the monsoon season of RCP scenarios and the danger situation of the heat of the city of Jaipur, are shown in the season-wise calculation. The correlation coefficient of ESI and environmental parameters of wet temperature, dry temperature, solar radiation, and relative humidity was obtained as 0.88, 0.96, 0.4, and −0.7, respectively, in a study by [32] Hajizadeh et al. (2016), which aimed to investigate the correlation between the environmental stress index (ESI) and WBGT index in a hot and dry climate.

As with the humidex, the lowest values of WBGT are observed for January month in the historical and future periods. The historical value of WBGT is 19.7, which will increase to 22.0 for RCP4.5 and 23.1 for RCP8.5 (Table 6). In the historical period, the high value of WBGT is obtained in the month of July, but it shifts to June for the future period. It is also observed that WBGT values are projected to decrease in the monsoon season (July to September) with the heat stress category of danger. The monthly pattern of values is similar for humidex, increasing from January to June/July and then further decreasing until December. Some cases of a shift from the existing caution condition to extreme caution condition in March, November, and December. The spatial variance of

WBGT in Jaipur city for the past and the future is explained in Figures 9–11. In Figure 9, the southern half of the city showed the greatest changes when compared to other places. Figures 10 and 11, which depict possible futures and determine the city's danger condition, show the same pattern. These show the monthly variation of WBGT values for a future period (both scenarios) compared with historical data. The indicator's value decreases throughout a three-month period from July to September, indicating a decline in indicators during the monsoon season. These areas came under the industrial zones and cover half of the city area. The green colour represents the area with low values, whereas the red colour represents high values.

Table 6. Average monthly variation in WBGT for historical and future periods.

	Historical	RCP4.5	RCP8.5	Historical	RCP4.5	RCP8.5
January	19.7	22.0	23.1	C	C	C
February	21.2	23.5	24.8	C	C	C
March	24.7	27.1	28.6	C	EC	EC
April	28.2	30.5	32.1	EC	EC	D
May	30.3	32.5	34.2	EC	EC	D
June	34.6	**35.8**	**37.3**	**D**	D	D
July	**37.3**	35.5	36.8	**D**	D	D
August	36.7	33.4	35.4	D	D	D
September	34.9	33.4	34.7	D	D	D
October	30.0	30.2	31.7	EC	EC	EC
November	25.5	26.2	27.5	C	C	EC
December	22.2	23.0	22.1	C	C	EC
Winter	21.0	22.8	23.3	C	C	C
Monsoon	**35.9**	**34.5**	**36.1**	**D**	D	D
Summer	27.7	30.0	31.6	EC	EC	EC
Autumn	27.8	28.2	29.6	EC	EC	EC

C—caution; EC—extreme caution; D—danger; ED—extreme danger.

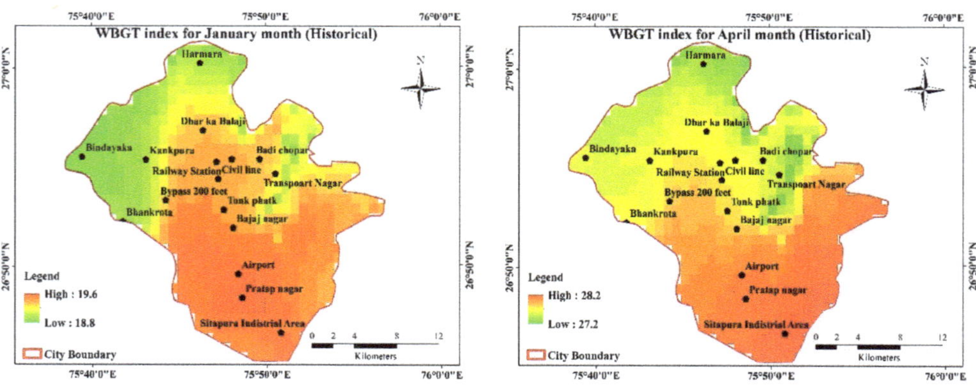

Figure 9. *Cont.*

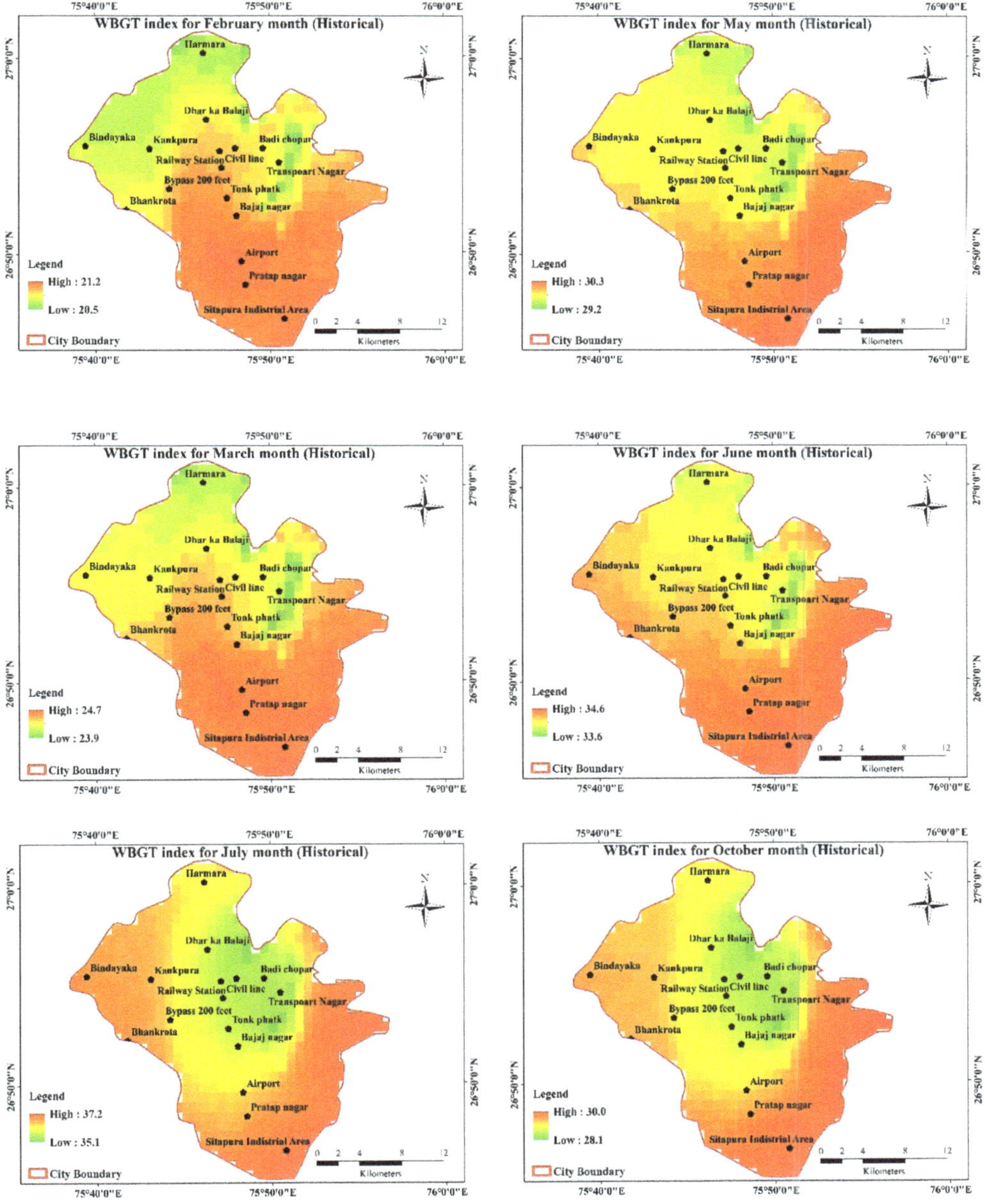

Figure 9. *Cont.*

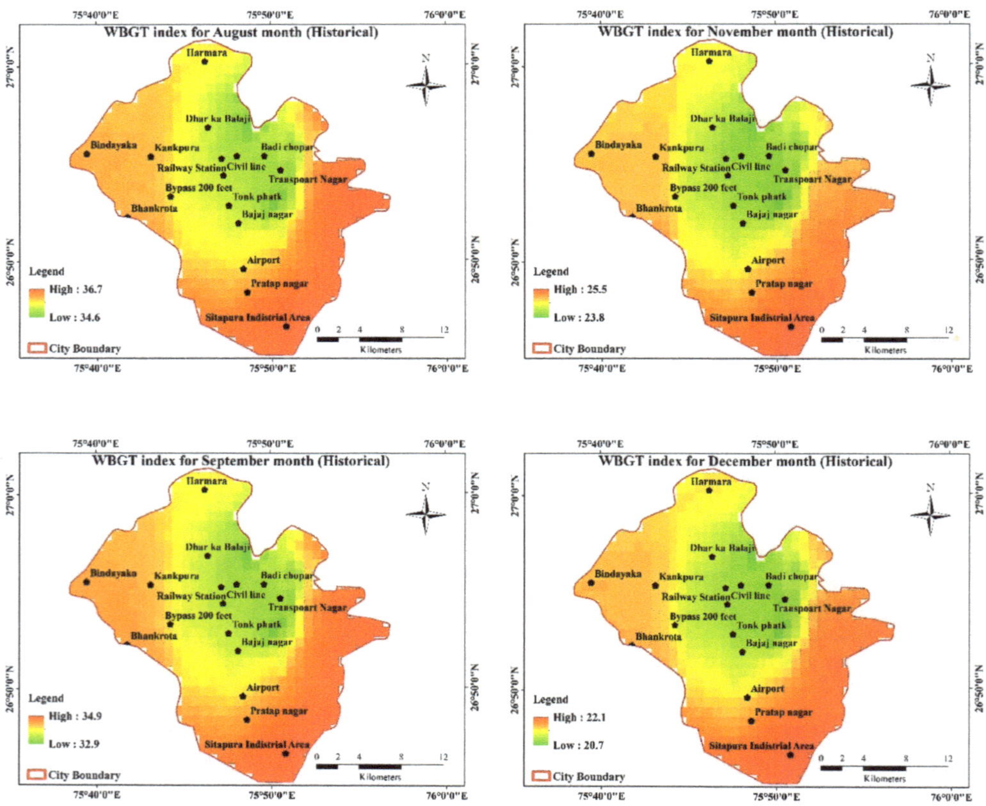

Figure 9. Spatial variation of WBGT for the historical period.

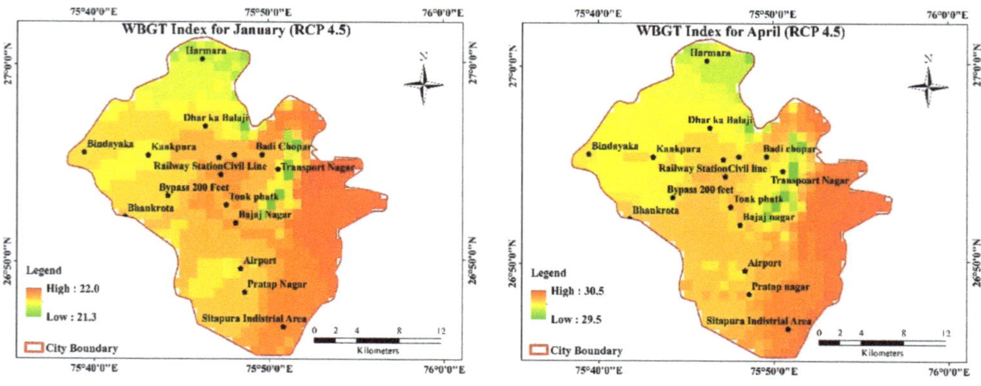

Figure 10. *Cont.*

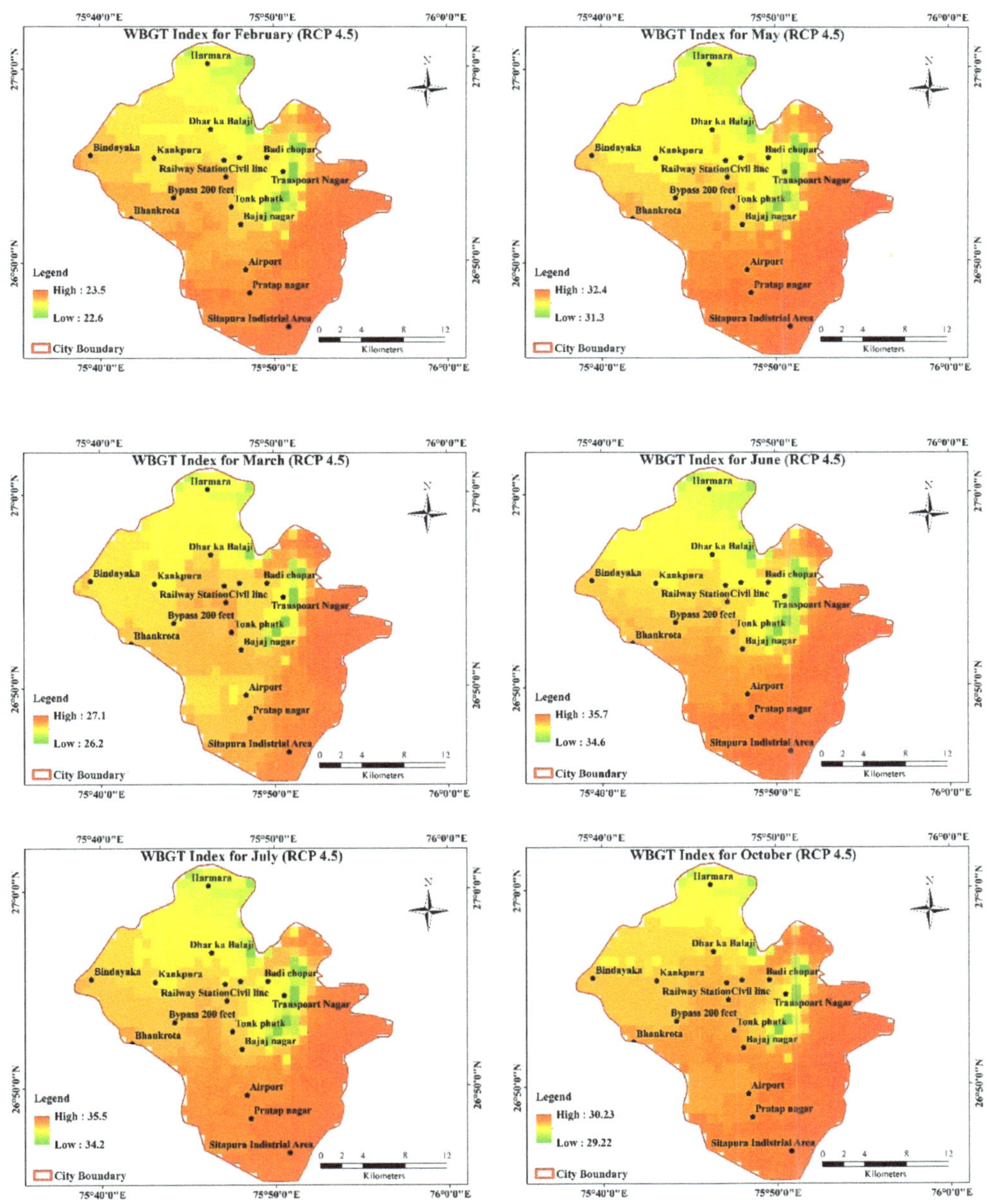

Figure 10. *Cont.*

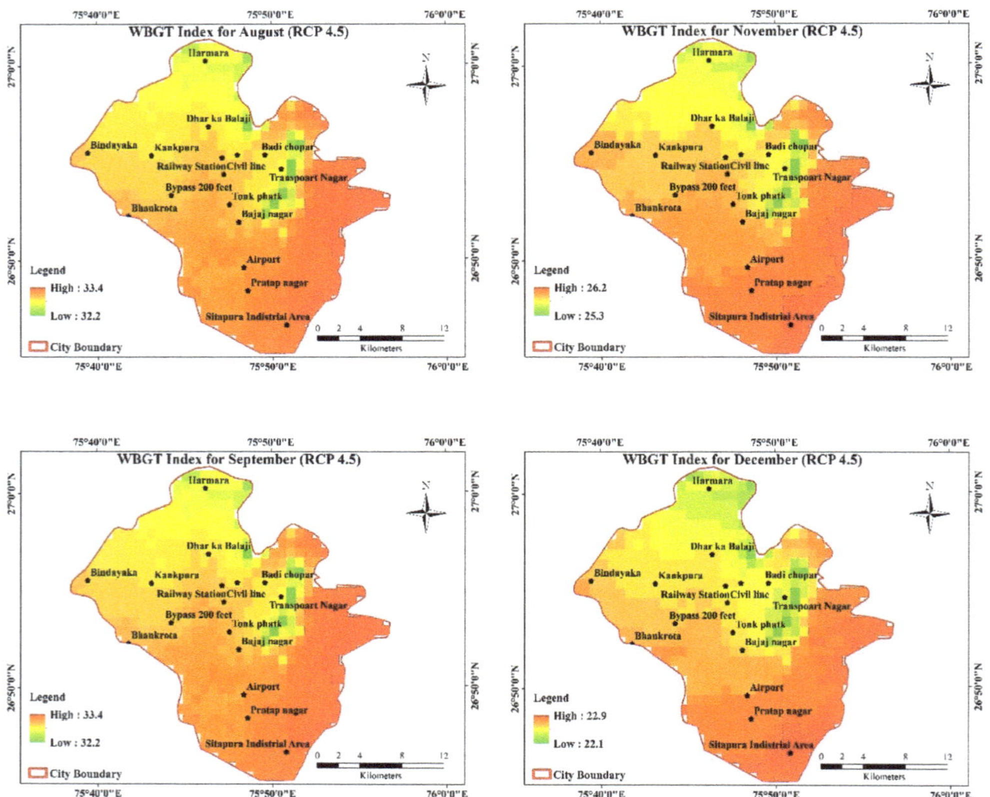

Figure 10. Spatial variation of WBGT for the future (RCP4.5) period.

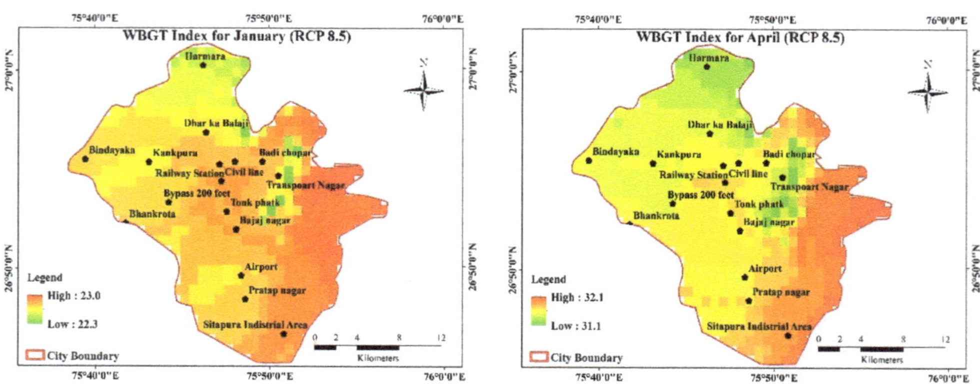

Figure 11. *Cont.*

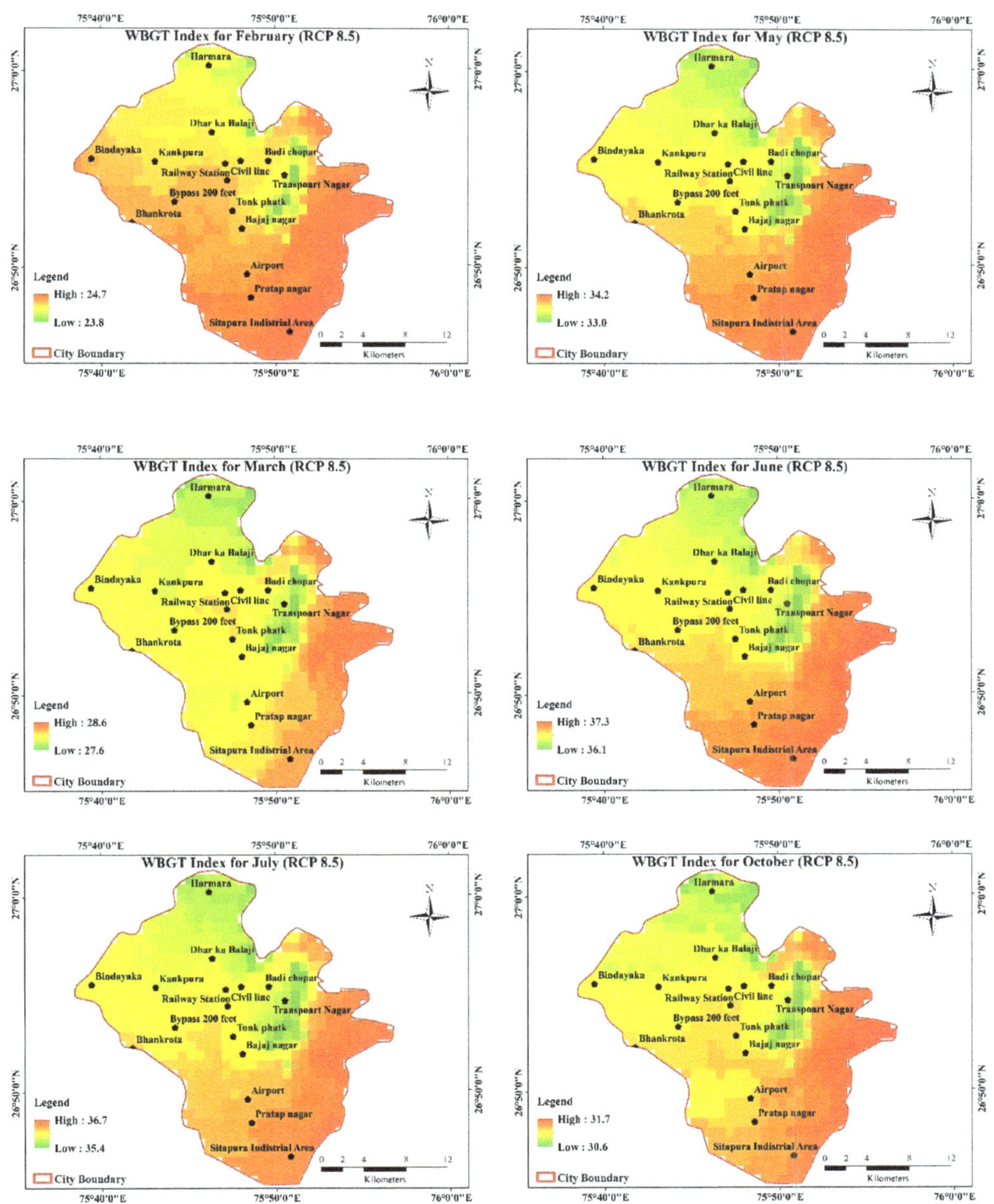

Figure 11. *Cont.*

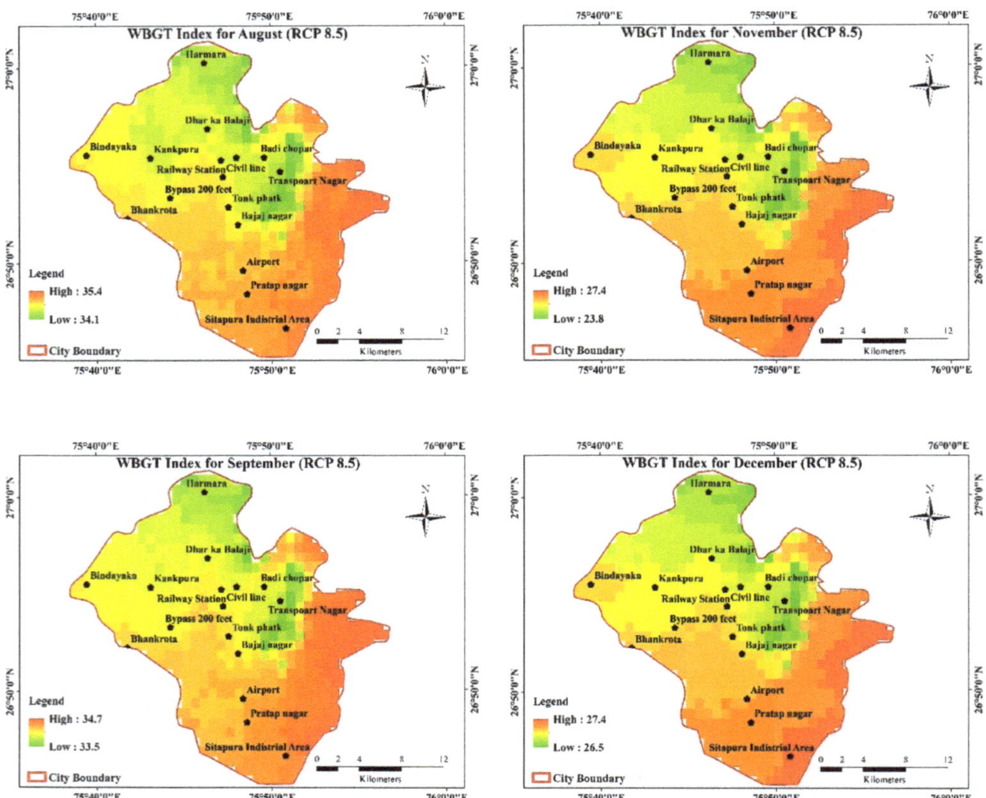

Figure 11. Spatial variation of WBGT for future (RCP8.5) period.

RCP8.5, which simulates a high emission scenario, consistently provides a large difference from RCP4.5. Figure 12 illustrates the disparity pattern, which is seen to be similar to the humidex indication. The study is being carried out to better understand how heat stress conditions vary seasonally. All four seasons' heat stress categories show little variation; however, the monsoon season shows a rise in the danger category. In the monsoon season, WBGT is at its highest; in the winter season, WBGT is at its lowest. An increasing value is found in the future period when compared to the historical period. Figure 13 shows the variation in monsoon WBGT for historical and future periods and the difference in indicator value within the city boundary. The WBGT is high in the summer and autumn season in the southeast direction and these changes are created due to the changes in land use pattern and expansion of urban areas.

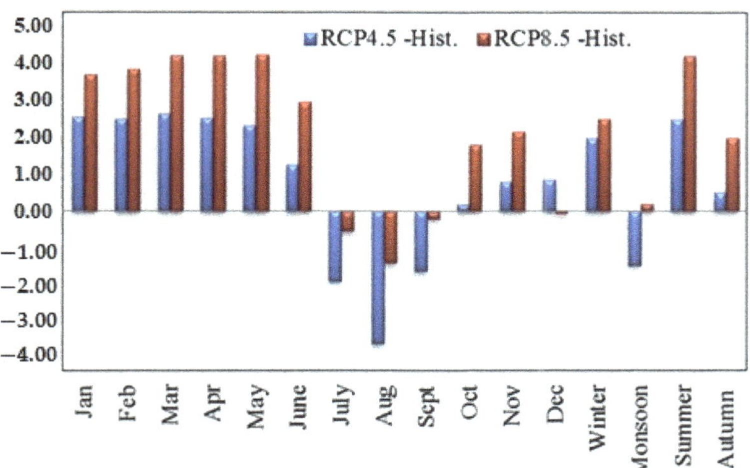

Figure 12. Difference in WBGT values for future scenarios (2050s).

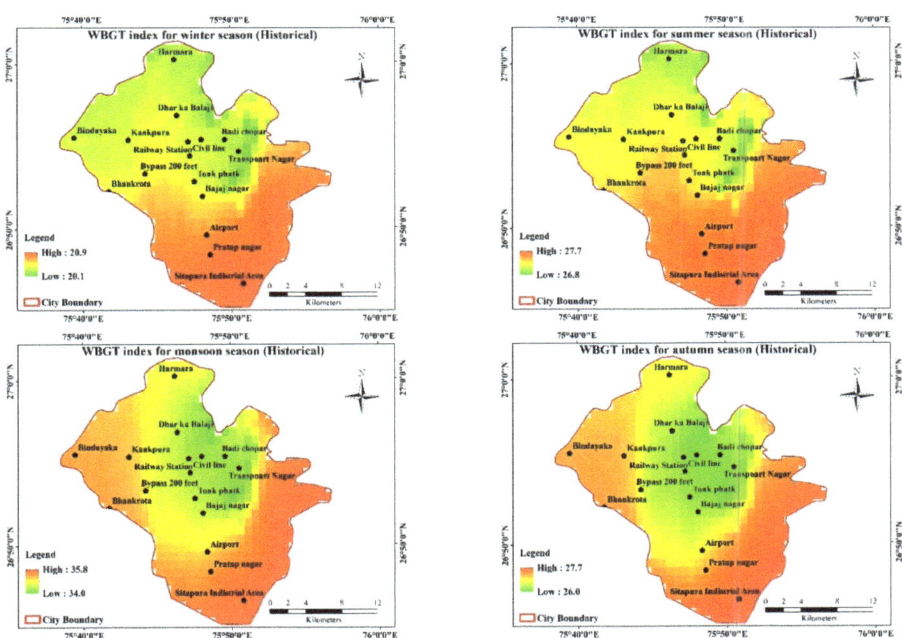

Figure 13. *Cont.*

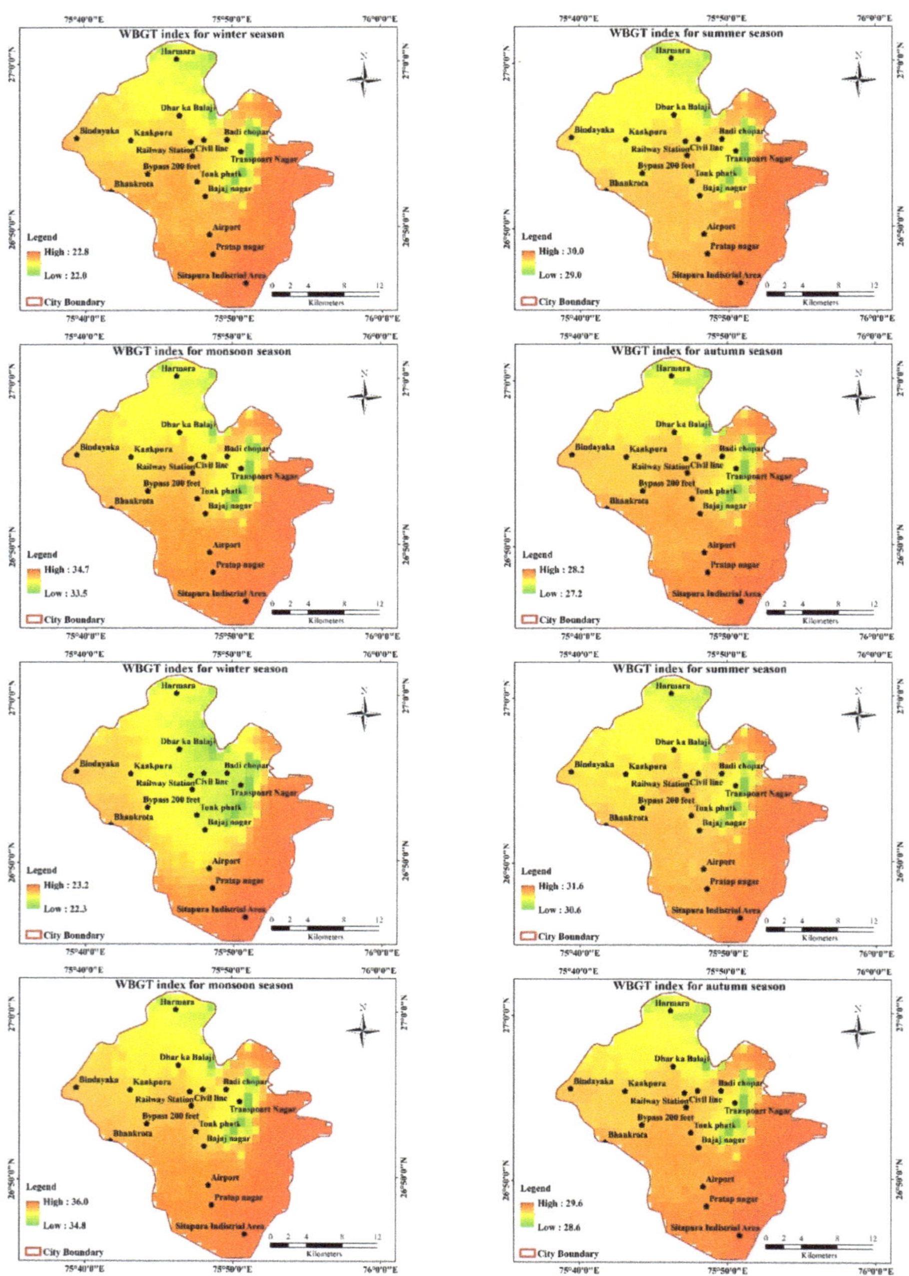

Figure 13. Seasonal map of the humidex index of the Jaipur city (Historical, RCP4.5, RCP8.5).

4.4. Normalized Difference Vegetation Index (NDVI)

In this study, NDVI was calculated for different periods: April 1993, April 2000, June 2010, and April 2015; NDVI values range from −1 to +1, different geographical features show the different NDVI values. These layers give different information through the bands and band 3 and 4 provides the vegetation with cover information of Jaipur city.

The extracted vegetation layer covers of NDVI were spatially compared with the colour composite image of Landsat-5 and Landsat-8 (TM and OLI) imagery. The range of NDVI in 1993 was −0.01 to 0.71, in 2000 was −0.019 to 0.63, and in 2010 was 0.04 to 0.56 of Landsat 5 TM imagery, and year 2015 shows the range of NDVI was −0.24 to 0.70 for the Landsat 8 OLI image of Jaipur city (Figure 14). The vegetation cover area utilizes solar radiation in the photosynthesis process and reduces the city's surrounding temperature.

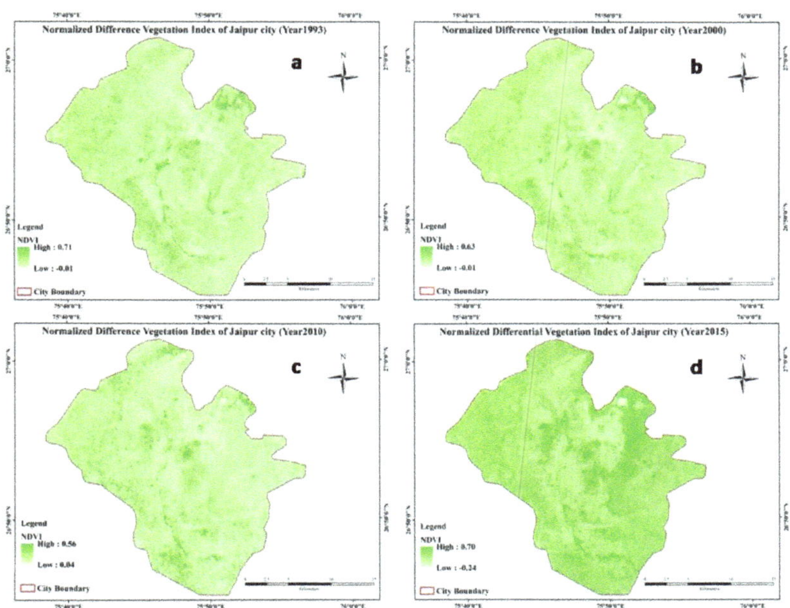

Figure 14. NDVI map of Jaipur city: (**a**) April 1993, (**b**) April 2000, (**c**) June 2010, and (**d**) April 2015.

4.5. Soil-Adjusted Vegetation Index (SAVI)

The study area is mainly classified into different types of land use. All these random samples are selected and the values of these cites are observed between NDVI and SAVI indices. The range of SAVI in 1993 was −0.005 to 0.50, in 2000 was −0.005 to 0.44, and in 2010 was −0.024 to 0.43 of Landsat 5 TM imagery, and 2015 shows the range of SAVI was 0.118 to 0.52 for the Landsat 8 OLI image of Jaipur city (Figure 15). This influence can be restricted using SAVI instead of NDVI. High NDVI and SAVI values were found in the buildup area.

On the other hand, there is a correlation between land use and NDVI data and data measured in meteorological stations in most research, including the current study, which is a significant reason for the efficiency of using this data for environmental issues. As a result, which can be derived indirectly using daily recorded metrological parameters in weather stations, it can be used to assess thermal conditions in Jaipur. Because evaluating environmental parameters for the calculation of heat stress indices is normally costly and time consuming, it is possible to alleviate this problem in environmental evaluations in open spaces by using daily recorded weather station data. Meteorological data has the advantage of being continuously recorded and providing a low-cost and comprehensive database for computing a variety of essential thermal indicators.

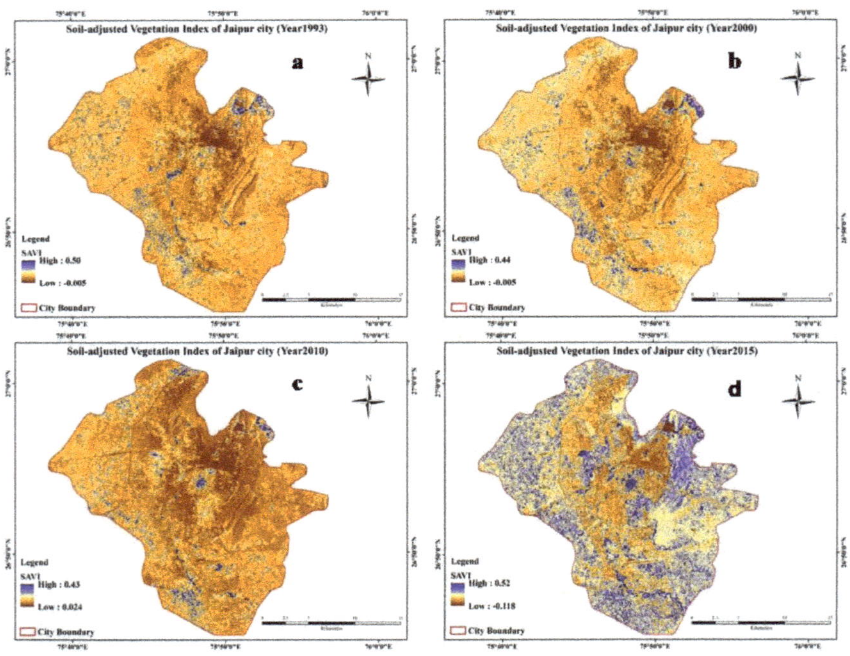

Figure 15. SAVI map of Jaipur city: (**a**) April 1993, (**b**) April 2000, (**c**) June 2010, and (**d**) April 2015.

5. Conclusions

The study's major goal was to predict the WBGT and humidex indexes for the past and future. This study demonstrates the expansion of urban land usage in Jaipur city from 1993 to 2015 using intermediate satellite images. The number of people living in cities has increased substantially in the last 23 years. The most prevalent type of land converted to urban areas is open terrain, followed by vegetation and hilly/rocky areas. The NDVI and SAVI indices are also used to determine the changes in land use patterns in the city and the amount of green space in the urban and peri-urban areas. The WBGT is highest during the monsoon season and lowest during the winter. When compared to the historical period, the future time shows an increase in value. Humidex's historical value has been 21.4, but it is projected to rise to 25.5 and 27.3 under the RCP4.5 and RCP8.5 scenarios. In May, the greatest humidex values were 39.5, 43.2, and 46.4 for the historical and two future RCP scenarios. The months of May and June are shown in the danger and extreme danger categories in the analysis. From January to May, the humidex values rise, then begin to fall until the month of December. It predicts that, with the exception of the monsoon season in the metropolis, discomfort levels will rise in the future. The findings indicate that the index's average value is increasing. Global warming or the absence of suitable conditions in these environments may be responsible for this trend. This will help in the identification of a better heat stress index for diverse situations and temperatures. There are some restrictions on the study; the distribution of indices studied throughout the different continents varies because the majority of studies undertaken in this field are focused on regions with hot climates. Application of WBGT and humidex indices has limited application in warmer climates as it shows a low level when the air temperature is in high range. The environmental heat index is preferred in occupational situations, but is not suitable at all work locations [33]. The possible reason for using these indices is comprehensiveness of the index for assessing the thermal stress conditions with limited data availability. However, it is a useful indices to understand the pattern of long-term change and warning purposes. The additional limitations of this analysis were the dearth of pertinent papers and the evidence provided in the articles. Appropriate protective strategies are required to prepare for the working population, which includes vulnerable persons whose occupational health and performance are harmed by heat stress.

Author Contributions: Conceptualization: S.C. and D.S.; methodology: S.C., D.S. and S.K.D.; formal analysis: S.C., D.S. and S.K.D.; investigation: S.C. and D.S.; writing—original draft: S.C. and S.K.D.; writing—review and editing: D.S., B.K.M. and R.D. All authors have read and agreed to the published version of the manuscript.

Funding: This is supported by Central University of Rajasthan and the Strategic Research Fund of the Institute for Global Environmental Strategies.

Institutional Review Board Statement: Not applicable.

Informed Consent Statement: Not applicable.

Data Availability Statement: Not applicable.

Conflicts of Interest: The authors declare no conflict of interest.

References

1. Field, C.B.; Barros, V.; Stocker, T.F.; Dahe, Q. *Managing the Risks of Extreme Events and Disasters to Advance Climate Change Adaptation: Special Report of the Intergovernmental Panel on Climate Change*; Cambridge University Press: Cambridge, UK, 2012.
2. Reiner, R.C., Jr.; Smith, D.L.; Gething, P.W. Climate Change, Urbanization and Disease: Summer in the City . . . . *Trans. R. Soc. Trop. Med. Hyg.* **2015**, *109*, 171–172. [CrossRef] [PubMed]
3. Joon, V.; Jaiswal, V. Impact of Climate Change on Human Health in India: An Overview. *Health Popul.-Perspect. Issues* **2012**, *35*, 11–22.
4. Marland, G.; Boden, T.A.; Andres, R.J. *CO2 Emissions. Trends: A Compendium of Data on Global Change*; Technical Report; Carbon Dioxide Analisys Center, Oak Ridge National Laboratory: Oak Ridge, TN, USA, 2000.
5. Rosenzweig, C.; Solecki, W.D.; Hammer, S.A.; Mehrotra, S. *Climate Change and Cities: First Assessment Report of the Urban Climate Change Research Network*; Cambridge University Press: Cambridge, UK, 2011.
6. Tan, J.; Zheng, Y.; Tang, X.; Guo, C.; Li, L.; Song, G.; Zhen, X.; Yuan, D.; Kalkstein, A.J.; Li, F.; et al. The Urban Heat Island and Its Impact on Heat Waves and Human Health in Shanghai. *Int. J. Biometeorol.* **2010**, *54*, 75–84. [CrossRef] [PubMed]
7. Dutta, D.; Rahman, A.; Paul, S.K.; Kundu, A. Estimating Urban Growth in Peri-Urban Areas and Its Interrelationships with Built-up Density Using Earth Observation Datasets. *Ann. Reg. Sci.* **2020**, *65*, 67–82. [CrossRef]
8. Gill, S.E.; Handley, J.F.; Ennos, A.R.; Pauleit, S. Adapting Cities for Climate Change: The Role of the Green Infrastructure. *Built Environ.* **2007**, *33*, 115–133. [CrossRef]
9. Kueh, M.T.; Lin, C.Y.; Chuang, Y.J.; Sheng, Y.F.; Chien, Y.Y. Climate Variability of Heat Waves and Their Associated Diurnal Temperature Range Variations in Taiwan. *Environ. Res. Lett.* **2017**, *12*, 074017. [CrossRef]
10. Nag, P.K.; Nag, A.; Sekhar, P.; Pandit, S. *Vulnerability to Heat Stress: Scenario in Western India*; National Institute of Occupational Health: Ahmedabad, India, 2009.
11. De, U.S.; Mukhopadhyay, R.K. Severe Heat Wave over the Indian Subcontinent in 1998, in Perspective of Global Climate. *Curr. Sci.* **1998**, *75*, 1308–1311.
12. van Oldenborgh, G.J.; Philip, S.; Kew, S.; van Weele, M.; Uhe, P.; Otto, F.; Singh, R.; Pai, I.; Cullen, H.; AchutaRao, K. Extreme Heat in India and Anthropogenic Climate Change. *Nat. Hazards Earth Syst. Sci.* **2018**, *18*, 365–381. [CrossRef]
13. Mazdiyasni, O.; AghaKouchak, A.; Davis, S.J.; Madadgar, S.; Mehran, A.; Ragno, E.; Sadegh, M.; Sengupta, A.; Ghosh, S.; Dhanya, C.T.; et al. Increasing Probability of Mortality during Indian Heat Waves. *Sci. Adv.* **2017**, *3*, e1700066. [CrossRef]
14. Ghalhari, G.F.; Heidari, H.; Dehghan, S.F.; Asghari, M. Consistency Assessment between Summer Simmer Index and Other Heat Stress Indices (WBGT and Humidex) in Iran's Climates. *Urban Clim.* **2022**, *43*, 101178. [CrossRef]
15. Mahgoub, A.O.; Gowid, S.; Ghani, S. Global Evaluation of WBGT and SET Indices for Outdoor Environments Using Thermal Imaging and Artificial Neural Networks. *Sustain. Cities Soc.* **2020**, *60*, 102182. [CrossRef]
16. Zare, S.; Shirvan, H.E.; Hemmatjo, R.; Nadri, F.; Jahani, Y.; Jamshidzadeh, K.; Paydar, P. A Comparison of the Correlation between Heat Stress Indices (UTCI, WBGT, WBDT, TSI) and Physiological Parameters of Workers in Iran. *Weather Clim. Extrem.* **2019**, *26*, 100213. [CrossRef]
17. Kakaei, H.; Omidi, F.; Ghasemi, R.; Sabet, M.R.; Golbabaei, F. Changes of WBGT as a Heat Stress Index over the Time: A Systematic Review and Meta-Analysis. *Urban Clim.* **2019**, *27*, 284–292. [CrossRef]
18. Aliabadi, M.; Jahangiri, M.; Arrassi, M.; Jalali, M. Evaluation of Heat Stress Based on WBGT Index and Its Relationship with Physiological Parameter of Sublingual Temperature in Bakeries of Arak City. *Occup. Med.* **2014**, *6*, 48–56.
19. Vatani, J.; Golbabaei, F.; Dehghan, S.F.; Yousefi, A. Applicability of Universal Thermal Climate Index (UTCI) in Occupational Heat Stress Assessment: A Case Study in Brick Industries. *Ind. Health* **2015**, *54*, 14–19. [CrossRef] [PubMed]

20. Gaspar, A.R.; Quintela, D.A. Physical Modelling of Globe and Natural Wet Bulb Temperatures to Predict WBGT Heat Stress Index in Outdoor Environments. *Int. J. Biometeorol.* **2009**, *53*, 221–230. [CrossRef]
21. Willett, K.M.; Sherwood, S. Exceedance of Heat Index Thresholds for 15 Regions under a Warming Climate Using the Wet-Bulb Globe Temperature. *Int. J. Climatol.* **2012**, *32*, 161–177. [CrossRef]
22. Murari, K.K.; Ghosh, S.; Patwardhan, A.; Daly, E.; Salvi, K. Intensification of Future Severe Heat Waves in India and Their Effect on Heat Stress and Mortality. *Reg. Environ. Change* **2015**, *15*, 569–579. [CrossRef]
23. Mathew, A.; Khandelwal, S.; Kaul, N. Investigating Spatio-Temporal Surface Urban Heat Island Growth over Jaipur City Using Geospatial Techniques. *Sustain. Cities Soc.* **2018**, *40*, 484–500. [CrossRef]
24. Chandra, S.; Sharma, D.; Dubey, S.K. Linkage of Urban Expansion and Land Surface Temperature Using Geospatial Techniques for Jaipur City, India. *Arab. J. Geosci.* **2018**, *11*, 31. [CrossRef]
25. Wang, Y.; Gao, J.; Xing, X.; Liu, Y.; Meng, X. Measurement and Evaluation of Indoor Thermal Environment in a Naturally Ventilated Industrial Building with High Temperature Heat Sources. *Build. Environ.* **2016**, *96*, 35–45. [CrossRef]
26. Adekunle, T.O.; Nikolopoulou, M. Winter Performance, Occupants' Comfort and Cold Stress in Prefabricated Timber Buildings. *Build. Environ.* **2019**, *149*, 220–240. [CrossRef]
27. Masterton, J.M.; Richardson, F.A. *Humidex: A Method of Quantifying Human Discomfort Due to Excessive Heat and Humidity*; Environment Canada, Atmospheric Environment: Toronto, ON, Canada, 1979.
28. Foody, G.M. On the Compensation for Chance Agreement in Image Classification Accuracy Assessment, Photogram. *Eng. Remote Sens.* **1992**, *58*, 1459–1460.
29. Ma, Z.; Redmond, R.L. Tau Coefficients for Accuracy Assessment of Classification of Remote Sensing Data. *Photogramm. Eng. Remote Sens.* **1995**, *61*, 435–439.
30. Monserud, R.A.; Leemans, R. Comparing Global Vegetation Maps with the Kappa Statistic. *Ecol. Model.* **1992**, *62*, 275–293. [CrossRef]
31. Batista, G.T.; Shimabukuro, Y.E.; Lawrence, W.T. The Long-Term Monitoring of Vegetation Cover in the Amazonian Region of Northern Brazil Using NOAA-AVHRR Data. *Int. J. Remote Sens.* **1997**, *18*, 3195–3210. [CrossRef]
32. Hajizadeh, R.; Mehri, A.; Jafari, S.; Beheshti, M.; Haghighatjou, H. Feasibility of Esi Index to Assess Heat Stress in Outdoor Jobs. *J. Occup. Environ. Health* **2016**, *2*, 18–26.
33. Golbabaei, F.; Asour, A.A.; Keyvani, S.; Kolahdoozi, M.; Mohammadiyan, M.; Ramandi, F.F. The Limitations of WBGT Index for Application in Industries: A Systematic Review. *Int. J. Occup. Hyg.* **2021**, *13*, 365–381. [CrossRef]

 *sustainability*

Article

Assessment of the Impacts of Spatial Water Resource Variability on Energy Planning in the Ganges River Basin under Climate Change Scenarios

Bijon Kumer Mitra [1,*], Devesh Sharma [2,*], Xin Zhou [1] and Rajarshi Dasgupta [1]

[1] Integrated Sustainability Center, Institute for Global Environmental Strategies, Kanagawa 240-0115, Japan; zhou@iges.or.jp (X.Z.); dasgupta@iges.or.jp (R.D.)

[2] Department of Atmospheric Science, School of Earth Sciences, Central University of Rajasthan, Rajasthan 305817, India

* Correspondence: b-mitra@iges.or.jp (B.K.M.); deveshsharma@curaj.ac.in (D.S.)

Abstract: Availability of water in the Ganges River basin has been recognized as a critical regional issue with a significant impact on drinking water supply, irrigation, as well as on industrial development, and ecosystem services in vast areas of South Asia. In addition, water availability is also strongly linked to energy security in the region. Hence, quantification of spatial availability of water resources is necessary to bolster reliable evaluation of the sustainability of future thermal power plants in the Ganges River basin. This study focuses on the risks facing existing and planned power plants regarding water availability, applying climate change scenarios at the sub-basin and district level up to 2050. For this purpose, this study develops an integrated assessment approach to quantify the water-energy nexus in four selected sub-basins of the Ganges, namely, Chambal, Damodar, Gandak, and Yamuna. The results of simulations using Soil and Water Assessment Tools (SWAT) showed that future water availability will increase significantly in the Chambal, Damodar, and Gandak sub-basins during the wet season, and will negligibly increase in the dry season, except for the Yamuna sub-basin, which is likely to experience a decrease in available water in both wet and dry seasons under the Representative Concentration Pathway (RCP) 8.5 scenario. Changes in the water supply-demand ratio, due to climate change, indicated that water-related risks for future power plants would reduce in the Chambal and Damodar sub-basins, as there would be sufficient water in the future. For 19 out of 23 districts in the Chambal sub-basin, climate change will have a moderate-positive to high-positive impact on reducing the water risk for power plants by 2050. In contrast, existing and future power plants in the Yamuna and Gandak sub-basins will face increasing water risks. The proposed new thermal power installations, particularly in the Gandak sub-basin, are likely to face serious water shortages, which will adversely affect the stability of their operations. These results will stimulate and guide future research work to optimize the water-energy nexus, and will inform development and planning organizations, energy planning organizations, as well as investors, concerning the spatial distribution of water risks for future power plants so that more accurate decisions can be made on the location of future power plants.

Keywords: water-energy nexus; spatial water variability; climate change; thermal power plant; Ganges River basin

Citation: Mitra, B.K.; Sharma, D.; Zhou, X.; Dasgupta, R. Assessment of the Impacts of Spatial Water Resource Variability on Energy Planning in the Ganges River Basin under Climate Change Scenarios. *Sustainability* **2021**, *13*, 7273. https://doi.org/10.3390/su13137273

Academic Editor: Miguel Amado

Received: 29 May 2021
Accepted: 24 June 2021
Published: 29 June 2021

Publisher's Note: MDPI stays neutral with regard to jurisdictional claims in published maps and institutional affiliations.

1. Introduction

Home to 600 million people, the Ganges is the most populous river basin in the world [1]. The Ganges River basin (GRB) is a strategically important river basin for all riparian countries, including Bangladesh, India, and Nepal, as more than 40% of people directly or indirectly relying on the water of this river for drinking, agriculture, energy generation purposes [2]. For instance, this river basin accounts for 25% of India's water resources, and more than 50% of irrigated areas in India are situated in this basin [1]. A vast amount of water is used for energy generation purposes. The Ganges River supplies water to several thermal power plants with more than 50 GW generation capacity [3]. Therefore, any changes in water availability in the Ganges will have paramount impacts on the development and wellbeing of the region.

Water resources in the GRB were once abundant, but are now under increasing stress, due to the growing demand. The Ganges is shown as water-stressed as per the Falkenmark water stress index with 1039 cubic meters of water per capita [4]. In this river basin, about 20% of people live without access to safe drinking water [5]. During the non-monsoon period, limited water resources hardly allow cropping to 1.3 times the net sown area [6]. Furthermore, climate change may exacerbate water stress, due to its impacts on hydrological dynamics in the GRB. Regional climate change model studies in the GRB predict an increase in annual mean temperature [7], and a rising trend of seasonal maximum and minimum temperatures [8]. This rise in temperature will lead to various dynamic changes, including a greater evapotranspiration loss [9], shrinking glaciers [10,11], and increased rainfall that will lead to more water flow, but with greater variability [12].

GRB is one of the hotspots of economic development in the region. It accounted for USD 700 billion of the GDP of India [1]. The river basin caters to a 40% share of the total electricity generation capacity in the region [3]. However, per capita energy consumption in riparian countries comes to 310 kWh in Bangladesh, 805 kWh in India, and 139 kWh in Nepal, which are far below the world average of 3130 kWh [13]. This leaves room for a step-up of the growth of the electricity sector in India for decades to come. Nevertheless, the electricity fuel mix in the region is dominated by coal and gas-based thermal power that accounted for 73% of the total electricity generation capacity [14,15]. Given conventional cooling technology, thermal power plants (TPP) require a large amount of water for cooling purposes. With the availability of indigenous coal and gas resources in India and Bangladesh, it is envisaged that future power generation will rely heavily on thermal sources. However, the sustainability of thermal power generation will be seriously affected by the climate-induced variability of water resources [16]. For example, drought events between 2013 and 2016 forced a shutdown of 14 major power plants, due to scarcity of cooling water, which incurred at least USD1.4 billion in potential revenue loss [17]. It implied that water availability for thermal power generation is in jeopardy, and this situation poses a serious operational risk for power plants in this river basin.

There are ample studies in the region covering the issues of direct use of water in agriculture, human habitat, and in other sectors. There are also certain studies in the field of energy use for water withdrawal focusing on pumping efficiency improvement, etc. However, there is little systematic literature looking at quantifying the interactions between energy infrastructure and spatial water availability under the impacts of climate change. In this context, the study narrates how water scarcity can be a major threat to energy security in the sub-basins of the Ganges River. To the best of our knowledge, this is a pioneer study of the GRB that deals with the water-energy nexus, particularly dealing with the water risks of existing and planned power plants, while considering the plausible climate change impacts on spatial water availability. The study aimed to assist informed decision-making related to power plant planning in the Ganges sub-basin in India, considering the impacts from spatial water resource distribution and long-term climate change.

2. Materials and Methods

2.1. Study Area

For this study, we selected four sub-basins of the Ganga River basin in India, namely, Yamuna, Chambal, Gandak, and Damodar. These sub-basins were choose based on the key features, including water supply, water demand and installed capacity of the existing thermal power plants, and the proposed location of the new thermal power plants. Figure 1 shows the location and characteristics of the selected sub-basins.

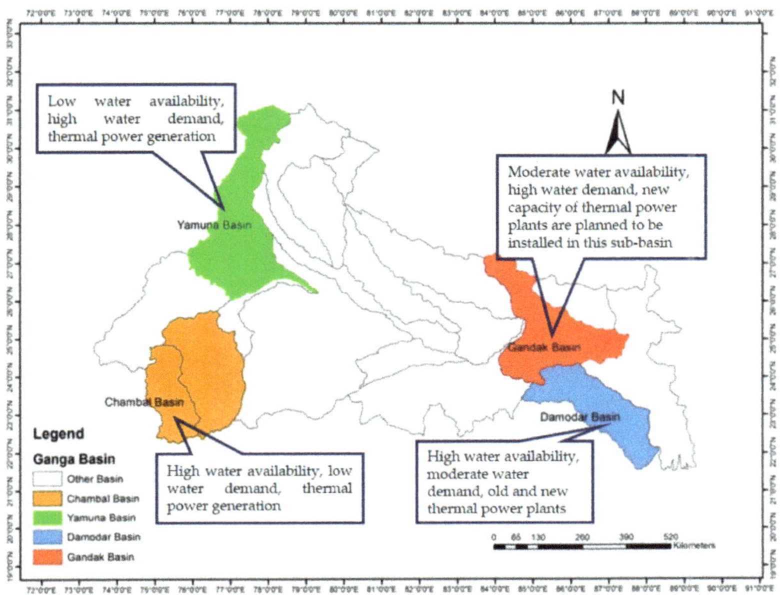

Figure 1. A map of the study areas with four selected sub-basins in India with the specific characteristics.

2.2. Methodological Framework

In this study, we took an integrated methodological approach that consists of the hydrological modeling of the spatial distribution of water availability, water demand assessment for non-energy sectors, collected water use intensity data for power generation through a field survey of 20 thermal power plants for estimation of water demand of energy generation in India. The overall framework is furnished in Figure 2.

2.3. Water Resource Variability Assessment

To classify the water resource availability, we used the Soil and Water Assessment Tool (SWAT) for assessing future water resource distribution at the sub-basin level. The advantages of using SWAT include; (i) it is suitable for ungauged water catchment and does not need calibration; and (ii) it simulates future water yield based on the physical data [16,18–21]. In the SWAT model, basins are considered as an agglomeration of multiple sub-watersheds. The sub-watersheds are further split into hydrologic response units (HRUs) characterized by similar land use, slope, and soil type, etc. For each HRU, the net hydrological balance is simulated based on precipitation, evapotranspiration, soil water, lateral sub-surface flow, and water yield [21,22].

The hydrological cycle is simulated by the SWAT model based on water balance Equation (1):

$$SW_t = SW_0 + \sum_{i=1}^{t} \left(P_{day} - Q_s - E_a - W_{seep} - Q_{gw} \right) \tag{1}$$

where SW_t is the final soil water content after t days (in mm); SW_0 is the initial soil water content (in mm) on day i; P_{day} is the amount of precipitation on day i (in mm); Q_s is the amount of surface runoff on day i (in mm); E_a is the amount of evapotranspiration on day i (in mm); W_{seep} is the amount of percolation and bypass flow exiting the soil profile bottom on day i (in mm); Q_{gw} is the amount of return flow on day i (in mm); t is the time (days).

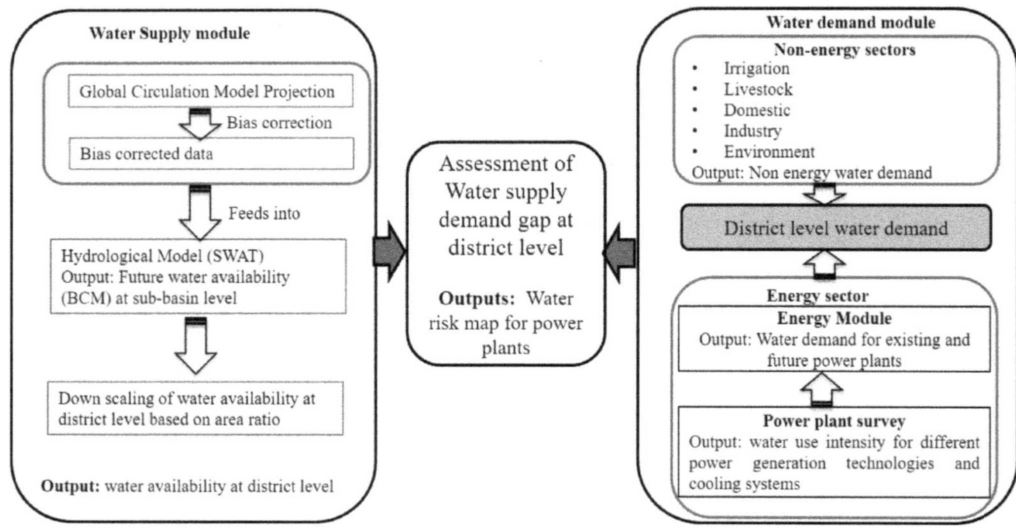

Figure 2. The methodological framework for assessing the impacts of spatial water resource variability on energy planning in the Ganges sub-basin in India under climate change scenarios.

Essential data, including the Digital Elevation Model (DEM), soil map, land use and land cover data, meteorological data, and climate model projections of temperature, and precipitation, were collected from various sources. For DEM, Shuttle Radar Topography Mission (SRTM) 90 m resolution was used [23]. Land use and land cover data from National Remote Sensing Centre (NRSC) and soil data were taken Food and Agriculture Organization (FAO) (available at http://www.fao.org/soils-portal/soil-survey/soil-maps-and-databases/harmonized-world-soil-database-v12/en/). Weather data, including precipitation of 0.5-degree grid data, the temperature of 1-degree grid data, relative humidity, solar radiation, wind grid weather data were taken from Global Weather Data for SWAT (available at http://globalweather.tamu.edu/).

A bias-correction technique based on the delta change method was employed to correct the bias of precipitation and temperature projections derived from the Global Circulation Model (GCM), consequently, minimize the uncertainty in future water availability projection. This bias-corrected projection was used to generate different climate change scenarios on water availability using the SWAT model. We used two climate scenarios, namely, the RCP 4.5 and RCP 8.5. To assess the climate change impacts on spatial water availability, the RCP 4.5 and RCP 8.5 climate scenarios from the MRI-CGCM3 model were used (available at http://pcmdi9.llnl.gov/). MRI-CGCM3 is selected considering its good performance at the basin scale for India [24]. MRI-CGCM3 is developed by the Meteorological Research Institute based on the earlier MRI-CGCM2 model. The simulation of the SWAT model provided water yield at the HRU level. The SWAT model simulated water yield of relevant HRU is used as a representative value for the districts located at the respective HRU. Then water yield was multiplied by the area of the district to estimate the water availability of each district.

2.4. Estimation of Water Demand

District level water demand for major water users was estimated, including domestic sector, irrigation, livestock, industrial use, environment water, and energy use for present and future scenarios, as mentioned below.

2.4.1. Water Demand for Domestic Use

Water demand for domestic use was calculated by multiplying the population of each district. As water use patterns in urban and rural significantly vary, domestic water demand for urban and rural populations calculated separately, and their cumulative sum represents water demand for domestic use of the respective district. For the present study, per capita, water demand of 150 L per capita per day (*lpcd*) and 70 *lpcd* was considered for urban and rural areas, respectively [25]. District-level population data is taken from the census (2001 and 2011). The following are the equations used for estimation of urban and rural domestic water demand:

$$D_{rural} = P_{rural} \times 70 \; lpcd \tag{2}$$

where D_{rural} denotes the water demand for the rural segment, and P_{rural} is the population in the rural area.

$$D_{urban} = P_{urban} \times 150 \; lpcd \tag{3}$$

where D_{urban} denotes the water demand for the urban segment, and P_{urban} is the population in the urban area.

2.4.2. Water Demand for Irrigation

For estimating water demand for irrigation, major cereal crops are identified for each of the four sub-basins. The reference evapotranspiration of each major crop, ET_0 (in mm/day), is estimated based on which the evapotranspiration of each major crop, ET_c, the amount of water demand of the crop under standard conditions, is calculated by Equation (4) as follows:

$$ET_c = ET_0 \times K_c \tag{4}$$

where K_c is the crop coefficient suggested by FAO for the main crop in each of the sub-basins.

Effective precipitation, $R_{effective}$, rainfall available for crop growth, is estimated based on the FAO/AGLW formula see Equation (5) by which total rainfall, R_{total}, is corrected by taking account of the losses (as a percentage of total rainfall), due to runoff and percolation [26].

$$\begin{aligned} R_{effective} &= 60\% \times R_{total} - 10, \text{ if } R_{total} <= 70 \text{ mm} \\ R_{effective} &= 80\% \times R_{total} - 25, \text{ if } R_{total} > 70 \text{ mm} \end{aligned} \tag{5}$$

Irrigation water demand is calculated by Equation (6).

$$D_{irrigation} = ET_c - R_{effective} \tag{6}$$

To estimate the total irrigation water requirement at the district level, national reference values were utilized [27]. To calculate the base year (2010) water demand, a reference factor of 1.45 was used, which was multiplied by the cereal irrigation water demand of 2010. Similarly, for calculating future irrigation water demand, different we used different multiplication factors (e.g., 1.003 for the 2020s, 1.007 for the 2030s, 1.168 for the 2040s, and 1.329 for the 2050s).

2.4.3. Water Demand for Livestock

We calculated the livestock water demand by multiplying the livestock population by the water use rate per head for different types of animals. Data on the district-level population of the livestock is collected from the census (2007 and 2012). It is, however, impossible to estimate the decadal growth for future livestock as the regular fluctuation are not known. Hence, we considered an increase of 10% in water demand on a decadal basis.

2.4.4. Water Demand for Industry

Water demand for the industrial sector was estimated as a percentage of urban and rural domestic water use, as recommended by the Central Pollution Control Board of India (1989).

$$D_{industry} = D_{rural} \times f_{rural} + D_{urban} \times f_{urban} \tag{7}$$

In which $D_{industry}$ denotes industrial water demand, D_{rural} is the rural domestic water demand, and D_{urban} is the urban domestic water demand, see Equations (2) and (3), and f_{rural} represents rural water use factors, which is considered 25%, and f_{urban} represents urban water use factors is considered 5% in this study.

2.4.5. Environmental Water Requirement

Environmental water requirement, i.e., the amount of water required to maintain ecological processes and biodiversity, is an important consideration for estimating future water demand. However, it is tough to estimate environmental water requirements, due to a lack of necessary data. In this study, environmental water requirement is estimated as 1.23% of the total water demand as recommended by the Central Water Commission (2015).

2.4.6. Energy Water Demand Estimation

To collect the water use intensity of different power generation technologies, including the different cooling systems, a power plant survey was conducted. During the power plant surveys, various information, including fuel types (*fuel*), installed capacity ($C_{install}$), power generation technologies (*tech*), cooling systems (*cool*), plant load factors (L), source of water, water use intensity ($I_{fuel, \, tech, \, cool}$), etc., were collected. Energy water demand (D_{energy}) is calculated using Equation (8):

$$D_{energy} = C_{install} \times 24 \times 365 \times L \times I_{fuel, \, tech, \, cool} \tag{8}$$

For estimation of future water demand from power generation, the study relied on the disclosed information of the total planned fuel mix by the Central Energy Authority (see Appendix A) and used Equation (8).

2.5. Water Risk Assessment for Future Power Generation

Based on the simulated future water resources and sectoral water demand at the district level for each sub-basin, the supply-demand ratio, defined as (supply-demand)/supply, is calculated for the present period (2010). Districts were classified into four classes of water risks based on the value of the supply-demand gap ratio, including highly water-stressed (<0.0), moderately water-stressed (0.0 to 0.5), no stress (0.5 to 1.0), and water surplus (>1.0). Then, future period changes in the supply-demand ratio in 2050 compared with the level in 2010 are calculated for each of the four sub-basins. The changes in the supply-demand ratio (in percentage) are classified into five levels to indicate the impact of climate change on future water risks under RCP 4.5 (see Table 1).

Table 1. Classification of the changes in the supply-demand ratio for assessing climate-induced water risk for future power plants.

Level of Changes in the Supply-Demand Ratio (%)	Colour	Description of Level of Effect
More than 25	Green	High positive
Between 5 to 25	Blue	Moderate positive
5 to −5	Yellow	Negligible/No change
Between −5 to −25	Brown	Moderate negative
less than −25	Red	High negative

3. Results and Discussions

3.1. Climate Parameters under RCP 4.5 and RCP 8.5

As per climate model scenarios, it is projected that there will be an increase in precipitation in all the four sub-basins irrespective of RCP 4.5 and RCP 8.5 scenarios, as shown in Table 2. The model suggested that overall, four sub-basins will increase in precipitation in the future with a more positive change in the mid-future period and then a decrease in the positive trend far into the future. This is a good sign as it will improve water availability in the sub-basin. The projection also revealed that all sub-basins will receive more precipitation under the RCP 8.5 scenario than the RCP 4.5 scenario. The Chambal basin will have a 9% increase in precipitation in the far future period to a 17% increase in the mid-future period under the RCP4.5 scenario. This increase varies from 11% in the far future period to 31% in the midfuture period. In the case of Yamuna, an increase of precipitation by 19% in the near future and 14% in the mid-future is predicted under RCP 4.5. However, the model predicts a slight decrease in precipitation in the far future under RCP 4.5. Under the 8.5 RCP scenario, precipitation will increase from 4% in the near future to a 26% increase in the midfuture period.

Table 2. Precipitation and evapotranspiration in four sub-basins under RCP 4.5 and RCP 8.5 scenarios (in mm).

	Chambal		Yamuna		Gandak		Damodar	
	P	E	P	E	P	E	P	E
Historical	854	443	744	452	1075	608	1464	712
RCP 4.5								
Near Future (2011–2040)	940	423	888	512	1253	653	1652	793
Mid Future (2041–2070)	1006	533	847	495	1233	649	1743	763
Far Future (2071–2100)	934	468	740	465	1127	624	1646	780
RCP 8.5								
Near Future (2011–2040)	1024	494	775	476	1175	649	1661	829
Mid Future (2041–2070)	1119	504	941	517	1318	665	1829	832
Far Future (2071–2100)	949	482	862	515	1335	661	1850	849

Note: P denotes precipitation; E denotes evapotranspiration.

It represents a positive change in water availability in the Yamuna sub-basin. In the Gandak sub-basin, there is also an increase in precipitation which varies between 6 to 23% under RCP 4.5, whereas between 13 to 34% under the RCP 8.5 scenario. All three future periods show an increase in precipitation. In the Damodar sub-basin, the percentage change in precipitation varies between 12 to 19% under RCP 4.5 scenario, whereas the increase is predicted to be between 13 to 26% under the RCP 8.5 scenario.

In all the four sub-basins, there is a projected increase in evapotranspiration (E), as presented in Table 2. In the Chambal sub-basin, there is a slight decrease in E in the near future, but it will increase in the mid-future and far-future under RCP 4.5. Under RCP 8.5, there will be an increase in E for all future periods. In the Yamuna sub-basin, E will increase under both RCP 4.5 and RCP 8.5 scenarios. Under the RCP 8.5 scenario ratio of E and P will increase in the far future period, which implies a reduction of water availability in the far future. In the Gandak sub-basin, the ratio of E and P will decrease under both the RCP 4.5 scenario and the RCP 8.5 scenario. Similar observations are made in the Damodar sub-basin as there is a decrease in E and P ratio f under both scenarios.

3.2. Water Availability Assessment under Climate Change Scenarios

Figure 3 presents the changes in the water yield in 2030 and 2050, respectively, as compared with the levels in the historical period (1976–2005) in four sub-basins under RCP 4.5 and RCP 8.5. Water yield is not evenly distributed throughout the year and will show significant seasonal variation, depending on the physical conditions, such as precipitation, evapotranspiration, and surface runoff, etc. Under RCP 4.5, all the sub-basins will show an increasing trend of water yield during the wet seasons in 2030 and 2050. Changes in the water yield in the wet season will vary from 8% in 2030 to 55% in 2050 in the Damodar sub-basin. Under RCP 8.5, Yamuna will face a negative change in the water yield in both 2030 and 2050, even during the wet season. In contrast, water yield will increase significantly in all other basins. The results show that RCP 8.5 will have more positive impacts on the water yield than RCP 4.5 in both the Chambal and Damodar sub-basins. The results imply that the total water yield will increase in Chambal, Damodar, and Gandak at all times, but with great seasonal variabilities. However, under the extreme climate scenario, water yield in the Yamuna will decrease.

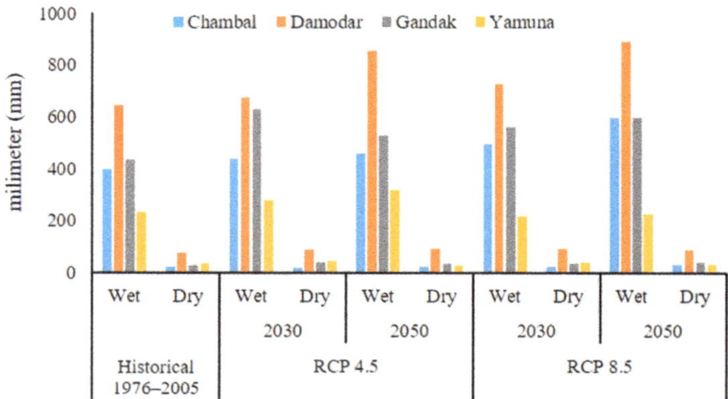

Figure 3. Seasonal water yield in four sub-basins under climate change scenarios in 2030 and 2050.

A positive effect of climate change on the water yield will result in an increase of available water in the four sub-basins (Figure 4). The results show that under both RCP 4.5 and RCP 8.5, Chambal will have the largest among of available water ranging from 47,823 MCM (million cubic meter) in 2050 under RCP 8.5 to 34,701 MCM in 2030 under RCP 4.5. Among four sub-basins, Yamuna will have the lowest volume of available water. The results imply that the increasing amount of available water will positively support water-intensive development, including thermal power generation, in the Chambal sub-basin.

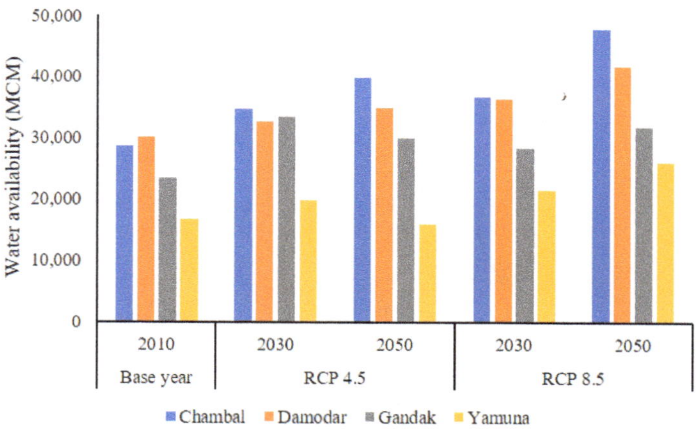

Figure 4. Future water availability in four sub-basins under climate change scenarios.

3.3. Water Demand Assessment

3.3.1. Water Demand of Non-Energy Sectors

Table 3 shows the water demand from five non-energy sectors in four sub-basins for the base period 2010 and the future period (2050). The results show that the future water demand from non-energy sectors will increase, due to population growth, industrial development, and irrigation requirements. Out of the four sub-basins, the Chambal sub-basin would have the least water demand, and the Yamuna sub-basin would have the most water demand in 2010. In all the four sub-basins, non-energy sectors water demand will increase significantly by 37% in Chambal, 34% in Damodar, 37% in Gandak, and 42% in Gandak in 2050. Although the rate of water demand increase will be high for domestic and industrial sectors, the share of irrigation water demand will dominate the total non-energy sector water demand followed by domestic water demand, which will continue until 2050. The Yamuna sub-basin will lead the highest water demand, including both irrigation water demand and the domestic water demand, among the four sub-basins, followed by the Gandak sub-basin.

Table 3. Water demand changes from non-energy sectors in four sub-basins under climate change scenarios in 2050 compared to 2010 (in MCM).

Water Demand	Base Period 2010	Change in 2050
Chambal		
Domestic	725	635
Industry	113	133
Livestock	189	88
Irrigation	8895	2927
Environment	125	47
Total	10,112	3828
Damodar		
Domestic	925	625
Industry	123	106
Livestock	200	93
Irrigation	20,281	6672
Environment	273	92
Total	22,099	7589
Gandak		
Domestic	1375	1441
Industry	139	163
Livestock	218	101
Irrigation	22,688	7464
Environment	306	116
Total	24,796	9285
Yamuna		
Domestic	2291	2728
Industry	446	621
Livestock	333	154
Irrigation	23,437	7711
Environment	333	140
Total	26,940	11,354

3.3.2. Water Demand of Energy Sector

A field survey was carried out to collect water use intensity data for existing power plants. Power plants were chosen based on the fuel types and technologies employed for the cooling systems, including open loop cooling systems, closed-loop cooling systems, and dry cooling systems. As such, it was observed that cooling technologies have major impacts on the energy sector's water demand. It was observed that the water use intensity in thermal power generation varies from 3.3 m^3/MWh with a closed-loop cooling system to 70 m^3/MWh with an open-loop cooling system. Furthermore, the requirement for water is much higher in coal-based power plants than gas-based power plants. Apart from gas-based power plants, the results revealed that all the thermal power plants under our survey exceeded the upper limit for regulated water use intensity (2.5 m^3/MWh) [28].

According to the CEA database, coal-based power plants dominate the majority of the existing installed thermal power capacity in four sub-basins (see Table A1). For example, 100% of the installed capacity is based on coal in the Damodar and Gandak sub-basin. As a result, energy-water demand is the highest in the Damodar sub-basin, followed by the Gandak sub-basin. Energy water demand will decrease in the Chambal and the Damodar sub-basins and will maintain a similar level in the Yamuna sub-basin; while it will substantially increase in the Gandak sub-basin. Among four sub-basins, the Damodar sub-basin has the largest thermal power capacity (17.9 GW). As a result, water demand for power generation is the highest among the four selected sub-basins. This situation will continue until 2030. In 2030, the water demand for thermal power generation will be more than 400 MCM (Figure 5). The Gandak sub-basin has the second-highest water demand for thermal power generation, and in 2030, the thermal power generation will require nearly 255 MCM of water. In contrast, estimates show that water demand for thermal power generation will reduce in the Chambal and Yamuna sub-basins by 2030.

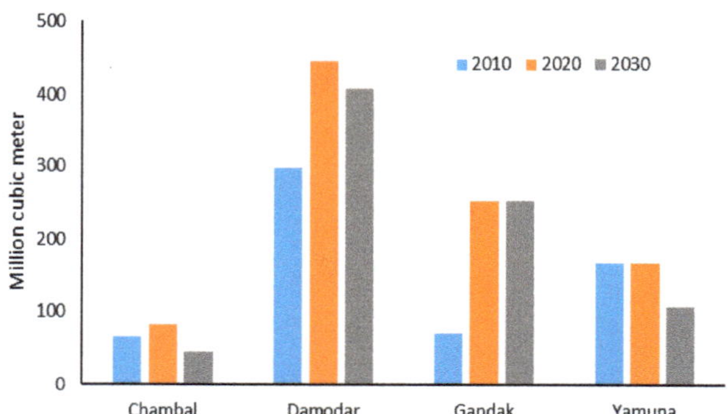

Figure 5. Water demand for existing power plants in four sub-river basins.

3.4. Assessment of Water Risks to Future Power Plants

Water supply-demand balance was calculated by subtracting total water demand (non-energy and energy-related water use, as shown in Table 3 and Figure 5) from the amount of available water in the future (Figure 4). Water supply-demand balance analysis revealed that the Chambal and Damodar sub-basins will have surplus water for all the periods. The estimate shows the amount of water surplus in the Damodar sub-basin will decrease over time.

In 2010, the surplus water volume in the Damodar sub-basin was 8072 MCM, and the estimate predicts a reduction by 15% in 2050. Water deficit will become more serious in both the Yamuna and Gandak sub-basins. In the Yamuna sub-basin, the water deficit will increase by 86%.

In Figure 6, water risk maps for 2010 showed that only a small part of the Chambal sub-basin located in the upper part of the area would face high water stress (in red), and some parts would have moderate water stress (in orange). The majority of the sub-basin would have a water surplus (in yellow). In the case of the Damodar sub-basin, districts at the upper catchments would have a water surplus. However, districts at the lower catchment areas would face moderate to high water risk, and many of the existing power plants are located in this part of the Damodar sub-basin. It indicates exiting power plants might face water shortage for operation. In the case of the Gandak and Yamuna, most of the districts would have moderate to high water risks.

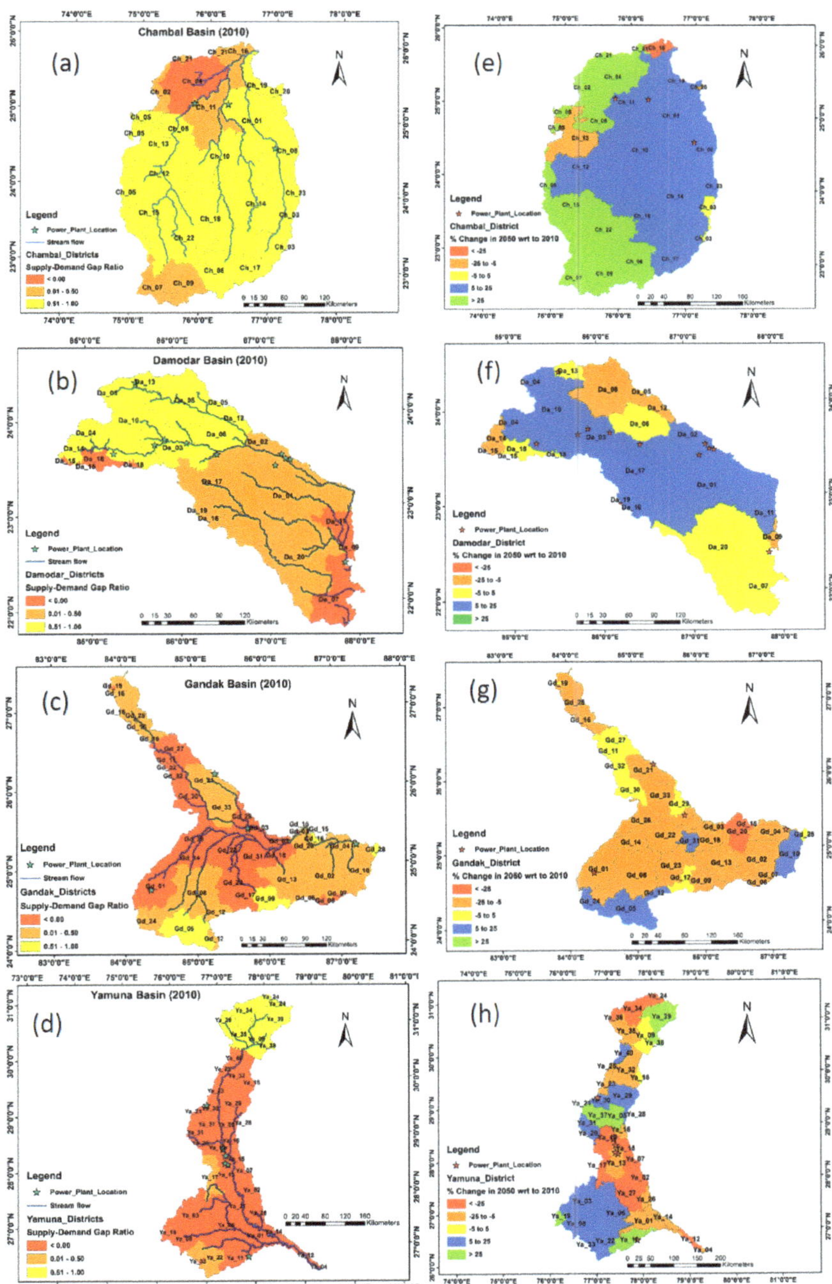

Figure 6. Water risks in 2010 (**a**–**d**) and the percent changes in the supply-demand ratio in 2050 in four sub-basins compared with 2010 (**e**–**h**).

Future water risks were assessed based on the percentage change of the supply-demand ratios for 2050 compared with the levels in the base year (2010) to comprehend future changes in water availability. The classification of water risks is shown in Table 1.

The results of the changes in the water supply-demand ratios at the sub-basin level indicate that water risks will reduce in the Chambal and Damodar sub-basins which will have surplus water in the future. Figure 6 shows that out of 23 districts in the Chambal sub-basin, climate change will have moderate positive to high positive impacts on reducing the water risks in 19 districts in 2050. However, four districts, including Bhopal (Ch_03), Neemuch (Ch_13), Sawai Madhopur (Ch_16), and Shivpuri (Ch_20), will face increasing water risks. Similarly, water risks will be reduced in the Damodar sub-basin. The number of districts with moderate positive impacts on reducing the water risks from climate change will increase to 10 in 2050. The results revealed that water risks for the existing and future power plants would decrease in the Chambal and Damodar sub-basins. In contrast, the existing and future power plants in the Yamuna and Gandak sub-basins will face increasing water risks in the future. Districts that will receive moderate negative impacts will increase to 19 in 2050. In the Gandak and Yamuna sub-basin, all the existing power plants and future power plants are located in the districts with increasing water risks. Particularly in the Gandak sub-basin, there will be a number of planned thermal power installations whose operations will face severe water risks. Please see Table A2 for the list of districts.

Existing or planned thermal power plants that are located in areas with high water risks may face serious water shortages, which will impact the stability of their operations. Districts with moderate to high positive impacts on the water risks from climate change in the Chambal and Damodar sub-basins can be considered as the appropriate locations for new electricity generation projects in the future.

4. Conclusions

Given the importance of water resources in the GRB for drinking water supply, energy generation, irrigation, and maintaining ecosystem services in South Asia, this study assessed the spatial variability of water resources under different climate change scenarios and the potential risks to future energy supply, regarding water availability. The study was conducted for four selected sub-basins. An integrated assessment method was developed, combining a single climate model (MRI-CGCM3), hydrological model (SWAT), and water demand projections for both non-energy sectors (domestic sector, agriculture, livestock, industry, and the environment) and the energy sector (thermal power generation). To gather primary data on water use intensity, we conducted a survey of thermal power plants in the four sub-basins and used the survey data to estimate the water demand for power generation in the four selected areas. The power plant survey results revealed that the water use intensity in coal thermal power plants varies from 3.3 m^3/MWh to 70 m^3/MWh, depending on power generation technologies, cooling technologies, as well as the quality of coal. An assessment of water availability under RCP 4.5 and RCP 8.5 shows that future water availability will increase in three sub-basins during the wet season except for the Yamuna sub-basin under RCP 8.5. Likewise, water demand will also steadily increase in four sub-basins, dominated by irrigation requirements and followed by domestic use. For the water demand from energy, of four sub-basins, Gandak will have the highest water demand for cooling down the coal power plants, followed by Damodar and Yamuna. The lowest water demand for future energy generation will be in the Chambal sub-basin as there is no plan to install new thermal power plants, although more water will be available in Chambal in the future. In the Gandak sub-basin, there will be a number of planned new thermal power installations in the coming years. However, changes in the water supply-demand ratio in 2050 relative to 2010 demonstrated that out of 33 districts in Gandak, 22 districts would face moderate negative impacts from climate change, which will worsen the water risks for the existing and future power plants. Similarly, thermal power plants in the Yamuna sub-basin will also face high water risks. These results will inform development and planning organizations, energy planning organizations, as well as investors, with regard to the spatial distribution of water risks for future power plants. All entities may consider this information to plan the location of future power plants in the districts with high water availability, thereby mitigating any conflicts with other water users and minimizing the risks of losing out on revenue related to the forced shutdown, due to water scarcity.

Author Contributions: Conceptualization, investigation, visualization, methodology, writing—review and editing; formal analysis, software, B.K.M., D.S., X.Z. and R.D.; writing—original draft preparation, B.K.M., D.S. and X.Z.; validation, resources, supervision, project administration, funding acquisition, B.K.M., D.S., X.Z. and R.D. All authors have substantially contributed for the development of this manuscript. All authors have read and agreed to the published version of the manuscript.

Funding: This research was funded by the Asia-Pacific Network for Global Change Research (APN) with the project reference number: ARCP2015-13CMY-Zhou. The publication cost is funded by IGES Strategic Research Fund-2020.

Acknowledgments: This study has been conducted from the project 'Assessment of Climate-Induced Long-term Water Availability in the Ganges Basin and Impacts on Energy Security in South Asia' funded by the Asia-Pacific Network for Global Change Research (APN) with the project reference number: ARCP2015-13CMY-Zhou. We would like to extend our thanks to the research assistants Sawtantra Dubey and Aashis Sapkota, for their supports to this study. We are also thankful to those power plants, which provided relevant information and detailed data to support our field surveys. Authors are grateful to three anonymous reviewers who have immensely helped in improving the manuscript and Emma Fushimi, IGES Japan for her kind support in proof reading and edits.

Conflicts of Interest: The authors declare no conflict of interest. The funders had no role in the design of the study; in the collection, analyses, or interpretation of data; in the writing of the manuscript, or in the decision to publish the results.

Appendix A

Table A1. List of existing and future thermal power plants in four Sub-basins.

Power Plants in Sub-Basins	Fuel	Capacity (MW)	Cooling Technology	Operation Status
Chambal				
Kota Super Thermal Power Station	Coal	1240	Wet-closed loop	Operational
Chhabra Thermal Power Plant	Coal	2320	Wet-closed loop	Operational
Anta Thermal Power Plant	Gas	419	Wet-closed loop	Operational
Damodar				
Koderma Thermal Power Station	Coal	1000	Wet-closed loop	Operational
Patratu Thermal Power Station I	Coal	880	Wet-closed loop	Operational
Patratu Thermal Power Station II	Coal	4000	Wet-closed loop	Planned
Bokaro B Thermal Power Station	Coal	500	Wet-closed loop	Planned
Tenughat Thermal Power Station	Coal	420	Wet-opened loop	Operational
Mejia Thermal Power Station	Coal	2340	Wet-closed loop	Operational
Kolaghat Thermal Power Station	Coal	1260	Wet-closed loop	Operational
Durgapur Steel Thermal Power Station	Coal	1000	Wet-closed loop	Operational
Santaldih Thermal Power Station	Coal	500	Wet-closed loop	Operational
Chandwa Power Project Phase I	Coal	1080	Wet-closed loop	Construction
Tori power plant Unit 1	Coal	1800	Wet-closed loop	Construction
Raghunathpur Thermal Power Station phase I	Coal	1200	Wet-closed loop	Construction
Gola power station Unit I and II	Coal	126	Wet-closed loop	Operational

Table A1. *Cont.*

Power Plants in Sub-Basins	Fuel	Capacity (MW)	Cooling Technology	Operation Status
Gandak				
Kahalgaon Super Thermal Power Station	Coal	2340	Wet-closed loop	Operational
Nabinagar Super Thermal Power Project	Coal	1980	Wet-closed loop	Construction
Kanti Thermal Power Station	Coal	610	Wet-closed loop	Operational
Banka Power Project Stage I (Unit 1 and 2)	Coal	2320	Wet-closed loop	Construction
Barh I power station	Coal	1980	Wet-closed loop	Construction
Barauni power station Unit 8	Coal	250	Wet-closed loop	Operational
Yamuna				
IPGCL-Gas Turbine Power Station	Gas	270	Wet-opened loop	Operational
NTPC- Faridabad Thermal Power Plant	Gas	430	Wet-closed loop	Operational
Panipat Thermal Power Station I and II	Coal	1360	Wet-closed loop	Operational
Panipat Thermal Power Station I and II Unit 9	Coal	800	Wet-closed loop	Planned
Dholpur Thermal Power Station	Gas	330	Wet-closed loop	Operational

Table A2. Name of districts in four sub-basins with district code.

Chambal		Damodar		Gandak		Yamuna			
District Code	District Name	District Code	District Name	District Code	District Name	District Code	District	District Code	District
Ch_01	Baran	Da_01	Bankura	Gd_01	Aurangabad	Ya_01	Agra		
Ch_02	Bhilwara	Da_02	Barddhaman	Gd_02	Banka	Ya_02	Aligarh	Ya_34	Shimla
Ch_03	Bhopal	Da_03	Bokaro	Gd_03	Begusarai	Ya_03	Alwar	Ya_35	Sirmaur
Ch_04	Bundi	Da_04	Chatra	Gd_04	Bhagalpur	Ya_04	Auraiya	Ya_36	Solan
Ch_05	Chittaurgarh	Da_05	Deoghar	Gd_05	Chatra	Ya_05	Baghpat	Ya_37	Sonepat
Ch_06	Dewas	Da_06	Dhanbad	Gd_06	Deoghar	Ya_06	Bharatpur	Ya_38	Tehri Garhwal
Ch_07	Dhar	Da_07	East Midnapore	Gd_07	Dumka	Ya_07	Bulandshahr	Ya_39	Uttarkashi
Ch_08	Guna	Da_08	Giridih	Gd_08	Gaya	Ya_08	Dausa	Ya_40	Yamuna Nagar
Ch_09	Indore	Da_09	Haora	Gd_09	Giridih	Ya_09	Dehra Dun		
Ch_10	Jhalawar	Da_10	Hazaribag	Gd_10	Godda	Ya_10	Delhi		
Ch_11	Kota	Da_11	Hugli	Gd_11	Gopalganj	Ya_11	Dhaulpur		
Ch_12	Mandsaur	Da_12	Jamtara	Gd_12	Hazaribag	Ya_12	Etawah		
Ch_13	Neemuch	Da_13	Koderma	Gd_13	Jamui	Ya_13	Faridabad		
Ch_14	Rajgarh	Da_14	Latehar	Gd_14	Jehanabad	Ya_14	Firozabad		
Ch_15	Ratlam	Da_15	Lohardaga	Gd_15	Khagaria	Ya_15	Gautam Buddha Nagar		
Ch_16	Sawai Madhopur	Da_16	Purba Singhbhum	Gd_16	Kushinagar	Ya_16	Ghaziabad		
Ch_17	Sehore	Da_17	Puruliya	Gd_17	Koderma	Ya_17	Gurgaon		
Ch_18	Shajapur	Da_18	Ranchi	Gd_18	Lakhisarai	Ya_18	Haridwar		
Ch_19	Sheopur	Da_19	Saraikela Kharsawan	Gd_19	Maharajganj	Ya_19	Jaipur		
Ch_20	Shivpuri	Da_20	West Midnapore	Gd_20	Munger	Ya_20	Jhajjar		
Ch_21	Tonk			Gd_21	Muzaffarpur	Ya_21	Jind		
Ch_22	Ujjain			Gd_22	Nalanda	Ya_22	Karauli		
Ch_23	Vidisha			Gd_23	Nawada	Ya_23	Karnal		
				Gd_24	Palamu	Ya_24	Kinnaur		
				Gd_25	Pashchim Champaran	Ya_25	Kurukshetra		
				Gd_26	Patna	Ya_26	Mahamaya Nagar (Hathras)		
				Gd_27	Purba Champaran	Ya_27	Mathura		
				Gd_28	Sahibganj	Ya_28	Meerut		
				Gd_29	Samastipur	Ya_29	Muzaffarnagar		
				Gd_30	Saran	Ya_30	Panipat		
				Gd_31	Sheikhpura	Ya_31	Rohtak		
				Gd_32	Siwan	Ya_32	Saharanpur		
				Gd_33	Vaishali	Ya_33	Sawai Madhopur		

References

1. World Bank. *The National Ganga River Basin Project*; World Bank: Washington, DC, USA, 2015. Available online: http://www.worldbank.org/en/news/feature/2015/03/23/india-the-national-ganga-river-basin-project (accessed on 20 January 2021).
2. Anand, J.; Gosain, A.K.; Khosa, R.; Srinivasan, R. Regional scale hydrologic modeling for prediction of water balance, analysis of trends in streamflow and variations in streamflow: The case study of the Ganga River basin. *J. Hydrol. Reg. Stud.* **2018**, *36*, 32–53. [CrossRef]

3. Sinha, D. Reviving the Ganga, at the Cost of Its Ecology! 2014. Available online: http://www.indiatogether.org/articles/ganga-river-waterway-reviving-and-impact-on-ecology-environment/print (accessed on 20 January 2021).
4. Gaur, A.; Amarasinghe, P. A river basin perspective of water resources and challenges 3. In *India Infrastructure Report 2011*; Oxford University Press: Oxford, UK, 2011.
5. Rasul, G. Water for growth and development in the Ganges, Brahmaputra, and Meghna basins: An economic perspective. *Int. J. River Basin Manag.* **2015**, *13*, 387–400. [CrossRef]
6. GoI (Government of India). *Ganges Basin*; Central Water Commission, Ministry of Water Resources: New Delhi, India, 2014. Available online: http://www.india-wris.nrsc.gov.in/ (accessed on 15 October 2020).
7. Moors, E.J.; Groot, A.; Biemans, H.; Scheltinga, C.T.V.; Siderius, C.; Stoffel, M.; Huggel, C.; Wiltshire, A.; Mathison, C.; Ridley, J.; et al. Adaptation to changing water resources in the Ganges basin, northern India. *Environ. Sci. Pol.* **2011**, *14*, 758–769. [CrossRef]
8. Shrestha, A.B.; Bajracharya, S.R.; Sharma, A.R.; Duo, C.; Kulkarni, A. Observed trends and changes in daily temperature and precipitation extremes over the Koshi river basin 1975–2010. *Int. J. Climatol.* **2017**, *37*, 1066–1083. [CrossRef]
9. Jeuland, M. Economic implications of climate change for infrastructure planning in transboundary water systems: An example from the Blue Nile. *Water Resour. Res.* **2010**, *46*. [CrossRef]
10. Maurer, J.M.; Schaefer, J.M.; Rupper, S.; Corley, A. Acceleration of ice loss across the Himalayas over the past 40 years. *Sci. Adv.* **2019**, *5*. [CrossRef] [PubMed]
11. Dyurgerov, M.B.; Meier, M.F. Twentieth century climate change: Evidence from small glaciers. *Proc. Natl. Acad. Sci. USA* **2000**, *97*, 1406–1411. [CrossRef] [PubMed]
12. Miller, J.D.; Immerzeel, W.W.; Rees, G. Climate Change Impacts on Glacier Hydrology and River Discharge in the Hindu Kush—Himalayas A Synthesis of the Scientific Basis. *Mt. Res. Dev.* **2012**, *32*, 461–467. [CrossRef]
13. World Bank Database. Electric Power Consumption (kWh per Capita). Available online: https://data.worldbank.org/indicator/EG.USE.ELEC.KH.PC (accessed on 20 January 2021).
14. CEA (Central Electricity Authority). All India Installed Capacity (in MW) of Power Stations. 2015. Available online: http://cea.nic.in/reports/monthly/installedcapacity/2015/installed_capacity03.pdf (accessed on 20 January 2021).
15. GOB (Government of Bangladesh). Power System Master Plan 2016 Summary. 2016. Available online: https://powerdivision.portal.gov.bd/sites/default/files/files/powerdivision.portal.gov.bd/page/4f81bf4d_1180_4c53_b27c_8fa0eb11e2c1/(E)_FR_PSMP2016_Summary_revised.pdf (accessed on 15 November 2020).
16. Gosain, A.K.; Rao, S.; Arora, A. Climate change impact assessment of water resources of India. *Curr. Sci.* **2011**, *101*, 356–371.
17. Luo, T.; Krishnan, D.; Sen, S. Parched Power: Water Demands, Risks, and Opportunities for India's Power Sector. Water Resources Institute. 2018. Available online: https://www.wri.org/publication/parched-power (accessed on 22 January 2021).
18. Mishra, H.; Denis, D.M.; Suryavanshi, S.; Kumar, M.; Srivastava, S.K.; Denis, A.F.; Kumar, R. Hydrological simulation of a small ungauged agricultural watershed Semrakalwana of Northern India. *Appl. Water Sci.* **2017**, *7*, 2803. [CrossRef]
19. Gosain, A.K.; Rao, S.; Srinivasan, R.; Reddy, N.G. Return-flow assessment for irrigation command in the Palleru river basin using SWAT model. *Hydrol. Process.* **2005**, *19*, 673–682. [CrossRef]
20. Kalcic, M.M.; Chaubey, I.; Frankenberger, J. Defining Soil and Water Assessment Tool (SWAT) hydrologic response units (HRUs) by field boundaries. *Int. J. Agric. Biol. Eng.* **2015**, *8*, 1. [CrossRef]
21. Arnold, J.G.; Srinivasan, R.; Muttiah, R.S.; Williams, J.R. Large area hydrologic modeling and assessment: Part I, model development. *J. Am. Water Resour. Assoc.* **1998**, *34*, 73–89. [CrossRef]
22. Srinivasan, R.; Ramanarayanan, T.S.; Arnold, J.G.; Bednarz, S.T. Large area hydrological modeling and assessment. Part II: Model application. *J. Am. Water Resour. Ass.* **1998**, *34*, 91–101. [CrossRef]
23. SRTM (Shuttle Radar Topography Mission). Global Digital Elevation Model. 2015. Available online: http://glcf.umd.edu/data/srtm/ (accessed on 5 June 2019).
24. Raju, K.S.; Nagesh Kumar, D. Ranking of global climate models for India using multi criterion analysis. *Clim. Res.* **2014**, *60*, 103–117. [CrossRef]
25. Van Rooijen, D.J.; Turral, H.; Biggs, T.W. Urban and industrial water use in the Krishna Basin, India. *Irrig. Drain.* **2009**, *58*, 406–428. [CrossRef]
26. Smith, M. *Manual for CROPWAT*; Version 5.2; FAO: Rome, Italy, 1988; 45p.
27. Amarasinghe, U.A.; McCornick, P.G.; Tushaar, S. India water demand scenarios to 2025 and 2050: A fresh look. In *Strategic Analyses of the National River Linking Project (NRLP) of India, Series 1: India Water Future: Scenarios and Issues*; Amarasinghe, U.A., Upali, A., Tushaar, S., Malik, R.P.S., Eds.; International Water Management Institute (IWMI): Colombo, Sri Lanka, 2009; pp. 67–83.
28. MOEFCC (Ministry of Environment Forests and Climate Change). S.O. 682(E). In *The Notification of the Government of India in the Ministry of Environment, Forest and Climate Change Vide Number S.O. 3305(E), Dated the 7th December, 2015. Gazette*; Government of India: New Delhi, India, 2015. Available online: http://www.moef.gov.in/sites/default/files/Thermalplantgazettescan.pdf (accessed on 20 January 2021).

 *sustainability*

Article

Evaluation of Qinghai-Tibet Plateau Wind Erosion Prevention Service Based on RWEQ Model

Yangyang Wang [1,2], Yu Xiao [1,2,*], Gaodi Xie [1,2], Jie Xu [3] , Keyu Qin [1,2], Jingya Liu [1], Yingnan Niu [1,2], Shuang Gan [1,2], Mengdong Huang [1,2] and Lin Zhen [1,2]

[1] Institute of Geographic Sciences and Natural Resources Research, Chinese Academy of Sciences, A11 Datun Road, Beijing 100101, China; wangyy.18b@igsnrr.ac.cn (Y.W.); xiegd@igsnrr.ac.cn (G.X.); qinkeyu@igsnrr.ac.cn (K.Q.); liujy.17b@igsnrr.ac.cn (J.L.); niuyn.19b@igsnrr.ac.cn (Y.N.); gans.17s@igsnrr.ac.cn (S.G.); huangmd.19s@igsnrr.ac.cn (M.H.); zhenl@igsnrr.ac.cn (L.Z.)
[2] University of Chinese Academy of Sciences, 19 A Yuquan Road, Shijingshan District, Beijing 100049, China
[3] School of Ecology and Nature Conservation, Beijing Forestry University, No. 35 Qinghua East Road, Haidian District, Beijing 100083, China; jiexu@bjfu.edu.cn
* Correspondence: xiaoy@igsnrr.ac.cn; Tel.: +86-10-6488-8157

Abstract: Ecosystem service research is essential to identify the contribution of the ecosystem to human welfare. As an important ecological barrier zone, the Qinghai-Tibet Plateau (QTP) supports the use of a crucial wind erosion prevention service (WEPS) to improve the ecological environment quality. This study simulated the spatiotemporal patterns of the WEPS based on the Revised Wind Erosion Equation (RWEQ) and its driving factors. From 2000 to 2015, the total WEPS provided in the QTP ranged from 1.75×10^9 kg to 2.52×10^9 kg, showing an increasing and then decreasing trend. The average WEPS service per unit area was between 0.72 kg m^{-2} and 1.06 kg m^{-2}. The high-value areas were concentrated in the northwest and north of the QTP, and the total WEPS in different areas varied significantly from year to year. The average retention rate of the WEPS in the QTP was estimated to be 57.24–62.10%, and high-value areas were mainly located in the southeast of the QTP. The total monetary value of the WEPS in the QTP was calculated to be between 223.56×10^9 CNY and 321.73×10^9 CNY, and the average WEPS per unit area was between 0.08 CNY m^{-2} and 0.13 CNY m^{-2}, showing a declining–rising–declining trend. The high-value areas gradually expanded to the west and east of the QTP. The slope was the most important factor controlling the spatial differentiation of the WEPS, followed by the landform type, average annual precipitation, and average annual wind speed, and human activities such as land-use change could improve the WEPS by returning farmland to grassland and desertification control in the QTP.

Keywords: wind erosion prevention service; revised wind erosion equation; geo-detector

Citation: Wang, Y.; Xiao, Y.; Xie, G.; Xu, J.; Qin, K.; Liu, J.; Niu, Y.; Gan, S.; Huang, M.; Zhen, L. Evaluation of Qinghai-Tibet Plateau Wind Erosion Prevention Service Based on RWEQ Model. *Sustainability* **2022**, *14*, 4635. https://doi.org/10.3390/su14084635

Academic Editor: Antonio Miguel Martínez-Graña

Received: 25 February 2022
Accepted: 4 April 2022
Published: 13 April 2022

Publisher's Note: MDPI stays neutral with regard to jurisdictional claims in published maps and institutional affiliations.

1. Introduction

The Qinghai-Tibet Plateau (QTP) is the most unique geographical unit in the world because of its high altitude and mountainous landforms. The QTP is considered an important ecological barrier in south-western China, as it supports various ecosystem services in the region and in neighboring countries [1,2]. Among these ecosystem services, wind erosion prevention service (WEPS) is very important because of the extreme climate and low vegetation coverage in most parts of the plateau [3–7]. The average wind erosion rate for the plateau as a whole reached the medium erosion standard in 2001 [8], and the area of desertification in the QTP accounted for 15.1% of the total region in 2015 [9]. Therefore, it is critical to assess the regional WEPS and analyze the influencing factors in order to forecast wind erosion and provide a scientific basis for desertification control [10].

The WEPS can be estimated by the difference between potential and actual sand erosion by ecosystems. Several models have been adopted to simulate sand erosion by ecosystems, including the sediment transport equation [11], the wind erosion equation

(WEQ) [12], the Texas erosion analysis model (TEAM) [13], the description model [14], the revised wind erosion equation (RWEQ) [15], the wind erosion prediction system [16], and the wind erosion stochastic simulator (WESS) [17]. Among these models, the RWEQ model is most commonly used for sand erosion simulations in the QTP. For example, Jiang [18] used the RWEQ model to evaluated wind erosion modules in Qinghai and showed that the total area of wind erosion in Qinghai Province accounts for more than half of the regional area, and the level of wind erosion damage in the Qaidam Basin is relatively serious. Huang [19] analyzed the WEPS of ecosystems in the Tibet Plateau from 1990 to 2010 by RWEQ, showing that the annual average soil wind erosion modulus, the average sand fixation service quantity, and the retention rate of the average annual WEPS in the Tibet Plateau were 1.58 kg m^2, 18.99 $\times$ 10^8 t, and 66.5%, respectively. Teng [10] adopted the RWEQ model to simulate the WEPS of the QTP from 2000 to 2015. However, studies on WEPS in this area have been limited, with most studies only involving local areas [20,21]. Accordingly, few studies have comprehensively described the WEPS and its dynamic characteristics over the whole QTP. As the terrain of the QTP is complex, the spatial pattern of sand erosion and WEPS may differ significantly between the east and west parts [9]. Few studies have examined wind erosion throughout the QTP or investigated the changes among different areas and ecosystems.

In recent years, many scholars have used multi-source data and methods such as the linear correlation model, multiple regression model, structural equation model, and constraint effect to investigate the factors (vegetation coverage, wind speed, precipitation, animal husbandry development, land use, etc.) driving the WEPS [10,22–27]. Latocha et al. showed that changes in land use have affected soil erosion to a larger extent than climate change in the Sudetes Mts. within the last 150 years [28]. Garbrecht et al. showed that warmer temperatures and reduced rainfall could exacerbate soil wind erosion in farmland areas in the southern Great Plains [29]. Li argued that the increase in the afforestation area is the main factor driving the improvement of WEPS in Inner Mongolia [30]. Sharratt showed that increase in the crop yield in the Colombian Plateau could reduce damage from soil wind erosion [31]. Few studies have conducted a driving factor analysis of the WEPS in the QTP. Teng [10] simulated the influences of climate changes and human activities on sand erosion and the WEPS in the QTP from 2000 to 2015 through comparative analysis of spatiotemporal dynamics. The above studies have helped us to understand the driving mechanism responsible for the spatial distribution pattern of WEPS, but there are still gaps in our knowledge on the interaction of different driving factors.

The geographic detector is a spatial statistical method based on spatial differentiation that is used to reveal the driving mechanism of a variable [32]. The geographic detector can analyze the interaction between two independent variables and determine the specific effect type while detecting the influences of variables with attributes of different types and values on the spatial distribution of the dependent variable. At present, the geographic detector is widely used in the analysis of the change process of spatial patterns in many fields such as meteorology, environmental pollution, ecology, human health, and regional planning [33–39]. Few studies have adopted the geographic detector to analyze the driving factors related to the WEPS. Zhang [40] conducted an exploratory study on the factors driving changes in the WEPS in Xilingol League using the geographic detector, and proved that the application of the geo-detector analysis on the WEPS is rational and scientific. The application of geographic detectors to analyze the factors driving changes in the spatial pattern of WEPS can account for the influences of type variables and numerical variables on service changes, analyze the interactions between various factors, and make up for the lack of analysis of factors driving the spatial distribution of the WEPS.

Therefore, this study adopted the RWEQ model to simulate the WEPS in the QTP from 2000 to 2015, estimated its monetary value using the restoration cost method, evaluated the differences in the material and monetary value of the WEPS among different cities and ecosystems, and analyzed the contributions and interactions of natural and human factors in the formation and evolution of the spatial pattern of the WEPS. This study can be used to improve the understanding of the spatial patterns of the WEPS in the QTP and provides crucial knowledge that can be used to make decisions about ecological restoration and protection to improve the WEPS in different ecosystems.

2. Methodology and Data

2.1. Study Area

The QTP is located in southwestern China, and involves five provinces (autonomous regions): Qinghai, Tibet, Sichuan, Xinjiang, Gansu, and Yunnan (Figure 1). In the western part of the QTP is the Pamir Plateau and the Karakoram Mountains, bordering Tajikistan and Pakistan, and in the southern part there are the Himalayas, bordering Nepal, Bhutan, and other countries. The QTP has an average elevation of more than 4000 m. The highest altitude is found on Mount Everest, 8844.43 m, whereas the Jinsha River is only 1503 m above sea level. The special geographical location of the QTP means that it has a unique plateau climate with a cold, hypoxic, and arid climate with strong radiation. The temperature is 15–20 °C lower than that of other areas at the same latitude. The complex terrain in this area also causes the climate to vary. The temperature decreases from southeast to northwest, and the average annual temperature can reach below −6 °C. Due to the blocking effect of the multiple high mountains, the annual precipitation decreases from south to north, and the annual precipitation is less than 50 mm. The total annual solar radiation on the QTP can be as high as 5400–7900 MJ·m^{-2}·year^{-1}, and the total annual sunshine hours are 2500–3200 h.

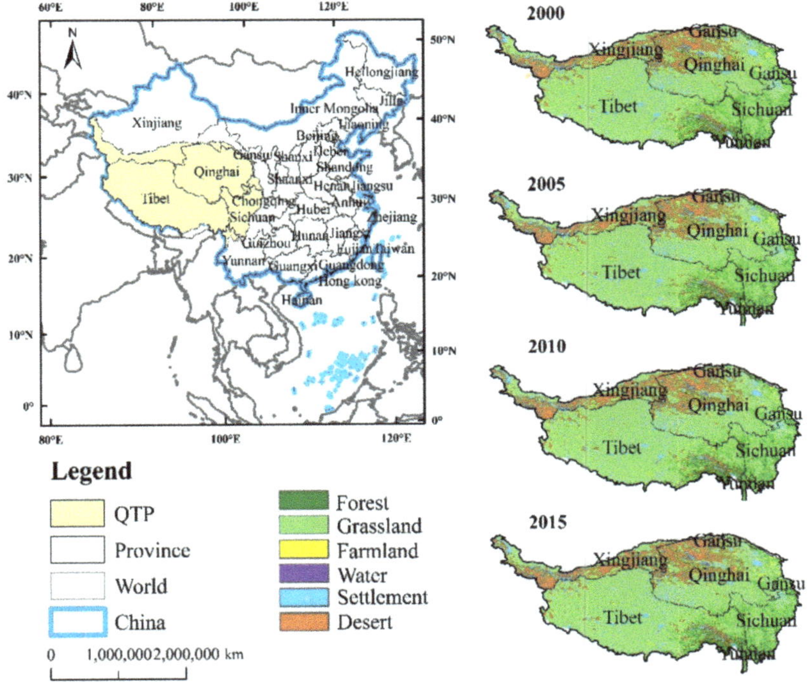

Figure 1. Location of the Qinghai-Tibet Plateau (QTP) and the land cover map from 2000 to 2015.

2.2. Method and Data

2.2.1. Calculation of WEPS

The WEPS represents the reduction of wind erosion caused by vegetation and is calculated as the difference between the potential wind erosion under bare soil conditions and the actual wind erosion under vegetation conditions. The potential and actual wind erosion are calculated with the RWEQ model [15,41]. The actual wind erosion is the level of soil erosion under the condition of vegetation coverage in a real scenario, whereas the potential wind erosion is the level of soil erosion under bare land conditions without vegetation. The WEPS was calculated as follows:

$$SL = \frac{2z}{s^2} \times Q_{max} \times e^{-\left(\frac{z}{s}\right)^2} \tag{1}$$

$$Q_{max} = 109.8 \times \left(WF \times EF \times SCF \times K' \times C\right) \tag{2}$$

$$s = 109.8 \times \left(WF \times EF \times SCF \times K' \times C\right)^{-0.3711} \tag{3}$$

$$SLR = \frac{2z}{s_r{}^2} \times Q_{rmax} \times e^{-\left(\frac{z}{s_r}\right)^2} \tag{4}$$

$$Q_{rmax} = 109.8 \times \left(WF \times EF \times SCF \times K'\right) \tag{5}$$

$$s_r = 109.8 \times \left(WF \times EF \times SCF \times K'\right)^{-0.3711} \tag{6}$$

$$G = SLR - SL \tag{7}$$

where SLR represents the potential wind erosion (kg m^{-2}); Q_{rmax} represents the maximum potential sediment transport capacity (kg m^{-1}); S_r represents the potential critical plot length (m); SL represents the actual wind erosion (kg m^{-2}); Q_{max} represents the maximum sediment transport capacity (kg m^{-1}); S represents the critical plot length (m); G represents the WEPS (kg m^{-2}); z represents the calculated downwind distance (m); and WF, EF, SCF, K', and C correspond to the factors of weather (kg m^{-1}), the soil erodibility fraction (%), soil crust factor (dimensionless), the soil roughness factor (dimensionless), and vegetation (dimensionless), respectively. The WF was calculated as follows:

$$WF = Wf \times \frac{\rho}{g} \times SW \times SD \tag{8}$$

$$Wf = u_2 \times (u_2 - u_1)^2 \times N_d \tag{9}$$

where Wf is the wind factor (m^3 s^{-1}); g represents the gravitational acceleration (m^2 s^{-1}); ρ represents the air density (kg m^{-3}); SW is the soil moisture factor (dimensionless); and SD represents the snow cover factor (dimensionless), which is the ratio of days with no snow cover to the total number of days studied. A snow cover depth less than 25.4 mm indicates no snow cover. The expression u_1 represents the threshold wind velocity at a height of 2 m (m s^{-1}). The threshold wind velocity for sand lands and sandy grasslands is 5 m s^{-1}; u_2 represents the monthly average wind velocity at a height of 2 m (m s^{-1}); and N_d represents the number of days with a monthly average wind velocity that exceeds the threshold wind velocity.

In this study, EF and SCF were assumed to be unchanged over time and were calculated using the soil raster data that were converted from a 1:1,000,000 soil data shapefile provided by the Harmonized World Soil Database (HWSD) developed by the International Institute for Applied System Analysis (IIASA) of the FAO. For the calculation, the classification criteria for soil texture taken from international data in the database were converted to American criteria to meet the factor calculation requirements by using the logistic growth model proposed by Skaggs et al. [42]. The values of EF and SCF were calculated as follows:

$$EF = 29.09 + 0.31 \times Sa + 0.17 \times Si + 0.33 \times \frac{Sa}{Cl} - 2.59 \times OM - 0.95 \times CaCO_3 \tag{10}$$

$$SCF = \frac{1}{1 + 0.0066 \times cl^2 + 0.021 \times OM^2} \tag{11}$$

where Sa, Si, Cl, OM, and $CaCO_3$ represent the content of sand, silt, clay, organic matter, and calcium carbonate (%), respectively.

The value of K' was calculated with the following equation:

$$K' = \cos \alpha \tag{12}$$

where α represents the slope gradient and was calculated by a digital elevation model (DEM) in ArcGIS. The value of C was computed as follows:

$$C = e^{-0.0483 \times SC} \tag{13}$$

$$SC = \frac{NDVI - NDVI_{min}}{NDVI_{max} - NDVI_{min}} \tag{14}$$

where SC represents the vegetation coverage (%); $NDVI_{max}$ represents the normalized difference vegetation index (NDVI) value of a highly vegetated grid; and $NDVI_{min}$ represents the NDVI value of a bare land grid. The values of $NDVI_{max}$ and $NDVI_{min}$ correspond to the NDVI values at cumulative frequencies of 95% and 5%, respectively.

2.2.2. Calculation of Retention Rate of WEPS

The retention rate of WEPS more accurately quantified the contribution of ecosystems on WEPS and avoided the impact of climatic factors on WEPS. The retention rate of WEPS was computed as follows:

$$F = \frac{G}{SLR} \tag{15}$$

where F is the WEPS retention rate.

2.2.3. Calculation of the WEPS Monetary Value

The monetary value of the WEPS can be measured as the cost required to restore sandy land. It was calculated with the following formula:

$$V_{sf} = \frac{SFQ}{\rho \cdot h} \times c \tag{16}$$

where V_{sf} is the WEPS value in CNY (CNY is the Chinese Currency); ρ is the soil bulk density (g cm^{-3}), which was 1.25 g cm^{-3} [43]; h is the thickness of the soil, which was calculated as 0.71 m [43]; and c is the average cost of the sand control project (CNY m^{-2}), based on the costs of afforestation and grass grid desertification control in The QTP (1.135 CNY m^{-2}).

2.2.4. Geographical Detector

The geographical detector (geo-detector) was mainly used to analyze correlations of the WEPS with the five selected impact factors and multiple impact factor interactions. This is because the geo-detector q values have clear physical meaning with no assumption of linearity, allowing us to objectively show that the dependent variable explains $100 \times q\%$ of the difference [44]. This study applied factor detectors and interaction detectors in geo-detectors to analyze the correlations between the WEPS and selected impact factors more comprehensively. The formula used to calculate for the value of q was

$$q = 1 - \frac{SSW}{SST} = \sum_{h=1}^{L} N_h \delta_h{}^2 \tag{17}$$

where $h = 1, \ldots, L$ is the stratification of variable Y or factor X, that is, the classification or partition; N_h and N are the number of units in the layer and the whole area, respectively; $\delta_h{}^2$ and δ^2 are the variance of Y with in the layer and in the whole area, respectively; and SSW and SST are sum of the variances within the layer and the total variance of the whole area, respectively.

Risk area detection was used to judge whether there was a significant difference in the average WEPS between the sub-areas of two influencing factors. This was tested using the statistic

$$t = \frac{\overline{Y}_{h=1} - \overline{Y}_{h=2}}{\left[\dfrac{Var\left(\overline{Y}_{h=1}\right)}{n_{h=1}} + \dfrac{Var\left(\overline{Y}_{h=2}\right)}{n_{h=2}} \right]^{1/2}} \tag{18}$$

where $\overline{Y}_h$ is the average WEPS of the sub-areas h, n_h is the number of samples in the sub-areas (h), and *Var* represents the variance.

Ecological detection was used to evaluate whether the influences of two influencing factors, X1 and X2, on the spatial differentiation of the windbreak and sand fixation were significant. This was done using the F statistic:

$$F = \frac{N_{X1}(N_{X2} - 1)SSW_{X1}}{N_{X2}(N_{X1} - 1)SSW_{X2}} = \frac{N_{X1}(N_{X2} - 1)\sum_{h=1}^{L1} N_h\sigma_h^2}{N_{X2}(N_{X1} - 1)\sum_{h=1}^{L2} N_h\sigma_h^2} \tag{19}$$

where N_{X1} and N_{X2} are the sample numbers of impact factors X1 and X2, respectively; SSW_{X1} and SSW_{X2} represent the sum of the variance within the layers corresponding to the impact factor stratification; and L_1 and L_2 represent the number of layers in impact factors X1 and X2.

Interactions occur between different factors. To evaluate whether the explanatory power of the dependent variable increases when the influencing factors work together, it was necessary to select the dominant factors. Interaction detection can identify interactions between different influencing factors, as showed by the following table (Table 1, Figure 2). Achievement of the geographic detector was operated with geo-detector software (http://geo-detector.cn/, accessed on 15 January 2022). The data needed to be separated and decentralized for the calculation due to their different spatial dimensions. ArcGIS software was used to extract the values of different data.

Table 1. Detecting factor indicators and class break values.

Factor	Class Break Value													
	1	2	3	4	5	6	7	8	9	10	11	12	13	14
Y	0.67	2.24	4.26	7.95	28.56									
X1	PL	PLA	H	SRM	MRM	BRM	EBRM	-	-	-	-	-	-	-
X2	Ls	SLs	Cs	Drs	Des	Ps	Sas	Wfs	Ses	Ms	As	Fa	Ur	Os
X3	2.06	4.5	7.13	9.76	12.58	15.58	18.76	22.34	27.03	47.68	-	-	-	-
X4	0.1	0.17	0.24	0.33	0.43	0.53	0.62	0.7	0.77	0.91	-	-	-	-
X5	2126.31	3506.49	4523.47	5685.72	6847.98	8010.24	9390.42	11,279.09	13,894.16	18,979.04	-	-	-	-
X6	−0.95	0.68	2.01	3.34	4.67	6.07	7.77	9.54	11.24	14.57	-	-	-	-
X7	0.64	1.07	1.41	1.70	2.00	2.33	2.64	2.94	3.31	5.04	-	-	-	-
X8	2050	2624	3110	3597	4038	4426	4761	5074	5458	8313	-	-	-	-
X9	1.00	2.00	3.00	4.00	>5.00	-	-	-	-	-	-	-	-	-
X10	1.00	2.00	3.00	4.00	>5.00	-	-	-	-	-	-	-	-	-

Y: grade of average WEPS (g km^{-2}), X1: landform types (PL: plains, PLF: platforms, H: hills, SRM: small rolling mountains, MRM: medium rolling mountains, BRM: big rolling mountains, EBRM: extremely big rolling mountains); X2: soil type (Ls: leaching soil, SLs: semi leaching soil, Cs: calcareous soil, Drs: dry soil, Des: desert soil, Ps: primary soil, Sas: semi-aqueous soil, Wfs: water-forming soil, Ses: saline earth, Ms: man-made soil, As: alpine soil, Fa: ferro-alumina, Ur: urban area, Os: other soil); X3: slope (°); X4: average annual NDVI, X5: average annual precipitation (mm), X6: average annual temperature (°C); X7: average annual wind speed (m·s^{-1}); X8: DEM (m); X9: average annual GDP (10^4 CNY km^{-2}); X10: population density (human km^{-2}).

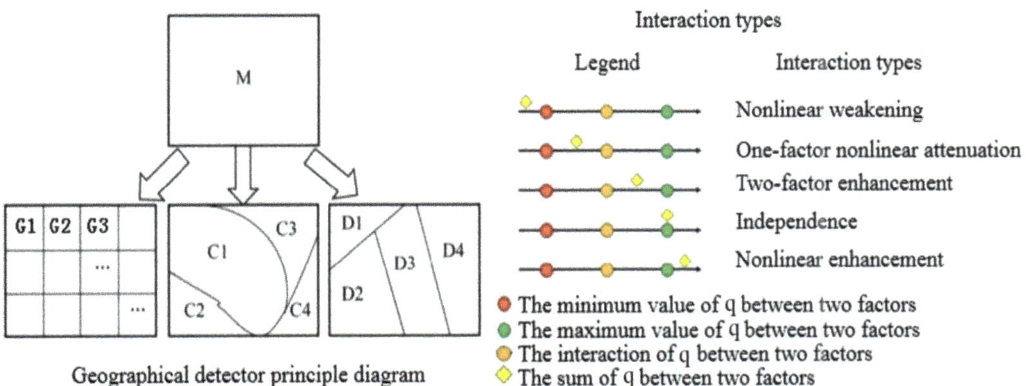

Figure 2. Diagram of the geographical detector principles and interaction types [32,40] M: the study area; G1, G2, G3: the dependent variables used for detection (average WEPS from 2000 to 2015 as Y factor); C1, C2, C3, C4: sub-areas of impact factor C; D1, D2, D3, D4: sub-areas of impact factor D; q: explanatory power of the impact factors on the WEPS.

Changes in the WEPS are simultaneously affected by natural factors such as climate, soil, and vegetation types, as well as by human activities such as grazing withdrawal and grazing prohibition (see previous research [39,40]). In this study, eight natural factors and two socioeconomic factors were selected as detection factors and used to analyze the factors driving change in the average annual WEPS in the QTP from 2000 to 2015 (Table 1). Based on data types and operability, the landform types and soil types of the QTP were divided into 7 and 14 categories, respectively. The numerical variables were classified by the natural discontinuity method in ArcGIS. For example, the population density and GDP per unit area were divided into 5 categories, and the remaining variables were divided into 10 categories (Table 1).

2.2.5. Data Source

Meteorological datasets, including data on the daily temperature, precipitation, wind speed, and solar radiation data from 2000 to 2015, were acquired from meteorological stations. Snow cover data with a 25-km resolution were obtained from a long-term snow depth dataset [45], which was accessed from the Cold and Arid Region Science Data Center (http://westdc.westgis.ac.cn, accessed on 15 January 2022). NDVI values (spatial resolution 500 m) were derived from the international scientific data mirror website of the computer network information center of the Chinese Academy of Sciences (http://www.gscloud.cn, accessed on 15 January 2022). The DEM data (spatial resolution 30 m), GDP data (spatial resolution 1 km), 1:1,000,000 landform data, and land cover data (spatial resolution 1 km) were derived from the Resource and Environment Data Cloud Platform of the Chinese Academy of Sciences (CAS) (http://www.resdc.cn, accessed on 15 January 2022). The 1:1,000,000 soil data shapefile was provided by the Harmonized World Soil Database (HWSD) developed by the International Institute for Applied System Analysis (IIASA) of FAO. During the calculation, the international soil texture classification criteria in the database were converted to American criteria to meet the factor calculation requirements by using the logistic growth model proposed by Skaggs [42]. The 1:4 million calcium carbonate content data were accessed from the Data Sharing Infrastructure of Earth System Science (http://www.geodata.cn, accessed on 15 January 2022). All of the above data were resampled to a 1 km spatial resolution.

3. Results

3.1. WEPS and Its Monetary Value in the QTP

From 2000 to 2015, the total *SLR* (potential wind erosion) in the QTP was computed to range from 56.36×10^8 t to 68.13×10^8 t. There was an increase of 5.67×10^8 t, and a trend of "rising–declining–declining" was shown. The *SLR* per unit area was calculated to be from 0 to 60.86 kg m^{-2}, and the average *SLR* per unit area was estimated to be from 2.40 kg m^{-2} to 2.95 kg m^{-2} (Figure 2). The *SLR* per unit area was higher in the western and northern parts, whereas it was lower in the southeastern parts. This shows that the western and northern areas of the QTP, which contain desert areas, are more threatened by wind erosion.

The total *SL* (actual wind erosion) in the QTP was calculated to be from 38.55×10^8 t to 43.39×10^8 t during the year (Figure 3). Compared to 2000, the total *SL* was 0.65×10^8 t lower in 2015. The *SL* per unit area was the highest in 2005, reaching 60.83 kg m^{-2}, and a continuous decline occurred from 2005 to 2015. The average *SL* per unit area was estimated to be 1.67 kg m^{-2}, 1.90 kg m^{-2}, 1.75 kg m^{-2}, and 1.65 kg m^{-2} in 2000, 2005, 2010, and 2015, decreasing by 0.02 kg m^{-2}. The *SL* per unit area was higher in the northern and western parts and lower in the eastern and southern parts. This shows that strong wind erosion occurred in the northern and western areas of the QTP. In addition, when comparing the SL and SLR in the western area, it can be seen that the vegetation cover significantly reduced the occurrence of wind erosion (Figures 1 and 3).

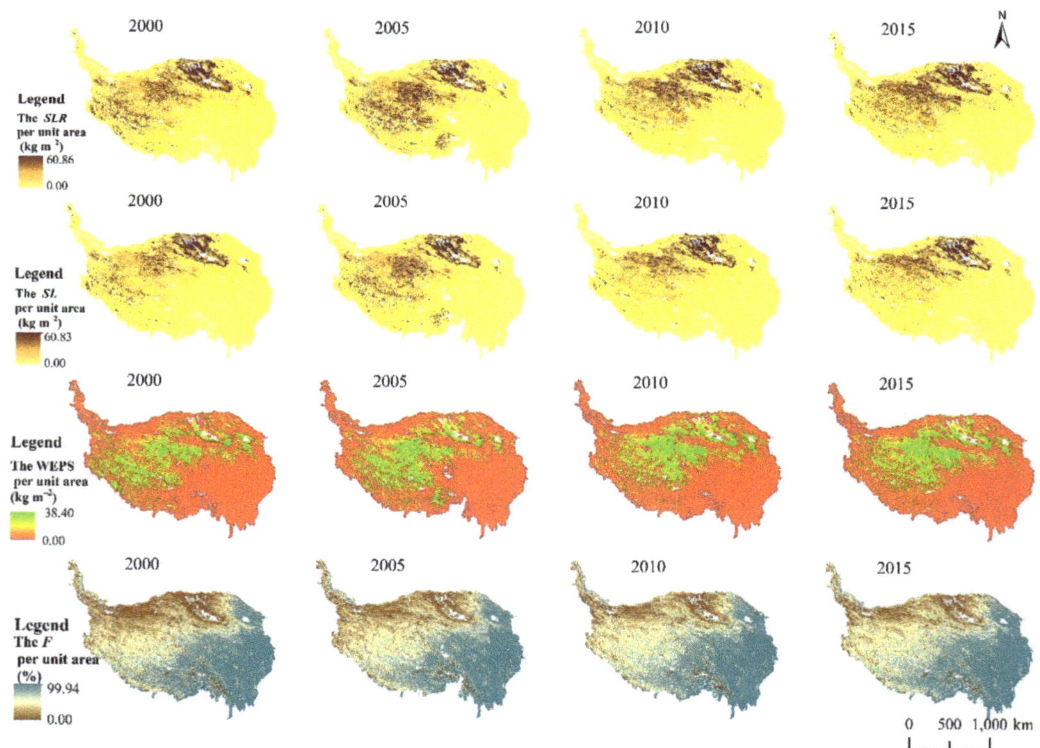

Figure 3. The amount of WEPS, SLR, SL, and F per unit area of the QTP from 2000 to 2015.

The total WEPS in the QTP was computed as 17.48×10^8 t in 2000 and 25.15×10^8 t in 2015, and the total monetary value was estimated to be 22.36×10^8 CNY in 2000 and 30.52×10^8 CNY in 2015. The WEPS decreased by 6.39×10^8 t in 2015 compared with the value in 2000, and its monetary value was reduced by 8.17×10^8 CNY. The WEPS per unit area was calculated to range from 0 to 38.40 kg m^{-2} in 2000 to 2015 (Figure 2), and the average WEPS per unit area was estimated to be from 0.09 kg m^{-2} to 0.13 kg m^{-2}. The WEPS per unit area was higher in the central and southwestern parts of the QTP, where grassland is the main land type, whereas it was found to be lower in the northern parts (Figures 1 and 3). It can be speculated that the grassland distribution area played a major role in preventing wind erosion in the QTP.

The average F value (retention rate of WEPS) in the QTP was between 57.24% and 62.10% from 2000 to 2015 (similar to that found in previous research [10]), showing an increasing trend. The F value was calculated to range from 0 to 99.94%. Different from the inter-annual variation in the WEPS, potential wind erosion, and actual wind erosion values, the retention rate of the WEPS in the QTP was highest in 2015 and lowest in 2000. Areas with higher WEPS retention rate were mainly located in the southeast where there is better vegetation coverage (Figure 1). A relatively obvious geographical distribution from northwest to southeast was shown.

3.2. WEPS and Its Monetary Value of Different Province and Cities in the QTP

Among the provinces in the QTP, Qinghai and the Tibet Autonomous Region were found to account for more than 90% of the WEPS. The total WEPS in Tibet showed the highest value, reaching 16.15 × 10^8 t in 2005, and its monetary value was calculated as 20.66 × 10^8 CNY. This was followed by Qinghai Province and Yunnan Province.

From 2000 to 2015, there were significant differences in the WEPS. The total WEPS in Xinjiang showed a gradual increasing trend over time, and the other four provinces showed a trend of "rising–declining". The Tibet Autonomous Region and Yunnan Province reached the highest total WEPS values in 2005, whereas in 2010, the maximum values occurred in Qinghai and Sichuan (Figure 4). The total WEPS in all provinces increased from 2000 to 2015, with Qinghai Province showing the largest increase of 3.55 × 10^8 t, followed by the Tibet Autonomous Region. It can be concluded that Tibet and Qinghai have greater WEPS concentrations.

In 2000, 2010, and 2015, the average WEPS per unit area was the highest for Qinghai province with values of 0.89 kg m^{-2}, 1.52 kg m^{-2}, and 1.42 kg m^{-2}, whereas in 2005, the Tibet Autonomous Region had the highest average WEPS per unit area, and the Yunnan province had the lowest values. From 2000 to 2015, the average WEPS per unit area in Sichuan province showed a gradual increasing trend over time, and that in Yunnan showed a trend of "rising-declining-rising". The other three provinces showed a trend of "rising-declining". However, the average WEPS per unit area increased for all provinces, with that in Qinghai increasing the most at 0.52 kg m^{-2}, followed by Xinjiang with 0.30 kg m^{-2}.

Among the 37 cities in the QTP, Nagqu was shown to have the highest total WEPS, accounting for 25–33% of the total. In 2015, its total WEPS amounted to 6.17 × 10^8 t. This was followed by Haixi, accounting for 16–22%. From 2010 to 2015, all cities showed a trend of "rising–declining". Compared with 2000, the total WEPS decreased in Shannan, Shigatse, Haibei, and Zhangye in 2015. Shigatse decreased by 0.54 × 10^8 t. Zhangye decreased the least, by only 0.06 × 10^8 t. The largest increase in the WEPS occurred in Haixi, where there was an increase of 1.91 × 10^8 t, followed by Naqu and Yushu, which increased by 0.19 × 10^8 t and 0.17 × 10^8 t, respectively.

In 2000, the average WEPS per unit area was the highest for Hainan Region at 1.33 kg m^{-2}, whereas during 2005 and 2015, Naqu reached the highest average WEPS per unit area with a range from 1.82 to 2.43 Guangyuan city had the lowest WEPS per unit area. From 2000 to 2015, the average WEPS per unit area in cities such as Mianyang, Ganzi, Ya'an, Yushu, Chengdu, Deyang, and Guangyuan showed a gradual increasing trend over time. In Shannan, Shigatse, Haibei, and Zhangye there was a volatile decreasing trend, whereas in other cities, a volatile growing trend was shown. Compared with 2000, the average WEPS per unit area in Yushu increased the most (0.71 kg m^{-2}), followed by Haixi with 0.64 kg m^{-2}. The average WEPS per unit area decreased the most in Shigatse (by 0.32 kg m^{-2}). Overall, the WEPS in most of the cities in the QTP improved.

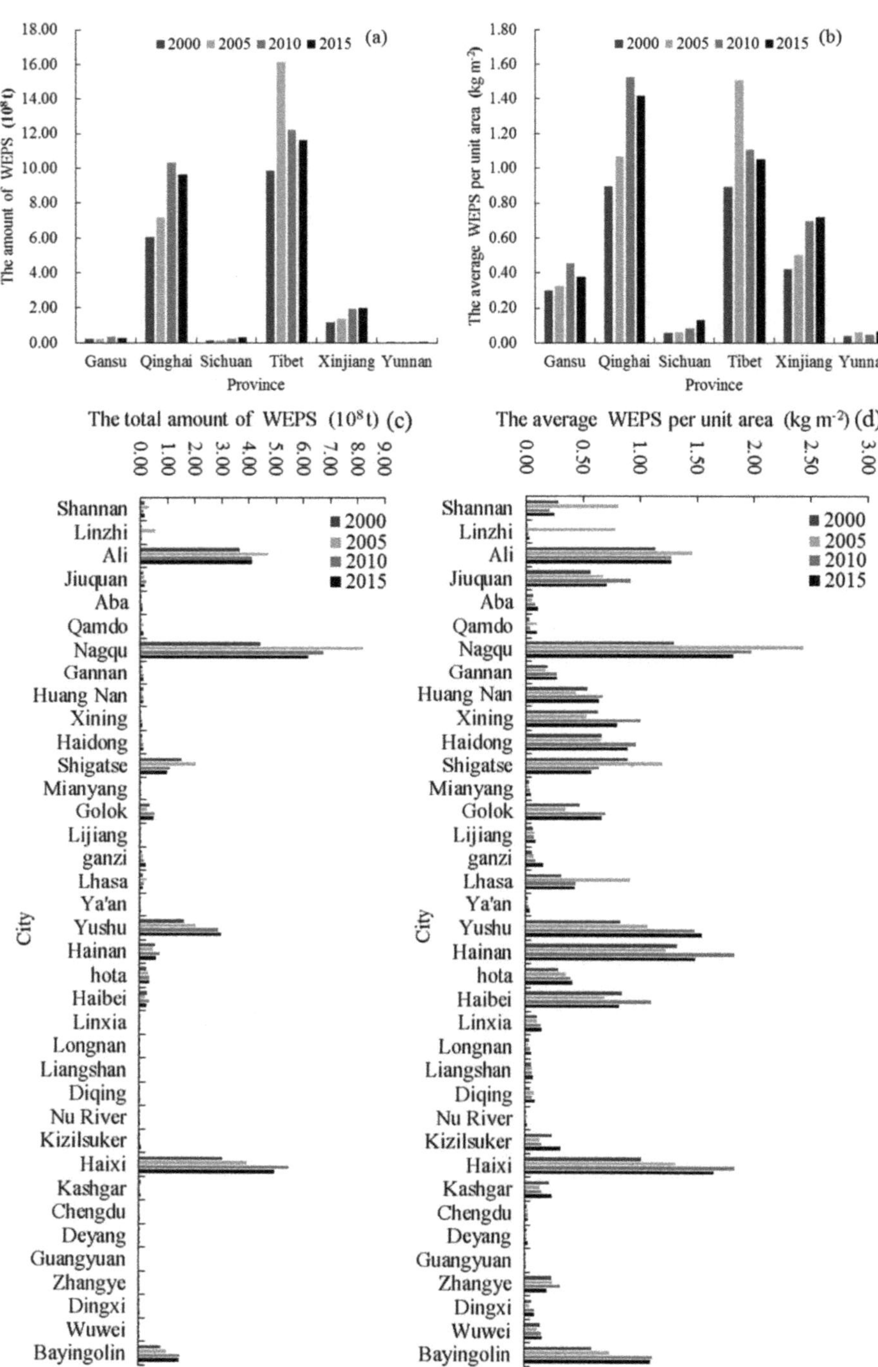

Figure 4. The WEPS in different provinces (**a**,**b**) and cities (**c**,**d**) in the QTP from 2000 to 2015.

3.3. WEPS and Its Monetary Value of Different Ecosystems in the QTP

Among the different types of ecosystems in the QTP, the grassland ecosystem has the highest WEPS, accounting for more than 68% of the total. The WEPS of the grassland ecosystem was estimated to range from 12.7×10^8 t to 18.29×10^8 t from 2010 to 2015. The monetary value was calculated to be from 16.24×10^8 CNY to 23.39×10^8 CNY. The settlement ecosystem had the lowest WEPS. In 2015, the average WEPS per unit area was the highest for the grassland ecosystem, 8.06 kg m^{-2}, followed by the settlement ecosystem, and the value for the forest ecosystem was the lowest. From 2000 to 2015, the total WPES in all ecosystems showed an increasing trend, among which the forest ecosystem increased the most, reaching 3.61×10^8 t, followed by the desert ecosystem (Figure 5).

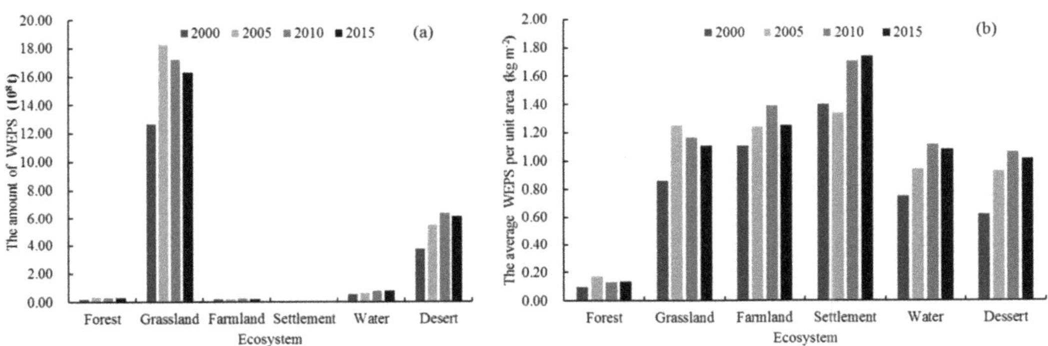

Figure 5. The total (**a**) and average (**b**) WEPS of different ecosystems in the QTP from 2000 to 2015.

3.4. Change of Spatial Pattern of WEPS in the QTP

From 2000 to 2015, the change in SLR per unit area in the QTP ranged from -27.74 kg m^{-2} to 15.29 kg m^{-2}, with an average change of 0.24 kg m^{-2}. An increase in SLR per unit area mainly occurred in the western areas of northeastern QTP, on the eastern edge, and in the central and northern parts of the area. Both the central and western parts showed a downward trend, especially the western part. The change in SL per unit area in the QTP ranged from -27.04 kg m^{-2} to 13.03 kg m^{-2}, with an average change of 0.02 kg m^{-2}. An increase in SL per unit area mainly occurred in northeastern QTP, on the eastern edge, and in the central and northern parts. Both the central and western parts showed a downward trend.

The change in WEPS per unit area in the QTP ranged from -14.85 kg m^{-2} to 20.40 kg m^{-2}, with an average change of 0.26 kg m^{-2}. Increases in the WEPS per unit area mainly occurred in the western part of northeastern QTP, on the eastern edge, and in the central and northern parts. Both the central and western parts showed a downward trend. The WEPS per unit area increased most significantly in Huzuoqi, Xinbalhuyouqi, and Chenbalhu in the west of Hulunbuir. In Zhenglan Banner in the southeast of Xilin Gol League, West Uzhumuqin in the northwest, the central and northern parts of Chifeng City, and the southeast of Ulanchabu City, the decline in the WEPS per unit area was more significant. From 2000 to 2015, the change in F per unit area in the QTP ranged from -89.17% to 97.89%, with an average change of 5.00%. A decrease in F per unit area mainly occurred in the western parts of northeastern QTP, on the eastern edge, and in the central and northern parts. Both the central and western parts showed an upward trend (Figure 6).

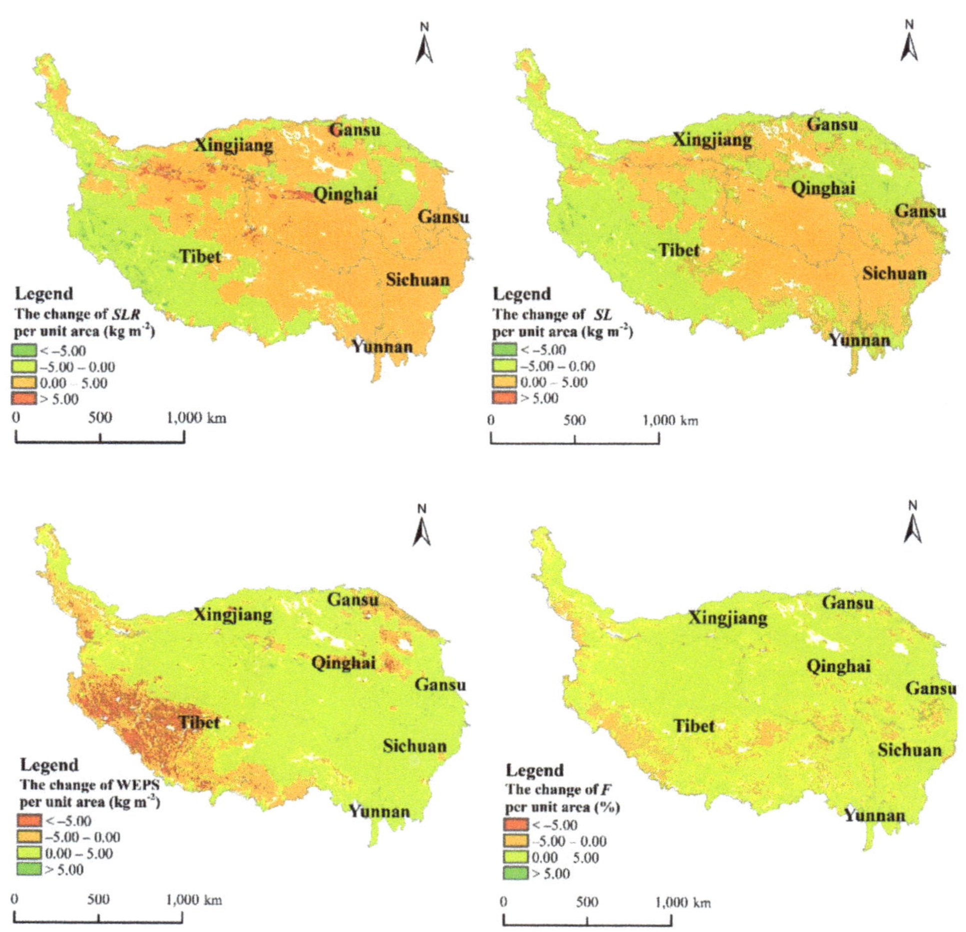

Figure 6. The changes in the WEPS, SLR, SL, and F per unit area in the QTP from 2000 to 2015.

3.5. WEPS in the QTP and the Detection of Its Driving Forces

This study set the grade of the average annual WEPS per unit area from 2000 to 2015 as the Y factor, and 23,853 samples were generated using geographic detectors to match and extract attribute values with other raster data layers (Figure 7). The landform type, soil type, slope, average annual NDVI, average annual precipitation, average annual temperature, average annual wind speed, DEM, average annual gross domestic product (GDP), and average annual population density were set as factors X1–X10 (Figure 8). The results show that the detected q values were as follows: slope (0.440) > landform type (0.275) > average annual precipitation (0.223) > average annual wind speed (0.186) > average annual NDVI (0.142) > average annual population density (0.124) > average annual GDP (0.076) > DEM (0.071) > average annual temperature (0.057) > soil type (0.053) (Table 2). Slope was found to be the most important factor controlling the spatial differentiation of the WEPS, followed by landform type, average annual precipitation, and average annual wind speed. The WEPS grade was not significantly affected by the average annual temperature or soil type. The influences of human factors, such as the average annual population density and average annual GDP, on the grade of the average annual WEPS tended to be less than the influences of most natural factors, which indicates that natural factors have dominated the formation of wind erosion in the area, but human activities have still had an impact on the WEPS in the QTP.

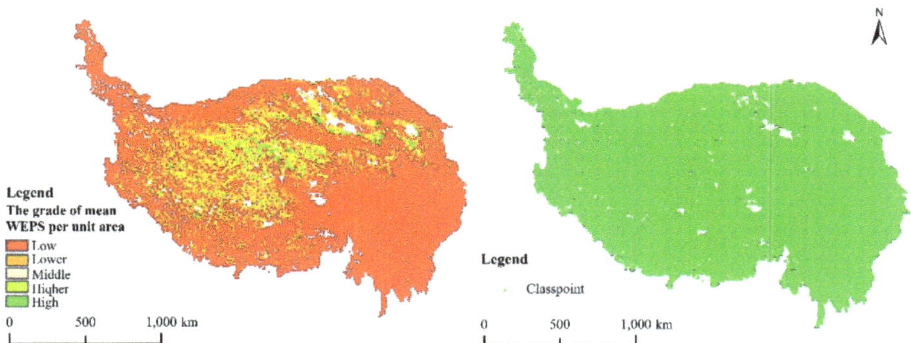

Figure 7. The grade of the average annual WEPS per unit from 2000 to 2015 and samples.

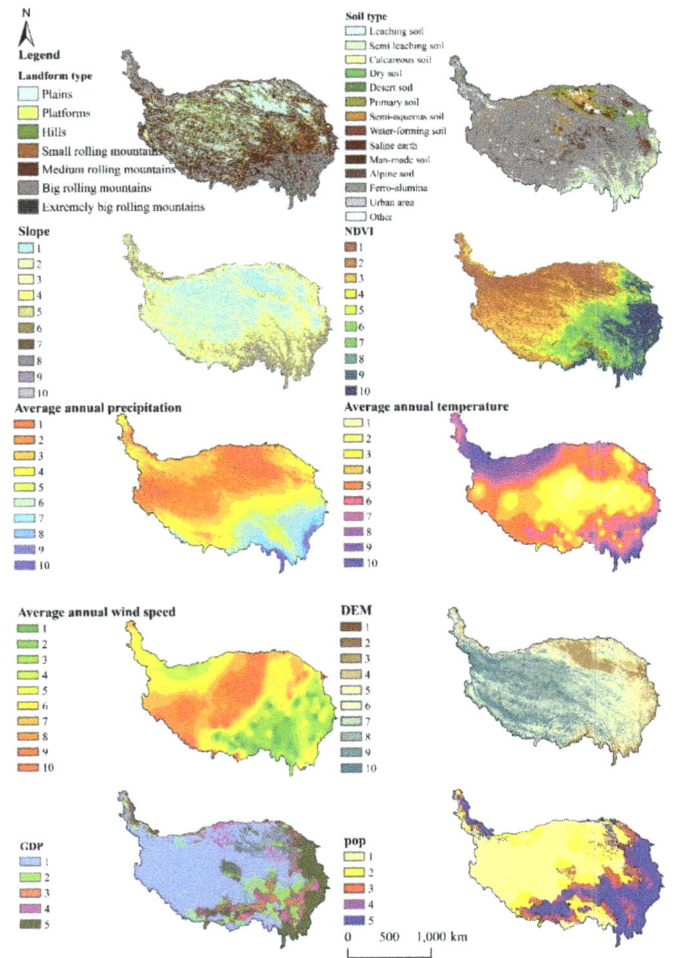

Figure 8. WEPS factors detected in the QTP.

Table 2. Result of factor detection.

Detected Factor	Landform Type	Soil Type	Slope	Average Annual NDVI	Average Annual Precipitation	Average Annual Temperature	Average Annual Wind Speed	DEM	Average Annual GDP	Average Annual Population Density
q statistic	0.276	0.053	0.440	0.142	0.223	0.057	0.186	0.071	0.076	0.124

The risk detector was used to detect the areas with the highest WEPS grade for each factor partition to reveal the internal mechanism responsible for the spatial pattern of the WEPS. The risk detection results show that, as the average annual wind increased, the average WEPS grade per unit area gradually increased. As the slope, average annual precipitation, average annual precipitation temperature, average annual GDP, and average annual population density increased, the average WEPS grade per unit area showed a trend of increasing and then decreasing. The highest WEPS grade occurred when the DEM ranged from 5074 m to 5458 m (Table 1, Figure 9). The average annual GDP and average annual population density trends show that appropriate levels of human activity can help to improve the WEPS grade, but excessive human disturbance in areas with moderate economic development and population density can lead to lower WEPS levels. However, the WEPS grade in the most economically developed areas was higher than that in the moderately developed areas, which could be due to the greater investment in, and influence of, environmental protection measures.

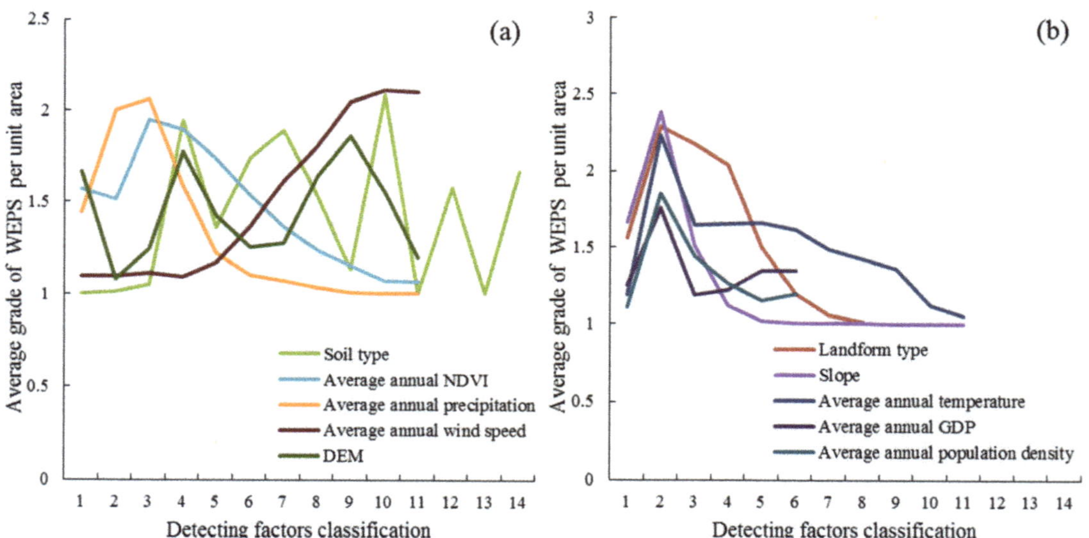

Figure 9. Average WEPS grade per unit area in relation to the classification of different detected factors. Note: (**a**) present the average WEPS grade per unit area in the classification of soil type, average annual NDVI, average annual precipitation, average annual wind speed, and DEM; (**b**) present the average WEPS grade per unit area in the classification of landform type, slope, average annual temperature, average annual GDP, and average annual population density.

The results for the detected factors (Table 3) show that the average annual precipitation, average annual GDP, and average annual population density were significantly different from all the other influencing factors. Landform type, slope, average annual NDVI, average annual wind speed, and average annual temperature showed significant differences with most of the other influencing factors, whereas the DEM showed no significant differences with most of the influencing factors.

Table 3. Statistical significance between detecting factors.

Detecting Factor	Landform Type	Soil Type	Slope	Average Annual NDVI	Average Annual Precipitation	Average Annual Temperature	Average Annual Wind Speed	DEM	Average Annual GDP	Average Annual Population Density
Landform type										
Soil type	N									
Slope	N	N								
Average annual NDVI	Y	N	N							
Average annual precipitation	Y	Y	Y	Y						
Average annual temperature	Y	Y	Y	Y	Y					
Average annual wind speed	N	N	Y	Y	Y	N				
DEM	N	N	N	N	Y	Y	N			
Average annual GDP	Y	Y	Y	Y	Y	Y	Y	Y		
Average annual population density	Y	Y	Y	Y	Y	Y	Y	Y	Y	Y

Y indicates a statistically significant difference (95% confidence level) between the two factors; N indicates no significant difference.

The multi-year average interaction detector results (Table 4) showed that each factor had an enhanced interaction effect on the WEPS of the QTP. Landform type, soil type, average annual NDVI, and average annual precipitation showed nonlinear enhancement relationships with other factors. Slope, average annual temperature, average annual wind speed, DEM, average annual GDP, and average annual population density showed nonlinear enhancement relationships with other factors. The two-factor enhancement between the slope and other factors was found to have a greater impact on the spatial variation of the WEPS grade, and the degree of influence (q value) on the WEPS grade exceeded 44%, which showed that the slope also strengthened the influences of other factors on the spatial distribution of the WEPS.

Table 4. Interaction type and q value between different detecting factors.

Detecting Factor	Landform Type	Soil Type	Slope	Average Annual NDVI	Average Annual Precipitation	Average Annual Temperature	Average Annual Wind speed	DEM	Average Annual GDP	Average Annual Population Density
Landform type	0.28									
Soil type	0.34	0.05								
Slope	0.46	0.50	0.44							
Average annual NDVI	0.39	0.19	0.57	0.14						
Average annual precipitation	0.38	0.25	0.55	0.26	0.22					
Average annual temperature	0.32	0.12	0.48	0.22	0.29	0.06				
Average annual wind speed	0.37	0.22	0.53	0.22	0.27	0.23	0.19			
DEM	0.32	0.13	0.46	0.22	0.28	0.14	0.26	0.07		
Average annual GDP	0.31	0.14	0.46	0.17	0.24	0.14	0.20	0.14	0.08	
Average annual population density	0.33	0.18	0.47	0.18	0.24	0.19	0.21	0.19	0.14	0.12

The shaded green indicates nonlinear enhancement interaction type between the two factors; the other indicates two-factor enhancement interaction type between the two factors.

However, the nonlinear enhancement relationship between the annual average temperature and other factors was found to have a high degree of influence on the spatial variation of the WEPS with a nonlinear enhancement for all factors. It can be concluded that although the single-factor effect of the annual mean temperature and the soil type on the spatial variation of the WEPS were weak, the interaction between the two factors had a higher level of influence on the WEPS than the two factors alone by changing the regional natural environment or human activities. This suggests that the sum of the individual effects of two factors indirectly affects the spatial variation of the WEPS.

4. Discussion

The change pattern of the WEPS was similar to that of the SLR (Figure 6), mainly because the WEPS is more dependent on natural factors, such as the slope, precipitation, and wind. As shown in previous studies, soil type played an important role in the spatial distribution pattern of the sand-stabilization service function [40], soil wind erosion rate was intensified significantly with increasing wind velocities by a power function [46], and the increase in rainfall reduced wind erosion [10]. It seems to be proven that climate changes are the leading factors that affect the soil wind erosion in a region overall, which is consistent with the finding of Li et al. [47].

Increasing fractional vegetation cover can effectively reduce the amount of soil erosion. People can change factors such as the NDVI by improving vegetation coverage and reducing the SL. From 2000 to 2015, a series of ecological conservation and restoration programmes and policies such as Grazing Forbidden Policy and the Grassland Ecological Protection Subsidies and Rewards Program have been implemented in the QTP, which have proven to be effective in reducing soil wind erosion [10]. In terms of changes in land use from 2000 to 2015, the amount of area translated from grassland ecosystem to desert ecosystem was the highest, followed by that from desert to grassland. The increase in the forest ecosystem area was mainly due to conversion from grassland and settlement areas (Figure 10). The change area mainly occurred in eastern part and southern edge of the QTP, where human activities are concentrated [48,49].

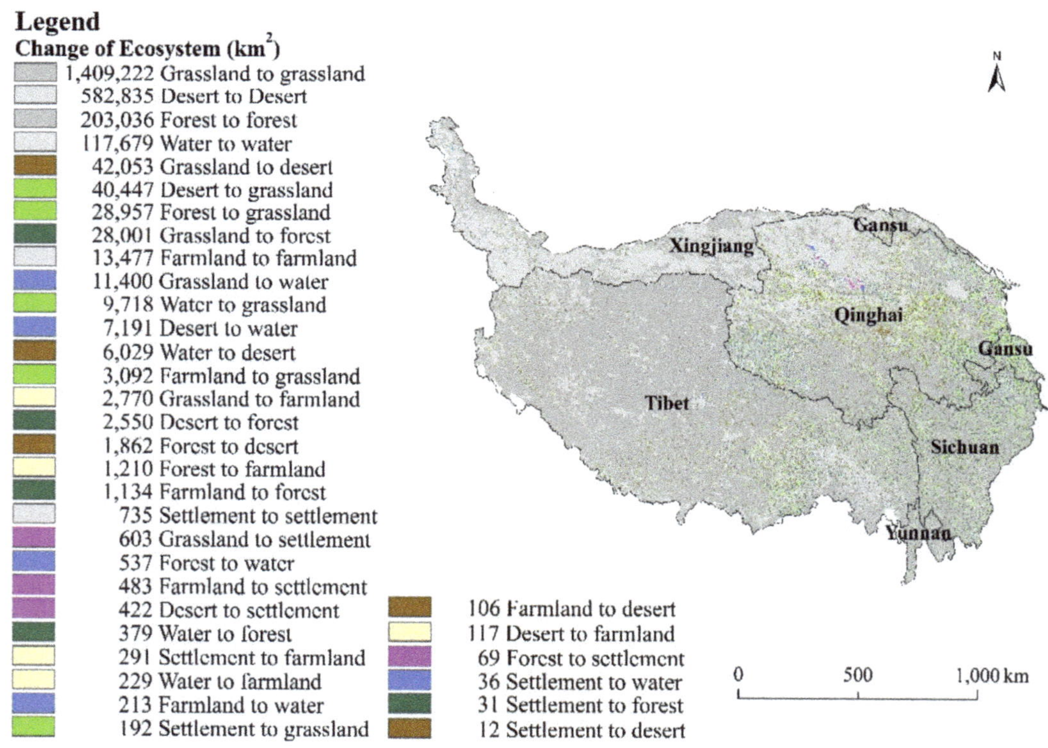

Figure 10. The area changes in ecosystems in the QTP from 2000 to 2015 (km^2).

Among the different types of land use change, the increase in the forest ecosystem area caused an increase in the WEPS, whereas the increase in the desert ecosystem area caused a decrease in the average WEPS in the QTP. The average WEPS per unit area only decreased in the area that was transformed from a water ecosystem to a settlement ecosystem (Figure 11). As same as the result shown in Chi's study [4], the land use/cover change (such as returning farmland to forest, shrub, and grassland) played a key role in reducing the soil wind erosion modulus and enhancing regional WEPS. It was found that the increase in grassland area played a certain role in improving the WEPS due to the control of desertification and the implementation of the returning farmland to grassland project. However, due to unsuitable land use, the resulting significant increase in the desert area is the main reason for the decrease in the total WEPS.

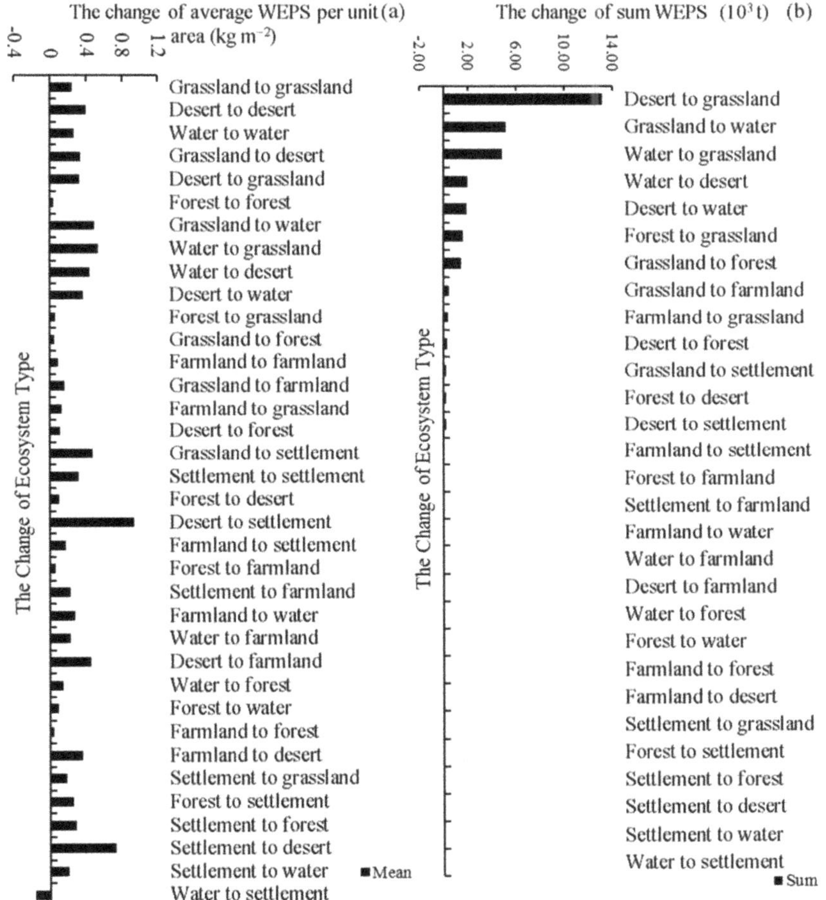

Figure 11. The change of mean and sum WEPS in different ecosystem in the QTP from 2000 to 2015.

Because land use change played an important role in WEPS change in QTP, reasonable land use management will help to promote regional ecological environment protection and sustainable development. In this study, grassland and desert changed significantly, over-grazing, logging, and crop planting have resulted in significant vegetation degradation and desertification, but the implementation of ecological protection projects has promoted the area of grassland and forest [50,51]. For further management, the grassland and desert in the eastern and south-eastern parts of the QTP could be restored to forests engineering practices, plantations, enclosures to limit human disturbances, etc., and the most widely adopted practice in the north and western part of the QTP should be natural restoration. In addition, water shortages in arid and semiarid regions were neglected and the conservation of water bodies should receive more attention.

5. Conclusions

This current study assessed spatial–temporal changes of the WEPS in the QTP from 2000 to 2015. The spatial dynamics of the WEPS were analyzed from the perspectives of various leagues and cities and ecosystems, and its driving forces such as contributions of influential factors to the WEPS and their interactions using geo-detector. The results indicate the following conclusions.

Areas with moderate and high soil erosion intensities were mainly distributed in the western and northern QTP, which contain distributed massive desert and sparse grassland. The annual mean WEPS showed an increasing trend from 2000 to 2015. Slope and weather factors, such as precipitation and wind speed, played leading roles in determining the spatial pattern and grade of the WEPS, and human activities showed an important factor in the variation of WEPS.

In addition, the implementation of projects, such as returning farmland to grassland and desertification control, increased the WEPS by increasing the grassland area. The increase in the desert ecosystem area due to unsuitable land use resulted in a significant reduction in the WEPS in the QTP.

However, there were still some limitations in our study. First, due to the lack of field observation data, this study did not consider the seasonal variation in the wind velocity threshold, thus the seasonal difference in the WEPS could not be analyzed. Second, further parameter localization in the determination of the erodibility boundary of the RWEQ model should be executed in the future to improve the accuracy of the simulation results. Finally, the study of the factors driving the WEPS using the geographic detector model showed the relationships among different factors, whereas the coupling effects of natural factors and human factors need to be studied further.

Author Contributions: Conceptualization, G.X. and Y.X.; methodology, J.X.; software, M.H.; validation, Y.W.; formal analysis, Y.W.; investigation, Y.W.; resources, S.G. and Y.N.; data curation, Y.W.; writing—original draft preparation, Y.W.; writing—review and editing, Y.X. and K.Q.; visualization, Y.X.; supervision, G.X., J.L. and L.Z.; project administration, Y.X.; funding acquisition, G.X. All authors have read and agreed to the published version of the manuscript.

Funding: This work was funded by the Strategic Priority Research Program of Chinese Academy of Sciences (XDA20020402), the National Natural Science Foundation of China (41971272), the Guangxi Science and Technology Major Project (AA20161002-3) and the Zhejiang Provincial Natural Science Foundation of China under Grant (LY18D010001).

Data Availability Statement: All data in this study were correctly referenced. Meteorological datasets, including data on the daily temperature, precipitation, wind speed, and solar radiation data from 2000 to 2015, were acquired from meteorological stations. Snow cover data with a 25-km resolution were obtained from a long-term snow depth dataset [45], which was accessed from the Cold and Arid Region Science Data Center (http://westdc.westgis.ac.cn, accessed on 15 January 2022). NDVI values (spatial resolution 500 m) were derived from the international scientific data mirror website of the computer network information center of the Chinese Academy of Sciences (http://www.gscloud.cn, accessed on 15 January 2022). The DEM data (spatial resolution 30 m), GDP data (spatial resolution 1 km), 1:1,000,000 landform data, and land cover data (spatial resolution 1 km) were derived from the Resource and Environment Data Cloud Platform of the Chinese Academy of Sciences (CAS) (http://www.resdc.cn, accessed on 15 January 2022). Achievement of the geographic detector was operated with geo-detector software [44] (http://geo-detector.cn/, accessed on 15 January 2022).

Conflicts of Interest: The authors declare no conflict of interest.

References

1. Xie, G.D.; Lu, C.X.; Leng, Y.F.; Zheng, D.; Li, S.C. Ecological assets valuation of the Tibetan Plateau. *J. Nat. Resour.* **2003**, *18*, 189–196. (In Chinese)
2. Li, X.W.; Li, M.D.; Dong, S.K.; Shi, J.B. Temporal-spatial changes in ecosystem services and implications for the conservation of alpine rangelands on the Qinghai-Tibetan Plateau. *Rangel. J.* **2015**, *37*, 31–43. [CrossRef]

3. Yang, C.; Geng, Y.; Fu, X.Z.; Coulter, J.A.; Chai, Q. The effects of wind erosion depending on cropping system and tillage method in a semi-arid region. *Agronomy* **2020**, *10*, 732. [CrossRef]

4. Chi, W.; Zhao, Y.; Kuang, W.; He, H. Impacts of anthropogenic land use/cover changes on soil wind erosion in China. *Sci. Total Environ.* **2019**, *668*, 204–215. [CrossRef] [PubMed]

5. Meng, N.; Yang, Y.-Z.; Zheng, H.; Li, R.-N. Climate change indirectly enhances sandstorm prevention services by altering ecosystem patterns on the Qinghai-Tibet Plateau. *J. Mt. Sci.* **2021**, *18*, 1711–1724. [CrossRef]

6. Wang, X.; Lang, L.; Yan, P.; Wang, G.; Li, H.; Ma, W.; Hua, T. Aeolian processes and their effect on sandy desertification of the Qinghai-Tibet Plateau: A wind tunnel experiment. *Soil Tillage Res.* **2016**, *158*, 67–75. [CrossRef]

7. Huang, K.; Zhang, Y.; Zhu, J.; Liu, Y.; Zu, J.; Zhang, J. The influences of climate change and human activities on vegetation dynamics in the Qinghai-Tibet Plateau. *Remote Sens.* **2016**, *8*, 876. [CrossRef]

8. Ping, Y.; Zhibao, D.; Guangrong, D.; Zhang, X.; Yiyun, Z. Preliminary results of using 137 Cs to study wind erosion in the Qinghai-Tibet Plateau. *J. Arid Environ.* **2001**, *47*, 443–452. [CrossRef]

9. Li, Q.; Zhang, C.; Zhou, N.; Shen, Y.; Wu, Y.; Zou, X.; Li, J.; Jia, W.; Wang, X. Spatial distribution of Aeolian desertification on the Qinghai-Tibet Plateau. *J. Desert Res.* **2018**, *38*, 690–700. [CrossRef]

10. Teng, Y.M.; Zhan, J.Y.; Liu, W.; Sun, Y.; Boappeah Agyemang, F.; Zhihui, L. Spatiotemporal dynamics and drivers of wind erosion on the Qinghai-Tibet Plateau, China. *Ecol. Indicators* **2021**, *123*, 107340. [CrossRef]

11. Bagnold, R.A.; Sandford, K.S.; Shaw, W.B.K. A further journey through the Libyan Desert (continued). *Geogr. J.* **1933**, *82*, 211–235. [CrossRef]

12. Woodruff, N.P.; Siddoway, F.H. A Wind Erosion Equation. *Proc. Soil Sci. Soc. Am.* **1965**, *29*, 602–608. [CrossRef]

13. Gregory, J.M.; Wilson, G.R.; Singh, U.B.; Darwish, M. TEAM: Integrated, process-based wind-erosion model. *Environ. Model. Soft.* **2004**, *19*, 205–215. [CrossRef]

14. Lindstrom Usda-Ars, M.J. A description of devices used in the study of wind erosion of soils. *Soil Sci.* **1985**, *140*, 313. [CrossRef]

15. Fryrear, D.W.; Bilbro, J.D.; Saleh, A. RWEQ: Improved wind erosion technology. *J. Soil Water Conserv.* **2000**, *55*, 183–189.

16. Hagen, L.J. Evaluation of the Wind Erosion Prediction System (WEPS) erosion submodel on cropland fields. *Environ. Model. Soft.* **2004**, *19*, 171–176. [CrossRef]

17. Van Pelt, R.S.; Zobeck, T.M.; Potter, K.N.; Stout, J.E.; Popham, T.W. Validation of the wind erosion stochastic simulator (WESS) and the revised wind erosion equation (RWEQ) for single events. *Environ. Model. Soft.* **2004**, *19*, 191–198. [CrossRef]

18. Jiang, L.; Xiao, Y.; Ouyang, Z.; Xu, W.; Zheng, H. Estimate of the wind erosion modules in Qinghai Province based on RWEQ model. *Res. Soil Water Conserv.* **2015**, *22*, 21–25. (In Chinese) [CrossRef]

19. Huang, L.; Cao, W.; Wu, D.; Gong, G. The temporal and spatial variations of ecological services in the Tibet Plateau. *J. Nat. Resour.* **2016**, *31*, 543–555.

20. Jiang, C.; Li, D.; Wang, D.; Zhang, L. Quantification and assessment of changes in ecosystem service in the Three-River Headwaters Region, China as a result of climate variability and land cover change. *Ecol. Ind.* **2016**, *66*, 199–211. (In Chinese) [CrossRef]

21. Zhang, C.; Gong, J.; Zou, X.; Dong, G.; Li, X.; Dong, Z.; Qing, Z. Estimates of soil movement in a study area in Gonghe Basin, north-east of Qinghai-Tibet Plateau. *J. Arid Environ.* **2003**, *53*, 285–295. (In Chinese) [CrossRef]

22. Zhang, H.B.; Peng, J.; Zhao, C.N.; Xu, Z.H.; Dong, J.Q.; Gao, Y. Wind speed in spring dominated the decrease in wind erosion across the Horqin Sandy Land in northern China. *Ecol. Indic.* **2021**, *127*, 107599. [CrossRef]

23. Jiang, C.; Yang, Z.; Wang, X.; Dong, X.; Li, Z.; Li, C. Examining the reversal of soil erosion decline in the hotspots of sandstorms: A non-linear ecosystem dynamic perspective. *J. Arid Environ.* **2021**, *186*, 104421. [CrossRef]

24. Jiang, C.; Liu, J.G.; Zhang, H.Y.; Zhang, Z.D.; Wang, D.W. China's progress towards sustainable land degradation control: Insights from the northwest arid regions. *Ecol. Eng.* **2019**, *127*, 75–87. [CrossRef]

25. Lyu, X.; Li, X.B.; Wang, H.; Gong, J.R.; Li, S.K.; Dou, H.S.; Dang, D.L. Soil wind erosion evaluation and sustainable management of typical steppe in Inner Mongolia, China. *J. Environ. Manag.* **2021**, *277*, 111488. [CrossRef] [PubMed]

26. Aubault, H.; Webb, N.P.; Strong, C.L.; Mc Tainsh, G.H.; Leys, J.F.; Scanlan, J.C. Grazing impacts on the susceptibility of rangelands to wind erosion: The effects of stocking rate, stocking strategy and land condition. *Aeolian Res.* **2015**, *17*, 89–99. [CrossRef]

27. Zhang, H.Y.; Fan, J.W.; Cao, W.; Harris, W.; Li, Y.Z.; Chi, W.F.; Wang, S.Z. Response of wind erosion dynamics to climate change and human activity in Inner Mongolia, China during 1990 to 2015. *Sci. Total Environ.* **2018**, *639*, 1038–1050. [CrossRef]

28. Latocha, A.; Szymanowski, M.; Jeziorska, J.; Stec, M.; Roszczewska, M. Effects of land abandonment and climate change on soil erosion—An example from depopulated agricultural lands in the Sudetes Mts., SW Poland. *CATENA* **2016**, *145*, 128–141. [CrossRef]

29. Garbrecht, J.D.; Nearing, M.A.; Steiner, J.L.; Zhang, X.J.; Nichols, M.H. Can conservation trump impacts of climate change on soil erosion? Anassessment from winter wheat cropland in the Southern Great Plains of the United States. *Weather Clim. Extrem.* **2015**, *10*, 32–39. [CrossRef]

30. Li, D.J.; Xu, D.Y.; Wang, Z.Y.; You, X.G.; Zhang, X.Y.; Song, A.L. The dynamics of sand—Stabilization services in Inner Mongolia, China from 1981 to 2010 and its relationship with climate change and human activities. *Ecol. Indicators* **2018**, *88*, 351–360. [CrossRef]

31. Sharratt, B.S.; Tatarko, J.; Abatzoglou, J.T.; Fox, F.A.; Huggins, D. Implications of climate change on wind erosion of agricultural lands in the Columbia plateau. *Weather Clim. Extrem.* **2015**, *10*, 20–31. [CrossRef]

32. Wang, J.F.; Xu, C.D. Geo-detector: Principle and prospective. *Acta Geogr. Sinica* **2017**, *72*, 116–134.

33. Polykretis, C.; Alexakis, D.D. Spatial stratified heterogeneity of fertility and its association with socio-economic determinants using Geographical Detector: The case study of Crete Island, Greece. *Appl. Geogr.* **2021**, *127*, 102384. [CrossRef]

34. Huang, J.X.; Wang, J.F.; Bo, Y.C.; Xu, C.D.; Hu, M.G.; Huang, D.C. Identification of health risks of hand, foot and mouth disease in China using the geographical detector technique. *Int. J. Environ. Res. Public Health* **2014**, *11*, 3407–3423. [CrossRef]

35. Shen, J.; Zhang, N.; Gexigeduren, H.B.; Liu, C.Y.; Li, Y.; Zhang, H.Y.; Chen, X.Y.; Lin, H. Construction of a GeogDetector-based model system toindicate the potential occurrence of grasshoppers in Inner Mongolia steppe habitats. *Bull. Entomol. Res.* **2015**, *105*, 335–346. [CrossRef]

36. Ren, Y.; Deng, L.Y.; Zuo, S.D.; Song, X.D.; Liao, Y.L.; Xu, C.D.; Chen, Q.; Hua, L.Z.; Li, Z.W. Quantifying the influences of various ecological factors on land surface temperature of urban forests. *Environ. Pollut.* **2016**, *216*, 519–529. [CrossRef]

37. Liang, P.; Yang, X.P. Landscape spatial patterns in the Maowusu (Mu Us) Sandy Land, northern China and their impact factors. *CATENA* **2016**, *145*, 321–333. [CrossRef]

38. Tian, F.; Liu, L.Z.; Yang, J.H.; Wu, J.J. Vegetation greening in more than 94% of the Yellow River Basin (YRB) region in China during the 21st century caused jointly by warming and anthropogenic activities. *Ecol. Indicators.* **2021**, *125*, 107479. [CrossRef]

39. Li, J.; Wang, J.; Zhang, J.; Liu, C.; He, S.; Liu, L. Growing-season vegetation coverage patterns and driving factors in the China-Myanmar Economic Corridor based on Google Earth Engine and geographic detector. *Ecol. Indic.* **2022**, *136*, 108620. [CrossRef]

40. Zhang, Y.; Xu, D.Y.; Wang, Z.Y.; Zhang, X.Y. The interaction of driving factors for the change of windbreak and sand-fixing service function in Xilingol League between 2000 and 2015. *Acta Ecol. Sinica* **2021**, *41*, 603–614.

41. Du, H.Q.; Dou, S.T.; Deng, X.H.; Xue, X.; Wang, T. Assessment of wind and water erosion risk in the watershed of the Qinghai-Qinghai-Tibet Plateau Reach of the Yellow River, China. *Ecol. Indicators* **2016**, *67*, 117–131. [CrossRef]

42. Skaggs, T.H.; Arya, L.M.; Shouse, P.J.; Mohanty, B.P. Estimating particle-size distribution from limited soil texture data. *Soil Sci. Soc. Am. J.* **2001**, *65*, 1038–1044. [CrossRef]

43. Lv, C.Q. Study on Soil Carbon Stock and Its Spatial Distribution, Influence Factors in Tibetan Plateau. Master's Thesis, Nanjing University, Nanjing, China, 2006.

44. Wang, J.F.; Li, X.H.; Christakos, G.; Liao, Y.-L.; Zhang, T.; Gu, X.; Zheng, X.Y. Geographical detectors-based health risk assessment and its application in the neural tube defects study of the Heshun region, China. *Int. J. Geograph. Inform. Sci.* **2010**, *24*, 107–127. [CrossRef]

45. Che, T.; Li, X.; Jin, R.; Armstrong, R.; Zhang, T. Snow depth derived from passive microwave remote-sensing data in China. *Ann. Glaciol.* **2008**, *49*, 145–154. [CrossRef]

46. Xie, S.B.; Qu, J.J.; Wang, T. Wind tunnel simulation of the effects of freeze-thaw cycles on soil erosion in the Qinghai-Tibet Plateau. *Sci. Cold Arid Reg.* **2016**, *8*, 187–195.

47. Li, J.; Ma, X.; Zhang, C. Predicting the spatiotemporal variation in soil wind erosion across Central Asia in response to climate change in the 21st century. *Sci. Total Environ.* **2020**, *709*, 136060. [CrossRef]

48. Sun, Y.; Liu, S.; Shi, F.; An, Y.; Li, M.; Liu, Y. Spatio-temporal variations and coupling of human activity intensity and ecosystem services based on the four-quadrant model on the Qinghai-Tibet Plateau. *Sci. Total Environ.* **2020**, *743*, 140721.

49. Wang, C.; Zhan, J.; Xin, Z. Comparative analysis of urban ecological management models incorporating low-carbon transformation. *Technol. Forecast. Soc. Chang.* **2020**, *159*, 120190. [CrossRef]

50. Li, Y.; Li, J.; Are, K.; Huang, Z.; Yu, H.; Zhang, Q. Livestock grazing significantly accelerates soil erosion more than climate change in Qinghai-Tibet Plateau: Evidenced from 137Cs and 210Pbex measurements. *Agric. Ecosyst. Environ.* **2019**, *285*, 106643. [CrossRef]

51. Pan, T.; Zou, X.; Liu, Y.; Wu, S.; He, G. Contributions of climatic and non-climatic drivers to grassland variations on the Tibetan Plateau. *Ecol. Eng.* **2017**, *108*, 307–317. [CrossRef]

 *sustainability*

Article

Herders' Perceptions about Rangeland Degradation and Herd Management: A Case among Traditional and Non-Traditional Herders in Khentii Province of Mongolia

Munguntuul Ulziibaatar [1,*] and **Kenichi Matsui** [2]

[1] Institute of Geography and Geoecology, Mongolian Academy of Sciences, Ulaanbaatar 15170, Mongolia
[2] Faculty of Life and Environmental Sciences, University of Tsukuba, Tsukuba 305-8572, Japan; matsui.kenichi.gt@u.tsukuba.ac.jp
* Correspondence: u.munguntuul@gmail.com; Tel.: +976-9908-8194

Citation: Ulziibaatar, M.; Matsui, K. Herders' Perceptions about Rangeland Degradation and Herd Management: A Case among Traditional and Non-Traditional Herders in Khentii Province of Mongolia. *Sustainability* **2021**, *13*, 7896. https://doi.org/10.3390/su13147896

Academic Editors: Kikuko Shoyama, Rajarshi Dasgupta and Ronald C. Estoque

Received: 2 May 2021
Accepted: 6 July 2021
Published: 15 July 2021

Abstract: Herders play essential roles in sustaining Mongolia's economy and rangeland conditions. As about 90% of Mongolia's livestock grazes on natural pasture, how herders manage it largely affects the future sustainability of the livestock industry. Since Mongolia transformed its grazing practices from communal management into loosely regulated household practices in 1990, overgrazing has become a growing concern. Considering this concern, this paper examines the extent to which traditional and non-traditional herders perceive pasture conditions and practice management. We conducted the questionnaire survey among 120 herders in Murun Soum of Khentii Province and asked about rangeland degradation and their coping strategies. To determine correlations between their perceptions/practices and sociodemographic characteristics, we conducted multiple regression analyses. We found that, overall, most herders identified rangeland conditions degrading and grass yield declining with less plant diversity and more soil damage by Brandt's vole. Herders' mobility and herd movement frequency have decreased since 1990, placing more strains on limited pasture areas. In coping with overgrazing, about 20% of the respondents had practiced traditional rangeland management, whereas many others had overlooked pasture conditions and increased goat production as the world's demand for cashmere rose. In response to our question about herders' future contribution of their traditional knowledge to sustainable rangeland management, traditional herders demonstrated their willingness to help local officials manage the pasture. This paper then explores how local administrations and herders may collaborate in the future.

Keywords: herder; rangeland degradation; perception; traditional rangeland management practices; Mongolia

1. Introduction

Open range livestock grazing practices in the Mongolian Steppes have faced tremendous economic, environmental, and social challenges from overgrazing, desertification, and pasture degradation since the 1990s [1,2]. Rangelands comprise more than 70% of Mongolia's territory, and about 90% of its livestock depends on them [3]. In the last three decades, however, 65% of Mongolia's vast rangelands have been at least partially degraded, and another 7% is completely degraded [4]. A satellite observation of vegetation optical depth (VOD) from 1988 to 2008 confirmed that all areas of the Mongolian steppes, the largest in the world, underwent significant vegetation reduction [2].

Rangelands largely belong to public land in Mongolia. A recent cashmere boom has led to a substantial goat population growth on public land [5]. Contrary to traditional pastoralists, contemporary herders have grazed their livestock in a fixed area with an increasing number of animals. FAO [6] predicted that population growth and changing lifestyle would increase pressure on rangeland in the future; at the same time, the world's meat consumption would increase from 229 million tons in 1999–2001 to 465 million tons in 2050. Considering these, the sustainability of rangelands in the Mongolian Steppes will face further sustainability challenges [4].

Several studies have identified the importance of herders' perceptions and observations as solution to mitigate these challenges [1,7–10]. Herders' observation of pasture conditions, for example, provides insights into ecological processes and changes [11–15]. Khwarae [9] suggested to incorporate community perceptions and practices into rangeland policies. The participation of traditional pastoralists in implementing these policies can help better sustain rangelands [16,17].

Considering these suggestions, this paper seeks to better understand herders' perceptions and knowledge concerning livestock management, grassland degradation, and other related environmental changes. Here we attempt to demonstrate how herders with various backgrounds perceived vegetation changes and responded to these changes. In addition, we intend to clarify the extent to which traditional knowledge still works for sustainably managing rangelands in Mongolia. Finally, we identify the degree of willingness among herders to participate in rangeland management in collaboration with local authorities.

Review of Livestock Sector and Rangeland Health in Mongolia

Mongolia's livestock industry experienced a turning point in the 1990s [3]. During this decade, Mongolia's livestock industry constituted about 30–40% of its GDP. Since 2000, this share has gradually declined, largely due to growing mining industries and periodic natural disasters that severely damaged grazing practices. As of 2018, the livestock industry produced 4.4 trillion tugrug (MNT) or 10.8% of GDP [18]. It accounts for 9.2%, or USD 642.9 million, of export income and employs 26.7% of the labor force, or 288,700 herders. In 2018, Mongolian herders produced 515,200 tons of meat, 902.4 million liters of milk, 33,000 tons of sheep wool, 10,900 tons of cashmere, 2000 tons of camel wool, and 17.4 million hides and skins. According to the National Statistics Office, the total number of five main livestock species in the country increased from 25 million in 1990 to 66.5 million in 2018, including 3.9 million horses, 4.4 million head of cattle, 459,700 camels, 30.6 million head of sheep, and 27.1 million goats [18].

Pasture degradation, a worldwide problem [10], is particularly serious in Mongolia [19]. In 1970, 90.2%, or 130 million hectares, of Mongolia's rangelands were covered with rich grass [20]. The percentage has dropped to 73.5% in 2018 [18]. Tseelei [21] estimated that Mongolia's rangelands are capable of having 25 million head of livestock, whereas 66.5 million head currently graze. According to the National Agency for Meteorology and Environmental Monitoring (NAMEM), which is responsible for monitoring rangeland health, as of 2015, 65% of the total rangeland was partially degraded and 7% is completely degraded with no possible chance of recovery. A UN study found that rangeland degradation costs Mongolia 368 billion tugriks (USD 135.7 million) per year [19].

In recent years, as we will discuss in the main discussion below, a distance for seasonal herd movement has decreased to the point where, in some areas, little or no traditional seasonal movement is observed. In making several field visits in rangeland, we observed that herders, especially young herders and newcomers, did not appear to know much about how rangelands can be cared for and sustained. Jamsran similarly observed this and argued that the "Tragedy of the Commons" [22] took hold of Mongolia's rangeland [23]. As the rangeland is considered to be the commons, the herders' and the government's responsibilities are somewhat unclear [24,25].

At the local level (Soum level), the Administration of Land Affairs, Geodesy and Cartography administers the Soum Annual Land Management Planning (SALMP) that regulates rangeland management throughout the country. As it has largely taken top-down approaches, local and traditional herders have had little room to participate in decision making, widening the gap between administrations and practitioners. The SALMP does not appear to have provided clear information for herders, either. Recommendations do not seem to reflect herders' perceptions [4].

2. Materials and Methods

2.1. The Study Area

In undertaking research for this paper, we selected one of the prominent grazing areas of Mongolia: Murun Soum (sub province) in Khentii Province. From our previous experience, we knew that this area largely depended on the livestock industry, and some traditional practitioners were there. This soum is located in eastern Mongolia, 311 km from Ulaanbaatar City and 27 km from Chinggis City, the capital of Khentii Province (Figure 1). The soum administration was established in 1923.

The average annual rainfall in this area is 160 mm. The average temperature ranges from −1 to 2 °C. The high temperature can reach 32.8 °C in June, and the coldest temperatures can be −48.2 °C in December and January. The total territory of Murun Soum is 218,000 km^2, and 90% of its territory is used as rangeland. A small portion of the land is used for crop production, including wheat (2500 hectares), potatoes, and other vegetables (90.1 hectares). Gravitational irrigation is practiced here by drawing water from the Murun River [26].

As of 2018, the total population of Murun was 1919, of which 468 were herders (280 herder households) who owned 179,800 livestock or 312,500 Sheep Forage Units (SFUs) (14,000 horses, 11,000 head of cattle, 91,800 head of sheep, and 63,000 goats) [18]. There are 642 head of livestock per household in Murun Soum, whereas the average livestock number per herder household in the nation was 352 in 2017. A total number of livestock increased from 97,000 in 2000 to 179,000 in 2018 (Figure 2). According to the National Statistical Office [27], the carrying capacity of pasture in this soum was 128,862 head of livestock or 233,966.7 Sheep Forage Units (SFUs), including 10,547 horses, 8891 head of cattle, 65,020 head of sheep, 43,963 goats, and 441 camels in 2018. This means that the number had already exceeded the carrying capacity of the soum by 33.6% as sheep equivalents. In particular, the number of goats increased dramatically from 3100 in 1970 to 63,000 in 2018 [28], largely due to the recent cashmere boom. The total human population in Murun Soum decreased from 2440 in 2000 to 1919 in 2018 [28].

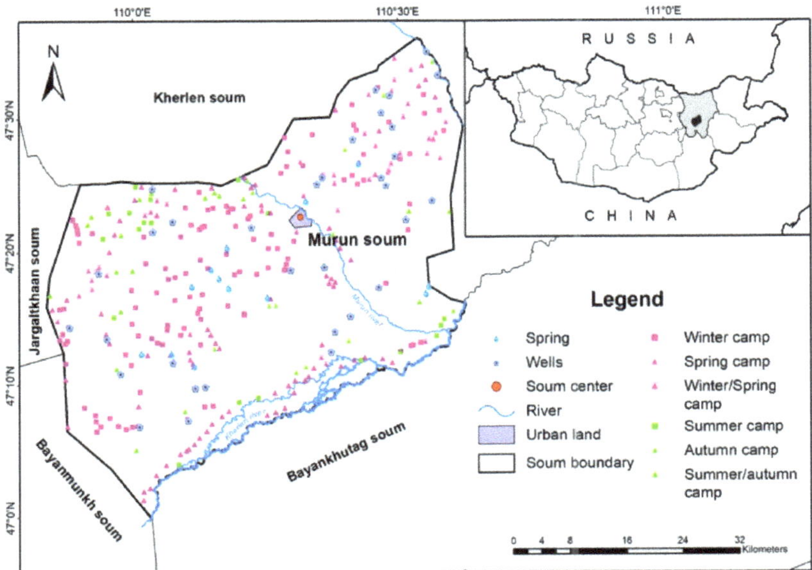

Figure 1. The Study Area Map. Source: Sustainable Livelihood II Project, 2010.

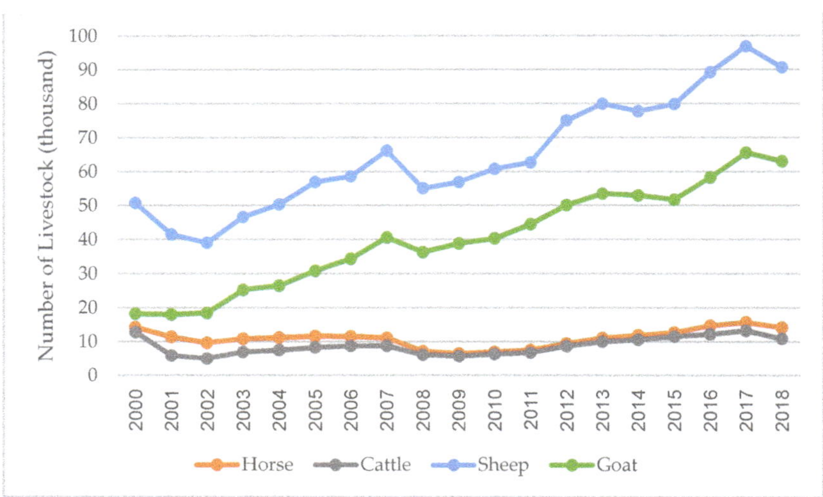

Figure 2. Number of Livestock in Murun Soum (2000–2018); Source: National Statistics Office, 2000–2018.

According to the 2013 Desertification Atlas of Mongolia [29], Murun Soum and its surrounding areas were affected by land degradation and desertification. Some areas were categorized under severe conditions due to intensive grazing near the Kherlen River, where summer camps were located. Gao's [30] remote sensing-based forage model (based on 50% utilization) showed that 91–100% of Murun Soum's rangeland forage was overused, although the National Rangeland Health Report [4] found that this rangeland condition had an 80–90% possibility to recover.

2.2. Data Collection and Analysis

The questionnaire survey was conducted from 15 December 2019 to 18 February 2020, in Mongolian. A simple random sampling method was used to select respondents. There are 468 herders (or 280 herder households) in Murun Soum, and we collected valid responses from 120 of them. As all respondents were actively engaged in herding in different parts of the Soum, it took us about two months to interview them. These herders were often camped in various remote locations and occasionally visited a Soum center where we met them.

This survey had five sections. The first section attempted to identify the sociodemographic characteristics of the respondents, including age, gender, education, household size, herding experience, the number of livestock, and monthly income. The second section consisted of multiple-choice questions, close-ended questions, and open-ended questions that were designed to understand grazing practices and observation about rangeland availability, water availability, and other grazing-related challenges. The third section had Likert-scale questions, multiple choice questions, and close-ended questions that aimed to understand respondents' perceptions about rangeland degradation and ecological changes. The next section consisted of multiple-choice questions, open-ended questions, and close-ended questions. These are to find the extent to which traditional rangeland management practices remained in the study area. The last section had a multiple-choice question, closed-ended questions, and five-point Likert-scale questions to identify respondents' needs and willingness to participate in rangeland management in the study area.

To better understand how the respondents observed the causes of rangeland degradation, we asked them to identify primary factors. We presented choices for the respondents based on past studies. There are several postulations about the causes of rangeland degradation among scholars. They found that rangeland degradation was caused by climate change and anthropogenic activities (e.g., increased goats) [25,31–36], lenient rangeland regulation or mismanagement [23], the disintegration of traditional practices and movement [31,33,37,38], and mining practices [32]. Bedunah and Angerer [39] argued that identifying the causes of rangeland degradation is rather complex and contentious as all stakeholders and environmental factors are interconnected. Middleton [40] claimed that overgrazing is not the only factor that accelerates rangeland degradation. Desertification and climate change also affected vegetation cover and soil conditions [4,41]. We adopted these factors in our question.

After collecting the data, a multiple regression analysis was conducted to find correlations between respondents' sociodemographic characteristics and their perceptions, including respondents' observations about pasture yield changes and future pasture degradation mitigation. Microsoft Excel was used for analyzing the data.

3. Results and Discussion

3.1. Sociodemographic Characteristics of the Respondents

The first part of our survey attempted to clarify the sociodemographic characteristics of the respondents (Table 1). The result shows that 61% of the respondents were 30 to 49 years old, showing signs of socioeconomic sustainability in herding activities for the next few decades. An additional 25% belonged to the 50–59 age group, including those who had knowledge about before and after 1990, a turning point from commune-based herding. In terms of gender, 79% of them were males. This means that primarily middle-aged males participated in the questionnaire survey. Regarding the household size, 51% had one to four family members, and the rest had more than five persons.

Table 1. Sociodemographic Characteristics of the Respondents.

	Category	Frequency	Percentage
	20–29	11	9%
	30–39	33	28%
Age	40–49	40	33%
	50–59	30	25%
	60 and above	6	5%
Gender	Male	95	79%
	Female	25	21%
	1–500	58	48%
Number of Household Livestock	500–1000	44	37%
	1001 and above	18	15%
	≤10	22	18%
	11–20	42	35%
Years of Grazing Experience	21–30	36	30%
	31–40	18	15%
	≥41	2	2%
	No formal	4	3%
	Primary	25	21%
	Lower Secondary	17	14%
Level of Education	Upper Secondary	53	44%
	Technical	13	11%
	Bachelor	8	7%
	Up to USD 71	47	39%
Monthly Income of Household	USD 72–179	22	18%
USD 1 = 2790 MNT(14 May 2020)	USD 180–358	40	33%
	USD 359 and above	11	10%
	Dairy products	31	26%
	Cashmere and wool	81	68%
Income Source	Meat	110	92%
	Salary (spouse)	12	10%

Even though most of the respondents were middle-aged, the respondents were well-experienced herders. We found that 35% of the respondents had herding experience of 11 to

20 years, 30% of 21 to 30 years, and 15% of 31 to 40 years. Looking at the level of education, the percentage of the respondents who had middle or upper secondary education was 44%. In addition, 21% had only up to primary education. In this study area, parents tend to send their teenage children to herders after completing their primary schooling. About 48% of the respondents had 200–500 head of livestock (categorized as middle wealth or lower-level household), and another 37% had 500–1000 head of livestock (wealthy level household). Those with more than 1000 head of livestock constituted 15%.

As for the monthly income, 39% of the respondents said that they earned less than 200,000 MNT (equivalent to USD 71.8 with the exchange rate on 26 April 2020), and 17% earned 200,001–500,000 MNT (USD 71.8–179.5). These income levels fell below Mongolia's per capita GDP in 2019 (USD 4295.28), according to the World bank data [42]. The rest had relatively livable incomes with 34% of the respondents having earned 500,001–1,000,000 MNT (USD 179.5–359) and 10% having more than 1,000,000 MNT per month. According to the National Statistics Office in 2019, the average monthly income per household in Mongolia was 1,343,428 MNT (USD 482.3). Selling meat was the most profitable source of income for them (92%), followed by cashmere and wool (68%). About 26% of the respondents were engaged in dairy product sales. The rest worked in administrative services at public offices.

3.2. Herders' Perceptions about Rangeland Conditions

To understand how herders observed current rangeland conditions, we first asked the respondents whether or not their rangeland area was sufficient for their livelihood. A half of them (50%) answered positively. Another half said that an excessive number of livestock had made it difficult to maintain the health of their rangeland. Next, we asked if the respondents found water availability sufficient in their grazing area. In response, 74% of the respondents said that it was adequate.

We then asked the respondents to choose listed challenges in livestock grazing (Figure 3). About 74% of the respondents chose rangeland degradation as the main challenge. Another 62% found a growing number of Brandt's vole problematic. Other challenges identified were drought (35%), decreasing plant species (34%), severe winter weather or so-called *dzud*, which killed a large number of livestock from starvation and cold (26%), and pasture dispute (16%). This result suggests that Murun herders faced two main problems: rangeland degradation and increasing Brandt's vole populations. Regarding the pasture dispute, we were told that some herders accidentally grazed someone else's pasture reserves for winter and spring without permission.

Brandt's vole, a rodent species, is considered the main threat to rangeland health and livestock breeding. It not only eats plants but also damages the roots by digging extended channels in the soil. A study found a connection between rangeland degradation and the growing number of Brant's vole in recent years in the study area [43]. Another study found that high density of Brand's vole's population caused vegetation loss [44]. However, this point appears to be contested among some scholars.

3.3. Herders' Perceptions about Rangeland Degradation and Ecological Changes

In this section, we asked the respondents about their observation of rangeland conditions. First, we asked them to evaluate the scale of changes (from "Severely degraded" to "Totally improved") in rangeland. The notable aspect of the result was that 42% of the respondents found pasture yield degrading and 30% found it moderately degrading. Another 12% found it severely degraded. On the other hand, those who found their pasture conditions somewhat improved consisted of only 5%. This result shows that about 95% of the respondents found their pasture more or less degrading.

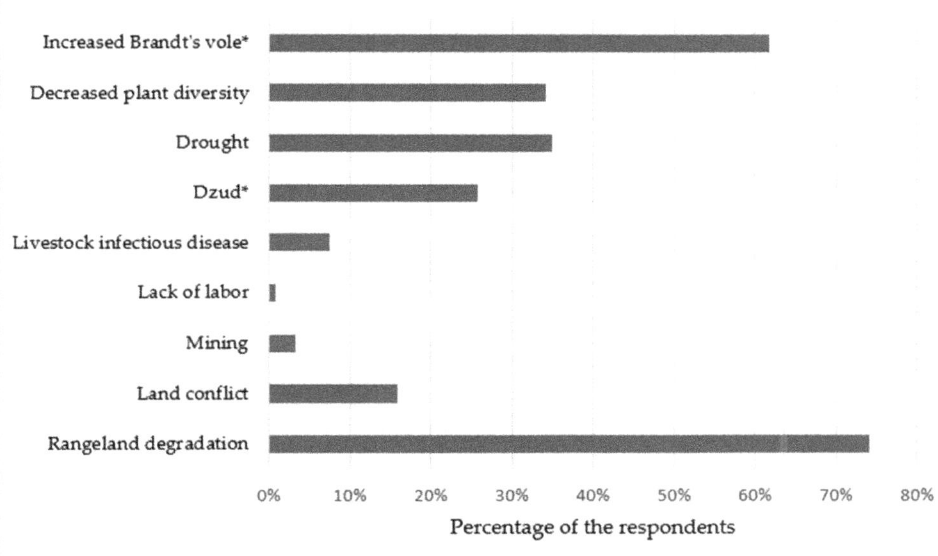

Figure 3. Murun Soum Herders' Perceived Challenges in Livestock Grazing. * *Dzud*, severe winter weather, killed a large number of livestock from starvation and cold.

To know more specifically about what aspects of rangeland conditions the respondents found their rangeland degraded, we administered ten indicators. Six of these focused on grass conditions: "Plants are less diverse"; "Yield has decreased"; "Poisonous plants were increased"; "Disappearing nutritious plant species"; "Plants are widely scattered"; and "Plants are getting shorter." In addition, two indicators focused on soil conditions: "Soil is getting drier" and "Soil is getting salinized." The other two were "Desertification has expanded" and "Brandt's vole increased" (Figure 4).

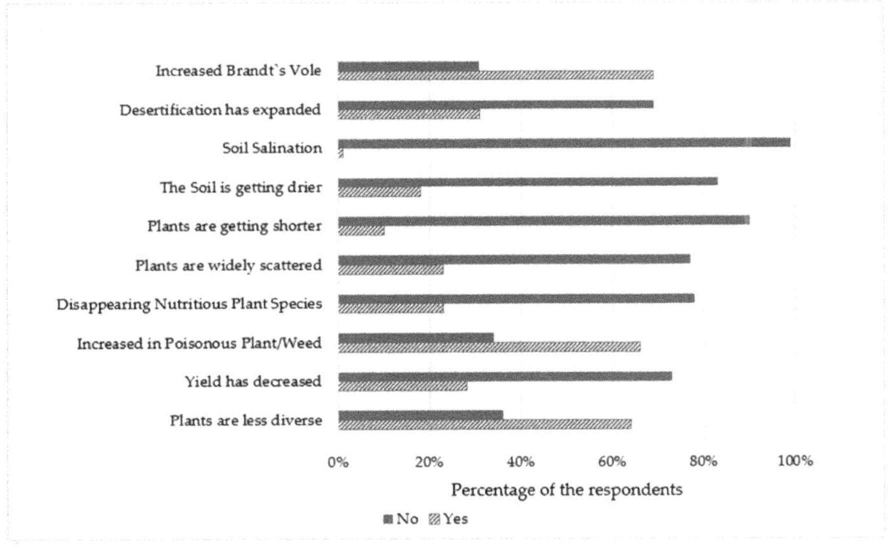

Figure 4. Identified Rangeland Degradation Indicators (Vegetation and Soil).

The result shows that more than 60% of the respondents identified the following three indicators: "Plants are less diverse," "Poisonous plants were increased," and "Brandt's vole increased." Regarding soil conditions, 17% of the respondents found it was getting drier, and only 1% found the soil salinized (1%). These results imply that the respondents mainly checked vegetation changes to assess pasture conditions.

Regarding the causes of degradation, the respondents found it mainly attributable to overgrazing (91%). They also identified reduced mobility in herding (92%), climate change (reduced amount of rain) (94%), increased Brandt's vole population (51%), increased livestock population (47%), pasture destruction by roads and vehicles (43%), weak pasture use regulation (36%), and an increase in goat numbers (18%) (Figure 5).

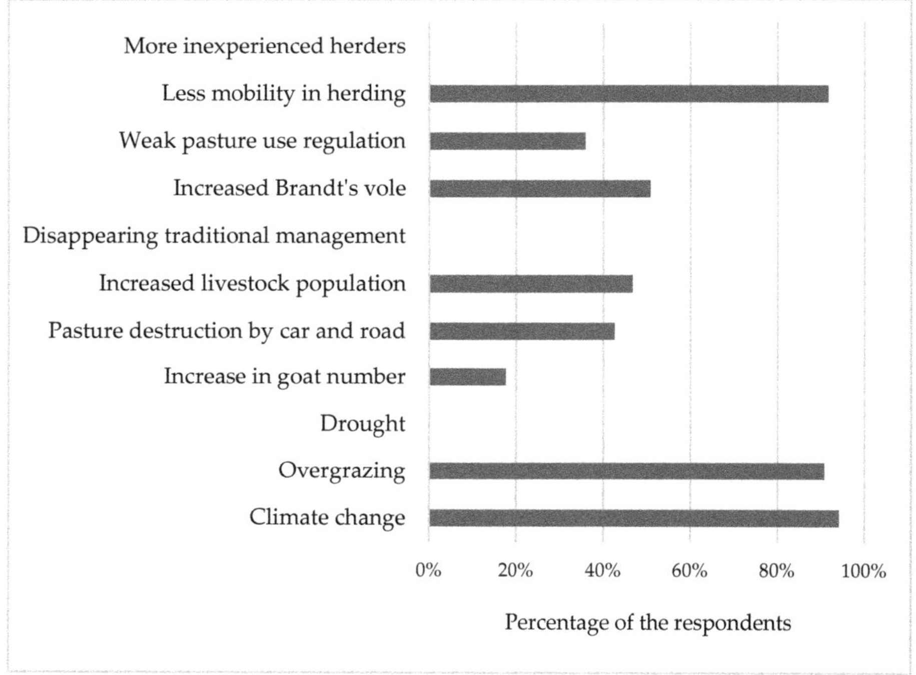

Figure 5. Observed Causes of Rangeland Degradation.

We then tried to understand respondents' willingness to limit the number of their goats in the future. The result shows that 51% did not intend to increase the goats. They may have considered the goat as the leading cause of overgrazing and pasture degradation. On the other hand, 15% of the respondents mentioned that they preferred to increase the number of their goats in coming years. Traditionally, herders managed the composition of the livestock. The proportion of goats was maintained within 18–20% of the total number of livestock to avoid pasture degradation and mitigate disaster impacts [23]. However, in Murun Soum, the number of goats rose sharply from 18,000 in 2000 to 63,000 in 2018 (35% of the total livestock) largely, due to growing international cashmere demand [45].

3.4. Effects of Herders' Socioeconomic Characteristics on Their Perception

We conducted a multiple regression analysis to better understand respondents' perceptions about observed changes in pasture yield. We paired respondents' perceptions with age, education, household size, grazing experience, number of household livestock, and monthly income. We found that education (p-value = 0.010) and gender (p-value = 0.008) were significantly correlated (Table 2). About 23% of the respondents were educated up to the tertiary level.

Table 2. Factors Determining Respondents' Observations about Pasture Yield Changes (from "Severely degraded" to "Totally improved").

Variable	Coefficients	Standard Error	Critical Value	p-Value	Lower 95%	Upper 95%
Intercept	2.704	0.706	3.829	0.000	1.305	4.103
Age	−0.008	0.017	−0.496	0.621	−0.041	0.025
Gender	−0.091	0.283	−0.320	0.008 *	−0.652	0.471
Education	−0.019	0.035	−0.541	0.010 *	−0.087	0.050
Household Size	0.042	0.073	0.574	0.567	−0.102	0.185
Grazing Experience (years)	0.027	0.017	1.642	0.103	−0.006	0.060
Household Livestock (number)	0.000	0.000	0.232	0.817	0.000	0.001
Number of People Engaged in Herding	0.064	0.112	0.573	0.568	−0.157	0.286
Monthly Income of Household	−0.161	0.087	−1.853	0.067	−0.332	0.011

* p-value < 0.05.

The respondents were asked about whether or not rangeland degradation would occur in the future. The result shows that 40% of the respondents said that the rangeland degradation would not happen in the future. On the other hand, 24% of the respondents thought it would worsen in the future. Herders with large herds tended to be confident that pasture degradation would be alleviated. Herders with small herds believed that large-scale grazing was to be blamed for pasture degradation.

Perceptions regarding the future mitigation of pasture degradation widely varied among the respondents. To understand why the respondents thought differently, we correlated their perceptions with such sociodemographic features as age, education, household size, grazing experience, the number of household livestock, and monthly income. We found that education (p-value = 0.004) and monthly income (p-value = 0.011) had significant correlations with their perceptions about future pasture degradation mitigation (Table 3). This implies that the respondents with higher education and income levels tended to be more optimistic about future mitigation

Table 3. Determinants of Respondents' Perception about Future Pasture Degradation Mitigation (From strongly agree to strongly disagree).

Variables	Coefficients	Standard Error	Critical Value	p-Value	Lower 95%	Upper 95%
Intercept	3.522	0.579	6.078	0.000	2.374	4.670
Age	0.007	0.014	0.543	0.588	−0.020	0.035
Gender	−0.249	0.232	−1.073	0.286	−0.710	0.211
Education	0.031	0.028	−1.099	0.004 *	−0.088	0.025
Household Size	0.016	0.060	0.275	0.784	−0.102	0.134
Grazing Experience (years)	−0.008	0.014	−0.609	0.544	−0.035	0.019
Household Livestock (number)	0.000	0.000	−0.349	0.728	−0.001	0.000
Number of People Engaged in Herding	−0.023	0.092	−0.252	0.802	−0.205	0.159
Monthly Income of Household	0.183	0.071	−2.575	0.011 *	−0.324	−0.042

* p-value < 0.05.

3.5. Herders' Perceptions of Traditional Rangeland Management Practices

In the next section, we attempted to understand how herders decide to move their livestock to new pasture areas. More than 80% of the respondents considered both grass and water availability (Figure 6). In addition, 66% of the respondents would move their herds to rest the pasture. Other reasons included avoiding severe weather (12%) and to match livestock numbers to the specific carrying capacity of a specific pasture (13%). The rain/water availability (5%) and disease (4%) had low responses.

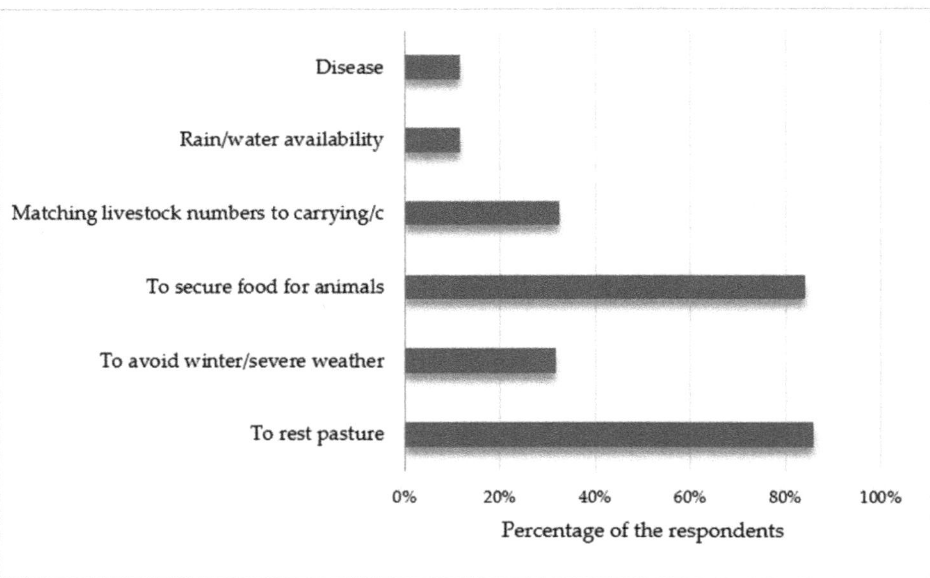

Figure 6. Murun Soum Herders' Reason to Move Their Livestock.

Past studies attributed the causes of pasture degradation and lowered livestock quality to the declining trend of herding movement and the frequency of changing pastures in Mongolia and Inner Asia in general [1,23]. Therefore, we first asked the respondents about the frequency of changing their pasture per year. In response, 3% of the respondents said to do so once or twice, 37% three to five times, 36% six to eight times, and 24% eleven times or more. Tumurjav [8] showed that traditional herders in steppes typically changed pastures seven to eight times a year. Considering this to be applicable to the study area, about 60% of the respondents still practiced traditional movement.

However, frequency is only one aspect of traditional herd movement. We attempted to know respondents' moving distance between camps. More than 74% of the respondents moved within 10 km each time during Spring–Summer and Summer–Autumn grazing seasons. In colder months, this distance became longer. Only 6% of the responding herders moved more than 30 km in spring and summer. Traditional herders in steppes typically moved 30 to 40 km [8]. Considering this, we found that more than 60% of the respondents moved their herds up to six times per year, but they tended to have fixed locations, and their moving distances were shorter than traditional practices.

Along with these movement questions, we attempted to identify rangeland management strategies and practices among the respondents in the past and the present. Here what we mean by the past is the so-called collective period or *negdel*, in which Marxist-based herding communes governed grazing activities within designated territories under the leadership of the General Committee and chiefs [46]. After 1991, communal governance was individualized with the introduction of the market economy.

Thus, we asked the respondents to select the practices they used to do during the collective period and then about current practices with multiple choices (Figure 7). The results show that before 1990, the respondents typically engaged in frequent seasonal movements (22%), *otor* (long-distance movement of herds) (20%), and reserving winter and spring pasture in summer (17%), matching livestock to the forage budget of the season (14%), herd splitting (13%), and rotational grazing management (10%). Fencing for hay in a certain pasture was widely practiced in other areas, but only 4% of the respondents did so in the study area.

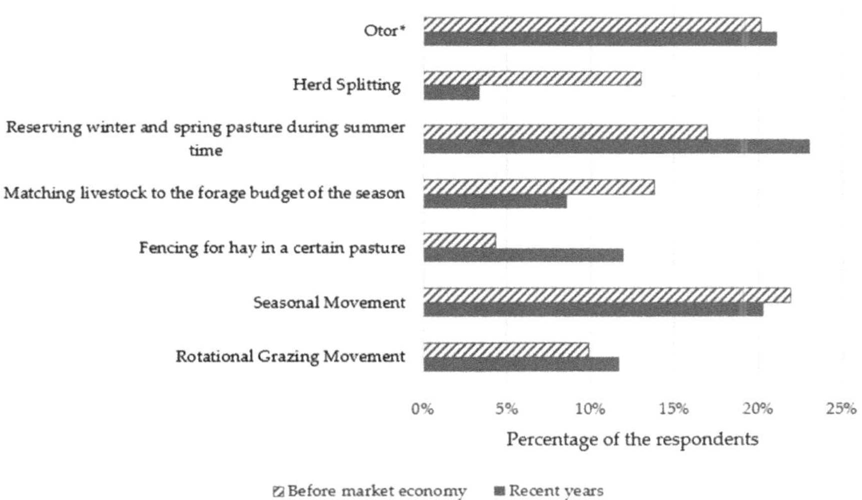

Figure 7. Comparison Between pre-1990 and Contemporary Traditional Rangeland Management Practices. * *Otor* means a long-distance movement of herds.

Concerning their current practices, the respondents, especially in the younger age groups, relied more on summer pasture reserves (23%), while *otor* (21%) and seasonal movement (20%) remained important among traditional herders. Practices of herd splitting (3%) and matching livestock to the forage budget of the season (9%) had become almost obsolete in the study area largely due to the collapse of communal governance [11].

Regarding *otor*, as it is related to movement distance, we added one more question to see if the respondents still practiced it. In response, 48% of the respondents said that they practiced it and moved their herds into neighboring soums, including Bayanmunkh, Bayan-Ovoo, Bayankhutag, and Umnudelger. By doing so, traditional herders aimed to fatten their animals.

3.6. Herders' Willingness to Participate in Rangeland Management

In the last section of our survey, we tried to understand how herders and government officials may cooperate for rangeland management in the present or the future. The respondents were first asked whether or not the soum government had helped them making herding decisions for better rangeland management. In response, 75% of the respondents said they made decisions without help. Only 8% of the respondents said that the soum government helped them manage. These results show an overall lack of cooperation between herders and the soum government in managing pastures. In the future, however, 47% of the respondents showed their willingness to support soum in managing pastures by using their knowledge and experience.

In addition, we attempted to understand respondents' willingness to improve their knowledge by either participating in training or receiving information. The result shows that 90% had not participated in training and workshop on rangeland management. However, 74% wanted to receive information about the carrying capacity of pasture for livestock and rangeland restoration. Some others wanted to stabilize sand dune expansion and land degradation by Brandt's voles.

To better understand the current state of collective pasture management, we asked the respondents about their involvement in a pasture user group, which has promoted local pasture management in the last ten years or so under the auspices of international organizations. We found that 59% of the respondents were not. Only 11% of the respondents were involved. The rest of the respondents were not sure about their involvement. This result possibly suggests that a communal organization, somewhat similar to *negdel* in the collective period, may not be able to exert influence over herding practices in the future.

We asked if the respondents were willing to limit the number of their livestock. About 65% of the respondents answered positively, but 18% found it impossible.

We attempted to understand the extent to which the local government enforces regulations to sustain the health of pastures. More than 65% of the respondents supported the pre-designed statement that the soum government should strictly observe regulations. Almost one-quarter of the respondents agreed that grazing fees should be introduced to herders for sustainable rangeland management. Those herders with relatively small numbers of livestock tended to believe that large-scale herders should pay higher grazing fees. Regarding the way rangeland should be managed sustainably in the future, 84% of the respondents agreed that herders should manage rangeland themselves. Jamsran [23] pointed out that herders had a certain level of distrust of local governments due to the past rangeland mismanagement.

After identifying respondents' willingness to work with the local government, we attempted to identify factors that influenced their willingness (Table 4). In doing so, we correlated their perceptions with some sociodemographic characteristics, such as age, education, household size, grazing experience, the number of household livestock, and monthly household income. The result shows that age (p-value = 0.015) and experience (p-value = 0.008) were particularly significant factors. This implies that those with more years of grazing experience expressed their willingness.

Table 4. Factors Influencing Respondents' Willingness to Help Soum Authorities Manage Rangeland.

Variable	Coefficients	Standard Error	Critical Value	p-Value	Lower 95%	Upper 95%
Intercept	1.731	0.484	3.580	0.001	0.773	2.689
Age	0.007	0.011	0.654	0.015 *	−0.015	0.030
Gender	0.145	0.194	0.748	0.456	−0.239	0.530
Education	0.043	0.024	1.799	0.075	−0.004	0.090
Household Size	−0.038	0.050	−0.773	0.441	−0.137	0.060
Grazing Experience (years)	0.003	0.011	−0.269	0.008 *	−0.025	0.019
Household Livestock (number)	0.000	0.000	0.194	0.847	0.000	0.000
Number of People Engaged in Herding	−0.143	0.077	−1.865	0.065	−0.294	0.009
Monthly Income of Household	0.098	0.059	1.656	0.100	−0.019	0.216

* p-value < 0.05.

4. Conclusions

This paper has examined Murun Soum herders' perspectives and knowledge about past and present livestock management, grassland degradation, and other related environmental changes. Our respondents, mainly middle-aged and elderly males, were veteran herders with almost 50% having more than 20 years of experience. Some had more than 40 years of experience. We also had the relatively young respondents and newcomers to this industry, especially those involved in cashmere production.

The results of our survey demonstrated that rangeland degradation and a growing number of Brandt's voles had posed the main threats to rangeland health in Murun Soum. About 95% of the respondents found that rangeland yields had somewhat declined. Their observation focused primarily on vegetation changes rather than soil conditions, although they did find damages to the rangeland soil by Brandt's voles.

Our study also found that the respondents moved their herds seasonally, reserved winter and spring pasture, and practiced *otor* for fattening their livestock. Comparing pre-1990 practices with contemporary ones, we found some remaining traditional practices, but these were declining. About a half of the respondents customarily practiced *otor* by crossing into neighboring soums, though the movement distance generally had become shorter in recent years.

Regarding the future participation in sustainable rangeland management in the study area, traditional herders showed their willingness to cooperate with soum authorities and other outsiders. Although more than 91% of the respondents depended on livestock outputs for livelihood, 65% showed their willingness to curb livestock numbers. Traditional small-scale herders, especially elderly ones among our respondents, were keenly aware of the importance of keeping vegetation diversity. They also expressed a need to maintaining the number of goats under the 18–20% range. Finally, we found that the post-1990 herding privatization has made an indelible mark on current herding practices, in which 84% of the respondents showed their interests in individual rangeland management rather than communal one.

Overall, it is possible to protect and improve rangeland health by adopting or enhancing traditional practices in Murun Soum if local and central governments, including researchers at public institutions, make a locally visible commitment to inviting traditional herders to participate in rangeland management. Thus, we argue that it is important to better understand traditional Mongolian herding methods in a given locality with regional socioeconomic situations.

Author Contributions: Conceptualization, K.M. and M.U.; methodology, K.M. and M.U.; software, M.U.; validation, M.U.; formal analysis, M.U. and K.M.; investigation, M.U.; resources, M.U.; data curation, M.U.; writing—original draft preparation, M.U.; writing—review and editing, K.M.; visualization, M.U.; supervision, K.M.; project administration, K.M.; funding acquisition, K.M. All authors have read and agreed to the published version of the manuscript.

Funding: This research was partially funded by the Project for Human Resource Development (JDS) Fellowship of the Japan International Cooperation Agency from 2018 to 2020.

Data Availability Statement: Not applicable.

Acknowledgments: We appreciate herders who shared their opinions and valuable comments with us in Murun Soum of Khentii Province, Mongolia.

Conflicts of Interest: The authors declare no conflict of interest.

References

1. Humphrey, C.; Sneath, D. *The End of Nomadism: Society, State, and the Environment in Inner Asia*; Duke University Press: Durham, NC, USA, 1999.
2. Liu, Y.Y.; Evans, J.P.; McCabe, M.F.; De Jeu, R.A.M.; van Dijk, A.I.J.M.; Dolman, A.J.; Saizen, I. Changing Climate and Overgrazing Are Decimating Mongolian Steppes. *PLoS ONE* **2013**, *8*, 6. [CrossRef] [PubMed]
3. Asian Development Bank. *Making Grasslands Sustainable in Mongolia Adapting to Climate and Environmental Change*; Asian Development Bank: Mandaluyong, Philippines, 2013; RPT136101.
4. NAMEM; MEGDT. *National Report on the Rangeland Health of Mongolia*; Government of Mongolia, National Agency for Meteorology and Environmental Monitoring, Green Development and Tourism: Ulaanbaatar, Mongolia, 2015.
5. Amgalan, O.; Avaadorj, D.; Batbuyan, B.; Batmandakh, A.; Binswanger, M.; Bolormaa, B.; Zagdsuren, Y. *Livelihood Study of Herders in Mongolia*; Mongolian Society for Range Management and Swiss Agency for Development and Cooperation, 2009; Available online: www.swiss-cooperation.admin.ch/mongolia/.../resource_en_,184798 (accessed on 22 April 2020).
6. FAO. *World Agriculture: Towards 2030/2050 Prospects for Food, Nutrition, Agriculture and Major Commodity Groups*; Food and Agriculture Organization of the United Nations: Rome, Italy, 2006.
7. Reed, M.S.; Dougill, A.J. Participatory Selection Process for Indicators of Rangeland Condition in the Kalahari. *Geogr. J.* **2002**, *168*, 224–234. [CrossRef]
8. Tumurjav, M. Traditional Animal Husbandry Techniques Practiced by Mongolian Nomadic People. In *Mongolia Today: Science, Culture, Environment and Development*; Badarch, D., Zilinskas, R.A., Eds.; Routledge: London, UK, 2003; p. 86.
9. Khwarae, G.M. Community Perceptions of Rangeland Degradation and Management Systems in Loologane and Shadishadi, Kweneng North, Botswana. Master's Thesis, Norwegian University of Life Sciences, Ås, Norway, 2006.
10. Ho, P.; Azadi, H. Rangeland Degradation in North China: Perceptions of Pastoralists. *Environ. Res.* **2010**, *110*, 302–307. [CrossRef] [PubMed]
11. Fernandez-Gimenez, M.E. The Role of Mongolian Nomadic Pastoralists' Ecological Knowledge in Rangeland Management. *Ecol. Appl.* **2000**, *10*, 1318–1326. [CrossRef]
12. Marin, A. Riders under Storms: Contributions of Nomadic Herders' Observations to Analyzing Climate Change in Mongolia. *Glob. Environ. Chang.* **2010**, *20*, 162–176. [CrossRef]

13. Berkes, F. *Sacred Ecology: Traditional Ecological Knowledge and Resource Management*; Taylor and Francis: London, UK, 1999.
14. Bruegger, R.A.; Jigjsuren, O.; Fernández-Giménez, M.E. Herder Observations of Rangeland Change in Mongolia: Indicators, Causes, and Application to Community-Based Management. *Rangel. Ecol. Manag.* **2014**, *67*, 119–131. [CrossRef]
15. Kristjanson, P.; Reid, R.S.; Dickson, N.; Clark, W.C.; Romney, D.; Puskur, R.; MacMillan, S.; Grace, D. Linking International Agricultural Research Knowledge with Action for Sustainable Development. *Proc. Natl. Acad. Sci. USA* **2009**, *106*, 5047–5052. [CrossRef] [PubMed]
16. Kgosikoma, O.; Mojeremane, W.; Harvie, B.A. Pastoralists' Perception and Ecological Knowledge on Savanna Ecosystem Dynamics in Semi-Arid Botswana. *Ecol. Soc.* **2012**, *17*, 11. [CrossRef]
17. Nakashima, D.; Galloway McLean, K.; Thulstrup, H.; Ramos Castillo, A.; Rubis, J. *Weathering Uncertainty: Traditional Knowledge for Climate Change Assessment and Adaptation*; McDonald, D., Ed.; UNESCO and UNU: Paris, France, 2012.
18. National Statistics Office. *Huduu Aj Ahuin Salbar 2018 [Agricultural Sector 2018]*. 2018. Available online: https://www.1212.mn/BookLibraryDownload.ashx?url=XAA%202018_last.pdf&ln=Mn (accessed on 15 June 2020).
19. Khishigbadam, G. Pasture Degradation Is Getting Serious. Available online: https://www.dnn.mn (accessed on 21 May 2020).
20. Banzragch, D.; Davaajamts, T. *Belcheer Ashiglakh Arga [Rangeland Use Method]*; State Press: Ulaanbaatar, Mongolia, 1970.
21. Tseelei, E. Manai Belcheer 25 Say Mal Tejeekh Ecologyn Chadavkhitai [Our Pasture Has the Ecological Capacity to Feed 25 Million Animals]. Available online: https://ikon.mn/opinion/1rwp (accessed on 20 June 2020).
22. Hardin, G. The Tragedy of the Commons. *Science* **1968**, *162*, 1243–1248. [CrossRef] [PubMed]
23. Jamsran, U. Involvement of Local Communities in Restoration of Ecosystem Services in Mongolian Rangeland. *Glob. Environ. Res.* **2010**, *14*, 79–86.
24. The World Bank. *Mongolia Environmental Monitor: Land Resources and Their Management*; The World Bank: Washigton, DC, USA, 2003.
25. Tseelei, E. Combination of Tradition and Contemporary Knowledge Lends a Solution to Mongolian Nomadic Herders. Available online: https://www.shareweb.ch/site/Agriculture-and-Food-Security/news/Documents/2015_12_mongolia.pdf (accessed on 14 June 2020).
26. *Murun Sumyn 90 Jilyn Oin Nom [Murun Soum 90th Anniversary Book]*; Narantuya, T.; Bayansuvd, B. (Eds.) Beijing, China, 2013.
27. Serjkhuu, S. *Mal Surgyn Toog Belcheeryn Daats, Nuhtsultei Uyalduulj Togtookh [Determine the Number of Livestock in Accordance with the Carrying Capacity of the Pasture]*. National Statistics Office. Available online: https://www.1212.mn/BookLibraryDownload.ashx?url=%D0%91%D1%8D%D0%BB%D1%87%D1%8D%D1%8D%D1%80-2017.pdf&ln=Mn (accessed on 10 May 2020).
28. National Statistics Office. Available online: www.1212.mn (accessed on 13 February 2020).
29. Nyamtseren, M.; Jamsran, T.; Sodov, K.; Doljin, D.; Zamba, B.; Erdenetuya, M. *Desertification Atlas of Mongolia*; Institute of Geoecology, Mongolian Academy of Sciences and Environmental Information Centre, Ministry of Green Development, Admon: Ulaanbaatar, Mongolia, 2013.
30. Gao, W.; Angerer, J.P.; Fernandez-Gimenez, M.E.; Reid, R.S. Is Overgrazing A Pervasive Problem Across Mongolia? An Examination of Livestock Forage Demand and Forage Availability from 2000 to 2014. In Proceedings of the Building Resilience of Mongolian Rangelands: A Trans-disciplinary Research Conference, Ulaanbaatar, Mongolia, 9–10 June 2015; pp. 35–41.
31. Ministry of Nature and Environment. *National Plan of Action to Combat Desertification in Mongolia*; Government of Mongolia Ministry of Nature and Environment: Ulaanbaatar, Mongolia, 1997.
32. Batjargal, Z. Desertification in Mongolia. In *RALA Report No. 200*; National Agency for Meteorology, Hydrology and Environment Monitoring: Ulaanbaatar, Mongolia, 1997.
33. Okayasu, T.; Muto, M.; Jamsran, U.; Takeuchi, K. Spatially Heterogeneous Impacts on Rangeland after Social System Change in Mongolia. *L. Degrad. Dev.* **2007**, *18*, 555–566. [CrossRef]
34. Addison, J.; Friedel, M.; Brown, C.; Davies, J.; Waldron, S. A Critical Review of Degradation Assumptions Applied to Mongolia's Gobi Desert. *Rangel. J.* **2012**, *34*, 125–137. [CrossRef]
35. Sheehy, D.P.; Damiran, D. *Assessment of Mongolian Rangeland Condition and Trend (1997–2009). Final report for the World Bank and the Netherlands-Mongolia Trust Fund for Environmental Reform*; World Bank and the Netherlands-Mongolia Trust Fund for Environmental Reform (NEMO): Ulaanbaatar, Mongolia, 2012.
36. Bayasgalan, B.; Mijiddorj, R.; Gomboluudev, P.; Oyunbaatar, D.; Bayasgalan, M.; Tas, A.; Narantuya, T.; Molomjamts, L. Climate Change and Sustainable Livelihood of Rural People. In *The Adaptation Continuum: Groundwork for the Future*; Devissher, T., O'Brien, G., O'Keefe, P., Tellam, I., Eds.; ETS Foundation: Leusden, The Netherlands, 2009.
37. Division of Cartography and Geographic Information System. *Mongolyn Undurlugyn Tsuljilt, Gazryn Burkhevch, Gazar Ashiglaltyn Orchin Ueyn Uil Yavtsyn Sudalgaa Tailan [Modern Study of Desertification, Land Cover and Modern Land Use Processes on Mongolian Plateau Report]*; The Mongolian Science and Technology Foundation: Ulaanbaatar, Mongolia, 2007.
38. Sternberg, T. Environmental Challenges in Mongolia's Dryland Pastoral Landscape. *J. Arid Environ.* **2008**, *72*, 1294–1304. [CrossRef]
39. Bedunah, D.J.; Angerer, J.P. Rangeland Degradation, Poverty, and Conflict: How Can Rangeland Scientists Contribute to Effective Responses and Solutions? *Rangel. Ecol. Manag.* **2012**, *65*, 606–612. [CrossRef]
40. Middleton, N. Rangeland Management and Climate Hazards in Drylands: Dust Storms, Desertification and the Overgrazing Debate. *Nat. Hazards* **2018**, *92*, 57–70. [CrossRef]
41. Sheehy, D.P.; Thorpe, J.; Kirychuk, B. Rangeland, Livestock and Herders Revisited in the Northern Pastoral Region of China. *USDA For. Serv. Proc.* **2006**, 62–82.

42. The World Bank. World Bank Open Data. Available online: https://data.worldbank.org/ (accessed on 10 February 2020).
43. Munkhnasan, T. *Fighting Management against Brand's Vole for Environmentally Friendly That Damaged Pasture*; Ministry of Food, Agriculture and Light Industry of Mongolia: Ulaanbaatar, Mongolia, 2019. [CrossRef]
44. Cui, C.; Xie, Y.; Hua, Y.; Yang, S.; Yin, B.; Wei, W. Brandt's Vole (Lasiopodomys Brandtii) Affects Its Habitat Quality by Altering Plant Community Composition. *Biologia* **2020**, *75*, 1097–1104. [CrossRef]
45. Pasotti, J. How Mongolia's Nomads Are Adapting to Climate Change. Deutsche Welle. 2017. Available online: https://www.dw.com/en/how-mongolias-nomads-are-adapting-to-climate-change/a-39310932 (accessed on 20 March 2020).
46. Goldstein, M.C.; Beall, C.M. *The Changing World of Mongolia's Nomads*; University of California Press: Berkeley, CA, USA, 1994.

sustainability

Article

Food Waste in Da Nang City of Vietnam: Trends, Challenges, and Perspectives toward Sustainable Resource Use

Ngoc-Bao Pham [1], Thu-Nga Do [2,*], Van-Quang Tran [3], Anh-Duc Trinh [4], Chen Liu [1] and Caixia Mao [1]

1   Institute for Global Environmental Strategies (IGES), 2108-11 Kamiyamaguchi, Hayama, Kanagawa 240-0115, Japan; ngoc-bao@iges.or.jp (N.-B.P.); c-liu@iges.or.jp (C.L.); caixia.mao28@gmail.com (C.M.)
2   Faculty of Energy Technology, Electric Power University, Hanoi 11900, Vietnam
3   Faculty of Environment, Danang University of Science and Technology, Da Nang 50608, Vietnam; tvquang@dut.udn.vn
4   Nuclear Training Center, Vietnam Atomic Energy Institute, Hanoi 11000, Vietnam; trinhanhduc@yahoo.com
*   Correspondence: dothu_nga2005@yahoo.com

Citation: Pham, N.-B.; Do, T.-N.; Tran, V.-Q.; Trinh, A.-D.; Liu, C.; Mao, C. Food Waste in Da Nang City of Vietnam: Trends, Challenges, and Perspectives toward Sustainable Resource Use. *Sustainability* **2021**, *13*, 7368. https://doi.org/10.3390/su13137368

Academic Editors: Kikuko Shoyama, Rajarshi Dasgupta and Ronald C. Estoque

Received: 3 June 2021
Accepted: 29 June 2021
Published: 1 July 2021

Publisher's Note: MDPI stays neutral with regard to jurisdictional claims in published maps and institutional affiliations.

Abstract: Food waste has become a critical issue in modern society, especially in the urbanized and fast-growing cities of Asia. The increase in food waste has serious negative impacts on environmental sustainability, water and land resources, and food security, as well as climate and greenhouse gas emissions. Through a specific case study in Da Nang City, Vietnam, this paper examines the extent of food waste generation at the consumption stages, the eating habits of consumers, food waste from households and service establishments, as well as prospects for the reuse of food waste as pig feed. The results of this study indicate that per capita food waste generation in Da Nang has increased from 0.39 to 0.41kg in 2016, 0.46 in 2017, and reached 0.52kg in 2018. According to the results of our consumer survey, 20% of respondents stated that they often generate food waste, 67% stated they sometimes do, and 13% stated they rarely do. Furthermore, 66% of surveyed households stated that their food waste is collected and transported by pig farmers to be used as feed for pigs. The use of food waste as feed for pigs is a typical feature in Da Nang. The study also found that there is a high level of consumer awareness and willingness to participate in the 3Rs (reduce, reuse, recycle) program, which was being initiated by the city government. In service facilities such as resorts and hotels, daily food waste reached 100–200 kg in large facilities and 20–120 kg in small facilities. This waste was also collected for use in pig farming. However, there has been a fall in demand for pig feed in line with a decrease in the number of pig farms due to the African swine fever epidemic that occurred during the implementation of this study. This paper suggests that there is a strong need to take both consumer-oriented waste prevention and waste management measures, such as waste segregation at source and introduction of effective food waste recycling techniques, to ensure that food waste can be safely and sustainably used as a "valuable resource" rather than "wasted."

Keywords: 3Rs program; landscape sustainability; municipal solid waste; pig farming; resource circulation; resource use efficiency; urban–rural nexus; zero-waste lifestyle

1. Introduction

On a global scale, it is estimated that about a third of the food produced for human consumption, or 1.3 billion tons, is lost at different stages of the food supply chain annually [1]. The environmental, social, and economic losses from food waste cost the global economy USD 2.6 trillion [2]. Environmental impacts occur from both food production and food waste management perspectives. The production stage requires 250 km^3 of water and 1.4 billion hectares of land, with about 30% of the agricultural land used to produce uneaten food [3]. Food loss and waste also emit 4.4 Gt CO_{2eq} of greenhouse gases (GHG), accounting for 8% of the total anthropogenic GHG emissions annually [4]. Looking at global food waste management, 58% of food waste from the distribution and marketing stages and 84% from the consumption stage is landfilled [5]. Like most developing countries in Asia, Vietnam still relies on open dumping as the main waste disposal method for waste management [6,7]. Landfill of food waste causes several health and environmental concerns, such as transmission of diseases [8] and GHG emissions [9].

Most studies on food waste are currently examining food waste in the context of developed economies. The per capita food waste in developed economies, such as in Europe and North America, is particularly high compared with other regions [10]. Moreover, the prevention of food waste generation is considered to be the most attractive option, as most waste can be prevented with proper measures [10]. Households in the United Kingdom discard about one third of the food they purchase, and 61% of this amount could have been consumed if the produce had been managed better [11]. Similarly, two thirds of the food waste from households in Germany could have been partly or completely avoided [12]. It has been shown that the average amount of per capita food waste generated in developed and developing countries is 107 kg/year and 56 kg/year, respectively [13]. Buchner et al. [14] also showed that there is a strong correlation between per capita food waste generation and income levels. For example, the amount of per capita food waste (kg/cap.day) in developed countries such as Australia, Germany, Singapore, the UK, and the United States comes to 0.25, 0.34, 0.40, 0.37, and 0.52, respectively [13]. On the other hand, the per capita food waste in countries such as Malaysia, China, Thailand, and Vietnam amounts to only 0.18, 0.14, 0.14, and 0.06, respectively. Nevertheless, in the context of urbanized cities in developing countries, food waste at the consumption stage in developing economies could be even higher than in developed countries [15,16], which indicates that addressing food waste in urbanized cities is an urgent issue in Asia.

Vietnam has made a remarkable journey from low to middle income in the past three decades. This rapid growth, along with industrialization and urbanization, has led to significant shifts in production and consumption patterns, as well as bringing about lifestyle changes, in line with increased consumerism and modified habits related to eating out [17]. As a result, there has been a significant increase in the amount of urban waste and food waste generated per capita, especially in large and rapidly developing cities like Da Nang City. This is having a detrimental impact on the environment and natural assets. It has been reported that food waste accounts for approximately 50–60% (or more in some cities) of the solid waste generated in urban areas, with this waste being eventually disposed of in landfills [18]. Existing research on food waste issues in Vietnam is dominated by waste management [9,19–22]. There are several waste management options identified, such as animal feeding, anaerobic digestion, composting, incineration, and landfilling [13]. Utilizing food waste for livestock feed is considered one major way to manage food waste in Vietnam. In Ho Chi Minh City, as much as 70% of the total food waste is used for livestock feeding [22,23]. Nevertheless, within the existing literature on food waste in Vietnam, little attention has been paid to the factors contributing to food waste generation at the different stages of the food supply chain, notably in the consumption stage, and so this paper aims to fill this research gap.

Through a specific case study conducted in Da Nang City, this paper aims to answer key questions on (i) the extent of food waste at the consumption stage of food supply chains in fast-urbanizing cities like Da Nang; (ii) habits of purchasing, preparing, and disposal of food waste by households and service establishments (e.g., resorts, hotels, and restaurants), as well as demand for food waste by pig farms on the outskirts of Da Nang; and (iii) the roles that reuse and recycling play in the reduction of food waste. The paper also initiates discussion on methodological issues of food waste relevant to the consumption stage of the food supply chains. In this study, food waste is defined as "any food, and inedible parts of food, removed from the goods supply chain to be recovered or disposed," as proposed by FUSIONS [24]. It does not include packaging waste, or food used for redistribution.

2. Materials and Methods

2.1. Description of the Study Site

Da Nang City is a port city located in central Vietnam. Da Nang is the fifth most populous city in the country, with a population of 1,080,700 as of 2018 and an area of 1285.4 km^2 [25]. Da Nang is subdivided into eight districts: six urban districts (Cam Le, Hai Chau, Thanh Khe, Lien Chieu, Nguyen Chieu, Ngu Hanh Son, and Son Tra) and two rural districts (Hoa Vang and Hoang Sa). It is further subdivided into one commune-level town, 14 communes, and 45 wards [26,27].

Municipal solid waste in Da Nang is generated from homes, hotels, restaurants, markets, shops, offices, schools, institutions, hospitals, airports, parks, and other commercial facilities. Municipal solid waste in the city is managed by the Da Nang Urban Environment Company [28]. The volume of municipal solid waste in Da Nang is projected to reach 1800 metric tons per day by 2025, 2400 metric tons per day by 2030, and 3000 metric tons per day by 2040. Currently, all municipal solid waste generated is treated at the Khanh Son landfill. The disadvantage of municipal solid waste disposal and treatment at this landfill is that it has become a hotspot for environmental pollution in Da Nang, and is an adverse, unfavorable use of land. Over time, costs for municipal solid waste collection and costs for treatment in this landfill have also increased, particularly to deal with odor, GHG emissions, and leachate that emerge from the landfill [29].

To collect data for this study, questionnaire surveys and key stakeholder interviews were conducted in three districts in Da Nang, namely, Hai Chau (urban district), Thanh Khe (urban district), and Hoa Vang (rural district), targeting local residents, owners of service facilities, and pig farmers, based on the characteristics of each district (Figure 1).

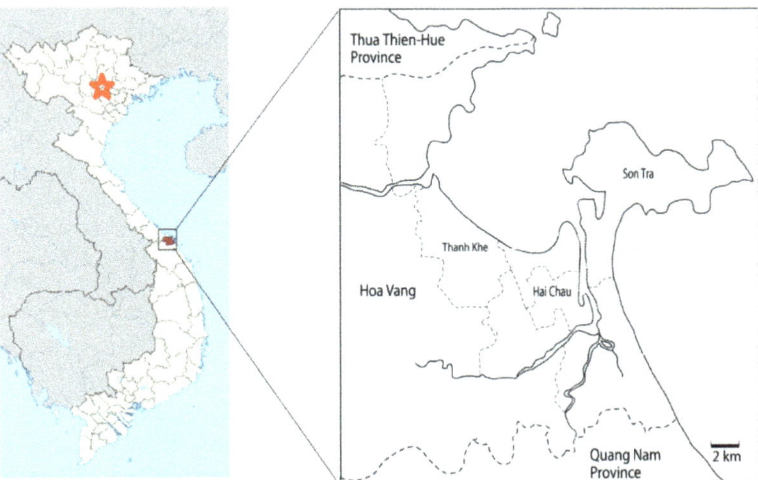

Figure 1. Location of survey area in Da Nang City, Vietnam.

2.2. Surveys—Questionnaire and Key Stakeholder Interviews

From 12 December 2018, to 1 January 2019, a survey on food habits and food waste segregation was conducted through interviews and questionnaires among 360 households in Hai Chau and Thanh Khe urban districts. This sample size was calculated using a formula developed by Yamane [30], setting a confidence level at 95%, and accounting for missing data. In order to understand food waste generation and demand, the research team interviewed owners of selected service establishments (resorts, hotels, and restaurants) and pig farmers (small-scale farms—less than 20 pigs, and medium-scale farms—less than 200 pigs) in Hoa Vang rural district from November to December 2019.

The survey was conducted through stakeholder interviews and questionnaires, which ensured a high recovery rate and minimized divergence due to respondent apathy. The questions asked in the household survey were based on the following points: (i) basic information about the respondents and their households; (ii) lifestyle habits (location of food purchases, daily meal preparation, and dietary habits); (iii) sorting and disposal habits of municipal and food waste; and (iv) knowledge of and willingness to participate in and implement the 3Rs (reduce, reuse, and recycle) program.

On the other hand, for swine producers, the survey focused on the demand for food waste, including (i) basic information on pig farms; (ii) food waste demand and collection methods; (iii) food processing methods for feeding pigs from food waste; and (iv) demand for pig feed from food waste in the near future.

3. Results and Discussion

3.1. Increase in Food Waste Fraction in Cities in Vietnam

Food waste-related studies have shown that the organic fraction of the waste tends to be higher in low-income countries and lower in high-income countries. Total organic waste tends to increase steadily with increasing affluence, at a more modest rate than the non-organic waste fraction. The organic rate in low-income countries is 64%, compared to 28% in high-income countries [31]. In many large cities in Vietnam, food residues account for a large proportion of the discarded municipal solid waste [32–37]. It has been reported that food waste in Ho Chi Minh City accounted for about 64.3–98.3% (average value of 81.6) by wet weight after three months of monitoring at Binh Thanh landfill in 2011 [36]. The composition of food waste at Ge Cat and Phuoc Hiep landfills in Ho Chi Minh City was reported to be 70.84% [35] and 61.3–68.9% [32] in 2011, and the proportion of food waste in the city's municipal waste increased to 83–88.8% in 2015 [13]. This suggests that the food waste component in municipal solid waste in Ho Chi Minh City has increased by several tens of percent. This is similar to the case in Da Nang. Kajima Corporation randomly sampled waste from garbage trucks arriving at the Khanh Son landfill in Da Nang and reported that food residues accounted for 43.7% of the wet weight of municipal solid waste [38]. Meanwhile, this data was reported by Da Nang DONRE [26,39] to be 56.85% in 2009. According to the Da Nang Department of Natural Resources, the percentage of biodegradable organics in municipal solid waste at the Khanh Son landfill was 66.71–74.65% [29,40], of which 80–90% was food waste for the period 2010–2014 [18]. This percentage of organic matter in solid waste in the city is similar to that published by Da Nang URENCO between 2007 and 2011, which was 68.47% [28]. This figure suggests that the organic and food waste components in Da Nang's municipal solid waste have been steadily increasing over the past decade.

3.2. Trends of Food Waste Generation in Da Nang City

Based on data collected from Da Nang DONRE [26], Da Nang UPI [29], and GSO [25], the municipal solid waste generation and corresponding food waste generation rates and solid waste generation rates were calculated and shown in Figure 2.

According to Figure 2, the population of Da Nang increased by 16.6%, from 926,000 in 2010 to 1,080,000 in 2018 [25]. On the other hand, the rate of increase in solid waste generation in the city during this period was much higher, calculated at 62.8%: from just 223,000 metric tons of municipal solid waste collected in 2010, this increased to 363,000 metric tons in 2018 [29]. Based on food waste composition data collected from previous reports, the generation rate per capita was calculated as 0.37 kg of food waste per day per capita in 2010, and this figure was maintained at 0.39–0.41 kg until 2016. Then in 2017, the figure rose to 0.46 kg, with a further increase to 0.52 kg in 2018.

Comparing this rate with other large cities (Hanoi, Ho Chi Minh City, Bangkok, and Phnom Penh) in Vietnam and other Southeast Asian developing countries [18,26,28,40,41], it shows that food waste generation in Da Nang is at a high level (Table 1).

As can be seen in the above table, the fraction of food waste in Da Nang's municipal waste was similar to that of Hanoi and Phnom Penh, but much lower than that of Ho Chi Minh City and much higher than that of Bangkok. It should be noted that although Bangkok's population was about nine times larger than that of Da Nang, the food waste generation rate per capita of the two cities was comparable in 2016. Similarly, the population of Ho Chi Minh City was about eight times larger than that of Da Nang, but the food waste generation rate per capital was only about 48% higher than that of Da Nang. This can be attributed to differences in lifestyles, solid waste management (food waste demand and disposal), people's attitudes, and local people's knowledge in Da Nang compared to other cities.

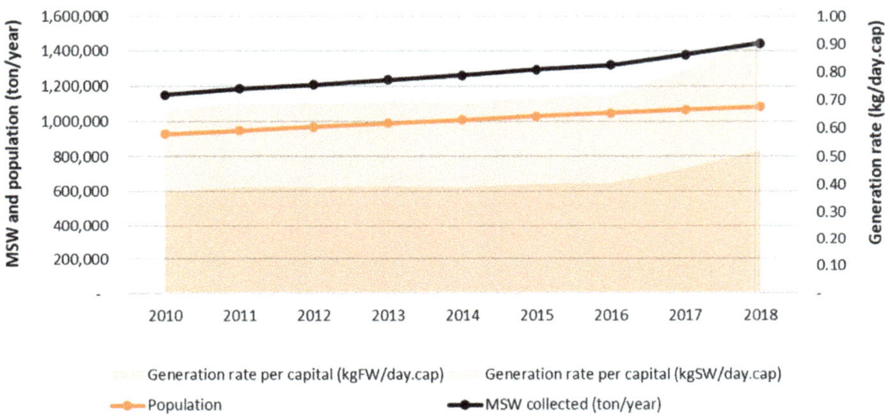

Figure 2. Municipal solid waste (MSW) and food waste generation in Da Nang City (2010–2018).

Table 1. Volume of municipal solid waste, composition, and ratio of food waste generation per capita in selected cities in Southeast Asia [18,26,28,40,41].

City	Da Nang	Hanoi	Ho Chi Minh City	Bangkok	Phnom Penh
Year	2018	2015	2018	2016	2015
Population (1000 persons)	1080	4371	8100	9166	1446
Food waste fraction (%)	57.3	57.3	65.7	47.6	51.9
Collected/landfilled municipal solid waste (tons/day)	994	4980	9400	10,130	1121
Food waste generation rate (kg/day.cap)	0.52	0.65	0.76	0.53	0.40

Moreover, looking at developed countries, based on the latest official data by the Tokyo Metropolitan Government, the average food waste per person per day was around 0.39 kg in 2012 [16]. Based on a report by the Natural Resources Defense Council, an average of 0.23 kg per person per day was wasted at home in the three US cities of Denver, Nashville, and New York [42]. WRAP reports the average food waste per capita per day was 0.42 kg in the UK in 2015 [43]. In addition, food waste generation per capita per day is 0.25 kg in Australia, 0.32 kg in Denmark, 0.27 kg in Sweden, 0.34 kg in Germany, 0.27 kg in South Korea, and 0.52 kg in the US [13]. Although the data are not directly comparable because of the different definitions of food waste and the scope of estimation, it is clear that food waste generation in Da Nang City is even higher than the average in many developed cities and countries.

3.3. Results of the Survey

3.3.1. Characteristics of the Selected Respondents

Table 2 summarizes the characteristics of households that participated in the survey.

Table 2. Characteristics of the respondents.

	Parameter	Number of Respondents
Gender	Male	119
	Female	241
Age	≤20	20
	21–30	60
	31–40	61
	41–50	64
	51–60	77
	>60	78
Occupation	Company employee	101
	Daily-base laborer	110
	Full-time housewife, pensioner	122
	Student (university, junior college, etc.)	27
Education	University, master's degree or higher	63
	Upper secondary school	261
	Lower secondary school	32
	Primary school	4
Household income	≤5 million VND/month	56
	5–10 million VND/month	172
	>10 million VND/month	132

As seen in Table 2, two thirds of the interviewees were women. Respondents' ages ranged widely from 18 to over 60 years of age, with more than half (57%) between 18 and 50 years of age. The level of education of the interviewees is also important in this study. About 90% of the respondents had a secondary education or higher. The occupation of the respondents was very diverse, with 28.1% working full time, 33.9% being retired and working at home, 30.6% working part time, and the remaining 7.5% being students. On the other hand, about 84% of the respondents had a medium to moderate income or higher.

3.3.2. Household Lifestyles Related to Food Waste Generation

Figure 3 shows the results of the survey on eating habits. Most respondents stated that they eat three meals a day. None of the respondents stated that they have no breakfast, and very few people answered that they have four meals a day.

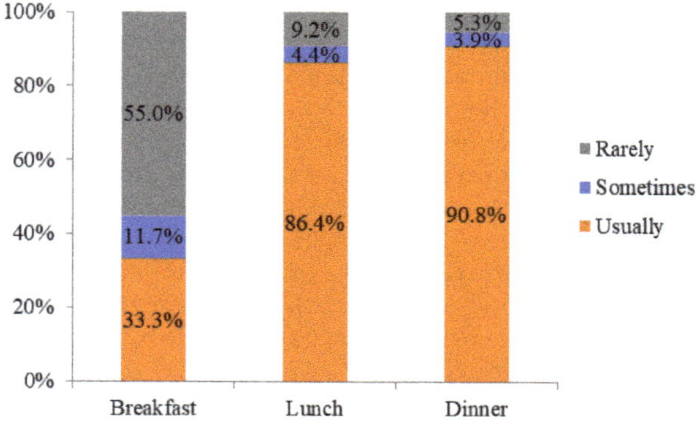

Figure 3. Frequency of eating meals at home.

These results are indicative of Vietnamese culture and traditions, with people usually eating lunch and dinner at home, four to five times a week. Conversely, more than half of the respondents reported that they rarely eat breakfast at home and prefer to eat out for breakfast. The survey results imply that Da Nang residents prefer to cook lunch and dinner at home rather than eating out. This trend is similar to Hanoi, indicating that most people tend to eat at home more often in both urban and rural areas [16,44]. Responses to the survey also showed that health and food safety is a key factor in decision-making on diet amongst consumers in Da Nang and Hanoi.

The results of the survey on places to buy food or ingredients for cooking food in Da Nang showed that the most common place is the traditional fresh market. Ninety-four people indicated that they choose both supermarkets and traditional fresh markets as the place to buy food or ingredients on a daily basis, whereas 239 people choose traditional fresh markets only. Very few households (19 households) purchase ingredients from either the supermarket or the hawker alone. A total of 242 respondents said that they are in the habit of going to the traditional fresh market every day, whereas the rest of the respondents stated that they go to the market three to six times a week. On the other hand, of the 108 people who choose to go to the supermarket to buy food, 91 of them buy food once or twice a week. This result indicates the traditional practice of Vietnamese people, whereby locals prefer to go to traditional fresh markets to buy ingredients for preparing food, may suggest the potential burden of food waste generated by the markets, e.g., expired foods, un-eaten parts of vegetables, fish, etc.

In the survey results on where people buy food, 360 people chose "fresh food" as their preferred choice. This figure is in line with the number of people who preferred to buy at fresh markets. Neither "prepared food" nor "canned food" were preferred by Da Nang households. This characteristic is also similar to the findings from the surveys in Hanoi [44]. They showed that residents in both rural and urban areas consume ready-made meals relatively infrequently. Buying fresh food and cooking it at home generates a lot of food waste from households. This is different from Bangkok, where local people eat out or consume ready-made meals frequently [16].

When asked how often they generate food waste, 20% stated "often" (four or more times a week), 67% stated "sometimes" (two to four times a week), and 13% stated "rarely" (two or fewer times a week). That means that about 87% of the respondents generate food waste more than twice a week. On the other hand, it was found that 66% of these households use food waste to feed pigs and just 2% use it to fertilize vegetables and for gardening. Food waste is collected and transported by pig breeders. In addition, 32% of households generate it with other waste. Compared to other large cities in Vietnam, a much higher percentage of food waste from households in Da Nang is used to feed pigs. This figure is comparable to other Asian countries, where there is a high demand for animal feed. For example, Japan and South Korea encourage reusing food waste to feed animals, with 33% and 81% of total food waste being used this way, respectively [45,46]. In contrast, the separation and collection of food waste are not practiced in many developing countries, and therefore almost all food waste is mixed with municipal solid waste (MSW), which is not then able to be purified and utilized for animal feed. The use of food waste to feed pigs is typical in Da Nang.

3.3.3. Food Waste Management at Service Facilities

Tourism is an important sector in Da Nang's economy, so it is essential to assess the part played by businesses, including resorts, hotels, and restaurants, in the total food waste generation in Da Nang City. This research carried out a survey of food waste in seven service establishments in urban Da Nang City. These included resorts (120 employees, 100–150 per day), hotels (large, 300 employees, 1200 per day) and restaurants (small, 10–35 employees, 50–80 per day). The results showed that the daily amount of food waste ranges from 100–200 kg for large facilities to 20–120 kg for small facilities; the highest daily food waste volumes are generated at the resorts (100–120 kg) and hotels (150–200 kg) targeted in the survey. Food waste from restaurants ranges from 40–80 kg per day. All of the facilities have a waste collection activity in their kitchens. This way of saving waste can generate income and provide support for pig breeders. Waste is collected from the service facilities in the early morning at 5 or 6 a.m. or late at night at 10 or 11 p.m., depending on the work schedule of the target facility. Since 2019, all types of food waste (including liquid forms) are put into 20–60 liter plastic containers and transported by bike or truck.

3.3.4. Food Waste Demand from Pig Breeders

A survey of farmers in the Hoa Khuong commune indicated that due to an outbreak of African swine fever [26], the number of farms and households raising pigs in 2019 decreased by 20–30%. In addition, their capacity also decreased from 80–200 pigs/year to 10–90 pigs/year in the 2018–2019 period. As a result, the demand for food waste decreased and pig breeders only collected it from rural areas. Liquid food waste generated by the service facility is filtered and discharged into the sewer system. Other food waste is mixed with other solid waste and transported to the landfill. Subsequently, the amount of food waste generated in urban areas increased and the amount of municipal solid waste at the Khanh Son landfill increased by 4% per day [26,47].

In fact, food waste could be safely used as pig feed by implementing the following two steps: removing impurities from food waste such as toothpicks, plastic bags, and clam shells, and cooking by adding more fiber and vegetables (which are available on farms) [19]. Existing literature suggests that food waste can be decontaminated for feeding using thermal treatment and preferably by complete sterilization [48]. Pig breeders use bikes to transport 60–240 kg of food waste for their pigs every day. This method could save 50% of the feed costs for pigs. In Vietnam, a farm raising 100–200 pigs can guarantee an annual income for a family with four members. According to Kato et al. [27,49], a large amount of food waste is collected by pig breeders in urban areas of Da Nang, with the amount collected dependent on the number of pigs on the farm and the number of collection trips per day. The study found that farmers collect about 26.3 metric tons of organic waste every day, which is equivalent to 4.1% of the domestic solid waste collected by the local government. Ninety-three percent of the farmers interviewed indicated that they would continue to use food waste for the next five years. Thus, all of the respondents confirmed that they would expand the size of their farms once the African swine fever outbreak is over. They also agreed on the method of collecting food waste at this time.

3.3.5. Assessment of Readiness for Food Waste Reduction, Reuse, and Recycling

This study assessed awareness and willingness to participate in the 3Rs (reduce, reuse, and recycle) for food waste reduction. The results are shown in Figure 4 and indicate a surprising percentage of respondents who had knowledge of the 3Rs and their application. Nearly half of the respondents (160 respondents) indicated that they did not know about the 3R initiative. Eighty-eight of the 360 respondents targeted were aware of the 3Rs but were not interested in applying them. Reasons cited included lack of time to separate waste, odor, and sanitation concerns, and not knowing how to separate waste. Some 105 respondents knew about the 3Rs, but only 85 people were implementing the 3Rs at home. Regarding the willingness of Da Nang residents to participate in the 3R initiative, 293 respondents, or about 82% of the total respondents, indicated that they were ready or willing to participate in the 3Rs (Figure 4b). However, about 18% of respondents either did not participate in the initiative or chose other options. A similar survey was conducted in Hanoi City, showing that the majority of respondents were willing to reduce food waste. Only a few indicated that they had no plans to reduce food waste [48]. The results from this survey suggested the importance of identifying potential ways for consumers to reduce food waste, using easier methods than current practices.

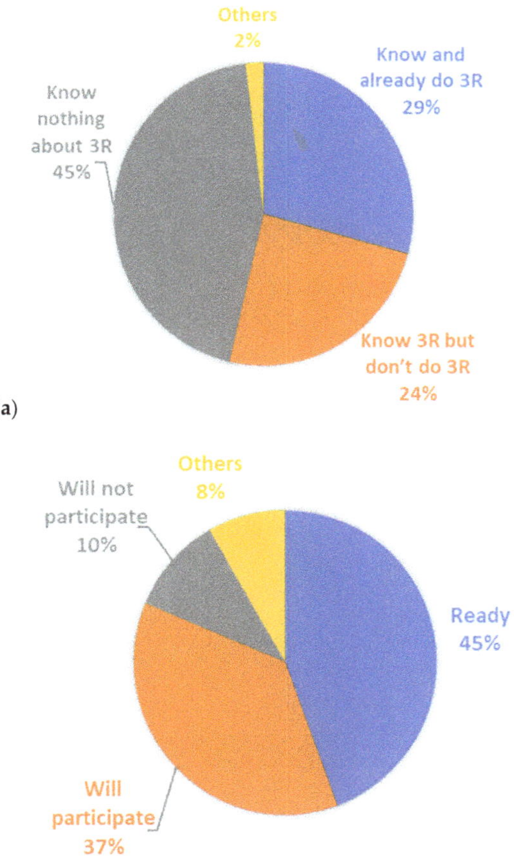

Figure 4. Accessing knowledge on the 3Rs (**a**) and readiness to apply the 3Rs (**b**) in Da Nang.

4. Way Forward and Policy Implications

The results of the survey carried out in Da Nang confirmed that due to the economic growth and urbanization, food waste in urbanized cities in Vietnam may be even higher than in developed countries, although the quantification methods differ. Thus, Da Nang needs to find a way to simultaneously address food waste generation reduction at the source and deal with food waste management issues.

The survey results show that consumers exhibit high awareness and willingness to apply the 3Rs to reduce food waste. Thus, several measures could be applied to target food waste reduction at the consumption stage. These could include awareness campaigns accompanied by toolkits to help consumers reduce their food waste. This type of guidance could encourage consumers to take practical steps such as improving their skills and knowledge in grocery shopping and planning, ordering smaller portions at restaurants, improving their knowledge about sell-by dates and labels, reusing leftovers and forgotten foods, and improving storage methods, thereby facilitating actions when eating at home and eating out [16].

From a waste management and landscape sustainability perspective, the survey results reveal that reusing food waste for pig farming is a major food waste management measure for food waste generated both in households and in service facilities. If the food waste is not reused, it has to be dumped outside the city, which degrades and devalues land. Therefore, reusing food waste is an economically and environmentally effective solution for food waste management. Nevertheless, due to the African swine fever outbreak [26], the number of pig farms fell in 2019, resulting in a decreased demand for food waste to use as feed for pigs in 2019. During the implementation of this study, livestock diseases have been brought under control, and households and farmers want to restart their farms to cater to food markets. However, local authorities have not started granting permits for pig re-farming in light of public health concerns. Meanwhile, research findings indicate that proper treatment of food waste using heating and fermentation technologies could effectively avoid contamination of feed for pigs, thereby ensuring stable management of food waste and reducing feed costs [19,50,51]. In accordance with Circular No. 21/2019/TT-BNNPTNT dated 28 November 2019, there have been several updates to the Articles of Law on Animal Husbandry regarding animal feed. Thus, from January 2020, it is now mandatory to carry out quality inspection when using food waste for animal feed. This new legislation is likely to increase the costs associated with feeding livestock, and so small and medium-sized pig farms seeking permission to re-farm are likely to be unable to afford to re-start and will disappear. A decrease in the number of pig farms will be a major problem. It has also been shown that pigs play a vital role in stable food waste management, as they can eat and digest different types of feeds [19,48,49]. Subsequently, if there is a decrease in the amount of food waste used for pig feed, food waste as a part of municipal waste will increase. Liquid food waste, such as fats and oils, flows into the sewer system, where anaerobic digestion causes clogging and produces foul odors. It will become difficult to classify, treat, and manage this increased amount of waste at the Khanh Son landfill [26].

At the policymaking level, national guidance states that the primary solutions for municipal solid waste management are reduction, reuse, and recycling (3Rs). Details and specific goals for municipal solid waste management in 2025 are as follows:

- Collect at least 90% of municipal solid waste and dispose of it at emission standards;
- Increase municipal solid waste reuse and recycling: Treatment process should be combined with energy or organic fertilizer creation;
- Less than 30% of municipal solid waste to be disposed of in landfills.

Meanwhile, Da Nang City has recently implemented waste segregation at source in the two central districts of Hai Chau and Thanh Khe as pilot models [29]. The rate of recycling, reuse, and energy and organic fertilizer production is expected to reach 85% by 2030. Looking at municipal solid waste management between 2015 and 2019 in the city area of Da Nang EPA [47], a municipal solid waste plan up to 2025 will be formulated to classify waste at source in the urban area of Da Nang. Households will be made aware of the importance of waste segregation at source and will be encouraged to participate. In addition, there are some challenges associated with municipal waste management in Da Nang City, including lack of collection and treatment facilities, making it impossible to segregate biodegradable organics. The city relies on farms in the suburbs for food waste collection. Therefore, the following key strategies are recommended to address the existing issues of municipal solid waste in general, and food waste management in particular: (i) supporting tools, equipment, and instructions on waste segregation at the source, particularly for food waste (biodegradable organics); (ii) investment in and upgrading of a secondary collection system and transfer stations for segregated municipal solid waste; (iii) support for research into pilots and models for recycling food waste into feed and organic fertilizer; (iv) additional provisions on classification and reuse; (v) supplementing the "unclassified—no collection" rule for food waste used as feed, municipal solid waste with observations, and penalties for violations; and (vi) strengthening of the urban–rural nexus since most pig farms in Da Nang are located in the rural district of Hoa Vang. This will facilitate more effective and sustainable food waste management, and ultimately prevent land degradation and contamination of the natural environment due to waste dumping in the urban vicinity. The strategies above with also enhance resource circulation and efficiency, and enable more sustainable resource use.

5. Conclusions

This case study on food waste in Da Nang City, Vietnam, has once again confirmed that food waste is being generated more and more at the consumption stage in urbanized cities in developing economies, even more so than in developed countries. Consequently, there is a strong need for Da Nang to formulate specific policies and regulations aimed at preventing and reducing food waste at source and in the consumption stage, while at the same time minimizing its negative impacts on the environment (including the reduction of greenhouse gas emissions and ameliorating climate change impacts). This requires appropriate collection, treatment, and recycling and disposal systems. This will facilitate the city in achieving Sustainable Development Goal (SDG) target 12.3, which aims to "halve per capita food waste at the retail and consumer levels, and reduce food losses along production and supply chains by 2030." Proper strategies will also help to reduce the burden on the existing Khanh Son landfill. Food waste is a cross-cutting issue, with strong interlinkages with other relevant issues such as food security, economic development, and environmental impacts. Therefore, it is also essential for Da Nang to consider establishing a food waste action plan, setting out a clear set of concrete actions aligned with other relevant policies (e.g., municipal solid waste management, food safety and hygiene, use of food waste for animal feed, agricultural development, consumer behavior, etc.). The plan could form part of the city's overall strategy to promote a circular economy approach through effective resource circulation and resource use efficiency.

It has been reported that households with less food waste tend to regularly practice the following five actions: (i) planning food purchase and use; (ii) less impulse buying; (iii) keeping track of food in stock; (iv) determining accurate portion sizes when cooking; and (v) using leftovers. Policy interventions can trigger changes in consumer behavior and enhance consumers' awareness and food management skills. Policy should target not only households, but also hotels and resort areas, as they are likely to have a significant impact on food waste reduction in the city.

This paper also suggests that Da Nang could address food waste issues from two perspectives—waste prevention and management, and landscape sustainability. In addition, the urban–rural nexus needs to be further strengthened, which will help to prevent land degradation and protect the natural environment while also enhancing resource circulation efficiency and sustainable resource use within the city. For waste prevention, the city could target consumers using awareness-raising campaigns and specific tips for food waste reduction for different occasions when eating at home or eating out. In terms of food waste management, the city could make it easier for consumers to carry out waste separation and introduce separate collection systems. In the meantime, a safeguard measure to ensure proper treatment of food waste for use as pig feed should be implemented. This will have cost-saving benefits for pig farmers and ensure proper assessment of waste management measurements.

This research is the first step to understanding food waste generation and management in Da Nang City. The data presented in this study may have various uncertainties due to the limited number of samples as well as limitations in self-reporting measures. Future research should aim to validate these self-reporting measures with more objective techniques for data collection. Studies could establish the links between food waste generation and food and waste systems to gain a more holistic view of food waste management. Additionally, more focus could be put on people's perspectives regarding food waste and their actual behavior, as well as their attitudes to current regulations. This type of study could be very useful in identifying potential and practical methods for consumers to apply the 3Rs more easily. These will consequently assist in achieving landscape sustainability and resource efficiency.

Author Contributions: Conceptualization, N.-B.P., T.-N.D., C.L. and C.M.; data curation, T.-N.D. and V.-Q.T.; formal analysis, N.-B.P., T.-N.D. and V.-Q.T.; funding acquisition, N.-B.P. and C.L.; investigation, N.-B.P. and V.-Q.T.; methodology, N.-B.P., T.-N.D., V.-Q.T. and C.L.; project administration, N.-B.P. and C.L.; supervision, T.-N.D., A.-D.T. and C.L.; visualization, T.-N.D., V.-Q.T. and A.-D.T.;

writing—original draft, N.-B.P., T.-N.D., V.-Q.T., A.-D.T. and C.M.; writing—review and editing, N.-B.P., T.-N.D., V.-Q.T., A.-D.T., C.L. and C.M. All authors have read and agreed to the published version of the manuscript.

Funding: This research was funded by the Institute for Global Environmental Strategies (IGES) under the Strategic Research Fund 2019–2020; and supported by the Environmental Research and Technology Development Fund (S16) of the Environmental Restoration and Conservation Agency of Japan under the project entitled: 'Policy Design and Evaluation to Ensure Sustainable Consumption and Production Patterns in Asian Region' (2016–2020).

Institutional Review Board Statement: Not applicable.

Informed Consent Statement: Not Applicable.

Data Availability Statement: Not Applicable.

Conflicts of Interest: The authors declare no conflict of interest.

References

1. Gustavsson, J.; Cederberg, C.; Sonesson, U.; Otterdijk, R.V.; Meybeck, A. *Global Food Losses and Food Waste*; FAO: Rome, Italy, 2011.
2. FAO. *Food Wastage Footprint: Full-Cost Accounting*. 2014. Available online: http://www.fao.org/3/i3991e/i3991e.pdf (accessed on 30 June 2021).
3. FAO. *Food Wastage Footprint: Impacts on Natural Resources*. 2013. Available online: http://www.fao.org/3/i3347e/i3347e.pdf (accessed on 30 June 2021).
4. FAO. *Food Wastage Footprint & Climate Change*. 2015. Available online: http://www.fao.org/documents/card/en/c/7338e109-4 5e8-42da-92f3-ceb8d92002b0/ (accessed on 30 June 2021).
5. Xu, F.; Yangyang, L.; Ge, X.; Yang, L.; Yebo, L. Bioresource Technology Anaerobic digestion of food waste—Challenges and opportunities. *Bioresour. Technol.* **2018**, *247*, 1047–1058. [CrossRef] [PubMed]
6. Idris, A.; Inanc, B.; Hassan, M. Overview of waste disposal and landfills/dumps in Asian countries. *J. Mater. Cycles Waste Manag* **2004**, *6*, 104–110. [CrossRef]
7. Luong, N.D.; Giang, H.M.; Thanh, B.X.; Hung, N.T. Challenges for municipal solid waste management practices in Vietnam. *Waste Technol.* **2013**, *1*, 6–9. [CrossRef]
8. Louis, G.E. A historical context of municipal solid waste management in the United States. *Waste Manag. Res.* **2004**, *22*, 306–322. [CrossRef] [PubMed]
9. Zhang, C.; Xu, T.; Feng, H.; Chen, S. Greenhouse Gas Emissions from Landfills: A Review and Bibliometric Analysis. *Sustainability* **2019**, *11*, 2282. [CrossRef]
10. Papargyropoulou, E.; Lozano, R.; Steinberger, J.K.; Wright, N.; Ujang, Z.B. The Food Waste Hierarchy as a Framework for the Management of Food Surplus and Food Waste. *J. Clean. Prod.* **2014**, *76*, 106–115. [CrossRef]
11. Ventour, L. *The Food We Waste*. Food Waste Report V2 2 (July). 2008, pp. 1–237. Available online: https://www.lefigaro.fr/assets/pdf/Etude%20gaspillage%20alimentaire%20UK2008.pdf (accessed on 30 June 2021).
12. Kranert, M.; Hafner, G.; Brarbosz, J.; Schneider, F.; Lebersorger, S.; Scherhaufer, S.; Schuller, H.; Leverenz, D. *A Search for the Amount of Food Waste and Recommendations to Decrease the Food Waste Rate in Germany*; Institute for Sanitary Engineering, Water Quality and Solid Waste Management (ISWA), University of Stuttgart: Stuttgart, Germany, 2012.
13. Dung, T.N.B.; Kumar, G.; Lin, C.Y. An overview of food waste management in developing countries: Current status and future perspective. *J. Environ. Manag.* **2015**, *157*, 220–229.
14. Buchner, B.; Claude, F.; Ellen, G.; John, R.; Gabriele, R.; Camillo, R.; Umberto, V. *Food Waste: Cause, Impacts and Proposal*. 2012. Available online: https://www.barillacfn.com/m/publications/food-waste-causes-impact-proposals.pdf (accessed on 30 June 2021).
15. Liu, C.; Hotta, Y.; Santo, A.; Hengesbaugh, M.; Watabe, A.; Totoki, Y.; Allen, D.; Bengtsson, M. Food waste in Japan: Trends, current practices and key challenges. *J. Clean. Prod.* **2016**, *133*, 557–564. [CrossRef]
16. Liu, C.; Mao, C.; Bunditsakulchai, P.; Sasaki, S.; Hotta, Y. Food waste in Bangkok: Current situation, trends and key challenges. *Resour. Conserv. Recycl.* **2020**, *157*, 104779. [CrossRef]
17. Ehlert, J. *Emerging Consumerism and Eating Out in Ho Chi Minh City, Vietnam: The Social Embeddedness of Food Sharing*; Routledge: Oxford, UK; New York, NY, USA, 2016.
18. MONRE. *National State of Environment Report: Solid Waste Management*; Ministry of Natural Resources and Environment of Vietnam (MONRE): Hanoi, Vietnam, 2017.
19. Huyen, L.T.T.; Cesaro, J.-D.; Duteurtre, G. Food Waste Recycling into Animal Feeding in Vietnam. In Proceedings of the Kick-Off Meeting of Blue Barrels Project, Hanoi, Vietnam, 30 November 2015.
20. Kawai, K.; Huong, L.T.M.; Yamada, M.; Osako, M. Proximate composition of household waste and applicability of waste management technologies by source separation in Hanoi, Vietnam. *J. Mater. Cycles Waste Manag.* **2016**, *18*, 517–526. [CrossRef]

21. Le, N.P.; Thu, T.; Nguyen, P.; Zhu, D. Understanding the Stakeholders' Involvement in Utilizing Municipal Solid Waste in Agriculture through Composting: A Case Study of Hanoi, Vietnam. *Sustainability* **2018**, *10*, 2314. [CrossRef]
22. Thi, N.B.D.; Tuan, N.T.; Thi, N.H.H. Assessment of food waste management in Ho Chi Minh City, Vietnam: Current status and perspective. *Int. J. Environ. Waste Manag.* **2018**, *22*, 111–123. [CrossRef]
23. Schneider, P.; Anh, L.H.; Wagner, J.; Reichenbach, J.; Hebner, A. Solid Waste Management in Ho Chi Minh City, Vietnam: Moving towards a Circular Economy. *Sustainability* **2017**, *9*, 286. [CrossRef]
24. Östergren, K.; Gustavsson, J.; Bos-Brouwers, H.; Timmermans, T.; Hansen, O.-J.; Møller, H.; Anderson, G.; O'Connor, C.; Soethoudt, H.; Quested, T.; et al. *Fusions Definitional Framework for Food Waste*. 2014. Available online: https://www.eu-fusions.org/phocadownload/Publications/FUSIONS%20Definitional%20Framework%20for%20Food%20Waste%202014.pdf (accessed on 30 June 2021).
25. GSO. *General Statistical Report*; General Statistical Office: Hanoi, Vietnam, 2019.
26. Danang DONRE. *Annual Environmental Report 2019. Subject Solid Waste Management*; Da Nang Department of Natural Resources and Environment (Danang DONRE): Da Nang, Vietnam, 2020.
27. Kato, T.; Pham, D.T.X.; Hoang, H.; Xue, Y.; Tran, Q.V. Food residue recycling by swine breeders in a developing economy: A case study in Da Nang, Viet Nam. *J.Waste Manag.* **2012**, *32*, 2431–2438. [CrossRef] [PubMed]
28. Danang URENCO. *Report of Municipal Solid Waste Management Status in the Period of 2007–2011*; Da Nang Department of Natural Resources and Environment (Danang DONRE): Da Nang, Vietnam, 2012.
29. Danang UPI. *Solid Waste Management Planning to 2030 and Vision to 2050*; Da Nang Urban Plan Institute (UPI): Da Nang, Vietnam, 2015.
30. Yamane, T. *Statistics, an Introductory Analysis*, 2nd ed.; Harper and Row: New York, NY, USA, 1967.
31. World Bank. *What a Waste—A Global Review of Solid Waste Management*. 2012. Available online: https://openknowledge.worldbank.org/handle/10986/17388 (accessed on 30 June 2021).
32. HCMC DONRE. *Report on Solid Waste Management in Ho Chi Minh City*; Ho Chi Minh City Department of Natural Resources and Environment (HCMC DONRE): Ho Chi Minh City, Vietnam, 2014.
33. CENTEMA. *The Report on Data Collection on Solid Waste Management in Ho Chi Minh City, Vietnam*; Center for Environmental Technology and Management (CENTEMA): Ho Chi Minh City, Vietnam, 2009.
34. Dung, K.M. Assessment of Effective Economic Environment—Proposed Feasibility Mining Scenarios after GÈ Cát Landfill Site Stops Receipting of Garbage. Bachelor's Thesis, Institute for the Environmental Science, Engineering and Management, Ho Chi Minh, Vietnam, 2015.
35. Nguyen, T.V. Solid waste separation at source: Necessary and sufficient condition for waste management in Ho Chi Minh. *Int. J. Environ. Sci. Sustain. Dev.* **2012**, *1*, 1–9.
36. Tran, D.T.M.; Le, T.M.; Nguyen, V.T. Composition and Generation Rate of Household Solid Waste: Reuse and Recycling Ability. *Int. J. Environ. Pollut.* **2014**, *4*, 73–81.
37. Nguyen, H.H.; Heaven, S.; Banks, C. Energy potential from the anaerobic digestion of food waste in municipal solid waste stream of urban areas in Vietnam. *Int. J. Energy Environ. Eng.* **2014**, *5*, 365–374. [CrossRef]
38. Kajima Corporation. *Project Design Document Form for Organic Waste Composting Project at Da Nang City, Vietnam*. United Nations Framework Convention on Climate Change. 2017. Available online: https://www.unescap.org/sites/default/files/Report_VN_Danang_WastewaterSolidWasteUrbanAgriculture_Kohlbacher_2015_0.pdf (accessed on 1 June 2021).
39. Danang DONRE. *Annual Environmental Report 2009. Subject Solid Waste Management*; Da Nang Department of Natural Resources and Environment (Danang DONRE): Da Nang, Vietnam, 2010.
40. MONRE. *Vietnam's 2011 National Environment Report—Solid Waste Section*; MONRE: Hanoi, Vietnam, 2011.
41. Phnom Penh Capital Administration. *Phnom Penh Waste Management Strategy and Action Plan 2018–2035*. 2018. Available online: https://www.iges.or.jp/en/pub/phnom-penh-waste-management-strategy-and/en (accessed on 1 June 2021).
42. NRDC. *Modeling the Potential to Increase Food Rescue: Denver, New York City, and Nashville*. 2017. Available online: https://www.nrdc.org/sites/default/files/modeling-potential-increase-food-rescue-report.pdf (accessed on 1 June 2021).
43. WRAP. *Courtauld Commitment 2025: Annual Review 2017–18*. 2019. Available online: https://wrap.org.uk/resources/report/courtauld-2025-review-201718 (accessed on 30 June 2021).
44. Chen, L.; Nguyen, T.T. Evaluation of Household Food Waste Generation in Hanoi and Policy Implications towards SDGs Target 12.3. *Sustainability* **2020**, *12*, 6565. [CrossRef]
45. Gen, I. Utilization of food wastes for urban/peri-urban agriculture in Japan. In Proceedings of the International Workshop on Urban/Peri-urban Agriculture in the Asian and Pacific Region, Tagaytay City, Philippines, 22 May 2006; pp. 85–97.
46. Kim, M.H.; Song, Y.E.; Song, H.B.; Kim, J.W.; Hwang, S.J. Evaluation of food waste disposal options by LCC analysis from the perspective of global warming: Jungnang case, South Korea. *Waste Manag.* **2011**, *31*, 2112–2120. [CrossRef]
47. Danang EPA. *Technical Report on Calculation and Predict of Leachate from Khanh Son Landfill*; Da Nang Environmental Protection Agency (Danang EPA): Da Nang, Vietnam, 2019.
48. Olivier, P.; Smet, J.D.; Hyman, T.; Pare, M. *Making Wastes Become to the Most Value Resources: Food Production, Fuel, Animal Feed and Fertiliser at Small Scale*, Translated version of the Management Committee of water and sanitary of Binh Dinh province; VIE0703511; People Committee of Binh Dinh Province: Binh Dinh, Vietnam, 2011.

49. Kato, T.; Tran, A.Q.; Hoang, H. Factors affecting voluntary participation in food residue recycling: A case study in Da Nang, Vietnam. *Sustain. Environ. Res.* **2015**, *25*, 93–101.
50. World Bank. *Assessment of Municipal Solid Waste and Industrial Hazardous Solid Waste Management in Vietnam: Option and Action to Implement National Strategy.* 2018. Available online: https://documents1.worldbank.org/curated/en/352371563196189492/pdf/Solid-and-industrial-hazardous-waste-management-assessment-options-and-actions-areas.pdf (accessed on 30 June 2021).
51. *Implementation Plan for Domestic Solid Waste Separation at Source in Da Nang until 2025*; Decision No. 1577/QD-UBND; Da Nang People's Committee (Danang PC): Da Nang, Vietnam, 2019.

 *sustainability*

Article

Towards Circulating and Ecological Sphere in Urban Areas: An Indicator-Based Framework for Food-Energy-Water Security Assessment in Nagpur, India

Bhumika Morey [1,*], Sameer Deshkar [1,*], Vibhas Sukhwani [2], Priyanka Mitra [2], Rajib Shaw [2], Bijon Kumer Mitra [3], Devesh Sharma [4], Md. Abiar Rahman [5], Rajarshi Dasgupta [3] and Ashim Kumar Das [5]

[1] Department of Architecture and Planning, Visvesvaraya National Institute of Technology (VNIT), Nagpur 440010, India

[2] Graduate School of Media and Governance, Keio University, Kanagawa Prefecture, Minato City 252-0882, Japan; vibhas@sfc.keio.ac.jp (V.S.); priyankamitra87@gmail.com (P.M.); shaw@sfc.keio.ac.jp (R.S.)

[3] Integrated Sustainability Center, Institute for Global Environmental Strategies (IGES), 2108-11 Kamiyamaguchi, Hayama, Kanagawa Prefecture, Minato City 240-0115, Japan; b-mitra@iges.or.jp (B.K.M.); dasgupta@iges.or.jp (R.D.)

[4] Department of Atmospheric Science, School of Earth Sciences, Central University of Rajasthan, NH-8, Bandarsindri, Ajmer 305817, India; deveshsharma@curaj.ac.in

[5] Department of Agroforestry and Environment, Bangabandhu Sheikh Mujibur Rahman Agricultural University, Gazipur 1706, Bangladesh; abiar@bsmrau.edu.bd (M.A.R.); ashimbsmrau@gmail.com (A.K.D.)

* Correspondence: bhumika.morey@gmail.com (B.M.); smdeshkar@arc.vnit.ac.in (S.D.)

Citation: Morey, B.; Deshkar, S.; Sukhwani, V.; Mitra, P.; Shaw, R.; Mitra, B.K.; Sharma, D.; Rahman, M.A.; Dasgupta, R.; Das, A.K. Towards Circulating and Ecological Sphere in Urban Areas: An Indicator-Based Framework for Food-Energy-Water Security Assessment in Nagpur, India. *Sustainability* 2022, 14, 8123. https://doi.org/10.3390/su14138123

Academic Editors: Marc A. Rosen and Bo Jiang

Received: 18 May 2022
Accepted: 1 July 2022
Published: 3 July 2022

Publisher's Note: MDPI stays neutral with regard to jurisdictional claims in published maps and institutional affiliations.

Abstract: The world's urban population is expected to nearly double by 2050, making urbanization one of the most disruptive developments of the 21st century. On a global-to-local scale, ensuring a secure and reliable supply of food energy and water (FEW) resources for all humans is a major challenge in such a scenario. While much attention has recently been focused on the concept of FEW security and the interactions between the three sectors, there is no universally acceptable framing of the concept due to the fact that latest studies are mainly focused on individual FEW sectors, with not much investigation into how they interact. This research aims to create a localized framework based on the principles of the emerging concept of the Circulating Ecological Sphere (CES), introduced by the government of Japan, for a limited number of security indicators and dimensions. It began with a thorough study of the relevant literature using the PRISMA method, identification of gaps in local indicators for urban areas in each of the existing frameworks, and the proposal of a new indicator framework that tackles collective FEW security in urban environments is made accordingly. The authors have applied a special mechanism for filtration of this literature dataset in the context of Nagpur City in accordance with data availability and case study context. To test the applicability of the indicator set, it has been applied to the specific case of Nagpur. Both online and offline surveys were conducted to collect data, and subsequently a weighted mean method was adopted to analyze the data and derive values for the indicator set.

Keywords: food-energy-water security; nexus; weighted mean method; indicator framework; circulating ecological sphere; Nagpur

1. Introduction

Food, energy, and water (collectively referred to as 'FEW' hereafter) resources are essential for the survival and advancement of human societies. Furthermore, they are also intricately linked in the form of a nexus. Water is used to produce both food and energy; energy is utilized to pump, purify, and distribute water, as well as to manufacture, harvest, store, transport, and cook food; and crops and food waste are increasingly being used as an energy source. Failure to effectively manage the resources in any of these FEW sectors can therefore lead to serious inefficiencies in FEW resource management. In recent years, there has also been a growing body of literature highlighting the close interlinkages and trade-offs between the three FEW sectors [1–4]. Several notable global agencies including, the World Economic Forum, World Water Forum, the Bonn 2011 Nexus Conference, and the World Water Week 2012, have called for an integrated approach to FEW security [3]. This concept highlights inextricable interactions between the three sectors [5]. A safe and dependable FEW system is one in which everyone has consistent access to clean water, energy, and healthy food [6].

There are also concerns about security in terms of the FEW nexus. Food security is defined as having consistent physical, social, and economic access to sufficient, safe, and nutritious food that fits dietary needs and food choices in order to live a healthy and active life. Access to reliable and affordable energy for cooking, heating, lighting, communications, and productive uses is referred to as energy security. Water security refers to the availability and accessibility of adequate and high-quality water for human and environmental usage [7]. Although the FEW security concept has gained considerable attention in recent times, there is no universally acceptable framing of the concept [8]. In parallel the emerging concept of a Circulating and Ecological Sphere (CES—previously known as RCES) is also being promoted by the government of Japan, resonating with the notion of FEW security. CES depicts a self-sufficient and decentralized society that utilizes and circulates regional resources, complementing and supplementing other regions based on its specific qualities, and that ensures sustainable, equitable, and efficient resource consumption in both urban and rural locations [9].

The shortage of FEW resources frequently accentuates conflicts and political instability, and threatens livelihoods worldwide [5]. Providing a secure and resilient supply of these resources to all humans is therefore a major challenge on a broad global-to-local scale [10]. As a backdrop to rapid urbanization trends and changing population dynamics, a shift of focus from global to local scale is therefore required to ensure FEW security and sustainable development. Urban areas are also increasingly being placed at the center of FEW nexus discussions due to the growing concentration of population and economic activities [11].

Today, ensuring FEW resource security is seen by many as an urgent problem that could threaten the lives of the urban population, if not effectively addressed. The world's urban population is predicted to nearly double by 2050, making urbanization one of the most disruptive developments of the 21st century [12]. Cities are increasingly concentrating populations, economic activities, social and cultural interactions, as well as environmental and humanitarian impacts, posing massive sustainability challenges in housing, infrastructure, basic services, food security, health, education, decent jobs, safety, and natural resources, to name a few [12]. The majority of the world's population is living in urban areas, and the share of the world's population living in urban areas is expected to increase from 55 percent in 2018 to 60 percent in 2030 [13]. Urban communities are sensitive and vulnerable to adverse climate conditions, and therefore FEW infrastructure systems are essential for contributing towards and maintaining the well-being of households [14].

Higher and more volatile energy and food prices have pushed natural resources to the forefront of the world agenda in recent years, while water scarcity has become a rising danger to industry, agriculture, and energy production. According to one prediction, global food, water, and energy demand will increase by 35, 40, and 50%, respectively, by 2030 [15]. Simultaneously, climate change will deteriorate the prospects for availability of these vital commodities [15].

Particularly for large developing countries like India with high population density, FEW nexus thinking is increasingly important, as such countries are highly water-stressed [16]. India is one of the world's largest irrigators [17] and irrigation accounts for the majority of the world's fresh water consumption. India is already struggling to fulfill domestic energy, food, and water demands, with 14% of the population living without electricity and roughly a third of the population cooking with conventional biofuels [18]. India has very little spare land or water, and it is one of the world's most vulnerable countries to climate change's effects. These problems will worsen in the coming decades [15]. Nagpur City, which is situated in the center of India, experiences all types of extreme weather conditions. Furthermore, the effect of urbanization and climate change are converging in threatening ways for the city [19].

In light of this challenge, it is essential to have a guiding framework when envisioning interventions in decaying communities from a sustainable standpoint that helps to coordinate existing policies while also exploring new possibilities based on local conditions. One such idea along these lines is based on the recent concept of CES. This research has been done with a good understanding of the concept of CES and focuses on analyzing urban areas in the context of a FEW nexus by following certain principles of CES. This study looks at the specific case of Nagpur City, in order to explore the relevance and application of the CES approach at a local level in terms of FEW security.

Urban areas largely rely on their surroundings for resource supply. The need for these resources depends on the components of urban systems. One major challenge is minimizing the use of these resources and developing a circular model [20]. Thus, the focus of this study is to develop a framework which can allow cities to assess their FEW security scenarios based on the CES concept. While existing studies on FEW security emphasize understanding and managing the interactions between the FEW sectors [5], most of them are focused on individual FEW sectors, with not much investigation into how they interact. The literature shows integrated studies of FEW security are limited on a global or regional scale. Study trends at the local level suggest that FEW resources are generally examined independently. There is only a small amount of research that focuses on how they are linked from a security viewpoint. These trends indicate a gap in the availability of urban level/local level indicators in all sectors, indicating the unique aspect of this study. Urban households consume considerable quantities of FEW resources to meet daily demand. A household is a unit of demand and it can also be the most appropriate unit for influencing consumption practices. A high portion of FEW consumption in cities can be attributed to household use [21]. Hence, the study focuses mainly on indicators that denote the situation at the household level.

To sustain FEW resources in the long term, and to promote human well-being and economic growth, there is a genuine need to address these underlying issues in an integrated manner [22]. Aiming to bridge this research gap, this study, together with the CES concept, works to systematically analyze the FEW security indicators and apply the indicator set at an urban scale. The three key objectives of this research are: (1) to review the existing indicators in FEW security domains; (2) to contextualize the existing indicators for the specific case of Nagpur City in India, and (3) to suggest a feasible framework for assessment of FEW security in Nagpur and test its applicability.

The remainder of the paper is structured as follows. Section 2 provides a brief overview of the FEW nexus with underlying security concerns for the specific case of Nagpur and the relevance of the indicator framework, identifying a gap in the research. Section 3 introduces the case study of Nagpur and briefly explains the FEW scenarios in the city. The research methodology is explained in Section 4. In Section 5, the urban level indicator framework is developed and applied to Nagpur City followed by a discussion of the results. Finally, in Section 6, FEW security for Nagpur is discussed and the key conclusions are made underlining the research limitations for the study.

2. Theoretical Background

This section provides an understanding of existing literature in the FEW domains. The Section is divided into three subsections. The first discusses the growing concerns in the domain of FEW security and the nexus-based challenges. In the second sub-section, the concept of CES is discussed in the context of the FEW nexus. The last section is about the relevance of an indicator framework for FEW security.

2.1. FEW Nexus and FEW Security Concerns

The multi-faceted challenges faced by human societies today, such as climate change, population growth, and urbanization, are simultaneously threatening the security of FEW resources at an accelerating rate [23]. These stressors combined with an ever-increasing demand for FEW resources are likely to further accelerate climate change and many other socio-economic issues [22] unless major policy reforms are implemented toward effective resource management. Specifically, the population in water-stressed countries is predicted to increase to four billion people by 2050 [24].

These complexities of interactions necessitate a comprehensive and in-depth understanding of the dynamics of global trends in urbanization [25] and its influence on natural resources. FEW systems are facing several threats to these propositions [26] and the unprecedented surge in urbanization and population growth rates is generating multiple impacts, affecting FEW demands. Sustainable natural resource management, in that regard, is crucial to ensure FEW security for a growing population, which could be conflicting while limited to a sectoral approach [5]. In the course of constructing a literature study, the authors arrived at several nexus-based indicators, that are expected to impact these resources for the specific case of Nagpur City. The indicators are discussed as follows:

2.1.1. Temperature

Climate change tends to increase existing and future risks associated with the management of resources [27]. The increase in average temperatures and rising demand for buildings with a more comfortable indoor environment lead to an increase in building energy consumption [28]. The impacts of climate change, which are driven by rising temperatures and changes in precipitation patterns, are expected to exacerbate the tension between the availability of and demand for resources. Rising average annual temperatures and precipitation can be considered the most significant climate change impacts for Nagpur City based on climate and geographical data. With increasing tendency, the average yearly temperature over the last two decades has been higher than the 50-year mean. The number of extreme heat events per year is also increasing, with 8 out of 10 years in the recent decade (2000–2019) indicating an above-average anomaly. In addition, based on seasonal statistics, the monsoon has seen a 16.4% rise in the occurrence of high temperatures between 2010 and 2019 [29].

2.1.2. Precipitation and Rainfall

Precipitation is one of the most critical indicators. The change in frequency and intensity of rainfall is often studied to understand the extent and magnitude of climate change, as rainfall has one of the most significant impacts on agriculture and food security. This is especially true for India, as the country's agriculture and, in turn, its economy is very dependent on seasonal rainfall. Thus the differences in global average temperatures and precipitation patterns (climate variability) have immense potential to affect agriculture in India. Therefore, it is important to examine and understand the spatio-temporal dynamics of this indicator to provide credible inputs to policymakers. Moreover, it is crucial to understand the changes in the precipitation pattern for Nagpur to understand the impact of climate change. The annual rainfall pattern over Nagpur has not changed significantly over the last 50 years. However, the seasonal distribution of rainfall demonstrates significant pattern changes that could have a negative impact on agriculture [29].

2.1.3. Urbanization and Population Growth

Due to encroachment and stress on existing natural resources, agricultural lands adjacent to cities face the brunt of this expansion. As a result, monitoring and controlling city expansion can help to sustainably manage resources. The availability of water changes as cities expand onto productive agricultural land. As a result, it is critical to keep an eye on this sprawl in order to spot places that are being overburdened by growth and make informed land-use decisions. The periphery of Nagpur has grown significantly representing a nearly 71% rise in all directions of the city. This growth was most noticeable along the Wardha, Bhandara, and Amravati highways. The eastern, northern, and southern segments of the city's periphery were dominated by agricultural regions that produced just one or two crops per year. Megaprojects like the Multimodal International Hub Airport at Nagpur (MIHAN) and the International Airport fueled urban growth in the city's eastern and southern outskirts. The area covered by scrubland (land dominated by bushes) decreased significantly, from 16.4% in 1991 to 5.5% in 2012. Scrublands are now restricted to the city's southern and northern reaches. Areas under double cropping and degraded scrublands were converted into residential or industrial areas during the urbanization process. The urban sprawl in Nagpur is due to inward migration to the city from villages, by people looking for better education, employment, recreation, and living standards [30].

2.2. Concept of Circulating and Ecological Sphere (CES)

The goal of CES is to create "a self-reliant and decentralized society where different resources are circulated within each region, leading to symbiosis and exchange with neighboring regions according to the unique characteristics of each region to re-discover regional resources and make optimum use of them in a sustainable manner" [31]. In line with recent global policy agreements such as the Sustainable Development Goals (SDGs) and the Paris Agreement, Japan's Ministry of the Environment introduced the concept of CES in its fifth Basic Environment Plan. The notion is rapidly being advocated as a foundation for future environmental policy in Japan and across the world [32]. The Government of Japan had previously developed the idea of a Regional Circular Sphere and adopted the 3R (reduce, reuse, and recycle) material recycling principles to build a sound material-cycle society in the second Basic Environment Plan. Notably, Japan's National Biodiversity Strategy (2012–2020) laid the foundation for creating a society that is in tune with nature [9].

It is primarily a policy approach that integrates three recognized principles: (1) a low-carbon society, (2) resource circulation, and (3) living in harmony with nature [33]. The CES reimagines the spatial dimension of human activities in order to maximize their range of activity, and to loudly promote a self-sufficient and decentralized society [34]. Furthermore, this approach could be implemented at all scales ranging from a community or municipal scale to a river basin or country scale. It begins at the local level, encouraging communities to examine their own potential to live in harmony with the environment by decreasing waste, producing renewable energy, and utilizing ecosystem services without harming them. It then scales up to create new value chains that complement local resources while organically connecting communities and regions to support one another. Finally, this new viewpoint optimizes the scale at which various human activities occur, reducing their environmental effect.

The idea of a Circulating and Ecological Economy (CEE) provides a theoretical foundation for the CES approach. Recycling, limiting resource use, and creating renewable energy sources are key to the ideals of a circular economy and low-carbon society. A circular economy focuses on "closing the loop", where the value of a resource is circulated as long as feasible and waste is minimized, if not eliminated [35]. It also emphasizes the utilization of renewable resources to accomplish the low-carbon society's supporting concept. A "low-carbon society" promotes low greenhouse gas emissions through converting to energy-efficient and low-carbon sources, as well as changing consumption patterns [36]. Overall, through using underused resources, both of these concepts contribute to the idea of resilient societies and rejuvenation. The third principle of CES as mentioned above focuses on a "society living in harmony with nature", which prioritizes economic recovery while safeguarding natural resources.

The CES idea emphasizes the importance of local renewable energy generation and decarbonization, as well as the investigation of innovative solutions to environmental concerns through stakeholder partnerships. CES may thus be viewed as a broad concept aimed at using circularity concepts in a collaborative manner that maximizes local strengths and resources in order to build resilience [37]. Hence, to ensure the successful implementation of this concept, research must be done on the creation of indicators and assessment methods that will measure and quantify progress, as well as identify the characteristics of Nagpur City that can be used to enhance sustainability and resilience.

2.3. Relevance of Indicator Framework in FEW Nexus

Indicators are measurable and they help to determine whether objectives have been achieved [38]. An indicator framework is an organized way to view data from different sources. It is a simple and concise way to present gathered data and help show the relevance and connection between different indicators. In a framework, data can be grouped or categorized and are often shown alongside detailed descriptions of associated measures and methods of calculation [39]. Systematically building a framework of indicators can help us to find gaps in the existing information [40]. Following this, as discussed on the interlinked nature of the FEW resources in previous sections, the authors have attempted to develop an indicator framework for urban FEW security to understand these sectors on the same platform in a quantitative way.

On similar lines, a review of recent studies from 2011 onwards (as summarized in Table 1) shows that significant research has been conducted in the domain of FEW security. However, there were very few integrated studies on FEW security at a global or regional scale [5,41–44]. On the other hand, the research trends at the local level reveal that FEW resources are being largely analyzed and addressed independently [24,45–52]. Amongst these, a very limited number of studies have focused on an interconnected security perspective. Most of the previous studies have analyzed FEW systems independently or examined two of the three FEW systems at a time [53]. These trends show that there is a gap in availability of urban level/local level indicators in all sectors as evident from Table 1

Table 1. Highlighting the focus of urban level indicators in FEW security domain.

Study	Regional/Global Level Indicators			Local/Urban Level Indicators		
	Food	Energy	Water	Food	Energy	Water
Arshad 2012	√					
Chuang 2013					√	
Nilsson et al. 2013						√
Flammini et al. 2014	√	√	√			
Rasul 2016	√	√	√			
Saladini 2018	√					
Jensen 2018						√
Adhikari 2018				√		
Prumbadia 2019					√	
Ibrahim et al. 2019				√		
Gesauldo 2019				√		
Markentonis 2019	√	√	√			
Putra 2020	√	√	√			
Casino-Loeza 2020				√		√
Haji et al. 2020				√		
Mabhaudi 2021				√		√

3. Case Study Context

3.1. About Nagpur

Nagpur is the location of the Zero Mile marker in India (location shown in Figure 1). Spread across a geographic area of 217.56 km^2, the city has an approximate population of 2.5 million. Due to its strategic central location in India, it is well connected to other major cities through vast transport networks. Besides its political and geographical significance, it is also a prominent educational hub and has a rich natural resource base [44,54]. Because of the region's enormous orange production, it is also well known as the "City of Oranges" in India. It is further projected to be one of the fastest-growing cities in the world from 2019 to 2035 with an average annual GDP growth rate of 8.41% [55].

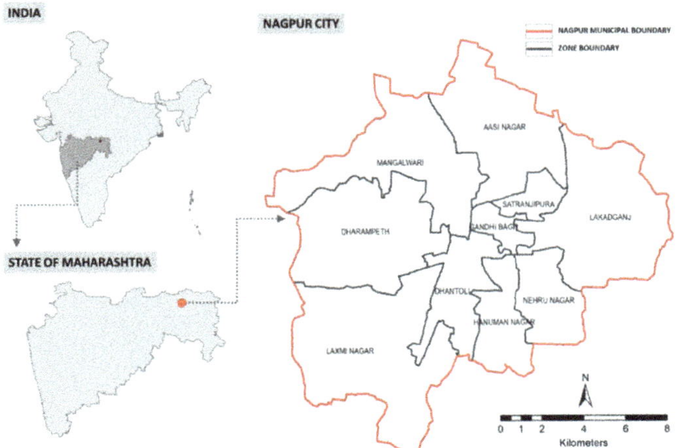

Figure 1. Location map of Nagpur city.

3.2. FEW Scenarios in Nagpur

Nagpur is not only the third-largest city in Maharashtra state but is also recognized as one of the fastest-growing urban agglomerations of India [56]. Historically, Nagpur is subject to a tropical climate with dry weather and experiences extreme weather conditions with significant consequences of climate variability [19]. It has witnessed severe water stress situations in recent years [54]. FEW sustainability can play a major role in ensuring the city develops in a resilient way [57], and Nagpur is still in the transformation phase. Meanwhile, the flow of FEW resources is separately governed and managed by different agencies [16,58]. Nagpur needs to leverage the interlinkages between FEW sectors to address this governance issue [44].

4. Research Methods

4.1. Literature Analysis

To identify the relevant research documents through the existing literature, this study followed the Preferred Reporting Items for Systematic Reviews and Meta-Analyses (PRISMA) guidelines [59]. In line with the PRISMA guidelines, search queries were mainly executed in the Elsevier's Scopus database, which is recognized to be the most comprehensive abstract and citation database of interdisciplinary literature (peer-reviewed), as well as additional high-quality online sources such as books and conference proceedings [60]. Figure 2 illustrates the results obtained through the Scopus search. A search query was first applied with the following keywords 'Food + Energy + Water + Security' in the "ALL" category. A total of 1910 papers were retrieved initially from Scopus using our defined search query. To filter out the documents focusing on indicators, the search query is then modified to 'Food + Energy + Water + Security + Indicator', through which 121 research documents were identified. Later, after manual screening, the literature database titles were extracted and the relevant literature was identified, irrespective of the study area, year of publication, type of document, etc. The next step was to shortlist the research documents, through which the 29 research documents were shortlisted for review. Other than the Scopus database search, a total of 15 other research documents and grey literature were also taken into consideration for this research.

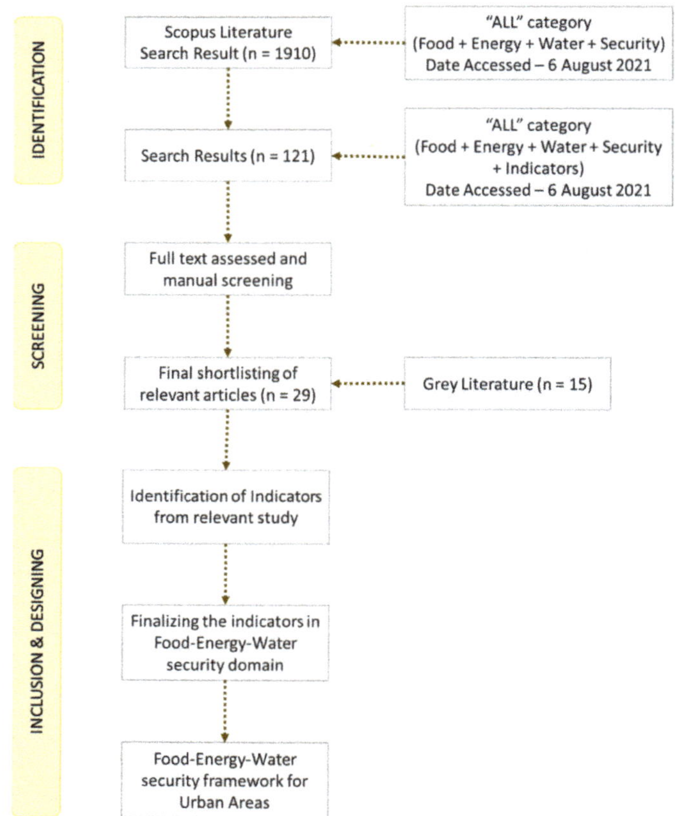

Figure 2. PRISMA flowchart for the study.

4.2. Framework Development

Based on the review of identified research documents, this research builds on four steps of analysis, as described below:

Step 1: To identify the key dimensions in the domain of FEW security. Table 2 shows the interconnected dimensions for the same.

Step 2: To identify the security indicators in FEW sectors. Table 3 shows the sector-wise security indicators.

Step 3: Formulation of an Indicator framework for the assessment of Urban FEW security.

Step 4: To collect the data primarily and secondarily.

Step 5: The final step was to assess this acquired data and conclude the assessment.

Figure 3 shows the methodology that was followed by the authors and which was instrumental in the assessment of FEW security for particular case studies. This indicator framework is believed ultimately to help authorities in the decision-making process in matters of governance for urban FEW security.

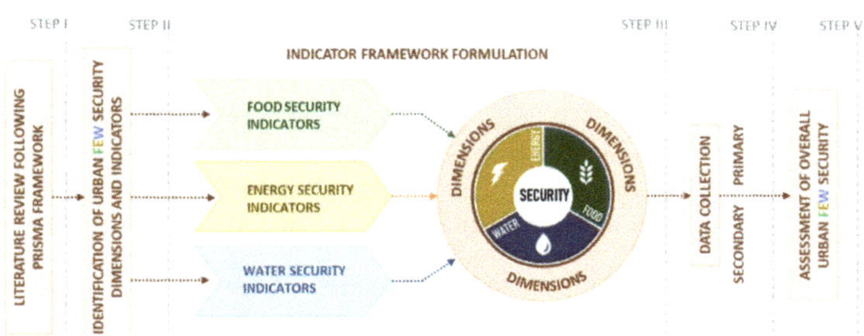

Figure 3. Methodology for development of framework.

4.3. Assessment for Nagpur City

4.3.1. Indicator Scoring and Weighted Mean Method

To avoid any bias in the scoring of each indicator, the assessment of indicators was done quantitatively. Thus, the framework can be applied to any case study irrespective of its character, location, and other defining peculiarities.

To calculate the scores for overall security based on all the indicators and for each dimension, the weighted arithmetic mean formula $W = \frac{\sum_{i=1}^{n} wi\,Xi}{\sum_{i=1}^{n} wi}$ (W = Weighted average, n = number of terms to be averaged, wi = weights applied to x values, Xi = data values to be averaged) was used, as it offers more nuance to each attribute.

4.3.2. Survey Method

To investigate the applicability of the framework for urban FEW security assessments, the study focused on the specific case study area of Nagpur City. Household-level surveys were conducted targeting city residents to gain an understanding of FEW security at the household level. In line with the 10 administrative zones of Nagpur, a stratified random sampling approach was adopted to cover households across the city. To ensure wider outreach across the city, offline surveys were conducted at the outset. In addition, an online questionnaire (through 'Google Forms') was circulated through Facebook and WhatsApp. When collecting the samples, the emphasis was more on collecting proportionate responses from all zones in the given time interval of 10 days (29 October 2021 to 7 November 2021). Given the context of the continuing COVID-19 pandemic and other concerns, the survey was performed over a 10-day time frame, as there was a perceived necessity for situation analysis in the defined time interval of 10 days. Since this was a pilot study authors decided to collect the samples with a confidence level of 85–90% which gives a z-score of 1.44–1.65. The survey was performed under the supervision of the authors; hence the margin of error was decided to be 5%. The sample size (SS) was calculated using formula $SS = \frac{\frac{z^2 \times p\,(1-p)}{e^2}}{\frac{z^2 \times p\,(1-p)}{e^2\,N} + 1}$

(N = population size, e = Margin of error, z= z-score), which came to be 208–273. The survey area does not cover the whole city but only the selected parcels for which we consider this sample size to be adequate. The online method was chosen particularly in view of the above mentioned COVID-19 issues. Even so, it was difficult to reach out to city residents and obtain a sufficient number of responses. Hence the authors do acknowledge that there was a rather limited number of samples collected for this study. After the defined period of 10 days, the online survey was turned off, and the survey responses were downloaded as a CSV file. A total of 243 responses were collected from the study area. Following this, all the samples collected through both online and offline modes were analyzed by the authors in Microsoft Excel.

5. Results

5.1. Formation of Indicator Framework

5.1.1. Review of Existing Indicators in FEW Domain

After filtering out relevant literature datasets using the PRISMA method, certain dimensions and indicators in the available literature were studied for all three sectors (FEW). Later, each of them was scrutinized fronting onto urban level assessment. The list of all these dimensions and indicators is shown in Tables 2 and 3.

Table 2. Identified dimensions in FEW security domain.

Dimensions	Source
Availability	[47,48,50,52]
Diversity	[24,51]
Quality	[24,47,48,61]
Service Sustainability	[24,47,48]
Capacity	[24]
Affordability	[24,48,52]
Health Risk	[24,61]
Regulations	[24,42,47]
Strategic Planning	[24,42]
Utilisation	[24,47]
Stability	[47]
Accessibility	[24,47,48,52,61,62]
Governance	[24,51,61]

Table 3. Sector wise indicators in FEW security domain.

Sector	Indicator	Source
Food	Overweight Children	[3,5,41,42,63]
	Cereal Yield	
	Food Supply Per Capita	
	Fertilizer Use/HA	
	Cereal Import Dependency Ratio	
	Prevalence of Undernourishment	[64]
	Low Per Capita Income	[3,47]
	Low and Unequal Distribution of Income	
	Poor and Highly Unstable Growth Performance especially in Agriculture	
	Unemployment and Underemployment	
	Low and Declining Farm Size	
	Inequalities in Land Distribution	
	Low Land Utilization	
	Social Discrimination	
	Population Growth	

Table 3. *Cont.*

Sector	Indicator	Source
	Access to Market	
	Poverty	
	Political Instability	
	Poor, Marginalized, Ethnic Group & Lower Caste Groups	
	High Infant Mortalities	
Energy	Access to Electricity	[3,5,41,42,45,50]
	Energy Use Per Capita	
	Electricity Consumption	
	Electricity from Hydroelectric	
	Access to Clean Fueltech for Cooking	
	Emission from the Energy Sector	
	Energy Intensity Level	
	Share of Renewable Energy in Total Primary Energy Supply (%)	
	Supply	[45,48]
	Excess Production	
	Demand	
Water	Population with Safe Drinking Water	[3,5,41,42]
	Population with Open Defecation	
	Water Resources Per Capita	
	Water Withdrawal Per Capita	
	Cultivated Area with Irrigation	
	Groundwater Extraction/Borewells	
	Water for Industrial Use	
	Municipal Water Withdrawal	
	Population using Basic Drinking Water Services (%)	[64]
	Population using Basic Sanitation Services (%)	
	Freshwater Withdrawal (% of Available Freshwater)	
	Scarce Water Consumption Embodied Imports	
	Groundwater Level	[65]
	Groundwater Recharge Rate	
	Groundwater Salinity	
	Groundwater pH	
	Type of Water Source	[46]
	Accessibility of Running Water	
	Household Water Cost	
	Water Consumption	
	Waterborne Disease	
	Rainfall	[27]
	Stream Flow	
	Relative Humidity	

Table 3. *Cont.*

Sector	Indicator	Source
	Ambient Water Quality	[50]
	Surface Water Availability	
	Groundwater Availability	
	Other Source Availability	
	Water Storage Capacity	
	Diversity of Water Sources	
	Water Quality	[3,14,24]
	Water Supply Capacity	
	Water Supply Coverage	
	Cost Recovery of Water Utilities	
	Water Tariff	
	Access to Sanitation	
	Water Contamination Incidents	
	Economic Loss Due to Water Pollution	
	Strategic Planning	
	Disaster Management	
	Regulation	
	Temperature	[66]
	Precipitation	

5.1.2. Indicator Framework for Urban FEW Security in Nagpur

Thereafter, dimensions and indicators were finalized for framing of the localized indicator set. Appendix A: Table 1 shows the indicator framework utilized for the assessment of FEW security. Each sector was divided into three dimensions, namely Availability, Accessibility, and Utilization, and further into indicators and sub-indicators. Lastly, the overall security dimension was added for all three sectors to understand the perception of the respondents. Most of the data (summarized in Appendix A: Table 1)) was collected through a primary source.

5.2. Application of the Indicator Framework for Nagpur City Context

A questionnaire survey was conducted in line with the finalized queries mentioned (Appendix A: Table 1) A total of 243 samples were collected from 10 zones of the city. The questionnaire also included some general queries like gender, zone of residence, income group, and household size, to understand the demographic character of the responses. As can be seen in Table 4, the maximum share of responses is obtained from three zones, namely Dharampeth, Laxminagar, and Hanuman Nagar. Overall, 52.18 percent of the respondents were male and 47.82 percent were female. The average household size of the samples collected was 4.2. The responses received were almost equal from all the income groups.

As can be seen in Figure 4, the dimension-wise final scores for all three sectors were mapped to understand the spatial scenario of FEW security for the city of Nagpur. Further, these sectoral securities with respect to three dimensions, namely availability, accessibility and utilization, are discussed in the following sections.

Table 4. Characteristics of the survey respondents.

Number of Surveys	243
Attribute	Response Percentage
Gender Group	
Male	52.26%
Female	47.74%
Other	0.00%
Household Size (Average Household size = 4.2)	
<3	31.69%
4 to 5	53.09%
6 to 8	12.76%
>8	2.47%
Area of Residence (Zone wise categorization of city area)	
Ashi Nagar	10.29%
Dhantoli	11.52%
Dharampeth	16.05%
Hanuman nagar	11.11%
Laxminagar	13.58%
Nehrunagar	7.82%
Gandhibagh	7.41%
Lakadganj	6.17%
Mangalwari	7.82%
Satranjipura	8.23%
Income Group (INR)	
<20,000	21.40%
20,000 to 40,000	21.40%
40,000 to 80,000	17.28%
80,000 to 1.25 lac	21.81%
>1.25 lac	18.11%

5.2.1. Food Security

1. Availability:

In terms of food availability, very few households in Nagpur city were reported to have their own kitchen garden or livestock, which undermines in-house food availability. At the same time, market food availability was perceived by city residents to be good. Compared with energy and water, food is less available in all 10 zones. Specifically for Gandhibagh, Mangalwari, Satranjipura, Hanuman Nagar, and Nehru Nagar, it is even less available in comparison with other zones. Overall, more consideration must be taken regarding the dimension of availability of food.

2. Accessibility:

Most of the respondents stated that they have a marketplace within 2 km of their household, thus indicating better physical access to food. Moreover, there was also moderately good economic accessibility for food as seen from the responses. In particular, economic accessibility for people in lower-income groups in the older areas was reported to be low. Most of them seemed satisfied with the quality and nutritional value of the food they have. Hanuman Nagar, Nehrunagar, Satranjipura Gandhibaug and Mangalwari zones are believed to be on verge of development for this particular dimension.

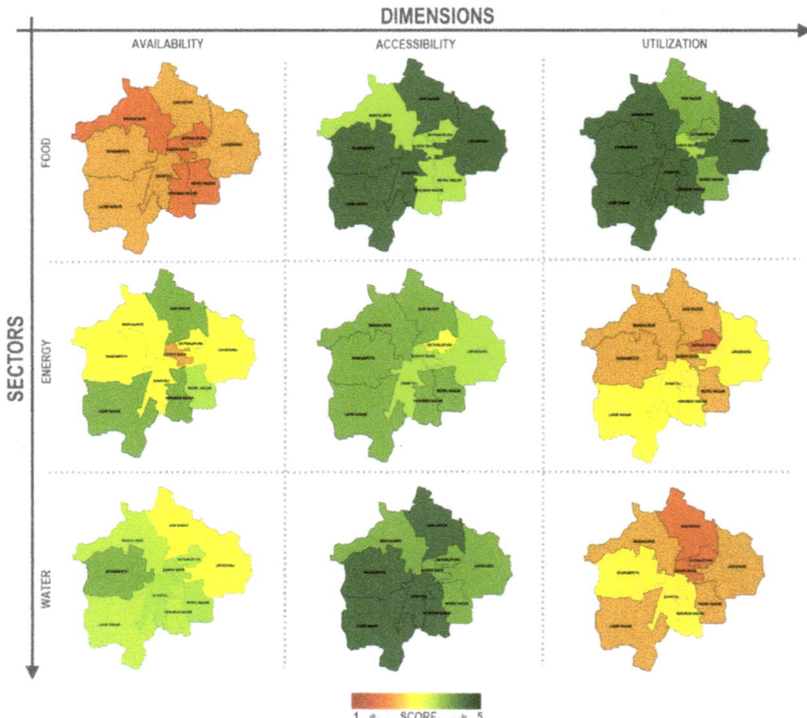

Figure 4. Final dimension wise FEW security scores for 10 zones of Nagpur city.

3.   Utilization:

The majority of respondents selected the energy-efficient cooking mode out of the available options. Overall, for all 10 zones, the dimension of utilization was better even compared with other dimensions of food security. Compared to water and energy utilization, the food sector elicited far better scores.

5.2.2. Energy Security

4.   Availability:

When energy availability is considered, the city of Nagpur was reported to have wide coverage with low power fluctuations. However, there were some fluctuations and power cuts, mostly observed in Dharampeth, Mangalwari, Gandhibaug, Satranjipura, Dhantoli and Lakadganj. Compared to other sectors, energy availability can be perceived as moderately good for all zones of Nagpur.

5.   Accessibility:

When energy accessibility is considered, economic access by the respondents to energy services was good, but their dependency on electricity for household appliances was at a maximum level. Figure 4 shows that energy accessibility can be considered moderate when compared to other sectors.

6.   Utilization:

Respondents were observed to be using the energy-efficient option for lighting, along with choosing more energy-efficient appliances when purchasing new products. However, city residents were not yet opting for renewable energy sources at the household level, which could enhance the city's energy security. The zone of Satranjipura is in significantly poor condition when compared to other zones. Overall, the dimension of utilization for energy needs more attention than availability and accessibility.

5.2.3. Water Security

7. Availability:

Most of the respondents, had an in-house storage facility for water, rather than availing themselves of groundwater sources. For the zones of Lakadganj and Ashi Nagar, the scores for water availability are comparatively low as most residents lacked both groundwater usage and in-house storage.

8. Accessibility:

The overall accessibility for water can be perceived as good, as residents were well covered by the water supply system. Residents did not need to spend much of their income on water tariffs, making water more accessible for all income groups. For all zones, it was observed that physical, as well as economic access, was good for all respondents.

9. Utilization:

From Figure 4, it can be observed that the dimension of utilization for water security is a matter of particular concern. While several respondents had very good access to the water supply, they were concerned about the quality of water they were using on daily basis. The city residents were mostly dependent on the municipal water supply system with very little diversity in terms of water sources. Households were still not practicing rainwater harvesting, despite the fact that rainwater could potentially be a good renewable water source for the city. Aashi Nagar, Satranjipuraand Gandhibaug zones showed significantly lower scores for water utilization, as most of the respondents from these particular zones are dependent on the municipal water supply and were concerned with the quality of water available to them. Most of the zones also have less diverse water sources for all-year-round use.

5.2.4. Overall FEW Security

Figure 5 shows the final weighted scores for each zone and for all three sectors. The sector-wise scores (overall security—indicator-based) are similar to the scores obtained from the survey results (overall security—respondent perspective). For the Aashi Nagar zone, this score is significantly less than the indicator-based score as the respondents perceive that the food security in their area is poor. Hence, the dimension of overall security for all three sectors could be crucial and justifiably fits the framework, as it gives a clear picture of how city residents' perceive FEW security in Nagpur.

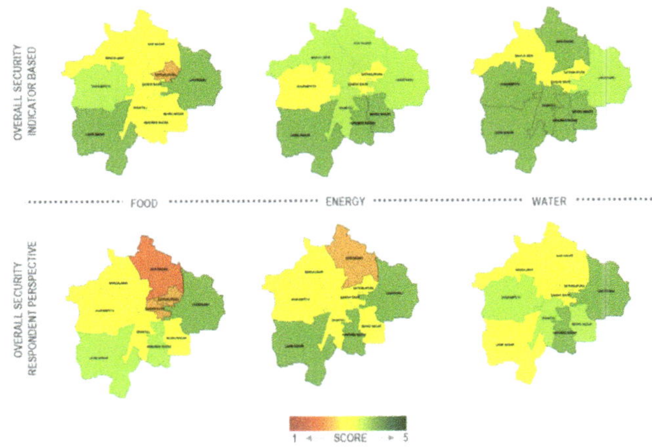

Figure 5. Overall security scores for 10 zones of Nagpur City.

6. Discussion and Conclusions

6.1. Discussion

This framework could be applied to support FEW security assessments for a city-level analysis. As discussed in Section 4, an indicator-based approach is preferred by the authors to analyze FEW security, as this can assist in analyzing resource security in a quantitative and non-biased way, irrespective of the selected case. A comprehensive understanding of FEW security issues and their importance is vital for rational decision-making. This study also covered issues that are less frequently examined at a local level, with urban related indicators included in the framework.

It was observed from the literature study that this kind of assessment on an urban scale has not been done so far (as evident from Table 1). Therefore, finalizing the assessment dimensions and indicators was an iterative and complex process. In the final framework, the indicators that the authors finalized were in the context of Nagpur City considering location, character, climate variability, and data availability. Thereby, when finalizing the framework as detailed in Section 5, a primary survey was conducted both offline and online. One challenge was that, because of certain restrictions due to the COVID-19, pandemic, many of the residents refused to engage in response, so the authors were unable to collect an equal number of samples from these 10 different zones of the city. At a later stage, the authors emphasized obtaining a proportionate number of responses from each zone. The limited sample size is thus acknowledged by the authors. As a future scope of this study, the authors might conduct similar surveys at different time intervals to perform a temporal analysis. Lastly, a qualitative approach was excluded from the study because different experts may interpret qualitative scales differently and the findings of the various criteria may be difficult to compare.

Analysis of FEW security was carried out to understand the current conditions of the city based on several dimensions, namely availability, accessibility, and utilization for FEW security. As evident in Figure 4 in the previous section, the indicators for food availability have a lower score as compared to other sectors. However, it is important to understand that cities predominantly rely on external areas for food. In that context, food security in Nagpur is reliant on rural areas of Nagpur district, and self-reliance for food availability is lower than for other dimensions, and also in comparison with other sectors. In should also be noted that the city is currently facing water scarcity issues. Thus, encouraging urban farming in the city will also have an impact on water and energy availability which currently score better than food availability. Secondly, when accessibility for FEW is considered, all three sectors show better scores. Lastly, the dimension of utilization concerning energy and water shows a lower score as compared to food. Household-level preparedness can be anticipated to be very low for all three sectors as in-house food availability scores are low, renewable energy sources are not made use of, people do not practice rainwater harvesting and there is not much diversity in terms of available resources. Overall this explains why residents must shift to utilization of more energy-efficient and renewable resources. Finally, if we look at the overall scenario (Figure 4), as compared to availability and utilization for all three sectors, the dimension of accessibility had better scores. Thus, the findings show a scenario where the population of Nagpur perceives the dimension of accessibility of FEW as "satisfactory" but the dimensions of availability and utilization as "moderate". Hence, it can be understood that Nagpur is not a resource-efficient city.

Rapid urbanization in Nagpur is bringing about a rise in population and, as a result, a rise in the demand for potable water. New power plants near the city have similarly increased the need for water. The year 2019 was indeed a wake-up call for the city. For the first time in several decades the Nagpur Municipal Corporation (NMC) supplied water on alternate days in mid-July when rains were delayed until late July in Nagpur. The area has also seen erratic precipitation in recent times. Such periods of scanty rainfall are especially concerning when it comes to food production in the surrounding rural areas, which supply most of the food to city residents. CES depicts a self-sufficient and decentralized society that utilizes and circulates resources, and as such the city should focus more on self-sustenance in terms of FEW resources, seeing as there is major reliance on the areas outside the city for FEW resources. As the concept of CES also emphasizes the necessity of cooperation and involvement for sustainable resource management, the level of awareness and willingness to support development policies is critical to the application of this concept [9]. Hence, to examine the level of awareness among residents, an overall security perspective was included as part of the survey. There seems to be a possibility for collaborative, bottom-up engagement on sustainability as a number of respondents are well aware of how crucial FEW security is.

Cities are at the forefront of managing change and are the driving force for action to reduce the use of resources by taking an integrated approach and planning. Not only are they the engines of the economy and home for their citizens, but municipalities also supply and control various public services to residents and businesses that influence the majority of resource use, and energy consumption [20]. Hence, to maintain these resources, it is essential to understand that these issues are intangibly linked, and thus must be addressed with a more integrated approach as mentioned above. In line with that, a sound indicator framework can be used as a management tool to assist governments to design implementation plans and policies in allocating FEW resources appropriately. This research therefore provides a method for assessing the state of FEW security in urban areas and helps them cope with upcoming challenges.

6.2. Conclusions

Given the character of competing developmental and environmental goals together with trends of rapid urbanization and population expansion in India, it has become necessary to explore the issues surrounding natural resource management while ensuring responsible use of all these resources. This study aimed to contribute an urban level assessment of collective FEW security by developing a localized framework for FEW security indicators and dimensions. Concurrently, its applicability for the specific case of Nagpur, India was also tested.

There have been a significant number of studies in the field of FEW security but these are generally limited to a global or regional scale. Study trends at the local level, on the other hand, suggest that FEW resources are examined and treated independently, with two of the three FEW systems examined at the same time. There have only been a small number of studies that focus on security among FEW sectors. Thus, the research set out in this study is unique among such studies.

The research began by identifying the relevant literature datasets, adopting the PRISMA method. Then a thorough analysis of the related literature was made, identifying gaps in local indicators for urban areas of each of the available frameworks, and proposing a new indicator framework that addresses the overall FEW security in urban areas.

A specific methodology for filtration of these indicators was applied by the authors considering the context of Nagpur City. Some indicators were necessarily left out by the authors in this study and these remaining indicators should also be investigated in the respective case study contexts. In addition, a qualitative approach could be considered including an expert consultation so as to develop a relevant indicator framework. Further practical testing of the framework would also provide more information about which FEW security indicators and dimensions should be addressed in the evaluation, as not all of the framework's elements are relevant in every scenario. On thing is certain and that is FEW nexus thinking and a CES approach require enormous political will. Currently both concepts are at a nascent stage, and operationalizing policy recommendations is likely to require much greater advocacy for effective policy change on both global to local scales. A systematic and thorough approach would be useful in foresight and risk management for that matter. The findings of the research, as discussed above, also suggest ways in which the different areas could enhance their self-reliance according to CES. However, the extent to which CES impacts development planning in Nagpur City will be determined by how receptive policymakers are to the concept.

This indicator framework is restricted to urban areas. A rural perspective might result in quite a different scenario from which several questions would emerge requiring further research to better translate them into integrated policy formulation. One research domain might be how to improve collaboration across different geographical scales, such as isolated municipalities and urban areas, intra-regional watershed areas, or at the global level. In this sense, it is critical to examine how to determine the most appropriate scales for resource consumption and circulation and how to effectively intervene at the respective scale of analysis. Another challenge is how to minimize the possible detrimental effects on FEW resources, such as the deployment of renewable energy facilities and nature conservation. Lastly, FEW nexus thinking can also explore collaboration between different resource management departments and policies, thereby encouraging the application of CES. All these challenges must be investigated via direct observation on the ground to better understand the complexity and challenges at different scales, thereby ensuring the transition to sustainability.

Author Contributions: All authors were involved in conceptualization, visualization, investigation, methodology, writing—review and editing; formal analysis, B.M., V.S., P.M. and A.K.D.; writing—original draft preparation, software, B.M.; validation, resource, supervision, project administration, funding acquisition, S.D., P.M., R.S., B.K.M., D.S., M.A.R. and R.D. All authors have substantially contributed for the development of this manuscript. All authors have read and agreed to the published version of the manuscript.

Funding: This research is supported by the Asia-Pacific Network for Global Change Research (APN) under the project titled "Developing Urban-Rural partnerships framework to mitigate climate-induced water availability impacts on Food, Energy and Water (FEW) security at the regional level" with the project reference number CRRP2020-05MY-Shaw, and by the Strategic Research Fund of the Institute for Global Environmental Strategies (IGES), Japan under the project titled "Circulating and Ecological Sphere Concept: Path Making for a Sustainable Transformation through Translating Global Goals into Local Actions".

Institutional Review Board Statement: Not applicable.

Informed Consent Statement: Informed consent was waived as no personal information was collected from respondents and only general opinion survey was conducted.

Data Availability Statement: Not applicable.

Acknowledgments: This study has been conducted under the project 'Developing Urban-Rural partnerships framework to mitigate climate-induced water availability impacts on Food, Energy, and Water (FEW) security at regional level' funded by the Asia-Pacific Network for Global Change Research (APN) with the project reference number CRRP2020-05MY-Shaw. This study is also supported by the Strategic Research Fund of the Institute for Global Environmental Strategies. The authors are grateful to three anonymous reviewers who have immensely helped in improving the manuscript, and to Emma Fushimi, IGES, Japan for her kind support in proof reading and edits.

Conflicts of Interest: The authors declare no conflict of interest.

Appendix A

Table 1. Indicator Framework and Criteria for Scoring of Indicators.

Dimension	Indicator	Question	Answer	Unit	Rating Criteria (Likert Scale)					Data Collection Source
					1	2	3	4	5	
				FOOD						
Availability	In-house Availability	Do you practice Kitchen Gardening/Urban Agriculture in your home?	Yes/No	% Respondents with answer = YES	0–20%	20–40%	40–60%	60–80%	80–100%	Primary
		Do you own any livestock, farm animals or poultry?	Yes/No	% Respondents with answer = YES	0–20%	20–40%	40–60%	60–80%	80–100%	Primary
	Market Availability	How much do you rate the market food availability?	Very Poor/Poor/Moderate /Good/Very Good	Option Chosen by Respondent	very poor	poor	moderate	good	very good	Primary
Accessibility	Physical Access	Do you have grocery store/food market within 2 km?	Yes/No	% Respondents with answer = YES	0–20%	20–40%	40–60%	60–80%	80–100%	Primary
	Economic Access	What % of your Income is spent on food?	0–100%	%	80–100%	60–80%	40–60%	20–40%	0–20%	Primary
	Quality of Food Available	How much do you rate the quality of food available to you?	Very Poor/Poor/Moderate /Good/Very Good	Option Chosen by Respondent	very poor	poor	moderate	good	very good	Primary
	Nutrition Adequacy	Do you have adequate access to nutritious food?	Yes/No	% Respondents with answer = YES	0–20%	20–40%	40–60%	60–80%	80–100%	Primary
Utilisation	Cooking Fuel	Which cooking option do you use?	LPG/Induction /Kerosene/Other	% Respondents using more eco-friendly and cost effective source	0–20%	20–40%	40–60%	60–80%	80–100%	Primary
Overall Security	Overall Food Security	How much do you rate the overall food security?	Very Poor/Poor/Moderate /Good/Very Good	Option Chosen by Respondent	very poor	poor	moderate	good	very good	Primary

Table 1. *Cont.*

Dimension	Indicator	Question	Answer	Unit	Rating Criteria (Likert Scale)					Data Collection Source
					1	2	3	4	5	
				ENERGY						
Availability	Power cuts	How frequently do you face power cuts in your area?	Daily/Once a Week/Once a Month/Rarely/Never	Option Chosen by Respondent	Daily	Once a week	Once a Month	Rarely	Never	Primary
	Power Fluctuation	Do you observe fluctuation in electricity supply now and then?	Yes/No	% Respondents with answer = NO	0–20%	20–40%	40–60%	60–80%	80–100%	Primary
	Service Coverage	What % of city/population covered under service supply?	—	—	0–20%	20–40%	40–60%	60–80%	80–100%	Secondary
Accessibility	Economic Access	What % of your income is spent on Electricity and other energy sources?	0–100%	%	80–100%	60–80%	40–60%	20–40%	0–20%	Primary
Utilisation	Electricity Consumption/Dependency on Electricity	Which electronic appliances do you use?	Refrigerator/TV/Geyser/Water Purifier/Induction/Laptop/Washing Machine/AC/Cooler/Microwave/Mixer/Other	Number of options selected by respondents	More than 8	8	6	4	2	Primary
	Lighting	Which Lights do you use?	CFL/LED/Halogen lamps/Flourescent lamps/Incandescent bulbs	% Respondents using most eco-friendly and cost effective source	0–20%	20–40%	40–60%	60–80%	80–100%	Primary
	Households using Renewable Energy Sources	Do you use solar panels for electricity supply?	Yes/No	% Respondents with answer = YES	0–20%	20–40%	40–60%	60–80%	80–100%	Primary

Table 1. *Cont.*

Dimension	Indicator	Question	Answer	Unit	Rating Criteria (Likert Scale)					Data Collection Source
					1	2	3	4	5	
	Energy Efficient Use	Which products do you prefer while purchasing?	Energy Efficient/Affordable	% Respondents with answer = Energy Efficient	0–20%	20–40%	40–60%	60–80%	80–100%	Primary
Overall Security	Overall Energy Security	How much do you rate the overall energy security?	Very Poor/Poor/Moderate /Good/Very Good	Option Chosen by Respondent	very poor	poor	moderate	good	very good	Primary
WATER										
Availability	Groundwater Availability	Do you have personal tubewell/common handpump?	Yes/No	% Respondents with answer = YES	0–20%	20–40%	40–60%	60–80%	80–100%	Primary
	Household Water Storage Availability	Do you have water storage tank (Over Head Tank/Under Ground Tank/Inhouse Storage)?	Yes/No	% Respondents with answer = YES	0–20%	20–40%	40–60%	60–80%	80–100%	Primary
Accessibility	Access to proper Sanitation Service	What % of city/population covered under service supply?	—	—	0–20%	20–40%	40–60%	60–80%	80–100%	Secondary
	Households with access to tap water supply	Do you have access to tap water supply?	Yes/No	% Respondents with answer = YES	0–20%	20–40%	40–60%	60–80%	80–100%	Primary
	Economic Accessibility	What % of your income is spent on water tariff?	0–100%	%	80–100%	60–80%	40–60%	20–40%	0–20%	Primary
Utilisation	Quality of Water Used	How much do you rate the quality of water available to you?	Very Poor/Poor/Moderate /Good/Very Good	Option Chosen by Respondent	very poor	poor	moderate	good	very good	Primary
	Diversity in Sources Used	Which water source do you use/is available to you?	Tap/Borewell/Tanker /Common Tap/Handpumps	Number of options selected by respondents	one	two	three	four	all	Primary

Table 1. *Cont.*

Dimension	Indicator	Question	Answer	Unit	Rating Criteria (Likert Scale)					Data Collection Source
					1	2	3	4	5	
	Use of Renewable Source	Do you practice Rainwater Harvesting at household/community level?	Yes/No	% Respondents with answer = YES	0–20%	20–40%	40–60%	60–80%	80–100%	Primary
Overall Security	Overall Water Security	How much do you rate the overall water security?	Very Poor/Poor/Moderate /Good/Very Good	Option Chosen by Respondent	very poor	poor	moderate	good	very good	Primary

References

1. Hoolohan, C.; Larkin, A.; McLachlan, C.; Falconer, R.; Soutar, I.; Suckling, J.; Varga, L.; Haltas, I.; Druckman, A.; Lumbroso, D.; et al. Engaging stakeholders in research to address water–energy–food (WEF) nexus challenges. *Sustain. Sci.* **2018**, *13*, 1415–1426. [CrossRef] [PubMed]
2. Scott, M.; Larkin, A. Geography and the water–energy–food nexus: Introduction. *Geogr. J.* **2019**, *185*, 373–376. [CrossRef]
3. Rasul, G. Food, water, and energy security in South Asia: A nexus perspective from the Hindu Kush Himalayan region☆. *Environ. Sci. Policy* **2014**, *39*, 35–48. [CrossRef]
4. Chen, Y.; Xu, L. Evaluation and Scenario Prediction of the Water-Energy-Food System Security in the Yangtze River Economic Belt Based on the RF-Haken Model. *Water* **2021**, *13*, 695. [CrossRef]
5. Putra, M.P.I.F.; Pradhan, P.; Kropp, J.P. A systematic analysis of Water-Energy-Food security nexus: A South Asian case study. *Sci. Total Environ.* **2020**, *728*, 138451. [CrossRef] [PubMed]
6. Ingram, J.S.I. A food systems approach to researching food security and its interactions with global environmental change. *Food Secur.* **2011**, *3*, 417–431. [CrossRef]
7. Nkiaka, E.; Okpara, U.T.; Okumah, M. Food-energy-water security in sub-Saharan Africa: Quantitative and spatial assessments using an indicator-based approach. *Environ. Dev.* **2021**, *40*, 100655. [CrossRef]
8. Zhang, X.; Vesselinov, V.V. Integrated modeling approach for optimal management of water, energy and food security nexus. *Adv. Water Resour.* **2017**, *101*, 1–10. [CrossRef]
9. Takeuchi, K.; Fujino, J.; Ortiz-Moya, F.; Mitra, B.K.; Watabe, A.; Takeda, T.; Jin, Z.; Nugroho, S.B.; Koike, H.; Kataoka, Y. *Circulating and Ecological Economy—Regional and Local CES: An IGES Proposal*; Institute for Global Environmental Strategies: Kanagawa, Japan, 2019.
10. McGrane, S.J.; Acuto, M.; Artioli, F.; Chen, P.Y.; Comber, R.; Cottee, J.; Farr-Wharton, G.; Green, N.; Helfgott, A.; Larcom, S.; et al. Scaling the nexus: Towards integrated frameworks for analysing water, energy and food. *Geogr. J.* **2019**, *185*, 419–431. [CrossRef]
11. Alfasi, N.; Fenster, T. Between the "Global" and the "Local": On Global Locality and Local Globality. *Urban Geogr.* **2009**, *30*, 543–566. [CrossRef]
12. UN. New Urban Agenda (Habitat 3). 2017. Available online: https://uploads.habitat3.org/hb3/NUA-English.pdf (accessed on 13 August 2021).
13. UN. Sustainable Cities, Human Mobility and International Migration. 2018. Available online: https://undocs.org/en/E/CN.9/2018/2 (accessed on 13 August 2021).
14. Dargin, J.; Berk, A.; Mostafavi, A. Assessment of household-level food-energy-water nexus vulnerability during disasters. *Sustain. Cities Soc.* **2020**, *62*, 102366. [CrossRef]
15. Ghosh, A.; Steven, D. India's Energy, Food, and Water Security: International Cooperation for Domestic Capacity. In *Shaping the Emerging World: India and the Multilateral Order*; Sidhu, W.P.S., Mehta, P.B., Jones, B.D., Eds.; Bbrookings Institution Press: Washington, DC, USA, 2013; p. 369.
16. Sukhwani, V.; Shaw, R. A Water-Energy-Food Nexus based Conceptual Approach for Developing Smart Urban-Rural Linkages in Nagpur Metropolitan Area, India. *IDRiM J.* **2020**, *10*, 16635. [CrossRef]
17. ICID. World Irrigated Area—2018. 2018. Available Online: https://www.icid.org/world-irrigated-area.pdf (accessed on 23 January 2022).
18. IEA. World Energy Outlook 2011. Paris. 2011. Available online: https://www.iea.org/reports/world-energy-outlook-2011 (accessed on 25 August 2021).
19. Sawadh, R.; Masram, B.Y.; Balamwar, S. Analysis of Past Behaviour of Climate Change on Basis of Multiple Parameters in Nagpur. *Zeichen J.* **2020**, *6*, 9.
20. European Environment Agency (EEA). *Technical, Urban Sustainability Issues—What Is a Resource-Efficient City?* European Environment Agency (EEA): Copenhagen, Denmark, 2015.
21. Hussien, W.A.; Memon, F.A.; Savic, D.A. An integrated model to evaluate water-energy-food nexus at a household scale. *Environ. Model. Softw.* **2017**, *93*, 366–380. [CrossRef]
22. Walker, S. Triple Threat: Water, Energy and Food Insecurity. Triple Threat Water, Energy Food Insecurity. 2020. Available online: https://www.wri.org/insights/triple-threat-water-energy-and-food-insecurity (accessed on 1 September 2021).
23. Zhuang, J.; Sayler, G.; Yu, G.; Jiang, G. Solving Shared Problems at the Food, Energy, and Water Nexus. *Eos* **2021**, *102*, 1739474. [CrossRef]
24. Jensen, O.; Wu, H. Urban water security indicators: Development and pilot. *Environ. Sci. Policy* **2018**, *83*, 33–45. [CrossRef]
25. Gragg, R.S.; Anandhi, A.; Jiru, M.; Usher, K.M. A Conceptualization of the Urban Food-Energy-Water Nexus Sustainability Paradigm: Modeling From Theory to Practice. *Front. Environ. Sci.* **2018**, *6*, 133. [CrossRef]
26. Shu, Q.; Scott, M.; Todman, L.; McGrane, S.J. Development of a prototype composite index for resilience and security of water-energy-food (WEF) systems in industrialised nations. *Environ. Sustain. Indic.* **2021**, *11*, 100124. [CrossRef]
27. Gesualdo, G.C.; Oliveira, P.T.; Rodrigues, D.B.B.; Gupta, H.V. Assessing water security in the São Paulo metropolitan region under projected climate change. *Hydrol. Earth Syst. Sci.* **2019**, *23*, 4955–4968. [CrossRef]
28. Bhatnagar, M.; Mathur, J.; Garg, V. Determining base temperature for heating and cooling degree-days for India. *J. Build. Eng.* **2018**, *18*, 270–280. [CrossRef]

29. NEERI. Environmental Status Report: Nagpur City. Nagpur. 2020. Available online: https://www.nmcnagpur.gov.in/assets/25 0/2021/06/mediafiles/Final_2020_ESR_-_AK.pdf (accessed on 5 September 2021).
30. Kar, R.; Reddy, G.O.; Kumar, N.; Singh, S. Monitoring spatio-temporal dynamics of urban and peri-urban landscape using remote sensing and GIS—A case study from Central India. *Egypt. J. Remote Sens. Space Sci.* **2018**, *21*, 401–411. [CrossRef]
31. Marome, W.; Rodkul, P.; Mitra, B.K.; Dasgupta, R.; Kataoka, Y. Towards a more sustainable and resilient future: Applying the Regional Circulating and Ecological Sphere (R-CES) concept to Udon Thani City Region, Thailand. *Prog. Disaster Sci.* **2022**, *14*, 100225. [CrossRef]
32. Ministry of the Environment Japan. Outline of the Fifth Basic Environment Plan. Available online: https://www.env.go.jp/policy/kihon_keikaku/plan/plan_5/attach/ref_en-02.pdf (accessed on 16 October 2021).
33. Institute for Global Environmental Strategies Homepages. Available online: https://archive.iges.or.jp/en/sdgs/sts.html (accessed on 16 October 2021).
34. Ortiz-Moya, F.; Kataoka, Y.; Saito, O.; Mitra, B.K.; Takeuchi, K. Sustainable transitions towards a resilient and decentralised future: Japan's Circulating and Ecological Sphere (CES). *Sustain. Sci.* **2021**, *16*, 1717–1729. [CrossRef]
35. Bačová, M.; Böhme, K.; Guitton, M.; Herwijnen, M.V.; Kállay, T.; Koutsomarkou, J.; Magazzù, I.; O'Loughlin, E.; Rok, A. *Pathways to a Circular Economy in Cities and Regions*; Interreg Europe Joint Secretariat: Lille, France, 2016.
36. Skea, J.I.M.; Nishioka, S. Policies and Practices for a Low-Carbon Society. *Clim. Policy* **2008**, *8*, S5–S16. [CrossRef]
37. Thapa, K.; Sukhwani, V.; Deshkar, S.; Shaw, R.; Mitra, B.K. Strengthening Urban-Rural Resource Flow through Regional Circular and Ecological Sphere (R-CES) Approach in Nagpur, India. *Sustainability* **2020**, *12*, 8663. [CrossRef]
38. CoastAdapt. Identifying Indicators for Monitoring and Evaluation. 2017. Available online: https://coastadapt.com.au/how-to-pages/identifying-indicators-monitoring-and-evaluation (accessed on 23 January 2022).
39. Government of Canada. What Is an Indicator Framework? 2017. Available online: https://health-infobase.canada.ca/datalab/indicator-framework-blog.html#:~{}:text=An%20indicator%20framework%20is%20an,and%20connec-tion%20between%20 different%20indicators (accessed on 24 January 2022).
40. Mainguet, C.; Baye, A. Defining a framework of indicators to measure the social outcomes of learning. In Measuring the Effects of Education on Health and Civic Engagement: Proceedings of the Copenhagen Symposium–© OECD 2006. 2006; p. 11. Available online: https://www.oecd.org/education/innovation-education/37425733.pdf (accessed on 24 January 2022).
41. Flammini, A.; Puri, M.; Pluschke, L.; Dubois, O. *Walking the Nexus Talk: Assessing the Water-Energy-Food Nexus in the Context of the Sustainable energy for All Initiative*; FAO: Rome, Italy, 2014.
42. Rasul, G. Managing the food, water, and energy nexus for achieving the Sustainable Development Goals in South Asia. *Environ. Dev.* **2016**, *18*, 14–25. [CrossRef]
43. Markantonis, V.; Reynaud, A.; Karabulut, A.A.; El Hajj, R.; Altinbilek, D.; Awad, I.M.; Bruggeman, A.; Constantianos, V.; Mysiak, J.; Lamaddalena, N.; et al. Can the Implementation of the Water-Energy-Food Nexus Support Economic Growth in the Mediterranean Region? The Current Status and the Way Forward. *Front. Environ. Sci.* **2019**, *7*, 84. [CrossRef]
44. Sukhwani, V.; Shaw, R.; Deshkar, S.; Mitra, B.K.; Yan, W. Role of Smart Cities in Optimizing Water-Energy-Food Nexus: Opportunities in Nagpur, India. *Smart Cities* **2020**, *3*, 1266–1292. [CrossRef]
45. Chuang, M.C.; Ma, H.W. An assessment of Taiwan's energy policy using multi-dimensional energy security indicators. *Renew. Sustain. Energy Rev.* **2013**, *17*, 301–311. [CrossRef]
46. Nilsson, L.M.; Berner, J.; Dudarev, A.A.; Mulvad, G.; Odland, J.; Parkinson, A.; Rautio, A.; Tikhonov, C.; Evengård, B. Indicators of food and water security in an Arctic Health context—Results from an international workshop discussion. *Int. J. Circumpolar Health* **2013**, *72*, 21530. [CrossRef]
47. Adhikari, S. Food security: Pillars, determinants and factors affecting it. *Retrieved Febr.* **2018**, *27*, 2021.
48. Prambudia, Y.; Rahmana, A.; Juraida, A. Framework for measuring urban energy security: Citizen perspective. *J. Pengemb. Kota* **2019**, *7*, 34–45. [CrossRef]
49. Ibrahim, H.; Uba-Eze, N.; Oyewole, S.; Onuk, E. Food Security among Urban Households: A Case Study of Gwagwalada Area Council of the Federal Capital Territory Abuja, Nigeria. *Pak. J. Nutr.* **2009**, *8*, 810–813. [CrossRef]
50. Mabhaudhi, T.; Nhamo, L.; Chibarabada, T.; Mabaya, G.; Mpandeli, S.; Liphadzi, S.; Senzanje, A.; Naidoo, D.; Modi, A.; Chivenge, P. Assessing Progress towards Sustainable Development Goals through Nexus Planning. *Water* **2021**, *13*, 1321. [CrossRef]
51. Bin Abdullah, F.; Iqbal, R.; Hyder, S.I.; Jawaid, M. Energy security indicators for Pakistan: An integrated approach. *Renew. Sustain. Energy Rev.* **2020**, *133*, 110122. [CrossRef]
52. Gasser, P. A review on energy security indices to compare country performances. *Energy Policy* **2020**, *139*, 111339. [CrossRef]
53. Liang, S.; Qu, S.; Zhao, Q.; Zhang, X.; Daigger, G.T.; Newell, J.P.; Miller, S.A.; Johnson, J.X.; Love, N.G.; Zhang, L.; et al. Quantifying the urban food–energy–water nexus: The case of the detroit metropolitan area. *Environ. Sci. Technol.* **2018**, *53*, 779–788. [CrossRef]
54. Deshkar, S. Resilience Perspective for Planning Urban Water Infrastructures: A Case of Nagpur City. In *Urban Drought*; Springer: Singapore, 2019; pp. 131–154. [CrossRef]
55. Holt, R. *Global Cities, Which Cities will Be Leading the Global Economy in 2035*; Oxford Economics: Oxford, UK, 2018.
56. NSSCDCL. NSSCDCL Homepages. Available online: https://nsscdcl.org/aboutnsscdcl.jsp (accessed on 15 March 2022).
57. Fernández, C.G.; Peek, D. Smart and Sustainable? Positioning Adaptation to Climate Change in the European Smart City. *Smart Cities* **2020**, *3*, 511–526. [CrossRef]

58. NIT. Nagpur Metropolitan Area Development Plan: 2012-32, Draft Development Plan Report. Nagpur Improvement Trust (NIT). 2015. Available online: https://www.nitnagpur.org/pdf/Metro_Region_DP.pdf (accessed on 21 September 2021).

59. Moher, D.; Liberati, A.; Tetzlaff, J.; Altman, D.G.; PRISMA Group. Preferred reporting items for systematic reviews and meta-analyses: The PRISMA statement. *PLoS Med.* **2009**, *6*, e1000097. [CrossRef]

60. Baas, J.; Schotten, M.; Plume, A.; Côté, G.; Karimi, R. Scopus as a curated, high-quality bibliometric data source for academic research in quantitative science studies. *Quant. Sci. Stud.* **2020**, *1*, 377–386. [CrossRef]

61. Paudel, S.; Kumar, P.; Dasgupta, R.; Johnson, B.; Avtar, R.; Shaw, R.; Mishra, B.; Kanbara, S. Nexus between Water Security Framework and Public Health: A Comprehensive Scientific Review. *Water* **2021**, *13*, 1365. [CrossRef]

62. Hua, T.; Zhao, W.; Wang, S.; Fu, B.; Pereira, P. Identifying priority biophysical indicators for promoting food-energy-water nexus within planetary boundaries. *Resour. Conserv. Recycl.* **2020**, *163*, 105102. [CrossRef]

63. Saladini, F.; Betti, G.; Ferragina, E.; Bouraoui, F.; Cupertino, S.; Canitano, G.; Gigliotti, M.; Autino, A.; Pulselli, F.; Riccaboni, A.; et al. Linking the water-energy-food nexus and sustainable development indicators for the Mediterranean region. *Ecol. Indic.* **2018**, *91*, 689–697. [CrossRef]

64. Cansino-Loeza, B.; Sánchez-Zarco, X.G.; Mora-Jacobo, E.G.; Saggiante-Mauro, F.E.; González-Bravo, R.; Mahlknecht, J.; Ponce-Ortega, J.M. Systematic Approach for Assessing the Water–Energy–Food Nexus for Sustainable Development in Regions with Resource Scarcities. *ACS Sustain. Chem. Eng.* **2020**, *8*, 13734–13748. [CrossRef]

65. Haji, M.; Govindan, R.; Al-Ansari, T. Novel approaches for geospatial risk analytics in the energy–water–food nexus using an EWF nexus node. *Comput. Chem. Eng.* **2020**, *140*, 106936. [CrossRef]

66. Liu, Q. Interlinking climate change with water-energy-food nexus and related ecosystem processes in California case studies. *Ecol. Process.* **2016**, *5*, 389. [CrossRef]

 *sustainability*

Article

Land Use-Based Participatory Assessment of Ecosystem Services for Ecological Restoration in Village Tank Cascade Systems of Sri Lanka

Sujith S. Ratnayake [1,2,*], Azeem Khan [1], Michael Reid [3], Punchi B. Dharmasena [4], Danny Hunter [5], Lalit Kumar [6], Keminda Herath [7], Benjamin Kogo [8], Harsha K. Kadupitiya [9], Thilantha Dammalage [1] and Champika S. Kariyawasam [2,8]

1   Applied Agricultural Remote Sensing Centre, School of Science and Technology, University of New England, Armidale, NSW 2351, Australia
2   Ministry of Environment, Battaramulla 10120, Sri Lanka
3   School of Humanities, Arts, and Social Sciences, University of New England, Armidale, NSW 2351, Australia
4   Faculty of Agriculture, Rajarata University of Sri Lanka, Mihintale 50300, Sri Lanka
5   The Alliance of Bioversity International and the International Center for Tropical Agriculture (CIAT), Via di San Domenico, 1, 00153 Rome, Italy
6   East Coast Geospatial Consultants, Armidale, NSW 2350, Australia
7   Faculty of Agriculture and Plantation Management, Wayamba University of Sri Lanka, Gonawila 60170, Sri Lanka
8   School of Environmental and Rural Science, University of New England, Armidale, NSW 2351, Australia
9   Natural Resources Management Centre, Department of Agriculture, Peradeniya 20400, Sri Lanka
*   Correspondence: rratnay2@myune.edu.au; Tel.: +61-421-077-454

Citation: Ratnayake, S.S.; Khan, A.; Reid, M.; Dharmasena, P.B.; Hunter, D.; Kumar, L.; Herath, K.; Kogo, B.; Kadupitiya, H.K.; Dammalage, T.; et al. Land Use-Based Participatory Assessment of Ecosystem Services for Ecological Restoration in Village Tank Cascade Systems of Sri Lanka. *Sustainability* **2022**, *14*, 10180. https://doi.org/10.3390/su141610180

Academic Editors: Ronald C. Estoque, Kikuko Shoyama and Rajarshi Dasgupta

Received: 10 June 2022
Accepted: 11 August 2022
Published: 16 August 2022

Publisher's Note: MDPI stays neutral with regard to jurisdictional claims in published maps and institutional affiliations.

Abstract: Village Tank Cascade System (VTCS) landscapes in the dry zone of Sri Lanka provide multiple ecosystem services (ESs) and benefits to local communities, sustaining the productivity of their land use systems (LUSs). However, there is a lack of adequate scientific research on the ESs of LUSs, despite the recent land use changes that have greatly impacted the provisioning of ESs. Collection of baseline ESs data is a pre-requisite for decision making on ESs-based ecological restoration and management of the VTCS. Thus, this study aimed at assessing ESs of the Mahakanumulla VTCS (MVTCS) located in the Anuradhapura district of Sri Lanka by using a participatory approach involving the integration of local knowledge, expert judgements and LUSs attribute data to assess the ESs. The methodology was designed to integrate the biodiversity and land degradation status of LUSs in a way that is directly linked with the supply of ESs. The study identified twenty-four ESs of the MVTCS based on community perceptions. The identified ESs were assessed as a function of LUSs to develop an ecosystem service supply (ESS) and demand (ESD) matrix model. The results reveal that the current overall ESD for regulating and supporting ESs is higher than the ESS capacity of MVTCS. The assessment also revealed that land degradation and biodiversity deterioration reduce the capacity to provide ESs. Downstream LUSs of the meso-catchment were found to be more vulnerable to degradation and insufficient to provide ESs. Further, the study established that ESs in the MVTCS are generated through direct species-based and biophysical-based providers. In addition, it emerged that social and cultural engagements also played an important role in association with both providers to generate certain types of ESs. Therefore, it can be concluded that VTCS ecological restoration depends on the extent to which integrated effort addresses the levels of ecological complexity, as well as the social engagement of communities and stakeholders. The results of this study provide a scientific basis that can inform future land use decision making and practices that are applicable to successful ESs-based ecological restoration and management of the VTCSs in the dry zone of Sri Lanka.

Keywords: village tank cascade system; ecosystem services; land use systems; ecosystem services mapping; ecosystem services trade-offs; ecosystem services-based ecological restoration

1. Introduction

In Sri Lanka, the Village Tank Cascade Systems (VTCSs) are considered unique social-ecological systems (SESs) that have also been recently recognised as one of the Globally Important Agricultural Heritage Systems (GIAHSs) attributed to complex hydrological and social-ecological relationships at different spatial scales [1]. Land use systems (LUSs) of VTCSs have evolved through unique biophysical-social interactions and are inexorably linked with rich biodiversity and ecosystem services (ESs) for their sustainability and resilience [2,3]. Ecosystem services provide the foundation for food and nutrition security to smallholder farming communities in the VTCSs by maintaining and improving ecosystem health and climate change resilience [4]. Over the last few years, there has been increased attention in the scientific literature to the value of ESs in SESs, leading to a better understanding of their contribution to human well-being and ecosystem health in the face of global environmental changes [5–8]. Moreover, after the Millennium Ecosystem Assessment (MEA) [8], the ESs concept has been broadened into a framework and increasingly adopted as a decision-making tool for the sustainable management of natural resources [9]. The VTCSs provide a model of a sustainable landscape in which human well-being was maintained in harmony with nature, despite environmental shocks during the past two millennia [10,11]. Thus, studying the ESs of the VTCSs provides insight into how people interact within a unique social-ecological system (SES) to obtain ESs for their survival and well-being, and helps guide efforts to maintain them into an uncertain future [12].

The Village Tank Cascade System (VTCS) comprises a mosaic of small-scale social-ecological LUSs, including smallholder farming systems and ecologically sensitive areas. About 23.3% of the agricultural lands (approximately 228,000 hectares of paddy lands) of the country are located in VTCSs, therefore sustainable maintenance of VTCSs is of great importance in enhancing the agricultural productivity of the country [13]. However, sustainability of the VTCSs will not be achieved if its ecosystem functions and ESs are not properly maintained and protected within the system. Some of the ESs associated with VTCSs have been identified and economically valued. Provisioning ESs are largely recognised at present, while other ESs—regulating, supporting and cultural services of the VTCSs—have not been adequately assessed and are mostly ignored [14–18].

A critical current challenge for maintaining the sustainability of VTCSs is how to increase their agricultural productivity in an environmentally sustainable way [12]. During the past two decades, devastating land use changes have caused biodiversity loss in VTCS landscapes, which have degraded ESs. Some of the land-use changes have resulted in the irreversible collapse of socio-ecological interactions and LUSs–ESs connections. [10,19–21]. Different LUSs may have different capacities for generating ESs [22]. Thus, the supply capacity variations of ESs across LUSs should be well understood in order to initiate sustainable strategies for ecosystem services-based ecological restoration and management of VTCSs. Increasing the ecological productivity of a VTCS through improving ESs could be one of the sustainable solutions for enhancing agricultural productivity [13,23]. In this context, a clear understanding of ESs and their associations with the land use system (LUS) attributes is important. However, research on VTCS ecology, its functions at different levels and the capacity of different ecological components to generate ESs is inadequate. For example, a review of published literature shows only two studies where the biodiversity of the VTCSs has been systematically quantified [24,25].

Ecosystem services assessment and mapping approaches have gained increasing attention, especially since 2000, in the global literature [26]. The essence of the ESs concept is described as the contribution of nature to human well-being [27,28]. Recent studies that have adopted participatory approaches to estimate and map ecosystem service supply and demand include those by Burkhard, Müller [29]; Burkhard, Kroll [30]; Casado-Arzuaga, Madariaga [31]; Burkhard, Kroll [32]; Jacobs, Wolfstein [33]; Burkhard, Kandziora [34]; Reyers, Biggs [35]; Paudyal, Baral [36]; Palomo, Felipe-Lucia [37]; Baral, Keenan [38]; Orsi, Ciolli [39]. Among others, Brück, Abson [9] noted that the majority of ESs assessments have focused on the overall value of ESs for society, rather than the intersection of ecosystem service (ES) values across space, time and user groups. Though there is an increasing trend of ESs assessment studies in Sri Lanka since 2000, research that quantifies the value of ESs through community engagements remains scant in the country [40]. Most ES bundles are generated in social-ecological systems as a result of the integrated outcomes of interacting social, ecological and cultural elements of the system [23,35,37,41–44]. Fisher, Turner [7] emphasised that people are the most important users and beneficiaries of ESs. Thus, participatory approaches that utilise peoples' perceptions could be more appropriate because the value of ESs is determined by the beneficiaries of the LUSs. On the other hand, most existing ES studies have focused on biophysical assessment and quantification of ecological and economic impacts due to land use changes in large-scale landscapes through advanced GIS and remote-sensing techniques [45,46]. Limited studies have taken into consideration the quantitative measurement of social perceptions that reflect the real beneficiary value of ESs [26,47] and other important ES components, such as biodiversity [26,45,48] and land health [49–51]. Therefore, a mixed approach, with the integration of social factors and local knowledge, is more effective for the assessment and mapping of ESs in a complex small-scale social-ecological system [52–54].

The present paper analyses the ESs of different LUSs of Mahakanumulla VTCS in Sri Lanka through community perceptions, expert estimations and biophysical indicators—species diversity and land degradation of LUSs. The specific objectives were to (i) identify and prioritise ESs; (ii) determine ES supply (ESS) and ES demand (ESD) varying across the LUSs; and (iii) map hotspots and bright spots of biodiversity, land degradation and ESS across the LUSs. The findings of this study will provide valuable and much-needed scientific information that could be applied to future land use decision making and practices in regard to ESs-based ecological restoration and management of the VTCSs in Sri Lanka.

2. Materials and Methods

2.1. Study Area

The Mahakanumulla Village Tank Cascade System (MVTCS) is located in the Anuradhapura district within the Malwathuoya River Basin of the north-central province of Sri Lanka (8°5′–8°15′ N and 80°20′–80°35′ E). There are 12 villages and 28 village tanks within the MVTCS, which covers a cascade area of about 4450 ha, with a water surface area of about 557 ha and an irrigation command area of about 758 ha. The population of the area is 3432 (47.8% male and 52.2% female) and 1193 households. There are ten farmer organisations within the MVTCS. The MVTCS landscape is characterised by a tropical monsoonal climate with a well-defined bi-modal rainfall pattern. The annual average rainfall of the area is 1445 mm and the average daily ambient temperature is 27 °C. The terrain of the MVTCS landscape is undulating and is characterised by slopes and valleys which determine the water movement. Three major soil groups (Reddish Brown Earths—Rhodustalfs (60%), Low Humic Gley—Tropaqualfs (30%) and Alluvials (10%)) are found in the area, which create different drainage conditions that provide favourable conditions for farmers to adopt three-fold traditional farming systems (lowland paddy, rain-fed upland and homestead gardens) in the MVTCS. The farming systems of the MVTCS are heavily fragmented and the majority of farm plots are less than two hectares. Intra-annual variation in rainfall (875 to 1875 mm) enables farmers to adapt two major cultivation seasons consisting of combinations of monsoonal and inter-monsoonal climatic seasons. These farming practices have symbiotic relationships with a rich array of biodiversity associated with traditional knowledge and the cultural practices of the area. The above social-ecological, hydrological and geomorphological features of the MVTCS contribute immensely to reducing climate stresses and maintaining sustainable food production. However, long-term land use changes and frequent intra-annual climate variations have impacted significantly the supply of ESs in the study area [12,13]. Altogether, five LUSs and ten major land use types (LUTs) could be identified from a 1:10,000 digital land use map of the study area obtained from the Land Use Policy Planning Department (LUPPD) of Sri Lanka (Table 1 and Figure 1).

Table 1. Land use system categories used for the ESs assessment.

Land Use System (LUS)	Land Use Type (LUT)	Code	Scale	Functions
Agricultural lands	Paddy	P	Macro	Irrigated paddy agro-ecosystem.
	Sparsely used crop land/Shifting cultivation (Chena)	SUCL	Macro	Rain-fed shifting cultivation with very few scattered trees.
	Seasonal crops	SC	Macro	Seasonal crop farming based on climatic seasons.
Forest lands	Dense forest	DF	Macro	Catchment forest (tropical dry mixed evergreen forest—habitat for wild animals).
	Open forest	OF	Macro	Secondary (sparse) forest trees and shrubs. Patches of Damana grasslands associated with tree vegetation.
	Scrub land	SL	Macro	Open areas with low vegetation, covered with small trees and shrubs—habitats for small wild species (amphibians, reptiles etc.).
	Forest plantation	FP	Macro	Dominant Acacia (*Acacia auriculiformis*) and monoculture Teak (*Tectona grandis*) plantation.
Water bodies	Tank/Minor reservoir	T/MNR	Macro	Village tanks. Four geometrical phases of the tank (dead storage, deep-phase, shallow-phase and high flood phase) provide habitats and support the survival of aquatic flora and fauna.
Rocky areas	Area with exposed rocks	RARE	Macro	Rocks and rock outcrops—habitat for few wild species (amphibians, reptiles, etc.).
Built-up areas	Home garden/Homestead	HG	Macro	Houses, home gardens with horticulture, vegetable and animal husbandry.
Micro-land uses (Ecological commons)	Upstream tree belt (Gasgommana)	UTB	Micro	Strip of trees found at the periphery of the tank bed. Functioning as a wind barrier, fish breeding habitat, silt filter, habitats for birds and small wild animals.
	Downstream reservation (Kattakaduwa)	DR	Micro	Diverse vegetation function as natural bio-filter to reduce salinity in seepage water before it reaches into the paddy fields. Habitat for many species.
	Upstream soil ridges (Isweti or Potaweti)	USR	Micro	Upstream earth ridges to prevent sediment inflow.

Table 1. *Cont.*

Land Use System (LUS)	Land Use Type (LUT)	Code	Scale	Functions
	Upstream water hole (Godawala)	UWH	Micro	Human-made water hole aims to trap sediment run-off and provides water to wild animals.
	Deep phase (Diyagilma)	DP	Micro	Central part of the tank bed. Various aquatic plants are grown in this area. Lotus and hydrilla species are dominant. Invasive aquatic plants such as water hyacinth, azolla, salvenia and water lettuce are also present.

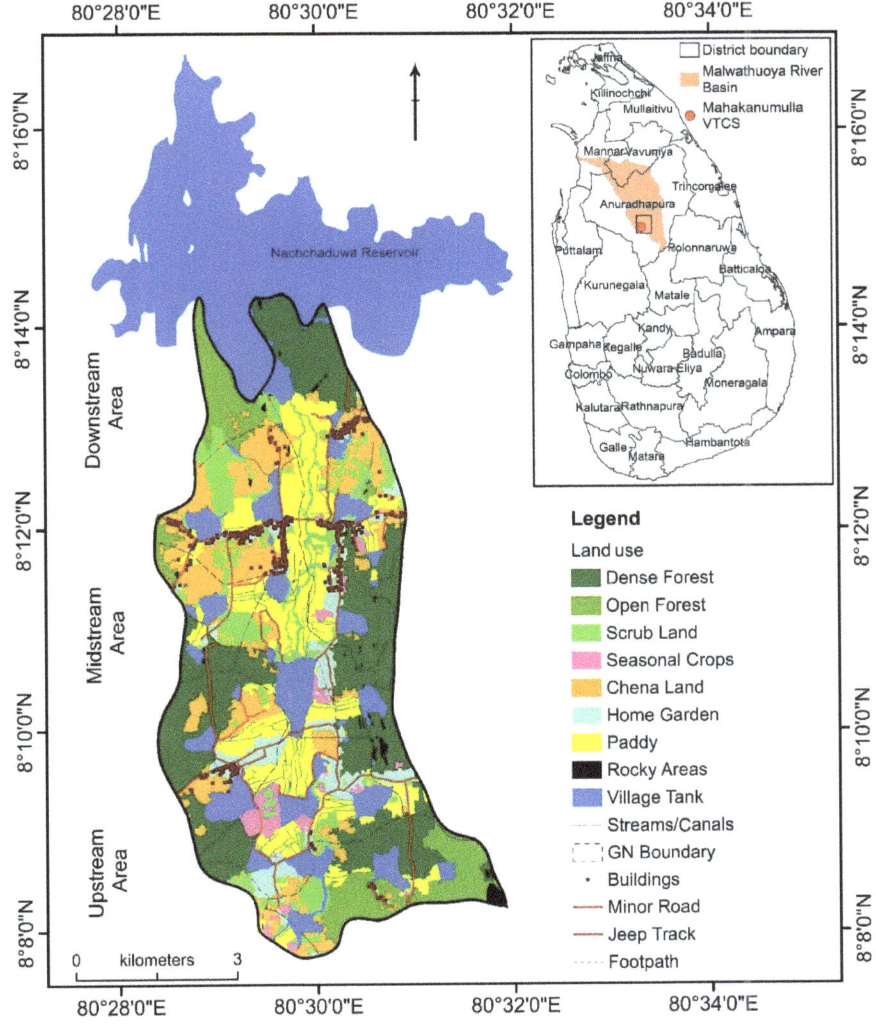

Figure 1. Location of the study area and major land use types.

2.2. Approach and Data Collection

The study utilised both qualitative and quantitative data collection methodologies. Participatory Rapid Appraisal (PRA) techniques were used to determine community perception of values they consider, while prioritising ESs and assessing the supply and demand of ESs provided by the LUSs of the MVTCS. An on-site field survey was carried out to assess the impact of land degradation on ESS of different LUTs. Biodiversity data were obtained from the biodiversity baseline survey of the MVTCS [55]. Land degradation and biodiversity data were integrated to map the ESS capacities of the MVTCS. Data were analysed using graphical and numerical summary measures and exploratory data analysis methods. On-site field survey data collection and participatory assessments were carried out with the support of the Natural Resources Management Centre (NRMC) of the Department of Agriculture and Wayamba University of Sri Lanka in December 2021 and May 2022, respectively.

2.3. Preparation of LUS Units Field Basemap

The LUSs digital map (1:10,000) layer (shapefile) of the study area was overlaid on the basemap of Google Earth imagery and the study area was selected as a Keyhole Markup Language (KML) file. On-screen digitising was undertaken by employing ArcMap (version 10.8.1) software from the Environmental Systems Research Institute, Redlands, California, USA, to demarcate fine-scale LUS units of the study area using the Google Earth image (KML file) accessed on 20 June 2021. In the basemap, each LUS unit was assigned a unique ID. The completed LUSs units field basemap was (i) uploaded as a KML file into Google Earth to use for navigation purposes through a smartphone to find the location of LUSs units and (ii) used as a reference basemap during on-site field assessment of ESs [56]. The methodological approach used in this study is elaborated in Figure 2.

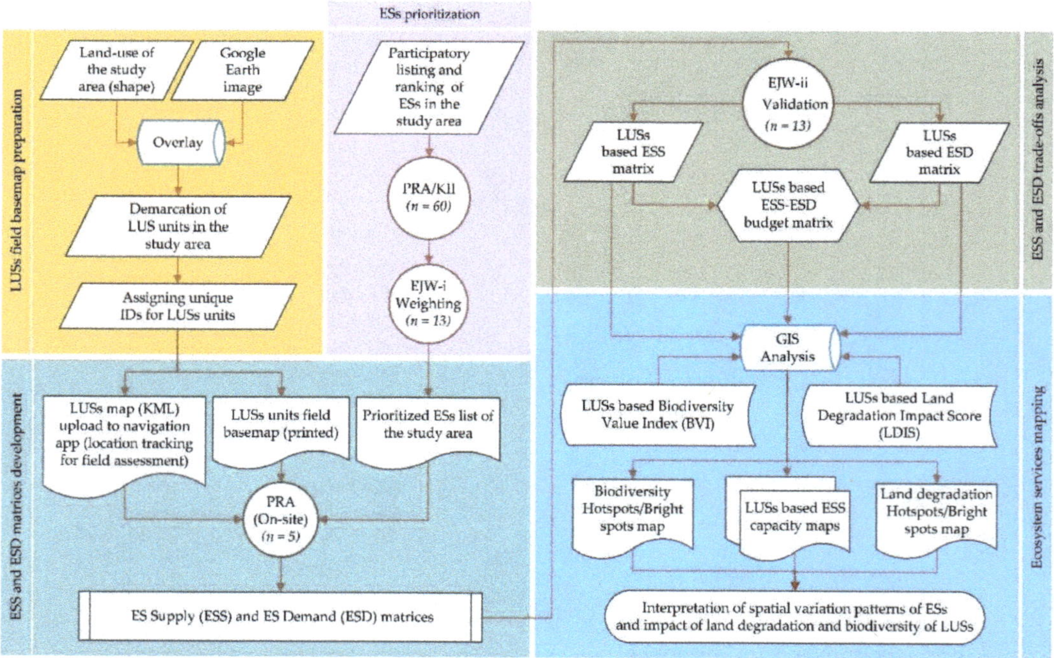

Figure 2. Methodological approach used to assess ecosystem services in the MVTCS. LUSs = Land Use Systems; ES = Ecosystem Service; KML = Keyhole Markup Language; PRA = Participatory Rapid Appraisal; KII = Key Informant Interviews; EJW = Expert Judgement Workshop.

2.4. Inventorying and Prioritisation of ESs

A review of the literature was first undertaken to screen and list out potential ESs associated with VTCS landscapes. In addition, participatory ESs screening was performed in the community to refine the initial list and prioritise ESs in the MVTCS. The prioritised ESs were organised under four main ESs categories: (i) Provisioning (P); (ii) Regulating (R); (iii) Cultural (C) services; and (iv) Supporting functions (S), based on the Common International Classification of Ecosystem Services (CICES—V5.1) [8,57].

The prioritisation of ESs was carried out through the adoption of a rapid participatory listing and ranking approach through PRA techniques [58,59]. The PRA was conducted by selecting community members of upstream, midstream and downstream areas of the MVTCS meso-catchment. An initial explanation of the ESs and their benefits to society was given by the facilitators. Community members were organised into groups and convened to identify ES indicators linked to different LUSs types; they were asked to score each ecosystem service based on its importance in providing benefits to society and to give reasons for the scores. Validated and analysed perception data were visualised using the 'ggplot2' package in R statistical software version 4.1.2 [60].

Often, the community members were uncertain how to convert their perception into a numerical score value, and therefore facilitators had to note down and convert their opinion on a scale of zero to five, where the higher the value, the higher the importance. The results were further validated through key informant interviews (KII), including government authorities and senior community members of the area. The first expert judgement workshop (EJW-i) was conducted to assign weights (%) for each ES based on their knowledge and experience. On-site biodiversity field survey data were analysed to verify the species-based ESs providers. Finally, on-site field verification exercises were conducted parallel to field data collection with the PRA members and experts to further verify the results.

2.5. Assessment of ESs' Supply and Demand

This study refers to the ESs' supply (ESS) capacity as the current potential of individual LUS units to provide different ESs bundles to the local beneficiaries. Ecosystem services' demand (ESD) refers to the amount of all ESs currently consumed or used in both the local area where they are generated or in areas outside local areas over a given period of time [30,53,61]. The study assessed 430 LUS units in the MVTCS which belonged to various macro and micro LUTs (Table 1). Ecosystem services' supply and demand values for particular LUS units were derived from on-site PRA exercises with key informants. The derived values (scores) were transferred to a pre-defined scale from zero to five—where the higher the value (score), the higher the demand or supply for the quantification of ESD and ESS for a particular LUS unit. All values were arranged in a matrix model to link ESs (y-axis) and the LUTs (x-axis). To reduce the uncertainties and improve the confidence level of perceived values by the community, a second expert judgement workshop (EJW-ii) was conducted to validate the ESS and ESD scores by integrating: (i) on-site observations to ensure real evidence and (ii) expert verification to reach a high level of scientific agreement on the final scores of the matrix model [62], thus increasing the scientific quality of the matrix model outcomes [63]. Validated and analysed perception data were visualised using the 'ggplot2' and 'fmsb' packages in R statistical software version 4.1.2 [60,64].

2.6. Mapping of ESS Capacity

Ecosystem service supply scores of the matrix were linked with the attributes of LUSs units' polygons using the 'unique code field' of the LUSs units as a common identifier field for the map production process. The mapping of ESs was carried out based on the values of the ESS matrix scores, Biodiversity Value Index (BVI) and Land Degradation Impact Score (LDIS) of each LUSs unit in the ArcMap version 10.8.1 GIS platform. The study used on-site biodiversity field survey data to derive the BVI by calculating the Simpson Diversity Index for shrubs, small plants, trees, crop plants, medicinal plants and aquatic flora and fauna for the LUTs level [65]. Land Degradation Impact Scores for each LUS unit were calculated based on land degradation field assessment survey data. The land degradation assessment adopted LUS-based LADA–WOCAT–QM (Land Degradation Assessment in Drylands–World Overview of Conservation Approaches and Technologies–Questionnaire for Mapping) approach for on-site assessment of the land degradation indicators in the MVTCS [50,66,67]. The study developed the LUTs and ESs matrix for a grid (100 m × 100 m)-based calculation of ESS capacity values. All ESS, BVI and LDIS values were normalised (rescaled) to be between 0 and 1 to develop raster maps of ESS capacity, biodiversity and land degradation hotspot and bright spot areas in the MVTCS landscape (Figure 3).

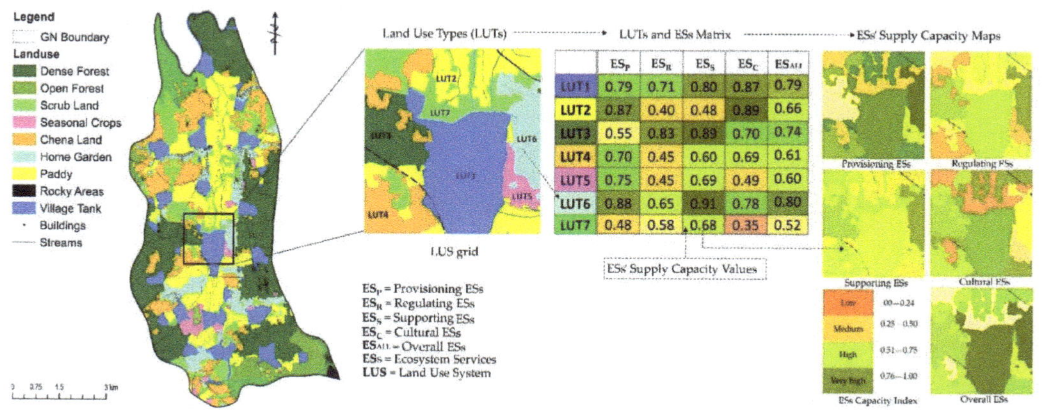

	ES_P	ES_R	ES_S	ES_C	ES_All
LUT1	0.79	0.71	0.80	0.87	0.79
LUT2	0.87	0.40	0.48	0.89	0.66
LUT3	0.55	0.83	0.89	0.70	0.74
LUT4	0.70	0.45	0.60	0.69	0.61
LUT5	0.75	0.45	0.69	0.49	0.60
LUT6	0.88	0.65	0.91	0.78	0.80
LUT7	0.48	0.58	0.68	0.35	0.52

Figure 3. An illustration of the process used to map ecosystem services.

3. Results

3.1. Informants of the PRA

Community assessments through PRA and KII involved 60 people (68% male and 32% female) who had different interactions with MVTCS, including the village community and local government officers. Expert judgement workshops involved 13 experts from NRMC, LUPPD, academia and research institutions. Among the community members who participated in the assessment, 88% were members and/or office bearers of the farmer organisations established in the MVTCS that were directly involved in the local governance of village tanks and associated livelihood activities in harmony with local government organisations. The age distribution of the informants ranged from 30 to 85 years. Most informants were seniors and had been living in the area for more than two generations, which indicates that they had firsthand information—key informants on the ecosystem benefits being received from the LUSs of the MVTCS. All informants were engaged in paddy cultivation in the MVTCS, while about 74% of them were found to be practicing upland farming, including shifting cultivation and home garden horticulture. The age distribution and the number of generations that the community members had been living in this area are shown in Figure 4.

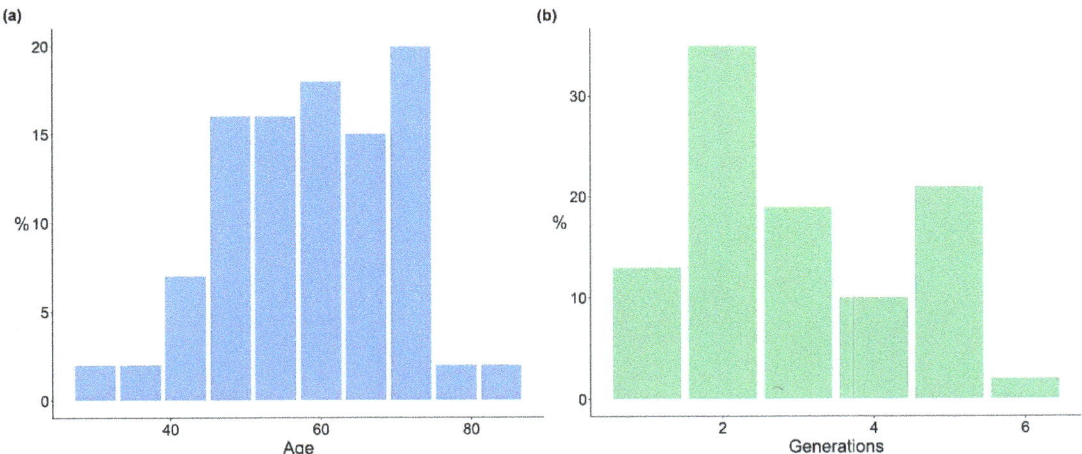

Figure 4. Age (**a**) and the generations (**b**) of the informants.

3.2. Establishment of ESs Priorities

This study revealed that the MVTCS landscape has provided a wide range of ES benefits to the local communities. Participatory listing of ESs by local community members identified twenty-four ESs—nine provisioning services, nine regulating services, four cultural services and two supporting functions. The findings also revealed that the community could identify key providers of ESs and had a strong relationship with biodiversity and biophysical elements of LUSs of the MVTCS. During the EJW-i, experts identified potential indicators to quantify the ESs identified in the participatory listing. The list of ESs perceived by local communities of the MVTCS is given in Table 2. The ranking of the ESs based on the community's perceived scores and experts' weighting values are provided in Figure 5.

Table 2. Summary of ecosystem services/functions and their key providers that emerged from the participatory rapid appraisal.

Ecosystem Service	Description	Key Providers	Potential Indicators Identified for ESs Quantification
Food production (P)	Cultivated food crops (paddy, cereals, lentils, pulses, vegetables, tubers and other seed crops).	- Crop species - Soil organisms - Biophysical elements	- Agrobiodiversity - Soil biodiversity - Crop productivity - Dietary diversity
Water for domestic use (P)	Capacity to provide clean water.	- Biophysical elements	- Soil health/land degradation - Pollution indicators - Groundwater recharge
Water for irrigation (P)	Capacity to provide water for agriculture.	- Biophysical elements	- Water productivity
Inland fisheries (P)	Edible fish species for food and nutrition.	- Fish species	- Biodiversity - Fish productivity - Dietary diversity
Livestock (P)	Reared livestock species for livestock products for food and nutrition. Provide organic manure for crop cultivation.	- Livestock species	- Livestock diversity - Livestock productivity - Dietary diversity - Soil organic matter content

Table 2. *Cont.*

Ecosystem Service	Description	Key Providers	Potential Indicators Identified for ESs Quantification
Fodder and grasses (P)	Existence of grazing lands (pastures) used in the diets of domestic herbivores.	- Plant species	- Biodiversity - Biomass productivity
Raw materials (P)	Plant species used as raw materials.	- Plant species	- Biodiversity - Biomass productivity
Medicinal plants (P)	Medicinal plants and materials.	- Plant species	- Biodiversity
Fruits and wild edibles (P)	Fruit species and wild edible plants.	- Plant species - Soil organisms	- Biodiversity - Soil biodiversity
Control of floods (R)	Capacity of LUSs to capture storm water and reduce runoff.	- Biophysical elements - Plant species of vegetative cover	- Land use, land degradation, soil health - Vegetative cover
Ground water recharge (R)	Capacity of LUSs to capture runoff water and enhance aquifer recharge.	- Biophysical elements	- Land use, land degradation, vegetative cover, soil health
Water purification (R)	Plant species capable to purify polluted water.	- Plant species	- Bioiversity—species richness
Local climate regulation (R)	Capacity of VTCS ecosystems to reduce negative effects of climate change and regulate air quality.	- Plant species	- Biodiversity—species richness - Ecological productivity/resilience
Global climate regulation (R)	Capacity of VTCS ecosystems to enhance carbon sequestering, and reduce GHG emissions.	- Biophysical elements	- Land use, vegetative cover, soil organic carbon - Ecological productivity/resilience
Pollination (R)	Capacity to maintain insects, birds and animal species as pollinators and seed dispersal animals that support crop pollination—food production and their contribution to gene flows and ecological restoration.	- Insect, bird, animal and host plant species	- Biodiversity—species richness - Habitat fragmentation - Pollution (soil, water, air)
Soil nutrient regulation (R)	Capacity of ecological components to maintain soil fertility and soil properties.	- Biophysical elements	- Land use, land degradation, vegetative cover, soil health
Soil erosion regulation (R)	Capacity of LUSs to provide soil retention, runoff control and reduce soil erosion.	- Biophysical elements	- Land use, land degradation, vegetative cover, soil health
Pests and diseases control (R)	Capacity to maintain of biological control agents to minimise incidence of pest and diseases outbreak.	- Insect, bird and vertebrate species	- Biodiversity—species richness - Pollution (soil, water, air)
Landscape diversity (S)	Capacity of LUSs to maintain ecologically and social-ecologically important habitats.	- Biophysical elements	- Landscape/habitat diversity - Landscape performance/resilience
Biodiversity (S)	Capacity of VTCS ecosystems to maintain globally and locally important biodiversity to support ecosystem processes and functions.	- Plant and animal species - Biophysical elements	- Biodiversity - Ecological productivity
Aesthetic and recreational values (C)	Capacity of landscape to provide areas of outstanding aesthetic beauty and quality. It provides environment for villagers and eco-travellers to relax—recreation and educational potentials.	- Plant and animal species - Biophysical elements	- Biodiversity - Landscape/habitat diversity
Traditional knowledge and values (C)	Existence of traditional knowledge systems and practices in the VTCS.	- Plant and animal species - Social elements	- Biodiversity - Biocultural diversity

Table 2. *Cont.*

Ecosystem Service	Description	Key Providers	Potential Indicators Identified for ESs Quantification
Cultural customary values (C)	Cultural traditions, customs and rituals connected with socio-cultural and ecological elements of the VTCS.	- Plant and animal species - Biocultural elements	- Biodiversity - Biocultural diversity
Spiritual and religious values (C)	Spiritual and religious customs associated with a sense of places in the VTCS environment.	- Biophysical elements - Biocultural elements	- Landscape/habitat diversity - Biocultural diversity

Note: P = Provisioning ES; R = Regulating ES; C = Cultural ES; S = Supporting ES functions.

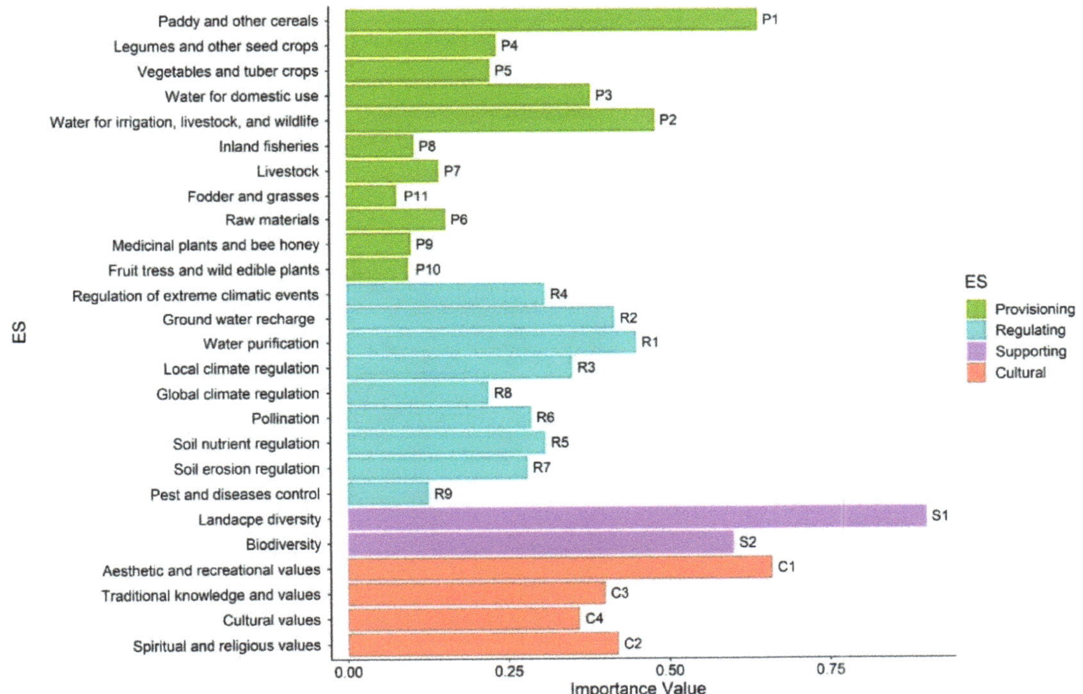

Figure 5. Community-perceived ranking of ecosystem services in MVTCS. ES = Ecosystem service; Importance value = Community perceived score; P = Provisioning ES; R = Regulating ES; C = Cultural ES; S = Supporting ES; Number (P1) = Rank within the ES category.

It was observed that the majority of food provisioning and regulating ESs identified are associated with direct species-based ESs providers (Table 2). Thus, further analysis of the biodiversity baseline survey data revealed that 276 plant species belonging to shrubs, small plants and trees, and 191 faunal species belonging to mammals, reptiles, birds, amphibians, land snails, butterflies and dragonflies, are found in various LUSs of the MVTCS. Many species are found to be common in several LUSs, while some species are multifunctional and multipurpose. Home garden LUSs record the highest number of plant species, while natural forest and downstream reservation areas accommodate many ecologically sensitive plant species important for providing regulating and supporting ESs. The baseline survey data also revealed that MVTCS is rich in agrobiodiversity, including 150 actively managed crop plant species bearing edible components of fruits, seeds, leaf, yam and bark. Most of these species are grown in home gardens and shifting cultivation (Chena) LUSs. Much of the diversity of crop species consists of crop landraces (110 landraces) maintained by the community. The distribution of plants, animals and crop species among different LUTs in MVTCS is visualised in Figure 6.

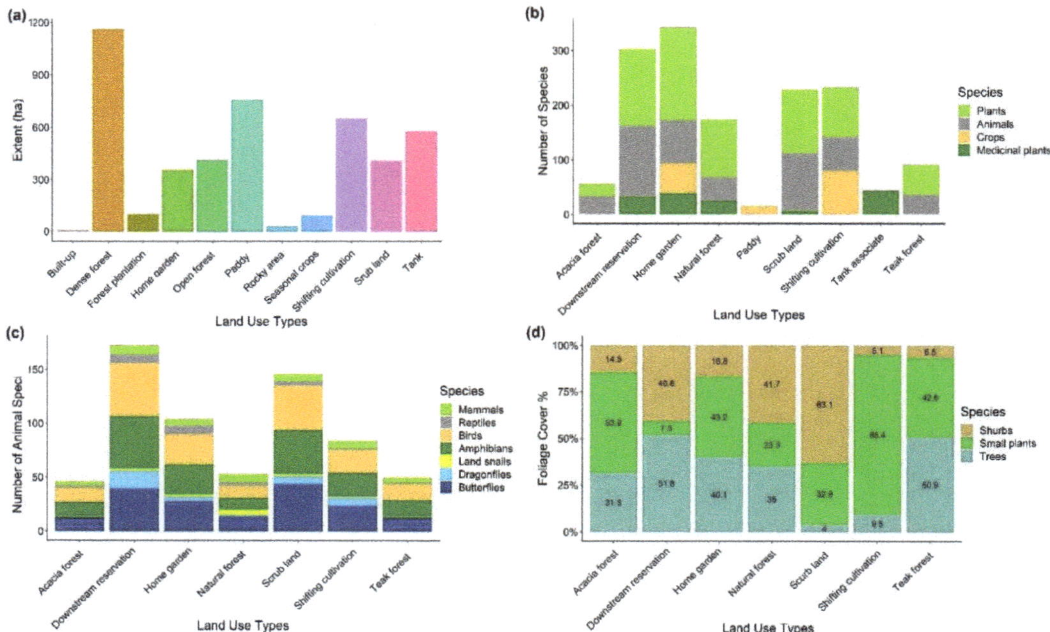

Figure 6. Land use types and species associations in the MVTCS: (**a**) land use types, (**b**) species occurrence (cumulative) in different habitats, (**c**) number of animal species and (**d**) foliage cover percentage of species.

3.3. Ecosystem Services' Supply and Demand

The current capacity to supply provisioning ESs ranges from zero to four. Home gardens, paddy fields and tanks are prominent in providing provisioning ESs. As far as the tanks are concerned, they still have some capacity to supply water for agriculture. The supply of ESs by shifting cultivation lands recorded low values. During on-site field verification, it was noticed that farmers practised agro-wells-assisted irrigation for their shifting cultivation for commercial purposes. To the extent that regulating ESs are valued, upstream tree belt (UTB), upstream soil ridge (USR), tanks and catchment forests were identified as ecologically important LUTs. The ESS matrix showed that most of the LUTs in the MVTCS have at least some capacity to maintain supporting ES functions. The tanks, paddy fields and home gardens are reported as the LUTs with the highest capacity to supply cultural ESs (Figure 7a). Demand for provisioning ESs ranged from zero to four. ESD matrix values indicated high demand for irrigation water from tanks to paddy fields. Compared to the demand for provisioning ESs, a fairly high demand for regulating ESs from all LUTs was recorded. There is a high demand from upstream soil ridge (USR) for regulating ESs. It was observed that the fairly high ESD values for regulating ESs were associated with food production LUTs. There were high ESD values that were observed to maintain biodiversity in the LUSs. Out of the four cultural ESs assessed, the most valued cultural ESs were the aesthetic and recreational services arising from particular LUTs (Figure 7b).

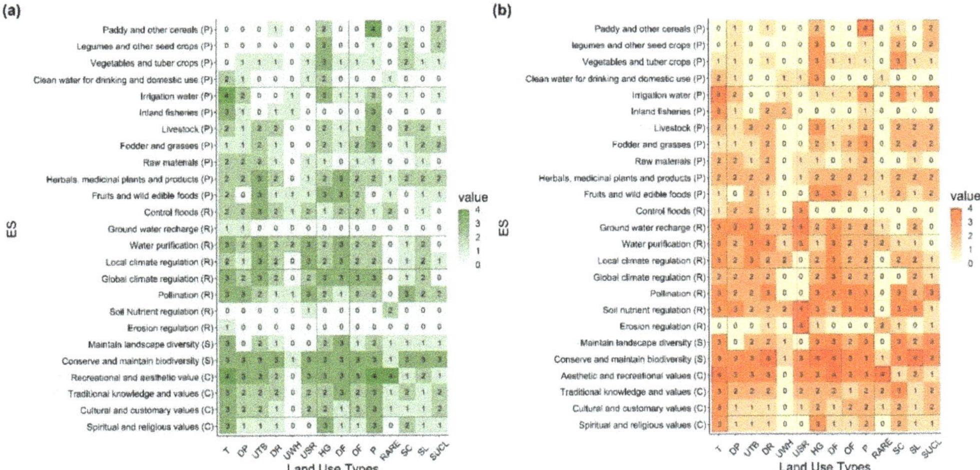

Figure 7. Ecosystem service supply (ESS) capacity evaluation matrix (**a**) and ecosystem service demand (ESD) evaluation matrix (**b**) based on different land use types in the MVTCS. T = Tank; DP = Deep-phase (Diyagilma); UTB = Upstream tree belt (Gasgommana); DR = Downstream reservation (Kattakaduwa); UWH = Upstream water hole (Godawala); USR = Upstream soil ridge (Iswetiya); HG = Home gardens; DF = Dense forests; OF = Open forests; P = Paddy; RARE = Area with exposed rocks; SC = Seasonal crop farming lands; SL = Scrublands; SUCL = Sparsely used crop lands (Chena); ES = Ecosystem service; P = Provisioning ES; R = Regulating ES; C = Cultural ES; S = Supporting ES.

The analysis of ESS and ESD trade-offs showed that the ESs' demand exceeds the supply capacity of the majority of the LUTs in the MVTCS (Figure 8). Negative values indicate occasions where demand exceeds supply (undersupply) and positive values indicate situations where supply exceeds demand (oversupply) (Figures 8c and 9a). Provisioning of all regulating ESs across all LUTs was recorded as undersupply (Figure 8c). There was a comparatively high undersupply of groundwater recharge and soil nutrient regulation ESs in the LUTs recorded (Figure 9b). In addition, the matrix analysis showed a decrease in pollinator services across most of the LUTs (Figure 9a). More importantly, LUTs that are central for generating supporting ES functions in the MVTCS recorded negative trade-off values (Figure 8c). It was observed that the natural forest LUS of the MVTCS has some supply capacity to provide all four ESs categories (Figure 8a). Further, ESD analysis revealed that the demand for tank-associated micro-land uses of upstream water holes (UWH), upstream soil ridges (USR) and downstream reservations (DR) was very high (Figure 8b). The field investigation revealed that the majority of micro-land uses were degraded.

3.4. Spatial Variation in ESS

This study produced ESS capacity maps in different LUTs in MVTCS to understand the spatial variation in current ESS capacities (Figure 10). The spatial pattern in ESS capacities across the MVTCS meso-catchment showed degradation of regulating and supporting ESs towards the downstream and valley bottom areas (Figure 10b,c). Overall, it emerged that the downstream and valley bottom LUSs of the meso-catchment have more capacity to provide provisioning ESs (Figure 10a), while midstream LUSs are important for providing cultural ESs linked with the rich biodiversity of these areas (Figure 10d). Based on the estimation of BVI and LDIS values, the study developed hotspot and bright spot maps of the spatial distribution of BVI and LDIS to understand the impact of biodiversity and land degradation on the ESS capacity of the MVTCS landscape (Figure 11). More hotspot areas impacted by land degradation on ESs were found in downstream and valley bottom areas of the meso-catchment (Figure 11b) and high-value biodiversity areas—bright spots were concentrated in the midstream areas (Figure 11a). Generally, ESS-rich, bright spot areas were found in the midstream areas, while more vulnerable areas of ESs degradation, hotspots, were found towards the downstream areas of the meso-catchment (Figure 11c).

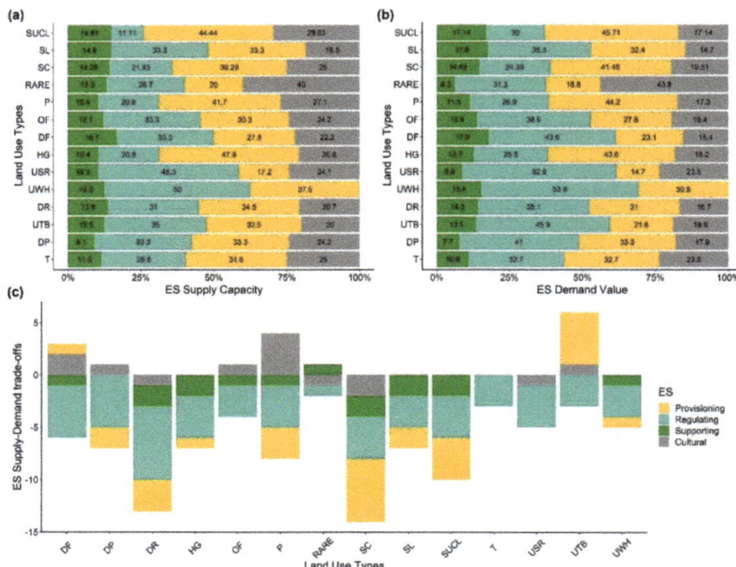

Figure 8. Percentage contributions of current supply (**a**), demand (**b**) and trade-offs (**c**) in four ecosystem service categories of different LUTs in the MVTCS based on community-perceived values. T = Tank; DP = Deep-phase (Diyagilma); UTB = Upstream tree belt (Gasgommana); DR = Downstream reservation (Kattakaduwa); UWH = Upstream water hole (Godawala); USR = Upstream soil ridge (Iswetiya); HG = Home gardens; DF = Dense forests; OF = Open forests; P = Paddy; RARE = Area with exposed rocks; SC = Seasonal crop farming lands; SL = Scrublands; SUCL = Sparsely used crop lands (Chena); ES = Ecosystem service.

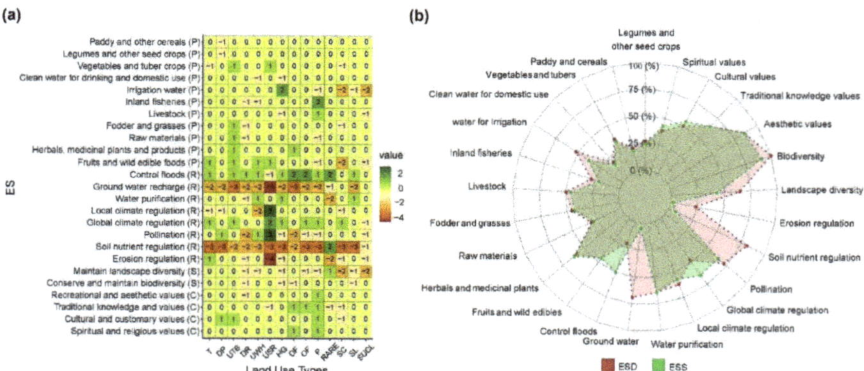

Figure 9. Ecosystem service supply (ESS) and ecosystem service demand (ESD) budget evaluation matrix (**a**); radar chart of ESS and ESD variations with different LUTs (**b**). T = Tank; DP = Deep-phase (Diyagilma); UTB = Upstream tree belt (Gasgommana); DR = Downstream reservation (Kattakaduwa); UWH = Upstream water hole (Godawala); USR = Upstream soil ridge (Iswetiya); HG = Home gardens; DF = Dense forests; OF = Open forests; P = Paddy; RARE = Area with exposed rocks; SC = Seasonal crop farming lands; SL = Scrublands; SUCL = Sparsely used crop lands (Chena); ES = Ecosystem service; P = Provisioning ES; R = Regulating ES; C = Cultural ES; S = Supporting ES.

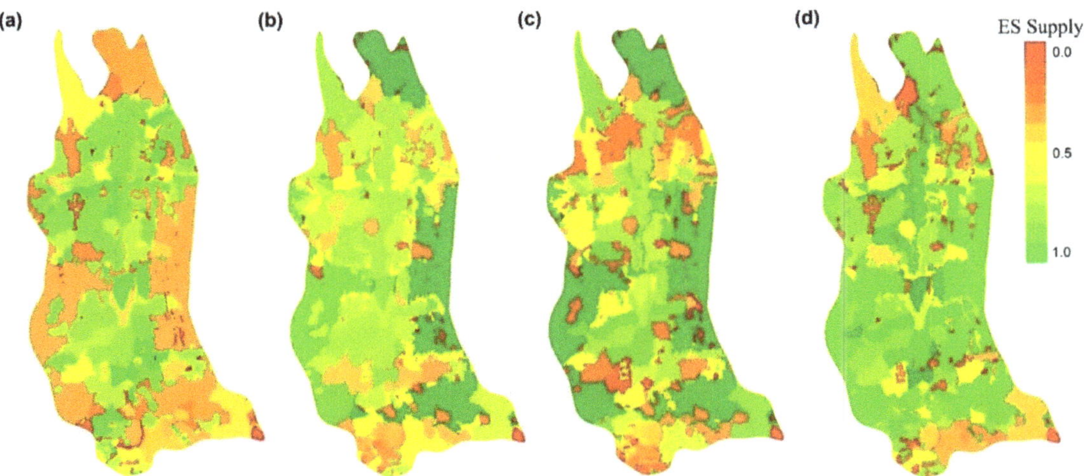

Figure 10. Ecosystem service supply capacities of different land use systems in the MVTCS: (**a**) provisioning, (**b**) regulating, (**c**) supporting and (**d**) cultural.

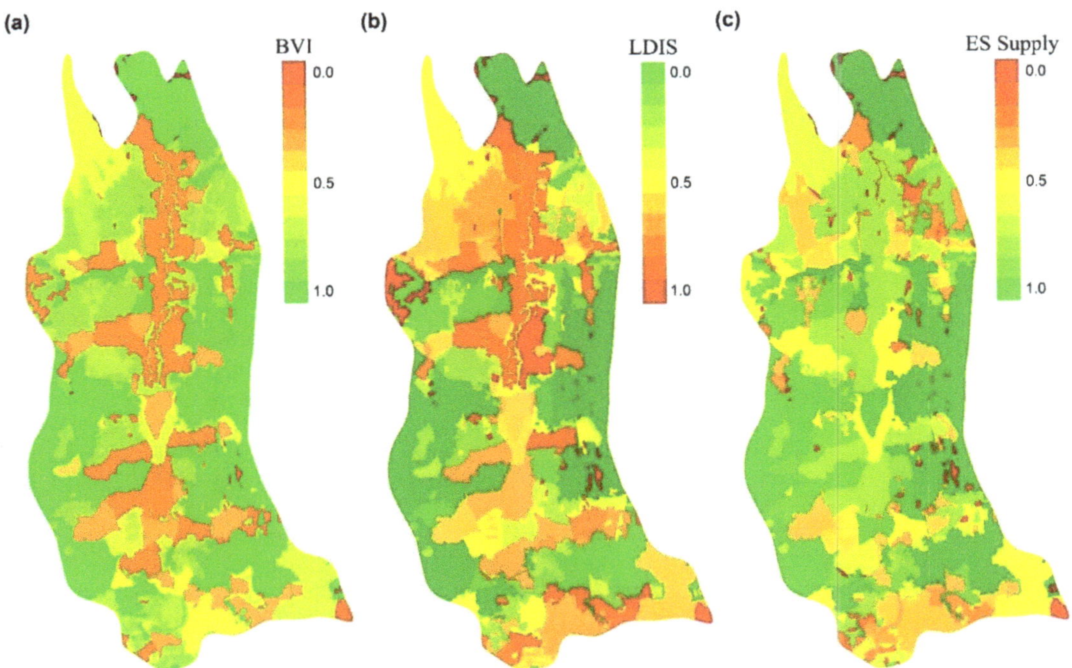

Figure 11. Ecosystem service supply hotspots and bright spots in the MVTCS: (**a**) biodiversity, (**b**) land degradation, and (**c**) overall ecosystem services supply. BVI = Biodiversity Value Index; LDIS = Land Degradation Impact Score.

4. Discussion

This study estimated the capacities of twenty-four ESs for the fourteen macro and micro-land use types identified in the MVTCS. The results show that the species-based ESs providers play an important role in generating ESs associated with biophysical elements of the land uses of the MVTCS. The majority of food provisioning ESs were established to be crop-based generated by interaction with biophysical elements of the LUSs in association with social engagements. The community prioritised food and water-related ESs as the straightforward ESS from the LUSs. Further, all regulating ESs identified had a substantial relationship with food and water provisioning in the MVTCS. This could be the reason the community placed considerable weight on most of the regulating services provided by the LUSs [13,68]. It indicated that the community perceptions appeared to favour their land use management practices and related ES benefits from the LUSs [31]. However, they ranked the role of ESs in pests and diseases control and global climate regulation at a lower level. This could be the reason the majority of community members failed to recognise the effectiveness of regulating ESs of the VTCS in terms of global climate change mitigation and biological control of pests and diseases.

Biodiversity and landscape diversity were identified as the key supporting ES functions from the MVTCS during the participatory listing. The study also found that all cultural ESs are generated as an integrated outcome of the biodiversity and social engagements of the VTCS. Further, it was observed that about 50% of the regulating ESs identified are associated with direct species-based ESs providers. This confirms the findings of [69–71] who established that, biodiversity and the species composition of LUSs is vital for maintaining the ecological balance of the VTCSs. Regaining the lost biodiversity in the VTCSs could be one of the most important steps in the ecological restoration of ESs that supports sustainable agricultural productivity. The significance of ESs restoration in global LUSs has been recognised by the United Nations (UN), declaring 2021–2030 as the Decade on Ecosystem Restoration [72,73]. Reassembling or re-creation of the original ecological components that once occupied VTCS–LUSs in the past is fundamental to ESs-based ecological restoration and management of VTCSs. This could involve assisting the recovery of degraded or fragmented ecological components of the VTCS land uses with a strong ecosystem structure to generate a variety of ecosystem functions and services [23,72,74–76]. However, it could be argued that the restoration of the ecological components/ESs to their past status is unattainable, given social-economic, political, institutional, technological and environmental changes in VTCSs, which in itself is a useful future research question.

Mapping of spatial variation of ESS across the MVTCS–LUSs showed a spatial distribution of areas of high ecological value and social-ecologically important land uses of the MVTCS, which is important in determining future land use and ESs trade-off decisions. Hotspot and bright spot maps of the study area demonstrate that overall ESs' generation capacity is influenced by biodiversity and the land degradation impacts of the land use. Further, it was observed that the degradation of ESs is high in the downstream and valley bottom areas of the meso-catchment. The high impact of land degradation and high social demand for provisioning ESs in the downstream LUSs could be one of the reasons affecting the downstream ESs. Impacts of land use change on biodiversity often lead to declines in ESs [77,78]. Thus, ESs supply-based land health indicators are important for the assessment of ESS capacity at the landscape level in the context of global change scenarios, such as climate, land use and socio-economic changes [79–83]. Therefore, this research provides an integrative, participatory bottom-up ESs assessment framework to apply in highly fragmented, diverse, small-scale SESs, based on their multifunctional LUSs. The resultant assessments can then inform future ecological restoration decision making and practices in the VTCSs. In contrast to conventional participatory ESs assessments, this study accounted for key ecological and land health indicators of the LUSs that directly link with the supply capacity of ESs.

Pre-testing of the participatory ESs screening questionnaire used in the study showed that prioritisation of ESs varied according to the community knowledge and the awareness of the ESs' use by the local communities. Participatory appraisal methods, although having both advantages and disadvantages, have been widely adopted in social-ecological research carried out in rural communities over the last two decades [84]. On the contrary,

ESs demand–supply matrices have some limitations associated with the need for more quantitative methods to estimate perception scores [62,85]. In addition, although on-site field assessment of ESs is straightforward and maybe more accurate, it takes more resources and time [36]. Therefore, to minimise such biases, initial awareness workshops for informants were conducted and the perception scores cross-validated through on-site field verifications and expert consultations. The diversity, complexity and fragmented nature of the arrangements of MVTCS–LUSs could create challenges to the production of quantitative species-based ESs data in a systematic manner.

Process-based biophysical models and toolkits, such as InVest—Integrated Valuation of Ecosystem Services and Tradeoffs, ARIES—ARtificial Intelligence for Ecosystem Services, MIMES—Multi-scale Integrated Models of Ecosystem Services, SolVES—Social Values for Ecosystem Services and MESH—Mapping Ecosystem Services to Human Well-Being, are popular for systematic modelling of ESs in data-rich, large-scale landscapes [53,78,86–90]. However, limitations were found in using such models for the assessment of ESs in small-scale SESs, such as VTCSs, due to the lack of baseline spatial data on biodiversity and land use systems. Thus, a complete survey of biodiversity associated with all LUSs, especially micro-land uses and aquatic environs, is required for a complete understanding of the vital relationships between land use and species interactions.

5. Conclusions

Ecosystem services are central to the existence and optimal multi-functioning of the LUSs of VTCSs. Thus, a comprehensive understanding of ESs is required to achieve optimum productivity of the VTCSs. This study demonstrated the use of the mixed-methods approach to assess, model and map ESs by integrating local knowledge and scientific estimations of land degradation and biodiversity indicators of LUSs of the VTCSs. The study introduced the process of combining biophysical data (biodiversity and land degradation) with social perception data into the ESs mapping process. Although the ESs mapping exercise is challenging due to some limitations, the study managed to map the spatial variation of ESs in the MVTCS using a bottom-up participatory data collection approach. Spatial variation of ESs' supply across LUSs revealed that the demand for all ESs was higher than the ESs' supply capacity of the MVTCS landscape. There was a fairly high demand for regulating and supporting ESs from all LUSs of the MVTCS. The study found that biodiversity plays an important role in generating ESs associated with LUSs. ESs maps revealed that loss of biodiversity and land degradation in the LUSs was directly linked to the overall ESS capacity. The success of the ecological restoration of VTCSs depends on the extent to which strategies address the diverse levels of cascade ecological complexity, as well as the social engagement of local communities. The study approach could be improved by integrating more biophysical and socio-economic data of the VTCSs that provide support to the successful ESs-based ecological restoration and management. Future research should focus not only on ESs assessments, but also consider their applicability to integrate potential ES indicators into ongoing ecological restoration planning processes in the VTCSs in Sri Lanka.

Author Contributions: Conceptualization, S.S.R.; Methodology, S.S.R.; Software, S.S.R. and H.K.K.; Data validation, S.S.R., P.B.D., K.H. and H.K.K.; Analysis, S.S.R., H.K.K. and K.H.; Visualization, S.S.R. and H.K.K.; Investigation, S.S.R.; Data curation, S.S.R., K.H. and H.K.K.; Writing—original draft preparation, S.S.R.; Writing—review and editing, S.S.R., A.K., M.R., P.B.D., D.H., L.K., B.K., C.S.K., K.H. and T.D.; Supervision, A.K., M.R., L.K., P.B.D., D.H. and T.D.; All authors have read and agreed to the published version of the manuscript.

Funding: This research was funded by an Australian Government Research Training Program (RTP) and Destination Australia Program (DAP) Scholarships through the University of New England (UNE), Australia, to the first author.

Institutional Review Board Statement: The participatory assessment undertaken in this study was conducted in accordance with the guidelines approval provided by the Human Research Ethics Committee of the University of New England, Australia (approval no: HE22-030, dated 3 May 2022).

Informed Consent Statement: Not applicable.

Data Availability Statement: Not applicable.

Acknowledgments: The first author thankfully acknowledges the Alliance of Bioversity International and the International Center for Tropical Agriculture (CIAT), Italy, for providing support for publishing this article. In addition, special thank goes to the Natural Resources Management Centre of the Department of Agriculture, Sri Lanka and Wayamba University of Sri Lanka for facilitating of the field data collection.

Conflicts of Interest: The authors declare no conflict of interest.

References

1. FAO. "Globally Important Agricultural Heritage Systems" (GIAHS). Available online: http://www.fao.org/giahs/en/ (accessed on 26 June 2021).
2. Jiao, Y.; Liang, L.; Okuro, T.; Takeuchi, K. Ecosystem services and biodiversity of traditional agricultural landscapes: A case study of the hani terraces in Southwest China. In *Biocultural Landscapes*; Hong, S.K., Bogaert, J., Min, Q., Eds.; Springer: Dordrecht, Germany, 2014; pp. 81–88.
3. Gu, H.; Sbramanian, S. *Socio-Ecological Production Landscapes: Relevance to the Green Economy Agenda*; United Nations University Institute of Advanced Studies (UNU-IAS): Yokohama, Japan, 2012; p. 66.
4. Wu, L.; Avishek, A. *Restoring the Ecological Foundation for Food Security: A Soil Organic Matter Perspective*; UNEP Policy Brief-12; Liu, J., Khan, A., Gommes, R.A., Oduor, A.M.O., Munang, R., Eds.; United Nations Environemnt Programme (UNEP): Nairobi, Kenya, 2013; p. 16.
5. Koohafkan, P.; Altieri, M.A. *Globally Important Agricultural Heritage Systems: A Legacy for the Future*; FAO—Food and Agriculture Organization of the United Nations: Rome, Italy, 2011; p. 41.
6. Lu, N.; Wang, M.; Ning, B.; Yu, D.; Fu, B. Research advances in ecosystem services in drylands under global environmental changes. *Curr. Opin. Environ. Sustain.* **2018**, *33*, 92–98. [CrossRef]
7. Fisher, B.; Turner, R.K.; Morling, P. Defining and classifying ecosystem services for decision making. *Ecol. Econ.* **2009**, *68*, 643–653. [CrossRef]
8. Reid, W.V.; Mooney, H.A.; Cropper, A.; Capistrano, D.; Carpenter, S.R.; Chopra, K.; Dasgupta, P.; Dietz, T.; Duraiappah, A.K.; Hassan, R.; et al. *Ecosystems and Human Well-Being-SYNTHESIS: A Report of the Millennium Ecosystem Assessment (MEA)*; Island Press: Washington, DC, USA, 2005; p. 137.
9. Brück, M.; Abson, D.J.; Fischer, J.; Schultner, J. Broadening the scope of ecosystem services research: Disaggregation as a powerful concept for sustainable natural resource management. *Ecosyst. Serv.* **2022**, *53*, 101399. [CrossRef]
10. Abeywardana, N.; Schütt, B.; Wagalawatta, T.; Bebermeier, W. Indigenous agricultural systems in the dry zone of Sri Lanka: Management transformation assessment and sustainability. *Sustainability* **2019**, *11*, 910. [CrossRef]
11. Chandrasekara, S.S.K.; Chandrasekara, S.K.; Gamini, P.H.S.; Obeysekera, J.; Manthrithilake, H.; Kwon, H.; Vithanage, M. A review on water governance in Sri Lanka: The lessons learnt for future water policy formulation. *Water Policy* **2021**, *23*, 255–273. [CrossRef]
12. Dharmasena, P.B. Cascaded Tank-Village System: Present status and prospects. In *Agricultural Research for Sustainable Food Systems in Sri Lanka*; Marambe, B., Weerahewa, J., Dandeniya, W.S., Eds.; Springer: Singapore, 2020; pp. 63–75.
13. Ratnayake, S.S.; Kumar, L.; Dharmasena, P.B.; Kadupitiya, H.K.; Kariyawasam, C.S.; Hunter, D. Sustainability of Village Tank Cascade Systems of Sri Lanka: Exploring cascade anatomy and socio-ecological nexus for ecological restoration planning. *Challenges* **2021**, *12*, 24. [CrossRef]
14. Vidanage, S.P.; Kotagama, H.B.; Dunusinghe, P.M. Sri Lanka's small tank cascade systems: Building agricultural resilience in the dry zone. In *Climate Change and Community Resilience*; Haque, A.K.K., Mukhopadhyay, P., Nepal, M., Shammin Md, R., Eds.; Springer: Singapore, 2022; pp. 225–235.
15. Vidanage, S.P. Economic Value of an Ancient Small Tank Cascade System in Sri Lanka. Ph.D. Thesis, University of Colombo, Colombo, Sri Lamka, 2020.
16. Vidanage, S.; Perera, S.; Kallesoe, M. *The Value of Traditional Water Schemes: Small Tanks in the Kala Oya Basin, Sri Lanka*; IUCN Water, Nature and Economics Technical Paper No. 6; IUCN—The International Union for Conservation of Nature: Colombo, Sri Lanka, 2005; pp. 1–76.
17. Dilhari, W.A.D.S.; Weerahewa, J. Valuing ecosystem services provided by Minor Village Tanks in the Dry Zone of Sri Lanka. In Proceedings of the Cascade Ecology and Management, University of Peradeniya, Faculty of Agriculture, Peradeniya, Sri Lanka, 17–18 September 2021.
18. Zubair, L. Modernisation of Sri Lanka's traditional irrigation systems and sustainability. *Sci. Technol. Soc.* **2005**, *10*, 161–195. [CrossRef]

19. Bhatta, L.D.; Chaudhary, S.; Pandit, A.; Baral, H.; Das, P.J.; Stork, N.E. Ecosystem service changes and livelihood impacts in the Maguri-Motapung wetlands of Assam, India. *Land* **2016**, *5*, 15. [CrossRef]

20. Anuradha, J.M.P.N.; Fujimura, M.; Inaoka, T.; Sakai, N. The role of agricultural land use pattern dynamics on elephant habitat depletion and human-elephant conflict in Sri Lanka. *Sustainability* **2019**, *11*, 2818. [CrossRef]

21. Ranagalage, M.; Gunarathna, M.H.J.P.; Surasinghe, T.D.; Dissanayake, D.; Simwanda, M.; Murayama, Y.; Morimoto, T.; Phiri, D.; Nyirenda, V.R.; Premakantha, K.T.; et al. Multi-decadal forest-cover dynamics in the tropical realm: Past trends and policy insights for forest conservation in dry zone of Sri Lanka. *Forests* **2020**, *11*, 836. [CrossRef]

22. Bhandari, P.; Mohan, K.C.; Shrestha, S.; Aryal, A.; Shrestha, B.U. Assessments of ecosystem service indicators and stakeholder's willingness to pay for selected ecosystem services in the Chure region of Nepal. *Appl. Geogr.* **2016**, *69*, 25–34. [CrossRef]

23. Timberlake, T.P.; Cirtwill, A.R.; Baral, S.C.; Bhusal, D.R.; Devkota, K.; Harris-Fry, H.A.; Kortsch, S.; Myers, S.S.; Roslin, T.; Saville, N.M. A network approach for managing ecosystem services and improving food and nutrition security on smallholder farms. *People Nat.* **2022**, *4*, 563–575. [CrossRef]

24. Marambe, B.; Pushpakumara, G.; Silva, P. Biodiversity and agrobiodiversity in Sri Lanka: Village Tank Systems. In *The Biodiversity Observation Network in the Asia-Pacific Region*; Nakano, S., Yahara, T., Nakashizuka, T., Eds.; Ecological Research Monographs, Springer: Tokyo, Japan, 2012; pp. 403–430.

25. Goonatilake, S.d.A.; Ekanayake, S.P.; Perera, N.; Wijenayake, T.; Wadugodapitiya, A. *Biodiversity and Ethnobiology of the Kapiriggama Small Tank Cascade System in Sri Lanka*; IUCN Programme on Restoring Traditional Cascading Tank Systems Technical Note 2; IUCN—The International Union for Conservation of Nature: Colombo, Sri Lanka, 2015; p. 189.

26. Zhang, X.; Estoque, R.C.; Xie, H.; Murayama, Y.; Ranagalage, M. Bibliometric analysis of highly cited articles on ecosystem services. *PLoS ONE* **2019**, *14*, e0210707. [CrossRef] [PubMed]

27. Díaz, S.M.; Settele, J.; Brondízio, E.; Ngo, H.; Guèze, M.; Agard, J.; Arneth, A.; Balvanera, P.; Brauman, K.; Butchart, S. The IPBES conceptual framework—Connecting nature and people. *Curr. Opin. Environ. Sustain.* **2015**, *14*, 1–16. [CrossRef]

28. Müller, F.; Fohrer, N.; Chicharo, L. The basic ideas of the ecosystem service concept. In *Ecosystem Services and River Basin Ecohydrology*; Chicharo, L., Müller, F., Fohrer, N., Eds.; Springer: Dordrecht, Germany, 2015; pp. 7–33.

29. Burkhard, B.; Müller, A.; Müller, F.; Grescho, V.; Anh, Q.; Arida, G.; Bustamante, J.V.J.; Van Chien, H.; Heong, K.L.; Escalada, M. Land cover-based ecosystem service assessment of irrigated rice cropping systems in southeast Asia—An explorative study. *Ecosyst. Serv.* **2015**, *14*, 76–87. [CrossRef]

30. Burkhard, B.; Kroll, F.; Nedkov, S.; Müller, F. Mapping ecosystem service supply, demand and budgets. *Ecol. Indic.* **2012**, *21*, 17–29. [CrossRef]

31. Casado-Arzuaga, I.; Madariaga, I.; Onaindia, M. Perception, demand and user contribution to ecosystem services in the Bilbao Metropolitan Greenbelt. *J. Environ. Manag.* **2013**, *129*, 33–43. [CrossRef] [PubMed]

32. Burkhard, B.; Kroll, F.; Müller, F.; Windhorst, W. Landscapes' capacities to provide ecosystem services-a concept for land-cover based assessments. *Landsc. Online* **2009**, *15*, 1–22. [CrossRef]

33. Jacobs, S.; Wolfstein, K.; Vandenbruwaene, W.; Vrebos, D.; Beauchard, O.; Maris, T.; Meire, P. Detecting ecosystem service trade-offs and synergies: A practice-oriented application in four industrialized estuaries. *Ecosyst. Serv.* **2015**, *16*, 378–389. [CrossRef]

34. Burkhard, B.; Kandziora, M.; Hou, Y.; Müller, F. Ecosystem service potentials, flows and demands-concepts for spatial localisation, indication and quantification. *Landsc. Online* **2014**, *34*, 1–32. [CrossRef]

35. Reyers, B.; Biggs, R.; Cumming, G.S.; Elmqvist, T.; Hejnowicz, A.P.; Polasky, S. Getting the measure of ecosystem services: A social–ecological approach. *Front. Ecol. Environ.* **2013**, *11*, 268–273. [CrossRef]

36. Paudyal, K.; Baral, H.; Burkhard, B.; Bhandari, S.P.; Keenan, R.J. Participatory assessment and mapping of ecosystem services in a data-poor region: Case study of community-managed forests in central Nepal. *Ecosyst. Serv.* **2015**, *13*, 81–92. [CrossRef]

37. Palomo, I.; Felipe-Lucia, M.R.; Bennett, E.M.; Martín-López, B.; Pascual, U. Chapter six -Disentangling the pathways and effects of ecosystem service co-production. *Adv. Ecol. Res.* **2016**, *54*, 245–283.

38. Baral, H.; Keenan, R.J.; Fox, J.C.; Stork, N.E.; Kasel, S. Spatial assessment of ecosystem goods and services in complex production landscapes: A case study from south-eastern Australia. *Ecol. Complex.* **2013**, *13*, 35–45. [CrossRef]

39. Orsi, F.; Ciolli, M.; Primmer, E.; Varumo, L.; Geneletti, D. Mapping hotspots and bundles of forest ecosystem services across the European Union. *Land Use Policy* **2020**, *99*, 104840. [CrossRef]

40. Sarathchandra, C.; Abebe, Y.A.; Wijerathne, I.L.; Aluthwattha, S.T.; Wickramasinghe, S.; Ouyang, Z. An overview of ecosystem service studies in a tropical biodiversity hotspot, Sri Lanka: Key perspectives for future research. *Forests* **2021**, *12*, 540. [CrossRef]

41. Hamann, M.; Biggs, R.; Reyers, B. An exploration of human well-being bundles as identifiers of ecosystem service use patterns. *PLoS ONE* **2016**, *11*, e0163476. [CrossRef]

42. Hamann, M.; Biggs, R.; Reyers, B. Mapping social–ecological systems: Identifying 'green-loop' and 'red-loop' dynamics based on characteristic bundles of ecosystem service use. *Glob. Environ. Chang.* **2015**, *34*, 218–226. [CrossRef]

43. Huntsinger, L.; Oviedo, J.L. Ecosystem services are social–ecological services in a traditional pastoral system: The case of California's Mediterranean rangelands. *Ecol. Soc.* **2014**, *19*, 8. [CrossRef]

44. Meacham, M.; Norström, A.V.; Peterson, G.D.; Andersson, E.; Bennett, E.M.; Biggs, R.; Crouzat, E.; Cord, A.F.; Enfors, E.; Felipe-Lucia, M.R. Advancing research on ecosystem service bundles for comparative assessments and synthesis. *Ecosyst. People* **2022**, *18*, 99–111. [CrossRef]

45. Ceaușu, S.; Apaza-Quevedo, A.; Schmid, M.; Martín-López, B.; Cortés-Avizanda, A.; Maes, J.; Brotons, L.; Queiroz, C.; Pereira, H.M. Ecosystem service mapping needs to capture more effectively the biodiversity important for service supply. *Ecosyst. Serv.* **2021**, *48*, 101259. [CrossRef]

46. Peng, H.; Hua, L.; Zhang, X.; Yuan, X.; Li, J. Evaluation of ESV change under urban expansion based on ecological sensitivity: A case study of three gorges reservoir area in China. *Sustainability* **2021**, *13*, 8490. [CrossRef]

47. Alamgir, M.; Pert, P.L.; Turton, S.M. A review of ecosystem services research in Australia reveals a gap in integrating climate change and impacts on ecosystem services. *Int. J. Biodivers. Sci. Ecosyst. Serv. Manag.* **2014**, *10*, 112–127. [CrossRef]

48. Oliver, I.; Parkes, D. *A Prototype Toolkit for Scoring the Biodiversity Benefits of Land Use Change*; NSW Department of Infrastructure Planning and Natural Resources: Parramatta, Australia, 2003; p. 40.

49. Smiraglia, D.; Ceccarelli, T.; Bajocco, S.; Salvati, L.; Perini, L. Linking trajectories of land change, land degradation processes and ecosystem services. *Environ. Res.* **2016**, *147*, 590–600. [CrossRef] [PubMed]

50. Bunning, S.; McDonagh, J.; Rioux, J. *Land Degradation Assessmentin Drylands (LADA): Manual for Local Level Assessment of Land Degradation and Sustainable Land Management*; FAO—Food and Agriculture Organization of The United Nations: Rome, Italy, 2016; p. 183.

51. Shepherd, K.D.; Shepherd, G.; Walsh, M.G. Land health surveillance and response: A framework for evidence-informed land management. *Agric. Syst.* **2015**, *132*, 93–106. [CrossRef]

52. Ji, Z.; Xu, Y.; Wei, H. Identifying dynamic changes in ecosystem services supply and demand for urban sustainability: Insights from a rapidly urbanizing city in Central China. *Sustainability* **2020**, *12*, 3428. [CrossRef]

53. Crossman, N.D.; Burkhard, B.; Nedkov, S.; Willemen, L.; Petz, K.; Palomo, I.; Drakou, E.G.; Martín-Lopez, B.; McPhearson, T.; Boyanova, K. A blueprint for mapping and modelling ecosystem services. *Ecosyst. Serv.* **2013**, *4*, 4–14. [CrossRef]

54. IUCN. *Tank Ecosystem Restoration of the Kapiriggama Small Tank Cascade System*; IUCN Programme on Restoring Traditional Cascading Tank Systems Technical Note 3; IUCN—The International Union for Conservation of Nature: Colombo, Sri Lanka, 2015; p. 29.

55. Ekanayake, S.P.; Ratnayake, S.S.; Sugathadasa, S.; Rajapakse, R.M. Chapter Three—Biodiversity baseline assessment. In *Baseline Assessment of the UNEP-GEF Managing Agricultural Landscapes in Socio-ecologically Sensitive Areas to Promote Food Security, Well-being and Ecosystem Health—Healthy landscapes Project*; Dharmasena, P.B., Kadupitiya, H.K., Ratnayake, S.S., Eds.; SACEP—South Asia Co-Operative Environment Programme: Colombo, Sri Lanka, 2020; pp. 151–219.

56. Kadupitiya, H.K.; De Silva, S.H.S.A.; Dilshani, D.G.S.; Weerasinghe, P. Use of smartphones for rapid location tracking in mega scale soil sampling. *Open J. Appl. Sci.* **2021**, *11*, 239. [CrossRef]

57. Haines-Young, R.; Potschin-Young, M. Revision of the common international classification for ecosystem services (CICES V5.1): A policy brief. *One Ecosyst.* **2018**, *3*, e27108. [CrossRef]

58. López-Marrero, T.; Hermansen-Báez, L.A. *Participatory Listing, Ranking, And Scoring of Ecosystem Services and Drivers of Change*; USDA Forest Service, Southern Research Station: Gainesville, FL, USA, 2011; p. 7.

59. Rey-Valette, H.; Mathé, S.; Salles, J.M. An assessment method of ecosystem services based on stakeholders perceptions: The Rapid Ecosystem Services Participatory Appraisal (RESPA). *Ecosyst. Serv.* **2017**, *28*, 311–319. [CrossRef]

60. Wickham, H. *ggplot2: Elegant Graphics for Data Analysis*; Springer: New York, NY, USA, 2016.

61. Syrbe, R.-U.; Grunewald, K. Ecosystem service supply and demand–the challenge to balance spatial mismatches. *Int. J. Biodivers. Sci. Ecosyst. Serv. Manag.* **2017**, *13*, 148–161. [CrossRef]

62. Jacobs, S.; Burkhard, B.; Van Daele, T.; Staes, J.; Schneiders, A. 'The Matrix Reloaded': A review of expert knowledge use for mapping ecosystem services. *Ecol. Modell.* **2015**, *295*, 21–30. [CrossRef]

63. Mastrandrea, M.D.; Mach, K.J.; Plattner, G.K.; Edenhofer, O.; Stocker, T.F.; Field, C.B.; Ebi, K.L.; Matschoss, P.R. The IPCC AR5 guidance note on consistent treatment of uncertainties: A common approach across the working groups. *Clim. Chang.* **2011**, *108*, 675–691. [CrossRef]

64. Nakazawa, M. Package 'fmsb'. Available online: https://cran.r-project.org/web/packages/fmsb/fmsb.pdf (accessed on 18 January 2022).

65. Simpson, E.H. Measurement of diversity. *Nature* **1949**, *163*, 688. [CrossRef]

66. García, C.L.; Teich, I.; Gonzalez-Roglich, M.; Kindgard, A.F.; Ravelo, A.C.; Liniger, H. Land degradation assessment in the Argentinean Puna: Comparing expert knowledge with satellite-derived information. *Environ. Sci. Policy* **2019**, *91*, 70–80. [CrossRef]

67. NRMC. *Assessment and Mapping of Land Degradation and Conservation: Kandy, Nuwara Eliya and Badulla Districts of Sri Lanka*; Rehabilitation of Degraded Agricultural Lands (RDAL) Project; NRMC—Natural Resources Management Centre, Department of Agriculture: Peradeniya, Sri Lanka, 2021; p. 63.

68. Dharmasena, P. Evolution of hydraulic societies in the ancient Anuradhapura Kingdom of Sri Lanka. In *Landscapes and Societies*; Martini, I., Chesworth, W., Eds.; Springer: Dordrecht, Germany, 2010; pp. 341–352.

69. Barral, M.P.; Benayas, J.M.R.; Meli, P.; Maceira, N.O. Quantifying the impacts of ecological restoration on biodiversity and ecosystem services in agroecosystems: A global meta-analysis. *Agric. Ecosyst. Environ.* **2015**, *202*, 223–231. [CrossRef]

70. Prober, S.M.; Byrne, M.; McLean, E.H.; Steane, D.A.; Potts, B.M.; Vaillancourt, R.E.; Stock, W.D. Climate-adjusted provenancing: A strategy for climate-resilient ecological restoration. *Front. Ecol. Environ.* **2015**, *3*, 65. [CrossRef]

71. Macfadyen, S.; Cunningham, S.A.; Costamagna, A.C.; Schellhorn, N.A. Managing ecosystem services and biodiversity conservation in agricultural landscapes: Are the solutions the same? *J. Appl. Ecol.* **2012**, *49*, 690–694. [CrossRef]
72. Fischer, J.; Riechers, M.; Loos, J.; Martin-Lopez, B.; Temperton, V.M. Making the UN decade on ecosystem restoration a social-ecological endeavour. *Trends Ecol. Evol.* **2021**, *36*, 20–28. [CrossRef]
73. Heger, T.; Jeschke, J.M.; Febria, C.; Kollmann, J.; Murphy, S.; Rochefort, L.; Shackelford, N.; Temperton, V.M.; Higgs, E. Mapping and assessing the knowledge base of ecological restoration. *Restor. Ecol.* **2022**, e13676. [CrossRef]
74. Clewell, A.F.; Aronson, J. *Ecological Restoration: Principles, Values, and Structure of An Emerging Profession*; Island Press: Washington, DC, USA, 2012.
75. Luck, G.W.; Harrington, R.; Harrison, P.A.; Kremen, C.; Berry, P.M.; Bugter, R.; Dawson, T.P.; De Bello, F.; Díaz, S.; Feld, C.K. Quantifying the contribution of organisms to the provision of ecosystem services. *BioScience* **2009**, *59*, 223–235. [CrossRef]
76. Gann, G.D.; McDonald, T.; Walder, B.; Aronson, J.; Nelson, C.R.; Jonson, J.; Hallett, J.G.; Eisenberg, C.; Guariguata, M.R.; Liu, J. International principles and standards for the practice of ecological restoration. *Restor. Ecol.* **2019**, *27*, S1–S46. [CrossRef]
77. Locatelli, B.; Lavorel, S.; Sloan, S.; Tappeiner, U.; Geneletti, D. Characteristic trajectories of ecosystem services in mountains. *Front. Ecol. Environ.* **2017**, *15*, 150–159. [CrossRef]
78. Liang, J.; Li, S.; Li, X.; Li, X.; Liu, Q.; Meng, Q.; Lin, A.; Li, J. Trade-off analyses and optimization of water-related ecosystem services (WRESs) based on land use change in a typical agricultural watershed, southern China. *J. Clean. Prod.* **2021**, *279*, 123851.
79. Ausseil, A.-G.; Dymond, J.R.; Kirschbaum, M.U.F.; Andrew, R.M. Assessment of multiple ecosystem services in New Zealand at the catchment scale. *Environ. Modell. Softw.* **2013**, *43*, 37–48. [CrossRef]
80. Li, B.; Chen, D.; Wu, S.; Zhou, S.; Wang, T.; Chen, H. Spatio-temporal assessment of urbanization impacts on ecosystem services: Case study of Nanjing City, China. *Ecol. Indic.* **2016**, *71*, 416–427. [CrossRef]
81. Wei, Y.; Wu, S.; Jiang, C.; Feng, X. Managing supply and demand of ecosystem services in dryland catchments. *Curr. Opin. Environ. Sustain.* **2021**, *48*, 10–16. [CrossRef]
82. Dharmasena, P.B. Environmental Richness in The Dry Zone Home Gardens in Sri Lanka. In Proceedings of the Eighth National Workshop on Multipurpose Trees, Kandy, Sri Lanka, 23–25 October 1997.
83. Potschin, M.; Haines-Young, R. Defining and measuring ecosystem services. In *Routledge Handbook of Ecosystem Services*; Potschin, M., Haines-Young, R., Fish, R., Turner, R.K., Eds.; Routledge: London, UK; New York, NY, USA, 2016; pp. 25–44.
84. Nyumba, T.O.; Wilson, K.; Derrick, C.J.; Mukherjee, N. The use of focus group discussion methodology: Insights from two decades of application in conservation. *Methods Ecol. Evol.* **2018**, *9*, 20–32. [CrossRef]
85. Schröter, M.; Remme, R.P.; Hein, L. How and where to map supply and demand of ecosystem services for policy-relevant outcomes? Letter to the Editor. *Ecol. Indic.* **2012**, *23*, 220–221. [CrossRef]
86. Villa, F.; Bagstad, K.J.; Voigt, B.; Johnson, G.W.; Portela, R.; Honzák, M.; Batker, D. A methodology for adaptable and robust ecosystem services assessment. *PLoS ONE* **2014**, *9*, e91001. [CrossRef]
87. Fang, Z.; Bai, Y.; Jiang, B.; Alatalo, J.; Liu, G.; Wang, H. Quantifying variations in ecosystem services in altitude-associated vegetation types in a tropical region of China. *Sci. Total Environ.* **2020**, *726*, 138565. [CrossRef] [PubMed]
88. Neugarten, R.A.; Langhammer, P.F.; Osipova, E.; Bagstad, K.J.; Bhagabati, N.; Butchart, S.H.M.; Dudley, N.; Elliott, V.; Gerber, L.R.; Arrellano, C.G. *Tools for Measuring, Modelling, and Valuing Ecosystem Services: Guidance for Key Biodiversity Areas, Natural World Heritage Sites, and Protected Areas*; IUCN-WCPA's Best Practice Protected Area Guidelines Series No 02; IUCN—The International Union for Conservation of Nature: Gland, Switzerland, 2018; p. 70.
89. Johnson, J.A.; Jones, S.K.; Wood, S.L.R.; Chaplin-Kramer, R.; Hawthorne, P.L.; Mulligan, M.; Pennington, D.; DeClerck, F.A. Mapping ecosystem services to human well-being: A toolkit to support integrated landscape management for the SDGs. *Ecol. Appl.* **2019**, *29*, e01985. [CrossRef] [PubMed]
90. Krkoška, L.E.; Harmáčková, Z.V.; Landová, L.; Pártl, A.; Vačkář, D. Assessing impact of land use and climate change on regulating ecosystem services in the Czech Republic. *Ecosyst. Health Sustain.* **2016**, *2*, e01210. [CrossRef]

sustainability

MDPI

Article

Payments for Watershed Ecosystem Services in the Eyes of the Public, China

Chunci Chen [1,2], Guizhen He [1,2,*] and Yonglong Lu [2,3]

[1] State Key Laboratory of Urban and Regional Ecology, Research Center for Eco-Environmental Sciences, Chinese Academy of Sciences, Beijing 100085, China; ccchen_st@rcees.ac.cn
[2] University of Chinese Academy of Sciences, Beijing 100049, China; yllu@rcees.ac.cn
[3] Key Laboratory of the Ministry of Education for Coastal Wetland Ecosystems, College of the Environment and Ecology, Xiamen University, Xiamen 361102, China
* Correspondence: gzhe@rcees.ac.cn; Tel.: +86-10-62844610; Fax: +86-10-62918177

Abstract: Recent decades have witnessed an increased development of schemes for payment for watershed ecosystem services (PWES). However, the public is usually excluded from PWES systems. Reliable and empirical research on PWES from the public perspective is scarce. Aiming to understand public perceptions, attitudes, participation, and responses to PWES, this paper investigated local residents living in the Yongding River watershed area through a face-to-face questionnaire survey. The results showed that the public had limited knowledge of PWES. The public was keen to be involved in PWES decision-making, but the current level of public participation was very low. Regarding willingness to pay (WTP) and willingness to accept (WTA), nearly 55% of the respondents supported paying the upstream residents for protecting the environment if they were beneficiaries in the downstream areas, while 85% of the respondents agreed to accept compensation if they were contributors to environmental improvement in the upstream areas. Although some of the respondents' daily lives were affected by the watershed environment, they were reluctant to pay, reflecting a sign of "free-riding". The regression analysis showed that public concerns, values, knowledge of PWES and the watershed environment, and demographic factors determined the WTP and WTA. The results of the contingent valuation method and opportunity costs method showed that the annual payment for headwater conservation areas (Huailai and Yanqing) ranged from CNY 245 to 718 million (USD 36 to 106 million). This study contributes to our limited knowledge and understanding of public sentiment and makes recommendations for improving public receptivity to PWES.

Keywords: public perceptions; payment for watershed ecosystem services; willingness to pay; willingness to accept; public participation

Citation: Chen, C.; He, G.; Lu, Y. Payments for Watershed Ecosystem Services in the Eyes of the Public, China. *Sustainability* 2022, 14, 9550. https://doi.org/10.3390/su14159550

Academic Editors: Kikuko Shoyama, Rajarshi Dasgupta and Ronald C. Estoque

Received: 30 May 2022
Accepted: 31 July 2022
Published: 3 August 2022

1. Introduction

Payment for ecosystem services (PES) can be identified as a voluntary transaction between ecosystem services users and providers that results in a conditional payment based on mutual agreement [1]. The PES seeks to fulfill both environmental and social goals such as environmental improvement and poverty reduction [2,3]. The National Forest Conservation Schemes in Costa Rica and Mexico and the Reducing Emissions from Deforestation and Degradation (REDD+) scheme have been widely reported as important PES practices [4,5]. There are currently more than 550 schemes around the world, with annual payments totaling over USD 36 billion [6]. In economic terms, PES can internalize the environmental externalities and create incentives for the landowner to ensure the provision of ecosystem services [7,8]. Kinzig et al. [9] highlighted that if PES is properly designed, it could eventually achieve win–win goals, creating special attractions in developing countries. Pagiola, Muradian, Wunder, and other prominent scholars have debated PES efficiency, arguing that too many goals could weaken PES efficiency and lead to trade-offs in ecosystem services [1,10,11]. Wunder et al. [12] indicated that environmental goals and non-environmental goals (welfare, equity, etc.) were both important. PES schemes are usually packaged as a series of social visions [13–15]. PES is considered to be the equivalent concept to ecological compensation in China and has been applied in different types of areas such as watersheds, forests, and wetlands [16–20].

Payment for watershed ecosystem services (PWES) is a voluntary scheme that offers cash payment or other benefits to ecosystem services providers in exchange for providing watershed services that help protect the watershed environment and water quality. PWES is a policy instrument to curb environmental deterioration, ensure water supply, prevent flooding, and maintain habitats for aquatic species and has received attention at both local and global scales over the past two decades [21–23]. There are currently 387 schemes, more than half of which are funded by governments, and nearly CNY 155.7 billion (USD 24.7 billion) was spent in 62 countries in 2015 [6]. PWES mechanisms are of three types. (1) Government-financed PWES refers to public finance rewards for landowners for promoting and protecting ecosystem services. As the government is the purchaser of the ecosystem services, the PWES scheme is implemented in a wide range of areas with uniform payment methods and standards. (2) Government- and user-financed PWES represents an institution that pools resources from multiple water users (private parties, government bodies, etc.) to pay upstream landowners for ensuring ecosystem services provision. (3) compliance PWES refers to quality trading and offsets. Ecosystem services (water) providers comply by paying upstream landowners for improving water quality in exchange for credits. China is the largest country to implement PWES; the total number of PWES schemes reached 69 in 2015, and the investment accounted for about 55% of total global investment. Water funds in South America presented the most rapid growth in the number of PWES schemes. More than 57 water funds were created in the past decade [6,24]. PWES schemes operate in different national contexts, with varying levels of government intervention through legislation and policy (Supplementary Table S1). Laws and supportive policies specific to PWES can establish regulatory oversight and provide funding for PWES schemes. The legal framework can influence the design and implementation of PWES. China's revised Environmental Protection Law (Supplementary Table S2) also emphasized the importance of PWES.

PWES has been high on China's environmental policy agenda since the late 1990s due to a series of floods and droughts. China's unique political and centralized power has allowed it to implement PES schemes at a tremendous scale and speed, improving ecological conditions and reshaping policy and the ecological landscape in a very short time [25]. The Sloping Land Conservation Program (SLCP) and the Natural Forest Conservation Program (NFCP) are the largest schemes in the world and play an important role in improving all ecosystem services (except for biodiversity habitat) and achieving additional purposes for rural development [26]. China dominates government-financed PWES schemes. The central government has issued guidelines for transboundary issues, encouraging provincial governments to agree on PWES schemes. In November 2020, the Regulation on Ecological Protection Compensation (Public Consultation Draft) was released, which is an incentive-based measure that adopts financial transfer payments or market transactions for environmental protection. Many provinces have established PWES schemes to protect watersheds. For example, Hebei Province issued a notice entitled Further Strengthening the Ecological Compensation for Water Quality at the Transboundary Section of Rivers. These practices have led to a series of provincial regulations regarding PWES in Hebei, Jiangxi, Zhejiang, and Jiangsu Provinces. A series of cases have been implemented for Xin'an River in Anhui and Zhejiang Provinces, Dongjiang River in Jiangxi Province, and Chaobai River in Beijing and Hebei Province [27–29]. There are four types of PWES in China. (1) Negotiated transactions among upstream governments and downstream governments. Under the jurisdiction of the higher-level government, upstream governments negotiate for and trade water resources, mostly using water rights trading, off-site development, and other ways to implement PWES schemes. (2) Jointly financed among governments. The upstream and downstream governments in the watershed co-finance schemes to protect the watershed environment, which has a two-way incentive for upstream and downstream governments. There are significant differences in economic development between upstream and downstream areas, and upstream environmental and economic behaviors affect the downstream demand for water resources. (3) Financial transfer payments among governments, including vertical transfers from the central government to local governments and horizontal transfers among local governments, which can fully mobilize the enthusiasm of local governments to strengthen environmental protection and pollution control. (4) Intergovernment compulsory withholding based on water-quality changes, mainly applied in transboundary PWES schemes. The government in the watershed must determine the payment subject and object and the water quality of the monitoring section. When the water quality is lower than the standard, the upstream government's compensation funds are withheld, and when the water quality is higher than the standard, the upstream government is rewarded with funds [30,31]. In 2019, the central financial budget for transfer payments for key ecological function zones was CNY 81.1 billion (USD 11.8 billion). China planned to establish a mechanism and full coverage of key fields and major ecological function zones by 2020. In 2021, China planned to establish PWES in the Yangtze River Economic Belt (YREB) to improve investment and payment mechanisms.

The rapid development of PWES schemes is not matched by sufficient theoretical research. PWES theory focuses on solving environmental externalities, including government regulation (based on a Pigouvian tax) and market incentives (based on the Coase theorem) [32], while the less frequent theoretical studies have addressed the cognitive and psychological dimensions of residents. The development of watershed conservation areas has long lagged behind that of non-conservation areas, and residents must make choices between improving livelihoods and protecting the environment. Determining whether PWES can achieve win–win goals (poverty alleviation and environmental improvement) requires a full study of residents' perceptions, values, participation, and behavior toward PWES [33]. However, residents are excluded from PWES management systems in most cases.

There are two forms of PWES management in China: (1) the vertical control form led by the central government and its functional departments and (2) the horizontal management form led by the local government and its departments (Figure 1). PWES management systems include PES formulation, water resources, environment, fisheries, natural resources, forests, wetlands, and finance. The National Development and Reform Commission (NDRC) is the decision-making authority for developing PWES policies, and other departments play supporting roles, e.g., financial departments manage funds, water resources departments manage water resources, and ecological environmental departments monitor water quality, etc.

PWES management in China is mainly led by government. Public participation in PWES schemes is still at the stage of conceptual advocacy, and no systematic participation mechanism has been formed. Previous studies illustrate strong personal connections to the environment, affected by identity, practical considerations, and livelihoods in the case of watershed protection, water shortages and reuse, risks of chemical parks and industry, the marine environment, climate change, etc. [34–40]. However, little attention has been paid to PWES from the public perspective. The existing studies are instructive but are typically focused on the assessment of PWES schemes, e.g., considering factors affecting public perceptions as part of the valuation of nonmonetary goods and services [41,42]. Regarding the gaps in PWES studies, we wish to examine public perspectives on PWES with a case study in the Yongding River watershed, a typical water-deficient region in northern China.

Note: NDRC, PDRC, MDRC, and CDRC: National, Provincial, Municipal, and County Development and Reform Commission, respectively; MWR: Ministry of Water Resources; RBC: River Basin Commission; WRMC: Water Resources Management Center; DPLR: Department of Policy, Law, and Regulation; DRLM: Department of River and Lake Management; RLCC River and Lake Conservation Center; DWRM: Department of Water Resources Management; PDWR, MDWR, and CDWR: Provincial, Municipal, and County Department of Water Resources, respectively; MEE: Ministry of Ecology and Environment; IB: Inspection Bureau; WEEMA: Watershed Ecological and Environmental Monitoring Administration; PDEE, MDEE, and CDEE: Provincial, Municipal, and County Department of Ecology and Environment, respectively; MARA: Ministry of Agriculture and Rural Affairs; BF: Bureau of Fisheries; PDARA, MDARA, and CDARA: Provincial, Municipal, and County Department of Agriculture and Rural Affairs, respectively; MNR: Ministry of Natural Resources; PDNR and NDNR: Provincial and Municipal Department of Natural Resources, respectively; NFGA, PFGA, MFGA, and CFGA: National, Provincial, Municipal, and County Forestry and Grassland Administration, respectively; WMD: Wetland Management Division; PWMD and MWMD: Provincial and Municipal Wetland Management Division, respectively; MF: Ministry of Finance; PDF: Provincial Department of Finance; MBF and CBF: Municipal and County Bureau of Finance, respectively.

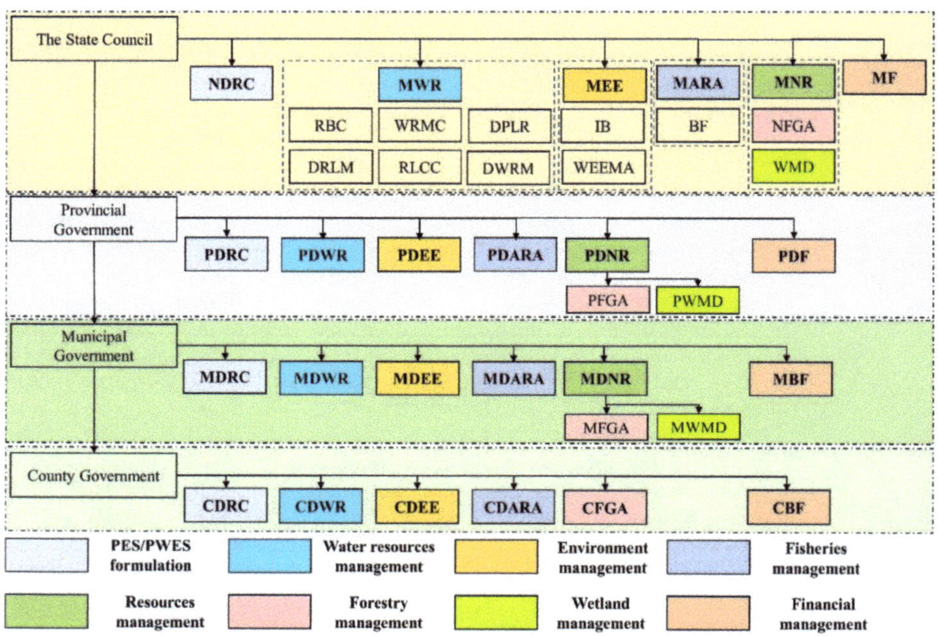

Figure 1. System of PWES-scheme-related departments and watershed management in China.

The Yongding River watershed has been affected by human activities for a long time. Reducing these pressures requires changes in the way residents interact with the watershed environment, which can be achieved through individual actions as well as governmental policies such as PWES schemes [43–45]. However, these efforts require public awareness of potential problems and support for mitigation actions [46,47]. To understand public perceptions, attitudes, and responses to PWES, survey methods are most often used [48–51]. In the past decade, research on public perceptions toward environmental issues has grown rapidly and has emerged as an important tool for policymakers and researchers. Public perceptions and responses to conservation schemes have recently been identified as emerging frontiers in research and have not yet been brought to the attention of decision-makers.

The aims of this study are therefore as follows: (1) to assess how the public views PWES, including the payment subject, object, method, and standard, and the sources of funds; (2) to identify the willingness to pay (WTP) and willingness to accept (WTA) of residents regarding PWES and the factors underlying their options; and (3) to evaluate how effective PWES is in implementation and how the public participates in PWES. To this end, face-to-face questionnaire surveys were conducted in the Yongding River watershed area. This study expands our limited knowledge and provides valuable information for policymakers via specific results on public knowledge and receptivity towards PWES. Combined with the background information from the public, the problems of PWES can be analyzed coherently so that specific recommendations can be made for improving PWES schemes. Section 2 introduces the research methods and data collection. Section 3 gives the results regarding public perceptions, attitudes, participation, and responses toward PWES. Section 4 presents the discussion and conclusions.

2. Methods and Data Collection

2.1. Study Area

The Yongding River watershed has a drainage area of 47,016 km^2. The Yongding River originates from the Sanggan River and the Yanghe River and flows into Guanting Reservoir, with a total length of 747 km (Figure 2a). The Yongding River and Guanting Reservoir were also important for safeguarding water resources and the watershed environment for the 2022 Winter Olympic Games. Both Hebei Province and Beijing City face a water crisis regarding development [52]. The competition for water resources between upstream (Hebei) and downstream (Beijing) is intensifying. To solve the conflicts, the governments have introduced some PWES schemes (e.g., PLDL) and policies in recent years, including water-quality target assessment and compulsory withholding of PWES in the Yongding River watershed (Figure 2b). In 2017, Beijing and Hebei jointly signed the Agreement on Compensation for Ecological Protection of the Water Conservation Area of the Chaobai River. The area was selected as a typical study watershed, considering the conflicts regarding water resources between the upstream (Hebei) and downstream (Beijing) areas and the high need for PWES.

2.2. Survey Design and Implementation

A survey was conducted through our designed questionnaire. The questionnaire was initially designed after data collection and preparation and was pre-tested in a series of rounds of face-to-face interviews with three experts and five residents in March 2017. The final questionnaire consisted of three parts (see the Appendix SA in the Supplementary Materials). Firstly, we investigated public values regarding watershed ecosystem services, public knowledge, and their preferences for the PWES subject, object, method, and standard. Secondly, we surveyed their WTP and WTA. Thirdly, we investigated public participation and the impact of PWES on livelihoods. The interviewees were residents living close to Yongding River and the Guanting Reservoir, based on assumption that these residents are stakeholders in, and are concerned about, the implementation of PWES. Four towns in Zhangjiakou City, Hebei Province and two towns in Yanqing District, Beijing were selected as survey sites (Figure 2). The survey sample size was determined based on the estimated overall proportion of simple random sampling, calculated at a 95% confidence level and with an absolute error of less than 4%. The required sample size was 601 for the survey locations with a total population of over 1 million (Supplementary Table S2).

A total of 637 interviewees were from these towns. The researchers randomly selected interviewees from among the residents of twenty villages in the four towns. Each respondent represented a family. Considering the high degree of social and demographic heterogeneity of this population, the sample was representative. The survey was carried out in March 2017 by six investigators, through face-to-face interviews. Interviewers took notes on binary, multiple-choice, and free-response answers.

2.3. Ecosystem Services and Opportunity Costs Evaluation

The contingent valuation method (CVM) was used for assessing the non-market value of ecosystem services. The CVM has its root in welfare economics and uses hypothetical, payment-based scenarios to trigger individuals' willingness to pay (WTP) or to accept compensation (WTA) for environmental improvement [53,54]. There remains a long-standing debate about the reliability of the CVM, with critics outlining several issues such as assumption bias and differences between WTP and WTA [55]. To avoid assumption bias, all investigators were trained prior to the survey to provide respondents with accurate descriptive information. The value of ecosystem services for the Yongding River watershed was estimated based on the following equations:

$$WTP = \sum_{i=1}^{k} AWTP_i \frac{n_i}{N} \tag{1}$$

$$WTA = \sum_{i=1}^{k} AWTA_i \frac{n_i}{N} \tag{2}$$

$$ES = \frac{N}{2}(WTP + WTA) \tag{3}$$

where WTP and WTA represent respondents' willingness to pay or to be paid and ES is the evaluation of ecosystem services by the public. AWTPi and AWTAi represent the WTP and WTA at level i, n_i is the corresponding number of respondents, and N is the total number of respondents.

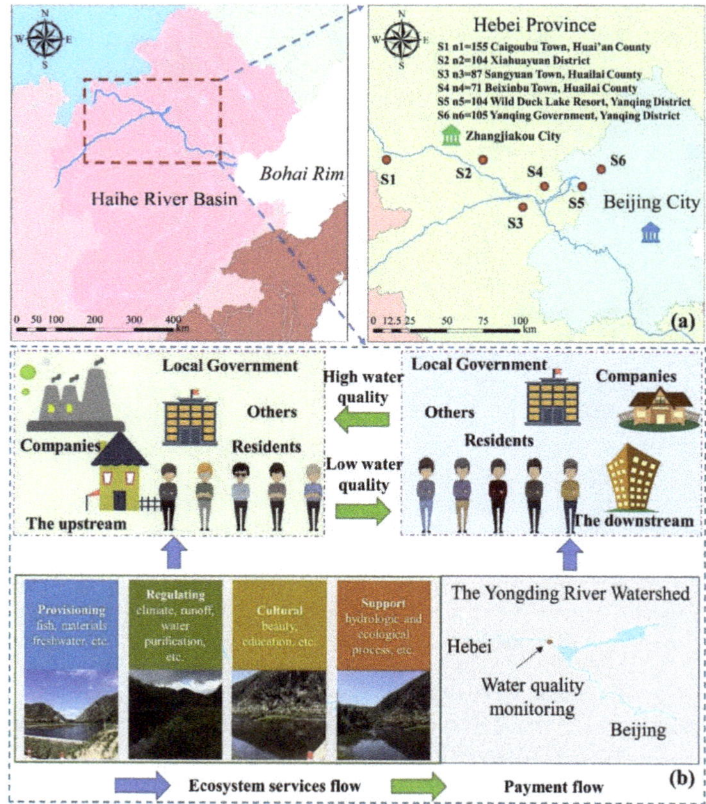

Figure 2. (**a**) Location of the Yongding River Basin and survey sites; (**b**) PWES mechanism in the Yongding River watershed, i.e., inter-government compulsory withholding pattern based on outbound water quality. Zhangjiakou provides water resources to Beijing. If the water quality of the monitored section does not reach the standard, Zhangjiakou's compensation funds are withheld. If the water quality of the monitored sector meets the standard, Zhangjiakou is rewarded with cash funds.

Then, we calculated the payment standard based on the opportunity costs model [56]. Huailai County and Yanqing District, where the Guanting Reservoir is located, are limiting industrial development to protect the watershed and therefore also limiting the livelihood improvement of residents. This is the opportunity cost in the upstream area and is mainly reflected in the difference in disposable income of residents in the restricted area (Huailai and Yanqing) and those in the unrestricted area (Zhangjiakou and Beijing). The annual opportunity costs are estimated based on the following equation:

$$OC = (T_C - T_S) \times N_t + (W_C - W_S) \times N_w \tag{4}$$

where OC is the opportunity cost of economic development, Tc and Wc are the disposable incomes of urban and rural residents in unrestricted areas, Ts and Ws are the disposable incomes of urban and rural residents in water-source areas, and Nt and Nw are the numbers of urban and rural residents.

2.4. Data Analysis

A total of 626 valid questionnaires were returned. Table 1 shows the social and demographic information of the respondents. Nearly 85% of respondents were under the university education level. The average age was about 44 years old. Nearly 65% lived within 10 kilometers of the watershed. Statistical analysis and data management were performed using SPSS 24.0. Descriptive analysis, correlation analysis, and logistic regression were applied to the survey data. The contingent valuation method was used to account for the value of the watershed.

Table 1. Socio-demographic information of respondents in the Yongding River watershed, China (n = 626).

Item	Status	Percentage (%)	Item	Status	Percentage (%)
Gender	Women	53.2		Farmer	48.4
	Men	46.8		Government employee	3.8
Average age	Years old	44.1		Enterprise employee	12.5
Education level	Primary school	17.6		Teacher	2.1
	Middle school	40.7		Researcher and scientist	0.6
	High school	26.5	Occupation	Self-employed	12.9
	University	14.7		Student	6.7
	Graduate	0.5		Retired	6.7
Annual income (thousand CNY)	≤10	23.8		NGO employee	0.6
	10.1–20	20.6		Media practitioner	0.5
	20.1–50	38.0		Others	5.1
	50.1–100	13.7	Family members	Average (persons)	3.6
	>100	3.9			

3. Results

3.1. Public Knowledge of Ecosystem Services and PWES

The Yongding River and the Guanting Reservoir hold great potential to provide ecosystem services for surrounding cities and communities. We listed ten types of ecosystem services, covering all four aspects of product provision services, regulation services, support services, and cultural services. Three-point scales were used for respondents to state their perceptions about the ecosystem services of the watershed, with scores of 1 to 3 representing unimportant to important. Average scores for these ten aspects were 2.87, 2.85, 2.87, 2.65, 2.78, 2.72, 2.77, 2.74, 2.70, and 2.75. The scoring for landscape value (entertainment, eco-tourism, etc.) was the lowest. The results indicated that about 68.5% to 87.8% of respondents thought that ecosystem services were important (Figure 3). Among them, soil and water conservation was perceived as an important ecosystem service by most respondents

(87.8%), followed by water purification (87.6% of the interviewees). However, the lowest number of respondents (68.5%) chose landscape value. The public often evaluates the importance of different ecosystem services for their purposes (e.g., recreation, scenery, etc.) rather than their ecological value. This poses a challenge for watershed protection due to potential differences between the ecosystem services protected and public preferences.

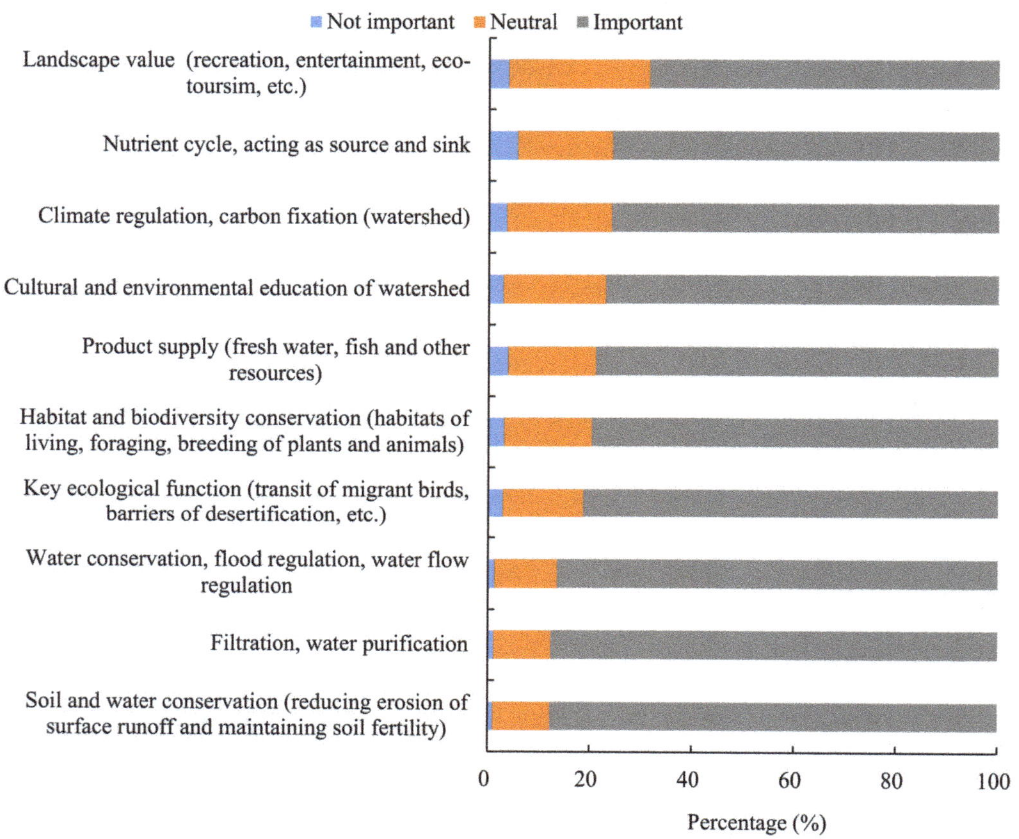

Figure 3. Public perceptions towards ecosystem services of the Yongding River, China (n = 626).

To measure how much knowledge the local people had acquired regarding PWES, we designed and asked questions about the concept and implementation of PWES. The results showed that most respondents (>70%) were unaware of watershed protection policies and PWES policies. Ordinal logistic regression analysis indicated that public occupation, place of residence, concerns, and education affected their knowledge of PWES (Table 2). The level of public concern about environmental issues contributed most to public knowledge about PWES. It is noteworthy that respondents in Beijing were more aware of PWES policies. When respondents were asked to prioritize management methods for solving environmental problems, they tended to indicate laws and regulations (35.1%), publicity, education, and information disclosure (21.4%), and economic methods (18.1%) as their top three priorities, with the least priority given to technical means. The results were within expectations, due to the existence of laws and regulations that can effectively reduce pollutant emissions [57]. Moreover, over half of the respondents (>50%) believed that they could play important roles in PWES decision-making and implementation. The Paddy Land to Dry Land (PLDL) scheme, which was implemented by Beijing, Zhangjiakou, and Chengde, was the PWES best known to respondents.

Table 2. Ordinal logistic regression analysis of public knowledge about PWES in the Yongding River watershed, China.

Factors	B	Sig.	Std. Error
Education level	0.206	0.043 *	0.102
Gender	−0.039	0.807	0.161
Age	0.012	0.730	0.036
Household income	0.119	0.161	0.085
Level of public concerns about environmental issues	0.403	0.000 **	0.085
Place of residence	−0.470	0.011 *	0.185
Relevance of occupation to environmental protection	0.109	0.038 *	0.053

Note: * $p < 0.05$; ** $p < 0.01$.

3.2. Public Perceptions of PWES

We designed and asked five questions about PWES, including payment subject, object, method, funding source, and standard. The responses are summarized in Figure 4. The PWES subject includes both the purchasers and providers of ecosystem services. The complexity of watershed environmental governance lies in the fact that not only does it involve externalities of residents' behavior but the watershed environment also has the characteristics of a public good when it exceeds a certain extent. The government assumes different roles when it manages PWES schemes in different ways. One way is when the government dominates the PWES and charges the beneficiaries and pays the providers of ecosystem services. The other way is when the beneficiaries of ecosystem services operate PWES in a market-oriented way, and the government undertakes tasks such as setting up the trading platform and releasing information. The majority (60.5%) of respondents thought that the central government should be the payment subject, while only 6% chose the beneficiaries, indicating that public perceptions toward PWES were that the government was the dominant player in PWES. Public opinion was not unanimous regarding payment objects, with 35.6% of respondents choosing poor and backward areas, followed by downstream areas suffering from pollution and damage, and only 19.2% choosing upstream areas. The public's expected payment methods were cash (54.0%) and ecological projects (25.4%). Regarding PWES funding sources, almost two thirds (66.3%) of respondents believed that funds should be raised by transfer payment from the central government to the local government. In response to the question on the payment standard, a range of responses was elicited. Nearly 36.1% of respondents thought the payment standard should meet personal basic needs, 23.2% considered that it should reach the average income level of the watershed, and the other 20.2% regarded the standard as covering direct economic losses caused by environmental protection. Respondents reported that the compensation funds should be used for the comprehensive improvement of the water environment and floating debris removal (19.3%), rural waste treatment and removal system construction (16.8%), and water-saving facilities construction (15.6%). Respondents believed that three main aspects of the benefits brought by PWES schemes were ensuring drinking water safety, protecting the environment and improving the quality of life, and guaranteeing food production security.

3.3. Public WTA and WTP for PWES Standard

Assessment of standards is one of the difficult issues in PWES research. Many methods have been proposed and applied to determine PWES standards, including evaluation of ecosystem services, opportunity cost, WTP and WTA methods, etc. [58]. When the payment of the PWES is greater than the opportunity cost and less than the ecosystem service value, it creates an effective incentive for both purchasers and providers of ecosystem services [59]. Although the value of watershed ecosystem services is difficult to measure, the existing value and social benefits of PWES schemes can be determined by applying the CVM [60,61].

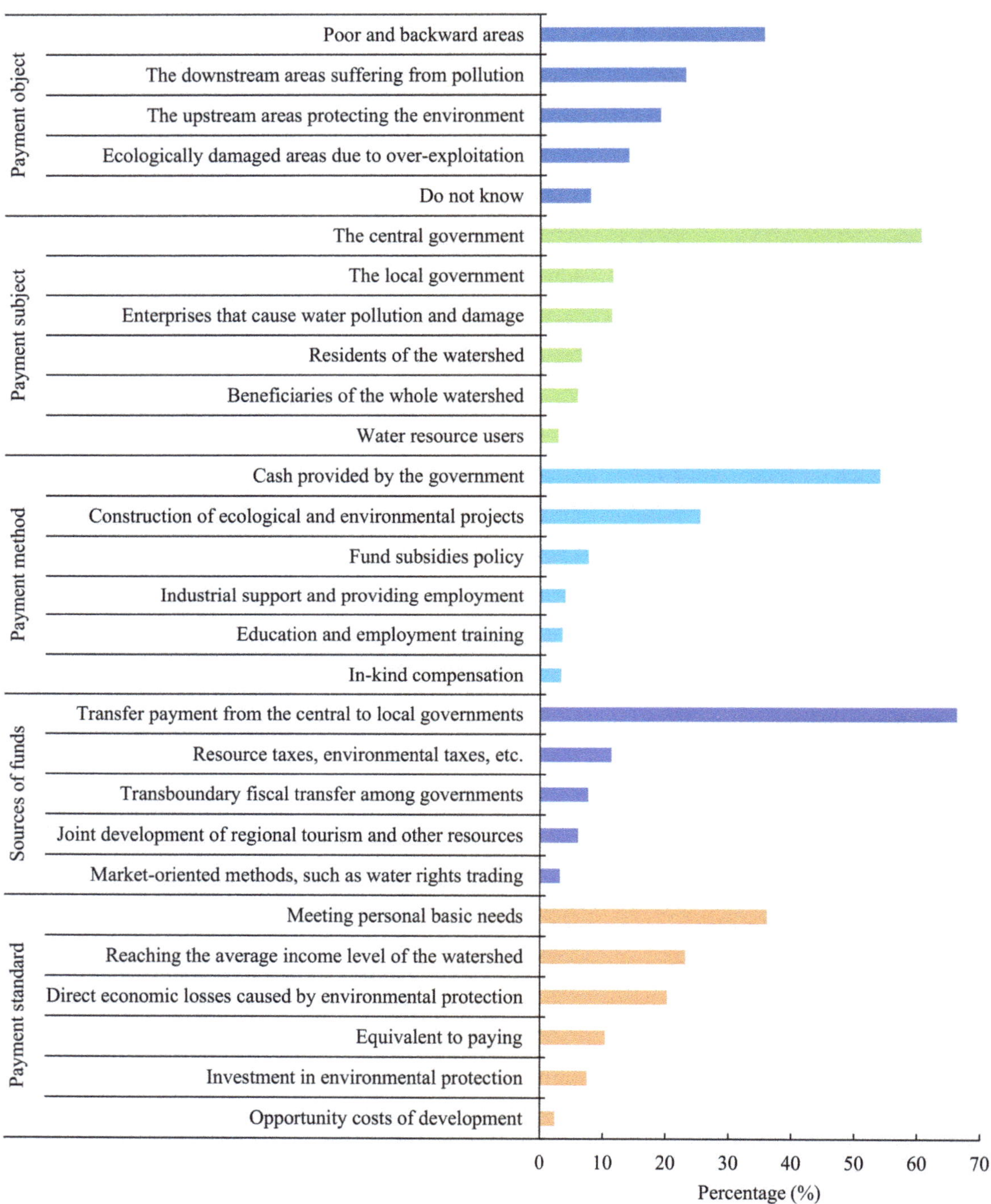

Figure 4. Public perceptions of PWES in the Yongding River watershed, China (n = 626).

Firstly, we investigated respondents' WTP and WTA. Over half of them (55.0%) indicated that they were willing to pay for environmental improvement upstream if they lived downstream and were the beneficiaries of the environmental improvement. Surprisingly, using living distance as a variable, the analysis showed that respondents living within 5 km of the Yongding River were significantly less willing to pay than those living further away ($p < 0.05$). Respondents preferred to pay in cash (43.0%), voluntary contributions (22.1%), and voluntary work (17.2%). Of the 282 respondents who were unwilling to pay, nearly 73% cited their low household income as the main reason. Some respondents (12.8%) said that individual payments were not sufficient, and the government should take responsibility for improving the watershed environment and funding watershed ecological restoration. The majority (85%) of respondents indicated that they were willing to accept compensation if they lived upstream and were contributors to environmental improvement. Nearly 70% expected cash compensation, and 15.2% wanted to be offered job opportunities. Conditional cash compensation is theoretically the best form of incentive; however, indirect non-cash compensation methods are more commonly used [62]. Although some of the interviewees felt that the watershed environment affected their daily lives, they were reluctant to pay for environmental goods, reflecting a sign of "free-riding". The results showed that the WTP and WTA were very different, with WTP and WTA mainly distributed in the interval of less than CNY 100 (USD 14.8) and over CNY 300 (USD 44.4), respectively (Supplementary Figure S1). Moreover, we found that the average annual values for WTP and WTA were CNY 96 (USD 14.2) and CNY 633 (USD 93.8), respectively. WTA was 6.6 times larger than WTP. The reasons for the difference between WTP and WTA were not only the social background of the residents but also their insufficient awareness of PWES. It is easy for them to overestimate their losses and underestimate their benefits, which leads to a higher WTA than WTP. Moreover, some studies on PWES point out that the tendency to punish ecologically damaging behaviors may also be a cause [63–66].

Then, we applied the CVM method to account for PWES standards, and the results showed that the annual payment was a total of CNY 245 million (USD 36.3 million) for Huailai County and Yanqing District, with CNY 131 million (USD 19.4 million) for Huailai County and CNY 114 million (USD 16.9 million) for Yanqing District (Figure 5). The reservoir area, as well as the upstream area of the Yongding River watershed had limited industrial development, in order to protect the environment; however, nearly 48% of respondents were not aware of that. We estimated the PWES standard under different scenarios based on opportunity costs. The results showed that the annual payments for Huaila County and Yanqing District were CNY 277 million (USD 41.0 million) and CNY 279 million (USD 41.3 million), based on the assumption of reaching the average income of other similar areas within the watershed. The result was close to the PLDL scheme (Figure 5). In 2018, Beijing paid CNY 322 million (USD 47.7 million) to Zhangjiakou City and Chengde City in Hebei Province to protect the Miyun Reservoir. The annual payments for Huaila County and Yanqing District were CNY 436 million (USD 64.6 million) and CNY 718 million (USD 106.3 million), based on the assumption of reaching the economic development level of other similar areas within the watershed. This result was significantly higher than the opportunity costs based on average income and the CVM-based assessment of ecosystem services.

Thirdly, we analyzed factors influencing WTP and WTA based on binary logistic regression. Public concerns, knowledge, perceptions, and demographic factors regarding PWES and environmental issues were incorporated into the model (Table 3). The results showed that higher education levels were associated with lower WTP and no significant change in WTA. Male respondents were more reluctant to receive compensation than females. The more they were concerned about environmental issues and the higher their income, the more they wanted to pay and the less they wanted to be compensated. In addition, respondents living in Hebei were willing to pay less than those in Beijing, while this was the opposite for WTA. Respondents who were more aware of PWES were more willing to pay, as were respondents who believed the environment had improved.

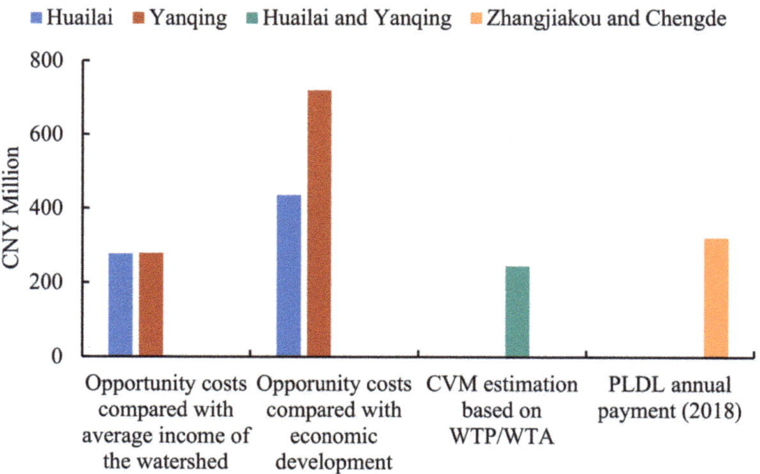

Figure 5. The PWES standard for the headwater conservation areas.

Table 3. BLR analysis of WTP in the Yongding River watershed, China.

Factors	WTP			WTA		
	B	Sig	Exp (B)	B	Sig	Exp (B)
Education level	−0.406	0.013 *	0.667	0.302	0.223	1.353
Gender	0.104	0.644	1.109	−1.016	0.004 **	0.362
Household income level	0.081	0.023 *	1.804	−0.377	0.036 *	0.686
Environmental improvement	0.771	0.000 **	2.162	0.442	0.064	1.556
Level of public concerns about environmental issues	0.256	0.043 *	1.292	−0.484	0.018 *	1.623
Place of residence	1.090	0.038 *	2.663	−1.312	0.046 *	0.269
Public knowledge about PWES	0.169	0.034 *	1.184	0.247	0.269	1.281

Note: * $p < 0.05$; ** $p < 0.01$.

3.4. Public Participation in the PWES Implementation

Public participation and involvement are prerequisites for promoting and developing China's water sustainability action [67]. This section of the survey was concerned with some issues regarding the PWES implementation. The overall responses on participation in some schemes were not very salient, with a total number of 90 responses, accounting for 14.4 % of respondents. Surprisingly, over half (55.8%) thought they had no direct relationship with watershed management. More than 37% of respondents thought their occupations had little relevance to environmental protection. Table 4 shows that public participation was mainly in the "Sloping Land Conversion Program". Only 41.1% indicated that officials (such as government staff, village officials, etc.) consulted their opinions and suggestions before and during the implementation phase. However, most respondents (77.7%) were still willing to participate in PWES schemes.

More than 90% thought they would like to support these schemes for improving the environment. The PWES method was mainly cash, according to the responses. Among respondents who participated in PWES schemes, almost all of them thought the PWES standard was not high. The current PWES schemes in the Yongding River watershed have not been effective in improving the income and living standards of the residents and the environment (Table 4). A majority of respondents would consider environmental impact and water consumption when choosing to engage in other occupations if the government stopped compensation schemes. Only 36.7% said they would continue to be engaged in agricultural production. Overall, these results indicated that public perceptions could generate impacts on behaviors and choices following participation in PWES schemes.

Table 4. Public participation in PWES schemes.

Question	Dominant Opinion	Percentage (%)
PWES scheme participation	SLCP	78.9
Whether to consult respondents	No	58.9
Desire to engage in decision-making	Yes	77.7
Whether to support or not	Yes	90.0
Reason for support	Can improve the environment	60.0
PWES methods	Cash or subsidy	92.2
PWES standard	Ordinary	60.0
Household income changes	Unchanged	72.2
Environmental improvement	Little improvement	48.9
Living standard improvement	Ordinary	46.7

4. Discussion and Conclusions

PWES schemes and policies have been implemented for a long time, since the Chinese government realized that deforestation posed serious water-quality and flooding threats [68,69]. This study reviewed the evolution and mechanisms of PES and PWES and assessed how the public viewed PWES and the factors underlying WTP and WTA. In the following section, we summarize the main findings of the above analysis.

4.1. WTP and WTA for PWES Standard

Watershed ecosystem services include water purification, soil and water conservation, flooding control, etc., which are difficult to measure, while the existing value of the social benefits can be determined by applying the CVM [70,71]. Many watershed-specific studies have also measured the public's WTP and WTA and calculated PWES standards based on the CVM. The study in the Heihe River watershed area showed that the residents had a strong willingness to participate in ecological management, and their WTP ranged from CNY 187 to 226 per year [72]. The WTP of the residents in the Weihe River watershed area was CNY 624 per year and the annual non-market value of watershed ecosystem services in Shaanxi Province was CNY 868 million [73]. The residents' WTP ranged from CNY 60 to 90 per year for the South–North Water Transfer Project in Zhengzhou. The residents' income level, education level, age, and preference had a significant impact on their WTP. Respondents thought that they had the right to use clean water, and their WTP had a strong self-interest motive. Changes in water quantity and quality affected their WTP [74]. Some of the residents thought that the watershed environment affected their daily lives, but they were reluctant to pay for public goods, reflecting a "free-riding" phenomenon that resulted in WTP being lower than WTA. A certain degree of wealth disparity between purchasers and providers of ecosystem services can help to increase the WTP/WTA ratio and potentially overcome the barriers posed by transaction costs. Thus, PWES schemes have a greater likelihood of success when they are established in an environment where there is a wealth disparity between purchasers and providers [64]. The PWES scheme should incorporate the non-market value of watershed ecosystem services, and the PWES standard should be explored in conjunction with the WTP and WTA.

4.2. Residents' Livelihoods in PWES Implementation

As the most important form of compensatory conservation, PWES has become a central component of the environmental protection strategy in China [75]. The upstream areas provide ecosystem services to the downstream areas and the downstream areas provide funds to the upstream areas. The better the water quality, the more the downstream areas pay to the upstream areas, and the worse the water quality, the more the upstream areas pay to the downstream areas. This is one of the most widespread forms of PWES practice.

Therefore, PWES with this two-sided feature is an incentive instrument to address water-use conflicts and transboundary pollution problems [76–78]. However, it has also increased the financial burden and subsequent social responsibility of the upstream areas because the opportunity costs of development restrictions in the upstream areas often outweigh the funds available from the PWES scheme. Nonetheless, the PWES funds are mostly used for pollution control and river regulation rather than for improving residents' livelihoods, which is the main reason for the low motivation for public participation in PWES implementation. Based on the survey, the public thought the current PWES schemes implemented in the Yongding River watershed were not effective in improving their livelihoods, incomes, and the watershed environment. Therefore, the PWES implementation should be enhanced so that both water quality and residents' livelihoods can be improved. Zheng et al. [29] pointed out that the PLDL scheme achieved a win–win goal, increasing water quantity while reducing nitrogen and phosphorus input. The benefits outweighed the costs for both the upstream and the downstream areas. PWES implementation requires attention to farmers' behaviors to maintain a win–win situation. The PLDL implementation also found that the input of pesticides and fertilizers to farmland increased significantly. We investigated the potential behavioral changes of residents after the cessation of the PWES. The results showed that many of them would probably resume their rural jobs, but most of them would consider environmental impacts and water consumption when choosing different occupations. This suggested that their behavior may change as a result of their involvement in PWES schemes. Therefore, PWES implementation should be more concerned with residents' livelihoods and behaviors. Future watershed management should pay more attention to the application of PWES schemes based on shared responsibilities between upstream and downstream areas [79].

4.3. Implication for Policy and Watershed Management

Over the past two decades, China has promoted PWES mechanisms to internalize environmental externalities from human activities in the context of integrated water resources management [17]. PWES is closely related to rural development and farmers' interests. To protect and restore ecosystems, watershed protection areas are inevitably constrained in terms of industrial and economic development. Therefore, PWES schemes play important roles in the sustainable development of poverty-removing areas, narrowing the gap between urban and rural development. PWES implementation, like other major ecological projects, is an important means for enhancing and improving ecosystem services. However, some types of ecosystem services (e.g., water-quality improvement) often come at the expense of other ecosystem services (e.g., agricultural production). Such trade-offs in ecosystem services are widespread and manifest at different spatial and temporal scales, thus creating uncertainty in the realization of ecological benefits [33]. Therefore, they should be considered in PWES management. PWES implementation is close to watershed management, which often transcends administrative jurisdictions. PWES schemes such as SLCP and PLDL are important components of sustainable watershed management and help to resolve water conflicts [80]. Moreover, transboundary PWES schemes represent a necessary strategy for horizontal cooperation between local governments for managing watersheds. There is a need to develop a transboundary PWES scheme in consultation with governments within watersheds and to take into account the interconnections between upstream and downstream areas, i.e., the ecological and social complexity [81].

The improvement of the PWES mechanism in China is inseparable from public participation. However, our analysis showed that the public was usually excluded from the PWES system, and public participation was at a low level. Members of the public can only passively accept PWES schemes and have little voice as participants, even though they believe they can play a large role in PWES development and implementation. Therefore, PWES management should be more inclusive, incorporating the public into the management system and providing the public with information on watershed governance, thus motivating people to participate in PWES management. Due to the weak ability and awareness of the public regarding access to information, certain strategies are still needed for policy advocacy and knowledge dissemination related to PWES [19]. The government plays an irreplaceable role in PWES implementation, and the public may place higher demands on the government's transparency and execution. Therefore, the government should improve public participation, disclose watershed governance information, and enhance transparency and fairness, thereby increasing public trust in the relevant authorities. These findings reveal the need for public participation in PWES, and the same approach can be applied to other regions of China. This study explored public perceptions, attitudes, participation, responses, and factors affecting the WTP and WTA, which were important for the future design and implementation of PWES and for providing practical insights into understanding public perceptions in other watershed areas. The results showed that the public had limited knowledge of watershed protection and PWES policies and was unaware of PWES mechanisms. Some of the respondents did not even know that upstream areas suffered economic loss as a result of protecting the environment. Therefore, the government and the media should take responsibility for disseminating information related to PWES policies. In addition, the government should help build public confidence in PWES via training, hearings, reports, and other means. The residents perceived the PWES standard as low, and their perceptions were closely related to their livelihoods. Although cash was the first choice for WTP and WTA, some would prefer to choose other ways to pay compensation (e.g., voluntary labor) or receive compensation (e.g., employment opportunities). Differentiated PWES standards and diversified PWES methods should be established in the Yongding River watershed area, considering residents' livelihoods, especially the affordability for low-income households and the preferred payment methods of the residents. Members of the public were willing to participate in decision-making regarding PWES and believed that they could play a very important role; however, the current level of public participation is very low. To mobilize public participation in watershed protection and PWES schemes, the government should make efforts to broaden the ways the public can participate and gradually institutionalize them. Public attitudes can take different forms such as active support or passive acceptance by different stakeholders and should be dealt with within the context of environmental psychology and social-science-oriented work.

Though some limitations exist, we intend to work further on this topic in the future. Firstly, the data were collected from two cities, with a relatively small sample size. Although the whole basin was not covered and the number of respondents was small, the respondents included a wide diversity of participants. More cases, covering the whole basin and giving a larger sample size, could contribute to better accuracy of the conclusion. Secondly, variables such as procedural justice, benefit, cost, and public norms are not within the scope of this study, which may also affect the full understanding of public attitudes and their WTP and WTA. In future work, we should include the above factors in the regression analysis. Thirdly, the scope of this study is limited to local responses at a specific time. Further studies will be conducted to examine the trend of support over time and the impacts of PWES on the livelihoods of the residents within the watershed area.

Supplementary Materials: The following supporting information can be downloaded at: https://www.mdpi.com/article/10.3390/su14159550/s1. Figure S1: The public's expected methods regarding WTP and WTA in the Yongding River watershed, China; Figure S2: Public willingness to pay and willingness to accept in the Yongding River watershed, China; Table S1: International legislative experience in PWES; Table S2: Relevant PWES policies and laws in China; Table S3: Sample size determination; Appendix SA: Questionnaire.

Author Contributions: Conceptualization, G.H.; methodology, C.C.; formal analysis, C.C.; investigation, G.H.; data collection, G.H.; writing—original draft preparation, C.C.; writing—review and editing, G.H. and Y.L.; supervision, G.H. and Y.L.; funding acquisition, G.H. All authors have read and agreed to the published version of the manuscript.

Funding: This research was funded by the National Natural Science Foundation of China (grant numbers 41877529 and U21A2041), the International Partnership Program of the Chinese Academy of Sciences (121311KYSB20190029), and Guangxi Special Fund Project on Innovation-Driven Development (Gui Ke AA20161004-04).

Institutional Review Board Statement: Not applicable.

Informed Consent Statement: Not applicable.

Data Availability Statement: Not applicable.

Conflicts of Interest: The authors declare no conflict of interest.

References

1. Wunder, S. Revisiting the concept of payments for environmental services. *Ecol. Econ.* **2015**, *117*, 234–243. [CrossRef]
2. Benra, F.; Nahuelhual, L.; Felipe-Lucia, M.; Jaramillo, A.; Jullian, C.; Bonn, A. Balancing ecological and social goals in PES design—Single objective strategies are not sufficient. *Ecosyst. Serv.* **2022**, *53*, 101385. [CrossRef]
3. Deng, X.; Yan, S.; Song, X.; Li, Z.; Mao, J. Spatial targets and payment modes of win–win payments for ecosystem services and poverty reduction. *Ecol. Indic.* **2022**, *136*, 108612. [CrossRef]
4. Collins, A.C.; Grote, M.N.; Caro, T.; Ghosh, A.; Thorne, J.; Salerno, J.; Mulder, M.B. How community forest management performs when REDD+ payments fail. *Environ. Res. Lett.* **2022**, *17*, 034019. [CrossRef]
5. Hook, A.; Laing, T. The politics and performativity of REDD+ reference levels: Examining the Guyana-Norway agreement and its implications for 'offsetting' towards 'net zero'. *Environ. Sci. Policy* **2022**, *132*, 171–180. [CrossRef]
6. Salzman, J.; Bennett, G.; Carroll, N.; Goldstein, A.; Jenkins, M. The global status and trends of Payments for Ecosystem Services. *Nat. Sustain.* **2018**, *1*, 136–144. [CrossRef]
7. Engel, S.; Pagiola, S.; Wunder, S. Designing payments for environmental services in theory and practice: An overview of the issues. *Ecol. Econ.* **2008**, *65*, 663–674. [CrossRef]
8. Hayes, T.; Murtinho, F.; Wolff, H.; López-Sandoval, M.F.; Salazar, J. Effectiveness of payment for ecosystem services after loss and uncertainty of compensation. *Nat. Sustain.* **2022**, *5*, 81–88. [CrossRef]
9. Kinzig, A.P.; Perrings, C.; Chapin, F.S.; Polasky, S.; Smith, V.K.; Tilman, D.; Turner, B.L. Paying for Ecosystem Services Promise and Peril. *Science* **2011**, *334*, 603–604. [CrossRef]
10. Pagiola, S. Payments for environmental services in Costa Rica. *Ecol. Econ.* **2008**, *65*, 712–724. [CrossRef]
11. Muradian, R.; Arsel, M.; Pellegrini, L.; Adaman, F.; Aguilar, B.; Agarwal, B.; Corbera, E.; de Blas, D.E.; Farley, J.; Froger, G.; et al. Payments for ecosystem services and the fatal attraction of win–win solutions. *Conserv. Lett.* **2013**, *6*, 274–279. [CrossRef]
12. Wunder, S.; Brouwer, R.; Engel, S.; Ezzine-de-Blas, D.; Muradian, R.; Pascual, U.; Pinto, R. From principles to practice in paying for nature's services. *Nat. Sustain.* **2018**, *1*, 145–150. [CrossRef]
13. Dasgupta, S.; Hamilton, K.; Pagiola, S.; Wheeler, D. Environmental Economics at the World Bank. *Rev. Environ. Econ. Policy* **2008**, *2*, 4–25. [CrossRef]
14. Vincent, J.R. Microeconomic Analysis of Innovative Environmental Programs in Developing Countries. *Rev. Environ. Econ. Policy* **2010**, *4*, 221–233. [CrossRef]
15. Shang, W.; Gong, Y.; Wang, Z.; Stewardson, M.J. Eco-compensation in China: Theory, practices and suggestions for the future. *J. Environ. Manag.* **2018**, *210*, 162–170. [CrossRef]
16. Pei, S.; Zhang, C.X.; Liu, C.L.; Liu, X.N.; Xie, G.D. Forest ecological compensation standard based on spatial flowing of water services in the upper reaches of Miyun Reservoir, China. *Ecosyst. Serv.* **2019**, *39*, 11. [CrossRef]
17. Lin, J.; Huang, J.; Hadjikakou, M.; Huang, Y.; Li, K.; Bryan, B.A. Reframing water-related ecosystem services flows. *Ecosyst. Serv.* **2021**, *50*, 101306. [CrossRef]
18. Feng, Y.; Zhu, A.; Liu, W. Quantifying inter-regional payments for watershed services on the basis of green ecological spillover value in the Yellow River Basin, China. *Ecol. Indic.* **2021**, *132*, 108300. [CrossRef]
19. Li, X.; Wang, Y.; Yang, R.; Zhang, L.; Zhang, Y.; Liu, Q.; Song, Z. From "blood transfusion" to "hematopoiesis": Watershed eco-compensation in China. *Environ. Sci. Pollut. Res.* **2022**, *33*, 49583–49597. [CrossRef]
20. Zhu, Q.; Tran, L.T.; Wang, Y.; Qi, L.; Zhou, W.; Zhou, L.; Yu, D.; Dai, L. A framework of freshwater services flow model into assessment on water security and quantification of transboundary flow: A case study in northeast China. *J. Environ. Manag.* **2022**, *304*, 114318. [CrossRef]
21. Farley, J.; Costanza, R. Payments for ecosystem services: From local to global. *Ecol. Econ.* **2010**, *69*, 2060–2068. [CrossRef]
22. Deng, Y.; Brombal, D.; Farah, P.D.; Moriggi, A.; Critto, A.; Zhou, Y.; Marcomini, A. China's water environmental management towards institutional integration. A review of current progress and constraints vis-a-vis the European experience. *J. Clean. Prod.* **2016**, *113*, 285–298. [CrossRef]
23. Retallack, M. The intersection of economic demand for ecosystem services and public policy: A watershed case study exploring implications for social-ecological resilience. *Ecosyst. Serv.* **2021**, *50*, 101322. [CrossRef]

24. Bremer, L.L.; Auerbach, D.A.; Goldstein, J.H.; Vogl, A.L.; Shemie, D.; Kroeger, T.; Nelson, J.L.; Benítez, S.P.; Calvache, A.; Guimarães, J.; et al. One size does not fit all: Natural infrastructure investments within the Latin American Water Funds Partnership. *Ecosyst. Serv.* **2016**, *17*, 217–236. [CrossRef]

25. Liu, J.; Li, S.; Ouyang, Z.; Tam, C.; Chen, X. Ecological and socioeconomic effects of China's policies for ecosystem services. *Proc. Natl. Acad. Sci. USA* **2008**, *105*, 9477–9482. [CrossRef]

26. Ouyang, Z.; Zheng, H.; Xiao, Y.; Polasky, S.; Liu, J.; Xu, W.; Wang, Q.; Zhang, L.; Rao, E.; Jiang, L.; et al. Improvements in ecosystem services from investments in natural capital. *Science* **2016**, *352*, 1455–1459. [CrossRef]

27. Feng, D.; Wu, W.; Liang, L.; Li, L.; Zhao, G. Payments for watershed ecosystem services: Mechanism, progress and challenges. *Ecosyst. Health Sustain.* **2018**, *4*, 13–28. [CrossRef]

28. Pan, X.; Xu, L.; Yang, Z.; Yu, B. Payments for ecosystem services in China: Policy, practice, and progress. *J. Clean. Prod.* **2017**, *158*, 200–208. [CrossRef]

29. Zheng, H.; Robinson, B.E.; Liang, Y.-C.; Polasky, S.; Ma, D.-C.; Wang, F.-C.; Ruckelshaus, M.; Ouyang, Z.-Y.; Daily, G.C. Benefits, costs, and livelihood implications of a regional payment for ecosystem service program. *Proc. Natl. Acad. Sci. USA* **2013**, *110*, 16681–16686. [CrossRef]

30. Jiang, K.; Zhang, X.; Wang, Y. Stability and influencing factors when designing incentive-compatible payments for watershed services: Insights from the Xin'an River Basin, China. *Mar. Policy* **2021**, *134*, 104824. [CrossRef]

31. Wang, J.F.; Hou, C.B. Study on implementation framework and compensation pattern of basin ecological compensation mechanism in China: From the perspective of compensation funds source. *China Popul. Resour. Environ.* **2013**, *23*, 23–29. (In Chinese)

32. Diswandi, D. A hybrid Coasean and Pigouvian approach to Payment for Ecosystem Services Program in West Lombok: Does it contribute to poverty alleviation? *Ecosyst. Serv.* **2017**, *23*, 138–145. [CrossRef]

33. Li, R.; Zheng, H.; O'Connor, P.; Xu, H.; Li, Y.; Lu, F.; Robinson, B.E.; Ouyang, Z.; Hai, Y.; Daily, G.C. Time and space catch up with restoration programs that ignore ecosystem service trade-offs. *Sci. Adv.* **2021**, *7*, eabf8650. [CrossRef] [PubMed]

34. Garcia-Cuerva, L.; Berglund, E.Z.; Binder, A.R. Public perceptions of water shortages, conservation behaviors, and support for water reuse in the U.S. *Resour. Conserv. Recycl.* **2016**, *113*, 106–115. [CrossRef]

35. Gu, Q.; Chen, Y.; Pody, R.; Cheng, R.; Zheng, X.; Zhang, Z. Public perception and acceptability toward reclaimed water in Tianjin. *Resour. Conserv. Recycl.* **2015**, *104*, 291–299. [CrossRef]

36. He, G.; Boas, I.J.C.; Mol, A.P.J.; Lu, Y. What drives public acceptance of chemical industrial park policy and project in China? *Resour. Conserv. Recycl.* **2018**, *138*, 1–12. [CrossRef]

37. He, G.; Chen, C.; Zhang, L.; Lu, Y. Public perception and attitude towards chemical industry park in Dalian, Bohai Rim. *Environ. Pollut.* **2018**, *235*, 825–835. [CrossRef]

38. Zhang, H.; Pang, Q.; Hua, Y.; Li, X.; Liu, K. Linking ecological red lines and public perceptions of ecosystem services to manage the ecological environment: A case study in the Fenghe River watershed of Xi'an. *Ecol. Indic.* **2020**, *113*, 106218. [CrossRef]

39. Gelcich, S.; Buckley, P.; Pinnegar, J.K.; Chilvers, J.; Lorenzoni, I.; Terry, G.; Guerrero, M.; Castilla, J.C.; Valdebenito, A.; Duarte, C.M. Public awareness, concerns, and priorities about anthropogenic impacts on marine environments. *Proc. Natl. Acad. Sci. USA* **2014**, *111*, 15042–15047. [CrossRef]

40. Lee, T.M.; Markowitz, E.M.; Howe, P.D.; Ko, C.-Y.; Leiserowitz, A.A. Predictors of public climate change awareness and risk perception around the world. *Nat. Clim. Change* **2015**, *5*, 1014–1020. [CrossRef]

41. Bremer, L.L.; Farley, K.A.; Lopez-Carr, D. What factors influence participation in payment for ecosystem services programs? An evaluation of Ecuador's SocioParamo program. *Land Use Policy* **2014**, *36*, 122–133. [CrossRef]

42. Chen, X.; Lupi, F.; He, G.; Ouyang, Z.; Liu, J. Factors affecting land reconversion plans following a payment for ecosystem service program. *Biol. Conserv.* **2009**, *142*, 1740–1747. [CrossRef]

43. Li, W.; Liu, J.; Li, D. Getting their voices heard: Three cases of public participation in environmental protection in China. *J. Environ. Manag.* **2012**, *98*, 65–72. [CrossRef]

44. Carr, G. Stakeholder and public participation in river basin management—An introduction. *WIREs Water* **2015**, *2*, 393–405. [CrossRef]

45. Zhang, Z.; Balay, J.W. How Much is Too Much?: Challenges to Water Withdrawal and Consumptive Use Management. *J. Water Resour. Plan. Manag.* **2014**, *140*, 01814001. [CrossRef]

46. Liu, J.; Ouyang, Z.; Miao, H. Environmental attitudes of stakeholders and their perceptions regarding protected area-community conflicts: A case study in China. *J. Environ. Manag.* **2010**, *91*, 2254–2262. [CrossRef]

47. Ross, V.L.; Fielding, K.S.; Louis, W.R. Social trust, risk perceptions and public acceptance of recycled water: Testing a social-psychological model. *J. Environ. Manag.* **2014**, *137*, 61–68. [CrossRef]

48. Buijs, A.E. Public support for river restoration. A mixed-method study into local residents' support for and framing of river management and ecological restoration in the Dutch floodplains. *J. Environ. Manag.* **2009**, *90*, 2680–2689. [CrossRef]

49. Khan, I.; Zhao, M. Water resource management and public preferences for water ecosystem services: A choice experiment approach for inland river basin management. *Sci. Total Environ.* **2019**, *646*, 821–831. [CrossRef]

50. Li, X.; Tilt, B. Public engagements with smog in urban China: Knowledge, trust, and action. *Environ. Sci. Policy* **2019**, *92*, 220–227. [CrossRef]

51. Gao, X.; Shen, J.; He, W.; Sun, F.; Zhang, Z.; Guo, W.; Zhang, X.; Kong, Y. An evolutionary game analysis of governments' decision-making behaviors and factors influencing watershed ecological compensation in China. *J. Environ. Manag.* **2019**, *251*, 109592. [CrossRef]
52. Jiang, B.; Wong, C.P.; Lu, F.; Ouyang, Z.; Wang, Y. Drivers of drying on the Yongding River in Beijing. *J. Hydrol.* **2014**, *519*, 69–79. [CrossRef]
53. Venkatachalam, L. The contingent valuation method: A review. *Environ. Impact Assess. Rev.* **2004**, *24*, 89–124. [CrossRef]
54. Lo, A.Y.; Jim, C.Y. Protest response and willingness to pay for culturally significant urban trees: Implications for Contingent Valuation Method. *Ecol. Econ.* **2015**, *114*, 58–66. [CrossRef]
55. Vassilopoulos, A.; Avgeraki, N.; Klonaris, S. Social desirability and the WTP–WTA disparity in common goods. *Environ. Dev. Sustain.* **2020**, *22*, 6425–6444. [CrossRef]
56. Roche, K.R.; Müller-Itten, M.; Dralle, D.N.; Bolster, D.; Müller, M.F. Climate change and the opportunity cost of conflict. *Proc. Natl. Acad. Sci. USA* **2020**, *117*, 1935–1940. [CrossRef]
57. Cheng, Z.H.; Li, L.S.; Liu, J. The emissions reduction effect and technical progress effect of environmental regulation policy tools. *J. Clean. Prod.* **2017**, *149*, 191–205. [CrossRef]
58. Sun, X.; Liu, X.; Zhao, S.; Zhu, Y. An evolutionary systematic framework to quantify short-term and long-term watershed ecological compensation standard and amount for promoting sustainability of livestock industry based on cost-benefit analysis, linear programming, WTA and WTP method. *Environ. Sci. Pollut. Res.* **2021**, *28*, 18004–18020. [CrossRef]
59. Wunder, S.; Engel, S.; Pagiola, S. Taking stock: A comparative analysis of payments for environmental services programs in developed and developing countries. *Ecol. Econ.* **2008**, *65*, 834–852. [CrossRef]
60. Li, C.; Peng, F.; Chen, H. Analysis of the Influencing Factors for Willingness to Pay of Payment for Ecosystem Services of River Basin: A Case of Changsha Reach of Xiang Jiang River Basin. *Econ. Gegrahy* **2012**, *32*, 130–135.
61. Zheng, H.; Zhang, L. Analysis of the people's willingness to pay for environmental services compensation and its influence factors in the Jinhua River Basin. *Resour. Sci.* **2010**, *32*, 761–767.
62. Muradian, R.; Corbera, E.; Pascual, U.; Kosoy, N.; May, P.H. Reconciling theory and practice: An alternative conceptual framework for understanding payments for environmental services. *Ecol. Econ.* **2010**, *69*, 1202–1208. [CrossRef]
63. Wang, Z.; Guo, L.; Zhang, J. Non-use value composition ratio and influencing factors in different attributes of resources based on cross-cases perspective. *Resour. Sci.* **2017**, *39*, 723–736. [CrossRef]
64. Wang, P.; Poe, G.L.; Wolf, S.A. Payments for Ecosystem Services and Wealth Distribution. *Ecol. Econ.* **2017**, *132*, 63–68. [CrossRef]
65. Xu, D.; Chang, L.; Hou, T.; Zhao, Y. Measure of watershed ecological compensation standard based on WTP and WTA: A case study in Liaohe River Basin. *Resour. Sci.* **2012**, *34*, 1354–1361.
66. Xu, D.; Tu, S.; Chang, L.; Zhao, Y. Interest conflict of River Basin ecological compensation based on evolutionary game theory. *China Popul. Dev. Stud.* **2012**, *22*, 8–14.
67. Yang, H.; Flower, R.J.; Thompson, J.R. Sustaining China's Water Resources. *Science* **2013**, *339*, 141. [CrossRef] [PubMed]
68. Sheng, J.; Qiu, W.; Han, X. China's PES-like horizontal eco-compensation program: Combining market-oriented mechanisms and government interventions. *Ecosyst. Serv.* **2020**, *45*, 101164. [CrossRef]
69. Liu, J.; Yang, W. Water Sustainability for China and Beyond. *Science* **2012**, *337*, 649–650. [CrossRef]
70. Chaikumbung, M.; Doucouliagos, H.; Scarborough, H. The economic value of wetlands in developing countries: A meta-regression analysis. *Ecol. Econ.* **2016**, *124*, 164–174. [CrossRef]
71. De Groot, R.; Brander, L.; van der Ploeg, S.; Costanza, R.; Bernard, F.; Braat, L.; Christie, M.; Crossman, N.; Ghermandi, A.; Hein, L.; et al. Global estimates of the value of ecosystems and their services in monetary units. *Ecosyst. Serv.* **2012**, *1*, 50–61. [CrossRef]
72. Shi, H.; Sui, D.; Wu, H.; Zhao, M. The influence of social capital on farmers' participation in watershed ecological management behavior: Evidence from Heihe Basin. *China Rural Econ.* **2018**, *1*, 34–45.
73. Shi, H.; Zhao, M. Willingness to pay differences across ecosystem services and total economic valuation based on choice experiments approach. *Resour. Sci.* **2015**, *37*, 351–359.
74. Zhou, C.; Li, G. The influencing factors for willingness to pay of payment for watershed services: A case of the water receiving area of Zhengzhou Cityof the Middle Route Project of the South-North Water Transfer Project. *Econ. Gegrahy* **2015**, *35*, 38–46.
75. Chen, C.; König, H.; Matzdorf, B.; Zhen, L. The Institutional Challenges of Payment for Ecosystem Service Program in China: A Review of the Effectiveness and Implementation of Sloping Land Conversion Program. *Sustainability* **2015**, *2015*, 5564–5591. [CrossRef]
76. Zhang, H.; Liu, G.; Wang, J.; Wan, J. Policy and practice progress of watershed eco-compensation in China. *Chin. Geogr. Sci.* **2007**, *17*, 179–185. [CrossRef]
77. Wang, H.J.; Dong, Z.F.; Xu, Y.; Ge, C.Z. Eco-compensation for watershed services in China. *Water Int.* **2016**, *41*, 271–289. [CrossRef]
78. Wang, X.; Shen, C.; Wei, J.; Niu, Y. Study of ecological compensation in complex river networks based on a mathematical model. *Environ. Sci. Pollut. Res. Int.* **2018**, *25*, 22861–22871. [CrossRef]
79. Qiu, J.; Shen, Z.; Huang, M.; Zhang, X. Exploring effective best management practices in the Miyun reservoir watershed, China. *Ecol. Eng.* **2018**, *123*, 30–42. [CrossRef]
80. Wei, C.; Luo, C. A differential game design of watershed pollution management under ecological compensation criterion. *J. Clean. Prod.* **2020**, *274*, 122320. [CrossRef]

81. Peng, B.; Chen, N.; Lin, H.; Hong, H. Empirical appraisal of Jiulong River Watershed Management Program. *Ocean Coast. Manag.* **2013**, *81*, 77–89. [CrossRef]

sustainability

Article

Scenario-Based Hydrological Modeling for Designing Climate-Resilient Coastal Water Resource Management Measures: Lessons from Brahmani River, Odisha, Eastern India

Pankaj Kumar [1,*], Rajarshi Dasgupta [1], Shalini Dhyani [2], Rakesh Kadaverugu [2], Brian Alan Johnson [1], Shizuka Hashimoto [1,3], Netrananda Sahu [4], Ram Avtar [5], Osamu Saito [1], Shamik Chakraborty [6] and Binaya Kumar Mishra [7]

[1] Natural Resources and Ecosystem Services, Institute for Global Environmental Strategies, Hayama, Kanagawa 240-0115, Japan; dasgupta@iges.or.jp (R.D.); johnson@iges.or.jp (B.A.J.); ahash@g.ecc.u-tokyo.ac.jp (S.H.); o-saito@iges.or.jp (O.S.)
[2] CSIR-National Environmental Engineering Research Institute (NEERI), Nehru Marg, Nagpur 440020, India; s_dhyani@neeri.res.in (S.D.); r_kadaverugu@neeri.res.in (R.K.)
[3] Graduate School of Agricultural and Life Sciences, University of Tokyo, Yayoi 1-1-1, Bunkyo-Ku, Tokyo 113-8657, Japan
[4] Department of Geography, Delhi School of Economics, University of Delhi, New Delhi 110007, India; nsahu@geography.du.ac.in
[5] Faculty of Environmental Earth Science, Hokkaido University, Sapporo 060-0810, Japan; ram@ees.hokudai.ac.jp
[6] Faculty of Sustainability Studies, Hosei University, Tokyo 102-8160, Japan; shamik.chakraborty.76@hosei.ac.jp
[7] School of Engineering, Pokhara University, Pokhara-30, Lekhnath 33700, Nepal; bkmishra@pu.edu.np
* Correspondence: kumar@iges.or.jp

Citation: Kumar, P.; Dasgupta, R.; Dhyani, S.; Kadaverugu, R.; Johnson, B.A.; Hashimoto, S.; Sahu, N.; Avtar, R.; Saito, O.; Chakraborty, S.; et al. Scenario-Based Hydrological Modeling for Designing Climate-Resilient Coastal Water Resource Management Measures: Lessons from Brahmani River, Odisha, Eastern India. *Sustainability* **2021**, *13*, 6339. https://doi.org/10.3390/su13116339

Academic Editor: Anastasios Michailidis

Received: 12 April 2021
Accepted: 31 May 2021
Published: 3 June 2021

Publisher's Note: MDPI stays neutral with regard to jurisdictional claims in published maps and institutional affiliations.

Abstract: Widespread urban expansion around the world, combined with rapid demographic and climatic changes, has resulted in serious pollution issues in many coastal water bodies. To help formulate coastal management strategies to mitigate the impacts of these extreme changes (e.g., local land-use or climate change adaptation policies), research methodologies that incorporate participatory approaches alongside with computer simulation modeling tools have potential to be particularly effective. One such research methodology, called the "Participatory Coastal Land-Use Management" (PCLM) approach, consists of three major steps: (a) participatory approach to find key drivers responsible for the water quality deterioration, (b) scenario analysis using different computer simulation modeling tools for impact assessment, and (c) using these scientific evidences for developing adaptation and mitigation measures. In this study, we have applied PCLM approach in the Kendrapara district of India (focusing on the Brahmani River basin), a rapidly urbanizing area on the country's east coast to evaluate current status and predict its future conditions. The participatory approach involved key informant interviews to determine key drivers of water quality degradation, which served as an input for scenario analysis and hydrological simulation in the next step. Future river water quality (BOD and Total coliform (Tot. coli) as important parameters) was simulated using the Water Evaluation and Planning (WEAP) tool, considering a different plausible future scenario (to 2050) incorporating diverse drivers and pressures (i.e., population growth, land-use change, and climate change). Water samples (collected in 2018) indicated that the Brahmani River in this district was already moderately-to-extremely polluted in comparison to the desirable water quality (Class B), and modeling results indicated that the river water quality is likely to further deteriorate by 2050 under all of the considered scenarios. Demographic changes emerged as the major driver affecting the future water quality deterioration (68% and 69% for BOD and Tot. coli respectively), whereas climate change had the lowest impact on river water quality (12% and 13% for BOD and Tot. coli respectively), although the impact was not negligible. Scientific evidence to understand the impacts of future changes can help in developing diverse plausible coastal zone management approaches for ensuring sustainable management of water resources in the region. The PCLM approach, by having active stakeholder involvement, can help in co-generation of the coastal management options followed by open access free software, and models can play a relevant cost-effective approach to enhance science-policy interface for conservation of natural resources.

Keywords: coast; Odisha; Brahmani River; climate resilience; water management; water quality; hydrological simulation; management plan

1. Introduction

Interactions between the land, ocean, and atmosphere in the coastal zone makes it highly dynamic in terms of structure and functioning [1]. Due to the high productivity of coastal areas, they are also among the most exploited and threatened geomorphic units and ecosystems on the earth [2]. The ease of access and abundant resources of coastal zones attracts diverse anthropogenic interferences, and because of the complexity of coastal zone management, has resulted in the misuse and abuse of these sensitive ecosystems [3]. More than 60% of the global population lives in coastal areas and low-lying deltaic zones. Hence, it is imperative to monitor the effects of diverse drivers affecting water quality in coastal areas [4]. Water security is key to ensuring and developing the overall resilience of coastal communities, where water security refers to the capacity of a population to safeguard sustainable access to adequate quantities of acceptable quality water for sustaining livelihoods, human well-being, and socio-economic development, as well as for ensuring protection against water-borne pollution and water-related disasters, and for preserving ecosystems in a climate of peace and political stability [5,6].

There is significant evidence that major drivers of change in coastal waterbody zones include rapid population growth, urban expansion/industrialization along the coastline, and climate variability (e.g., changing frequency and intensity of extreme weather events) [7,8]. Cumulative impacts of natural and anthropogenic activities can drive changes in the land-use/land cover in coastal zones, which in turn affects the services provided by the local ecosystems (e.g., their capacity for water provisioning and regulating surface water and groundwater) [9,10]. These key factors can cause a negative effect on the coastal ecology and economy of coastal communities, leading to eutrophication, contamination of seafood, land subsidence and/or salinization of underground aquifers, building up of solid wastes including non-biodegradables such as plastics, and chemical contamination from industrial and agricultural products such as pesticides and fertilizers [11–16]. The situation is further exacerbated by the lack of water governance and inadequate infrastructure in many developing countries [17,18]. Recognizing these concerns, the United Nations Sustainable Development Goals (SDGs) highlight the necessity of clean water for achieving different goals related to the environment (e.g., Goal 14, Life below water; Goal 15, Life on Land) and human well-being (e.g., Goal 6, Clean water and sanitation, Goal 12, Responsible consumption and production, Goal 13, Climate Action, Goal 2, Zero hunger) [19].

Sustainable management of the coastal zone requires accurate information on the various aspects that affect it, e.g., information on the local coastal habitats, coastal processes, natural hazards and their impacts, water quality, and living resources [1]. Effective management practices also depend on indigenous/local knowledge and suitable responses from concerned government agencies. For water resources, understanding how water and solutes move within and across the land-sea interface, and groundwater-freshwater interactions, are necessary for managing resources in a way that promotes ecological, societal and human well-being [20,21]. However, most coastal water resource management studies have focused on groundwater salinization and contaminant transport in the coastal aquifers, while assessing the status quo of river water quality from inland areas is still not well explored, especially for developing and underdeveloped countries [22]. This is one reason why hydrogeological/hydrogeochemical studies of coastal aquifers (e.g., to decipher their hydrological processes) have received greater scientific and societal interest in recent years.

Coastal aquifers, especially shallow ones, often experience intensive saltwater intrusion because of natural and anthropogenic processes [23]. A summary of key drivers affecting coastal aquifers, their effects, and counter measures to manage these precious resources is shown in Figure 1. One of the major requirements of planning coastal protection works is addressing the processes of erosion, deposition, sediment transport, flooding, and sea-level changes that is continuously altering the shorelines [24]. Along these lines, Jha et al. [25] demonstrated the application of a coastal water quality index (CWQI) and Geographical Information System (GIS)-based overlay mapping technique as one of the effective integrated approaches that is widely used to demarcate healthy and polluted areas. Keesstra et al. [26] mentioned holistic and integrated land-use management practices that can be used to achieve and localize SDGs related to water, food, health, and climate change. More importantly, water governance has an important role to play in all inclusive, holistic/integrated land-use management. Water governance is a particularly important issue when it comes to coastal aquifer systems because of their vulnerability to saltwater intrusion. A summary of the list of key drivers affecting coastal aquifers, their effects, and counter measures to manage these precious resources is shown in Figure 1.

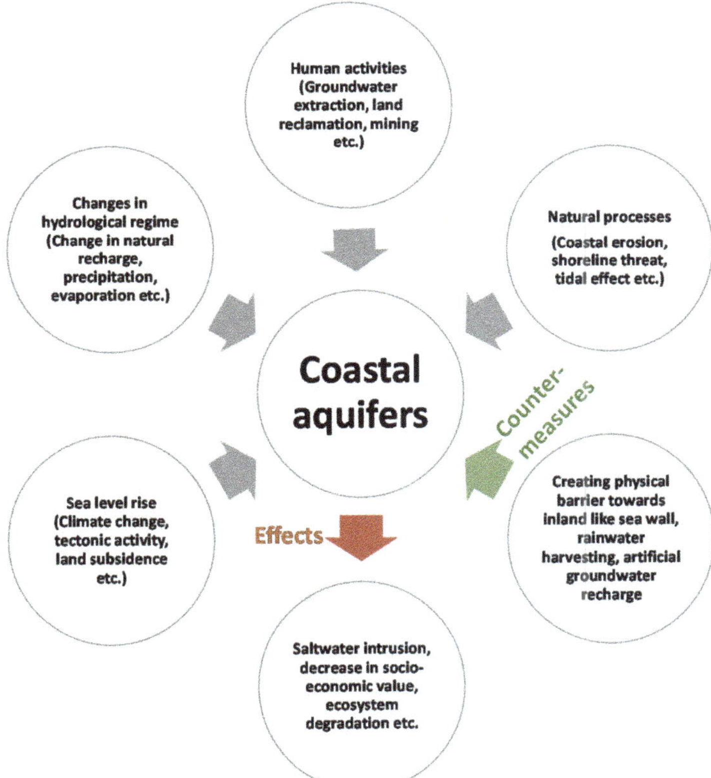

Figure 1. Drivers affecting coastal aquifer and their management options.

India, with its long coastline, is endowed with some of the world's most extensive and diverse mangrove ecosystems. Management of these ecosystems faces several challenges including lack of awareness of their benefits, poor governance, and inadequate management policies [27]. Hence, it is necessary to integrate local socio-economic understanding with the knowledge database of the coastal zone to appropriately understand the impact of anthropogenic activities and climate vulnerabilities [27–29]. The government of India has implemented Integrated Coastal Zone Management (ICZM) all across its coastal states for resource conservation, sustainable development, and pollution control. However, several constraints, ranging from insufficient scientific information, lack of guidelines, lack of sufficient baseline information and weak social basis, ambiguity of project activities, poor local governance, rapid economic reforms, lack of scientific forecasting, undue favors to coastal zone development followed by ineffective implementation, and enforcement are foreseen threats to these sensitive ecosystems [30].

The east coast of Odisha in India is facing particularly diverse drivers of change in the coastal zone viz., shrimp culture, aquaculture, typhoons/cyclones, sea-level rise, climate change, changes in river water quality and quantity, and land-use land cover changes, resulting in damage to its important mangrove ecosystems [31]. Kadaverugu et al. [27] highlighted changes in the local hydrological regime (i.e., reduction of clean water from upstream areas) as the most important factor, which will determine the salinity exposed to the mangroves. Additionally, this area is very important considering the Bhitarkanika National Park and many adjoining high value ecosystems host several mangrove species, saltwater crocodiles, and diverse floral and faunal components of high conservation value. Therefore, it is very important to monitor the river water quality of Odisha at regular intervals to develop science-based coastal management options in a timely manner. However, on the other hand, this area is very poorly investigated and hence calls for comprehensive scientific attention on an urgent basis.

Considering this information need, in this study we used the Participatory Coastal Land-Use Management (PCLM) approach to design future scenarios and simulate future water quality in Odisha, with the goal of aiding water resource management planning of the Brahmani River in the coastal aquifers of Odisha. The PCLM approach integrates scientific evidence for upscaling policy interventions, especially from Integrated Coastal Zone Management (ICZM) programs, and helps in developing appropriate preparedness plans and mitigation measures against the changes the region is facing. Such successful use of PCLM approach for water resource management and making land use more resilient to rapid global change in the case of the Philippines is well documented in Brian et al. [32].

2. Study Area

Kendrapara is one of the densely populated districts of India's Odisha state, and is situated on the central coastal plain zone of the state, on the eastern coast of India (Figure 2). The district has an area of 2644 km^2, and topographically, the region is flat, with an average elevation of 13 m above the mean sea level. The annual average rainfall of the region is 1556 mm which is realized in around 53 days, and nearly 80% of rainfall occurs in the monsoon season between June to September [33]. The region experiences a tropically hot and humid climate with the hot summers and cold winters. Summer is from March to May with the maximum temperature of 42.04 °C, while winter is from October to February with the minimum temperature of 12.8 °C.

The rivers Mahanadi, Brahmani, and Baitarani, along with their distributaries, define the drainage systems of the district, majorly comprising anastomosing drainage patterns [34]. Other rivers in the district include Karandia, Gobari, Brahmani, Kani, Baitarani, Birupa, Kharasrota, Paika, Chitroptala, and Hansua [35]. The delta region formed by the rivers Brahmani and Baitarani has underlain alluvial deposits, which resulted in fertile agricultural land. The soils in the district are predominantly of tertiary and recent alluvium deposited by the rivers. The thickness of alluvium of the Mahanadi delta is more than 600 m [36]. The coastlines of the district are dominated by Aeolian and sandy soils especially in the Rajnagar and Rajkanika blocks of the district. In general, 50% of the soil type is of entisols followed by inceptisols, water bodies, and mudflats, and the geographical area has a slope within 0–3% [37].

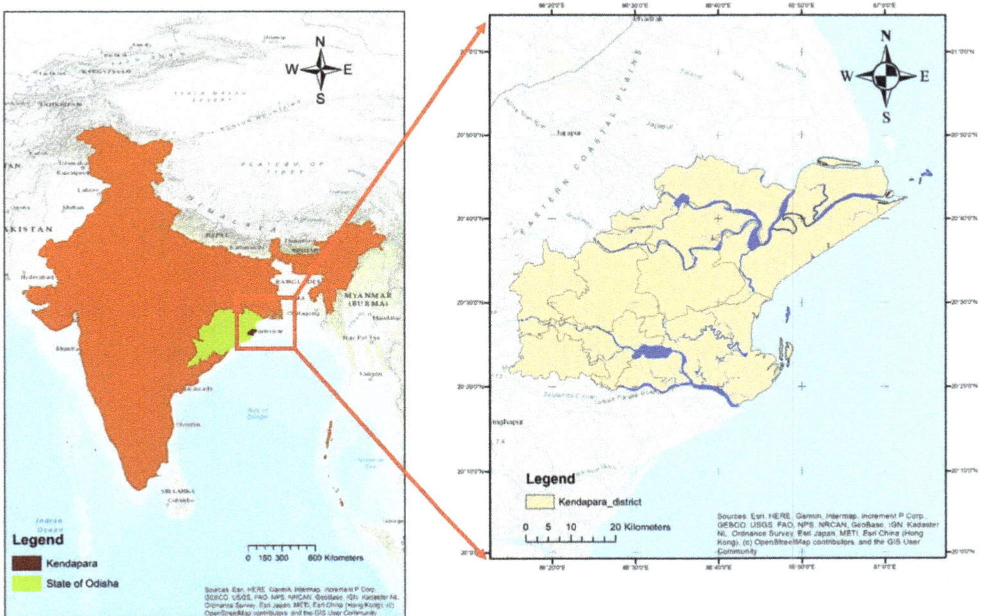

Figure 2. Study area map.

The Kendrapara district has a population of about 1.4 million as per the 2011 census [38], with a density of 545 people per square km. Nearly 94% of the population resides in rural areas of the district, and agriculture is the mainstay of the locals with 70% dependence on agriculture. Nearly 63% of the geographical area is under-cultivated with a cropping intensity of 194% [37]. Despite having enough water resources, this region is facing waterlogging and increasing salinity of cropland as the major concerns related to agriculture, mainly due to a lack of necessary irrigation structures, and proper management solutions in place [37]. Additionally, the region experiences frequent flooding from the dense network of rivers due to the geographical disadvantage owing to flat terrain and high rainfall intensity [35,37]. The district also has rich groundwater resources that fall under the safe category according to the Central Ground Water Board classification, having an overall stage of the groundwater development of 52.97% [38]. The pre-monsoon groundwater-level depth varies in the range of 1.65 to 5.43 m below ground level, while the post-monsoon level varies from 0.11 to 4.90 m below ground level [38]. The Brahmani River is very important for the Kendrapara district considering its invaluable contribution for water demand for different industries especially mining industries in the upstream region as well as managing rich species diversity among mangrove forests near the downstream area. Therefore, understanding the current status as well as future projection of river water quality is very crucial for sustainable socio-economic development.

3. Materials and Methods

The methodology adopted for this work is shown in Figure 3.

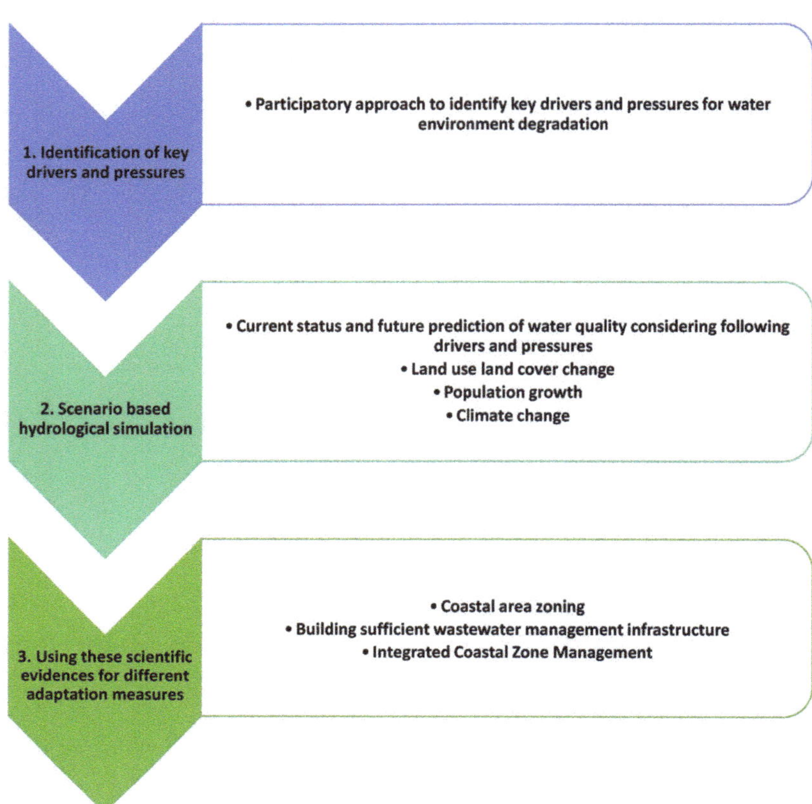

Figure 3. Flowchart for the study methodology.

3.1. Key Informant Interview

First, to identify key drivers and pressures which affect water quality in this study area, key informant's interviews (KII) were conducted with 18 resource persons who hold strong knowledge of the study area, including representatives from the national and state water ministry (n = 3), the agriculture ministry (n = 3), NGOs (n = 4), community members (n = 2), scholars and researchers (n = 6) (Figure 4). They were asked to summarize the different drivers and their intensity, which are affecting the land use of coastal zones. Here, we have employed the STEEP (sociological, technological, economical, environmental, and political factors) approach for driver classification through impact-uncertainty matrix [39]. At first, they discussed the main drivers and then, all of the informants prioritized different factors according to their level of impact and uncertainty on water resources on a Likert scale. For impact, level '5' corresponded to the highest impact and '1' represented the lowest impact, respectively. For uncertainty, level '1' corresponds to very certain whereas level '5' corresponds to highly uncertain. Previously, this method has been successfully used for scenario planning in coastal areas (e.g., Dasgupta et al. [39]).

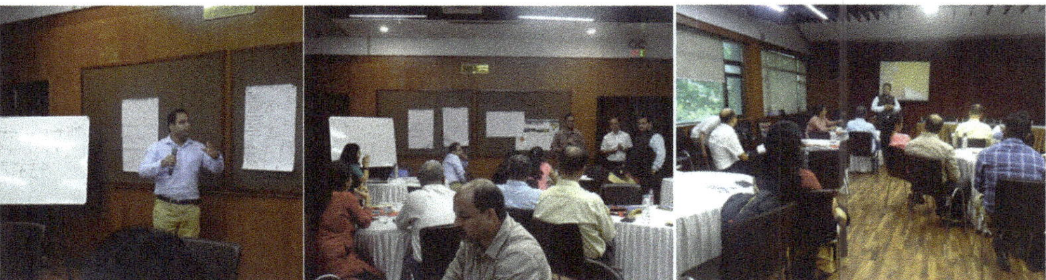

Figure 4. Key informant interview.

The analysis of the interviews and respective scoring indicated that among different drivers, sociological factors (i.e., population growth, awareness about water management, etc.) were the most significant, while political factors (like absence of law pertaining to freshwater exploitation, land allocation for aquaculture, etc.) were the least significant. The order for different drivers found was sociological > environmental > technological > economical > political. Digging further into the different key components of these factors, it was found that population growth, land-use land cover change, and climate change are the key factors responsible for water stress in this region.

Hence, we set up a numerical model to quantify the current status as well as future prediction of water resources using different scenario analyses.

3.2. Hydrological Model
3.2.1. Model Set-Up

In this study, the WEAP model was used to simulate streamflow and river water quality variables. The reason to select WEAP model for this simulation is because it is freely available to the developing nations and hence, its replicability by the local stakeholders will be easier. In addition, WEAP model functionality enables to generate various scenario building, necessary to answer "what-if" questions for the policy and decision makers. The schematic diagram for the problem domain was developed using different shapefiles such as administrative boundary of the study area and drainage network (developed with ArcGIS), and inbuilt links and nodes from WEAP. Here, the whole study area is divided into smaller catchments based on the topographical, hydrological, and confluence points, and climatic characteristics of the river basin. Different hydrological modules, namely rainfall-runoff (simplified coefficient method), irrigation demands only (simplified coefficient method), and rainfall-runoff (soil moisture method) are available in the WEAP platform to simulate various components of the hydrological cycle including the catchment's potential evapotranspiration, infiltration, and streamflow. However, based on the data availability, the rainfall-runoff method (simplified coefficient method), is used for the catchment simulation in this study [40].

The Streeter–Phelps model within WEAP was used to estimate pollution concentrations in water bodies. Simulation of oxygen balance in a river within this model is governed by two processes, i.e., consumption by decaying organic matter and reaeration induced by an oxygen deficit. The removal of BOD from water is a function of water temperature, settling velocity, and water depth as shown by Equations (1) and (2):

$$BOD_{final} = BOD_{init} exp^{\frac{-k_{rBOD}L}{U}} \tag{1}$$

$$k_{rBOD} = k_{d20}^{1.047\ (t-20)} + \frac{v_s}{H} \tag{2}$$

where BOD_{init} = BOD concentration at the top of the reach (mg/L), BOD_{final} = BOD concentration at the end of the reach (mg/L), t = water temperature (in degrees Celsius), H = water depth (m), L = reach length (m), U = water velocity in the reach, v_s = settling velocity (m/s), k_r, k_d and k_a = total removal, decomposition and aeration rate constants (1/time), k_{d20} = decomposition rate at reference temperature (20° Celsius).

3.2.2. Data Requirement for Model Set-Up

Future prediction (for the year 2050) of Brahmani River water quality was made using a scenario analysis to assess different possible water management plans. Key datasets used for the modeling were domestic wastewater discharge, past river water quality at different monitoring stations, population, rainfall, temperatures, river cross-section, river length, river discharge, land-use/land cover etc. List of data used to run the model and their sources is shown in Table 1.

Table 1. List of data used in this study with their source.

S. No.	Parameters	Time Interval	Scale	Source
1	Population	2008	Yearly	Census of India [41]
		2011–2050	Yearly	UNDESA [42]
2	Water quality (BOD, Tot. Coli)	2008–2019	Quarterly	Odisha Water Resource Board [43]
3	River cross section, streamflow	2008–2019	Quarterly	Odisha Water Resource Board [43]
4	Rainfall, Temperature	1984–2018 (Past data)—Average gave a value for current condition for year 2018	Monthly	Indian Meteorological Department [44]
		2021–2070 (Future data)—Average gave a value for future condition for year 2050		IPCC [45]
5	Land-use land cover map	2008 and 2018	Yearly	LANDSAT [46]
		2050		Land change modeler [47]

For water quality parameters, we have used biochemical oxygen demand (BOD) and Total coliforms collected at three points on the Brahmani River. Data for both water quality parameters were analyzed by the Odisha water resources board following the standard method of water quality analysis from APHA [48]. The reason to choose these two parameters is that this was the only available dataset at a regular basis for the simulation period. The principal reasons for selecting sampling locations were the accessibility of water samples, saving cost, and observing the effect of urbanization at approximately equidistant throughout the river length passing through the watershed. For hydrological modeling, three catchment areas in the Brahmani River watershed, which experienced inter-basin transfers, were considered. Pollutant transport from a catchment accompanied by rainfall-runoff is enabled by ticking the water quality modeling option. During non-rainy days, pollutants accumulate on the catchment surfaces and reach water bodies through surface runoff.

Regarding future precipitation data, two different Global Climate Models (GCMs) (MRI-CGCM3 and MIROC5) and Representative Concentration Pathway (RCP) (4.5 and 8.5) output were used after downscaling and bias correction. We have evaluated the average value of the change in monthly average precipitation for both RCP 4.5 and 8.5 to evaluate the climate change on water quality. The whole study area is divided into nine demand sites for estimating the effect of population growth and its associated domestic wastewater discharge on river water quality status. Primarily, these demand sites denote the population of different administrative units (blocks in this case) lying on either side of the Brahmani River within our study area. Future population in these demand sites were estimated by ratio method using UNDESA projected growth rate [42]. In the absence of exact information on the total domestic wastewater production, the daily volume of domestic wastewater generation per person considered for this study was 130 L [49]. Land-use land cover map was generated using LANDSAT 8 OLI satellite image and future map was generated using land change modeler; detailed information of this work was submitted for a peer-reviewed article and is under review at present [50]. Simulated water quality result is compared with Class B, i.e., class for outdoor bathing, a standard set for surface water quality desired by the state of Orissa. The local government set a goal to achieve Class B for the ambient water quality. Besides, it is relatively easier to achieve compared to Class A, i.e., water for drinking purposes. For Class B, the standard value for BOD and Total coliform are <3 mg/L and <500 CFU/100 mL respectively. The whole simulation process is divided into three phases: (a) model set-up and data input, (b) calibration and validation, and (c) future simulation using scenario analysis. The schematic diagram for the model set-up is shown in Figure 5.

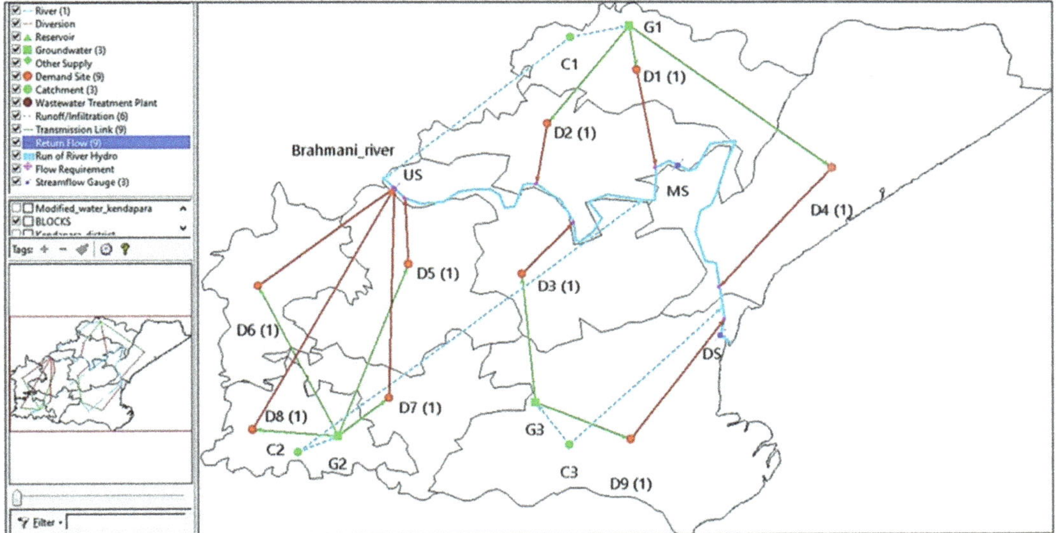

Figure 5. Schematic diagram for model set-up.

4. Results

4.1. Future Prediction of Key Factors Considered for Hydrological Simulation

Three key factors, i.e., land-use land cover change, climate change, and population growth were considered for the hydrological simulation. The future values of all the parameters were calculated for the target year 2050 and the result is shown in Figure 6. Figure 6a, showing the result for land-use land cover change, indicates that a significant amount of changes will occur for vegetation and built-up categories. The built-up area will be increased (by 96%) at the expense of a decrease in the area of vegetation (by 22%) mainly. Figure 6b shows the result for comparison between observed and downscaled data for monthly average rainfall using two different GCMs and two RCPs. The annual precipitation projected for future scenarios using the GCM downscaling method is almost the same with the current observed data. However, slight differences in the precipitation values are observed at a monthly scale. In the present study, we have considered these marginal changes in the precipitation values and studied whether they impart any significant variations in the river water quality. Figure 6c shows the result for population projection in the study area using the UNDESA growth rate. Looking at the result, it is found that the population for the year 2050 will be increased by 80% when compared to that of the year 2018.

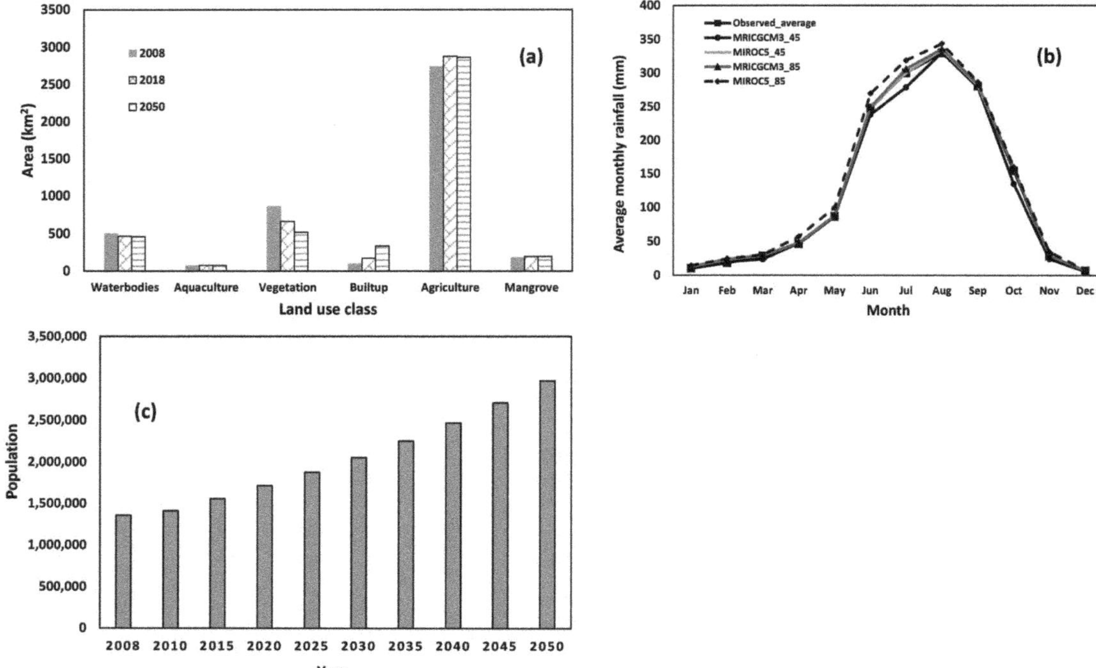

Figure 6. Future predicted value for different factors: (**a**) Land-use land cover; (**b**) Monthly average precipitation; (**c**) Population of the study area, considered in the hydrological simulation.

4.2. Model Performance Evaluation and Future Simulation

The WEAP model performance credibility was tested through calibration and validation of the data. It is essential to check the performance credibility of models, especially complex environmental models before projecting the future scenarios. The whole WEAP module is primarily divided into two parts, i.e., water quality component and hydrological component. In the present study, the model valuation was carried out using the BOD and streamflow information for water quality and hydrological components, respectively. Trial and error method were employed for the model calibration, and the best-fit results were obtained. The variables—effective precipitation and the runoff/infiltration were adjusted in the model simulations to match with the measured monthly stream flow data for the year 2018 (Table 2). The adjusted best-fit values of these variables were 94% and 55/45, respectively. Similarly, the best-fit values for the water quality components—BOD and Total coliform were 2.5 mg/L and 220 MPN/100 mL, respectively.

Table 2. Summary of parameters and steps used for calibration.

Parameters	Initial Value	Steps	Final Calibrated Value
Effective precipitation	100%	±0.5%	94%
Runoff/infiltration ratio	50/50	±5/5	55/45
Headwater quality (BOD—2 mg/L)	2 mg/L	±0.5/0.5	2.5
Tot. coliform (MPN/100 mL)	100 MPN/100 mL	±10/10	220

Results show a significant relation between the simulated and observed values of BOD and streamflow for the years 2018 and 2019, respectively. The reason for selecting 2018 and 2019 for validating BOD and streamflow, respectively, was regular availability of observed data without any gap. The validation of the model in terms of both BOD and streamflow is presented in bar diagrams of Figure 7a,b, respectively. BOD was simulated with a strong correlation with a correlation coefficient (R^2) = 0.86 having an average error of around 11% (Figure 7a). Streamflow for the Brahmani River at midstream was also simulated with a significant degree of correlation (R^2) = 0.81 with an average error of 13% for most of the months during 2019 (Figure 7b). Once validation is done with a satisfactory result, future water quality simulation was done.

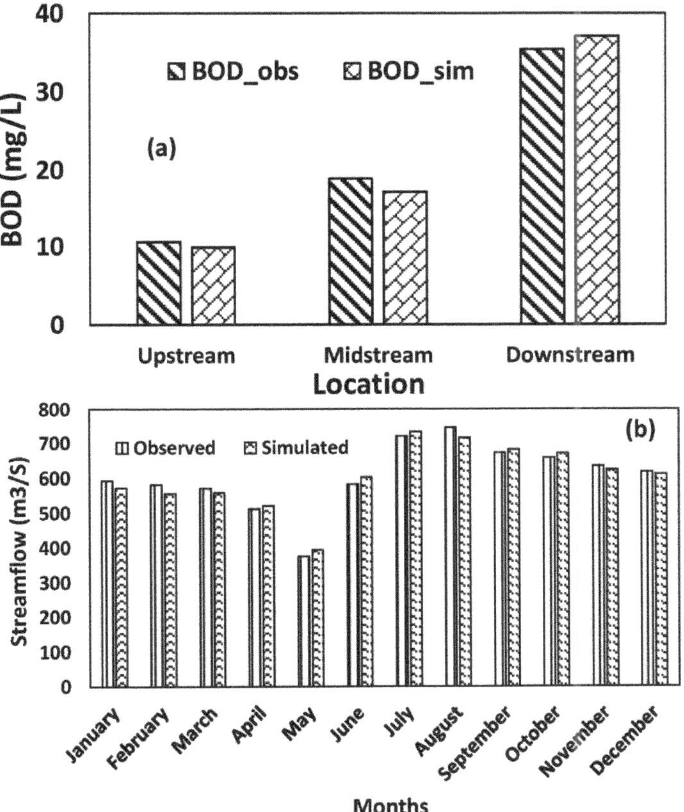

Figure 7. Validation of the model output by comparing simulated and observed (**a**) average biochemical oxygen demand (BOD) values for different locations for the year 2018 and (**b**) average monthly river discharge for year 2019 at Brahmani River midstream.

The impact assessment on the river water quality and flow was carried out by considering three main drivers of change viz. population growth, climate change, and land-use land cover change. A combination of these drivers was considered to obtain the following 'what-if' situations viz. (a) with only population growth; (b) population growth and with a moderate climate change; (c) population growth with an extreme climate change; (d) combination of population change, moderate climate change and land-use land cover change, and lastly (e) combination of population change, extreme climate change and land-use land cover change. In all these scenarios, the impact on the river was assessed without considering the adaptation measures. The obtained results through hydrological modeling were compared with that of national guidelines (Figure 8) outlined by the Central Pollution Control Board (CPCB) for Class B category of river quality (i.e., swimmable category for which BOD <3 mg/L and Tot. coli <500 CFU/100 mL).

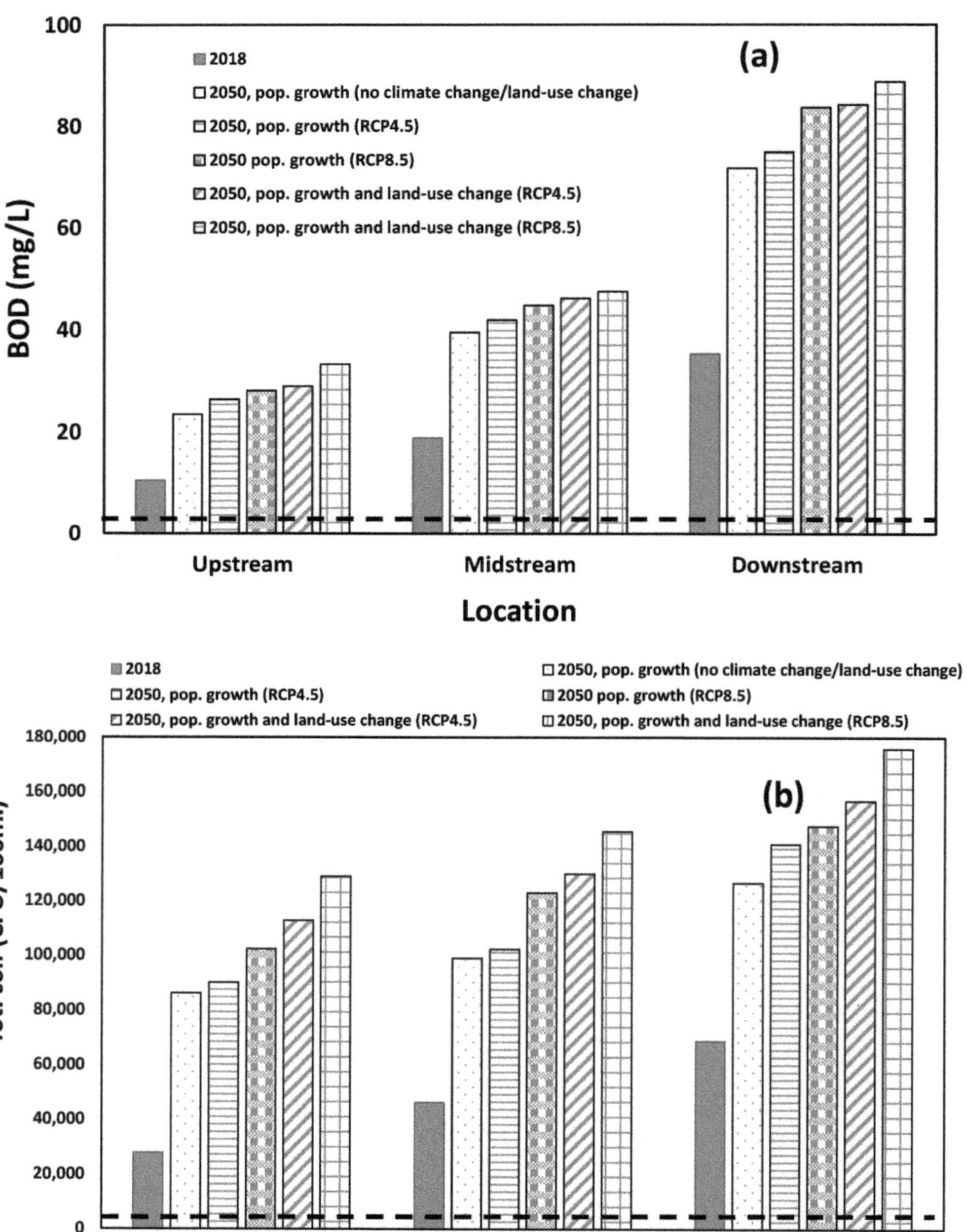

Figure 8. Results show simulated water quality parameters: (**a**) BOD; (**b**) Tot. coli for future scenario without adaptation measures.

The model simulated concentration of BOD for the year 2018 varies from 10.6 to 35.4 mg/L considering all the scenarios. When compared with the desired concentration of BOD from Class B category of river quality classification, it can be said that all the simulation values throughout the year falls under the moderately-to-extremely polluted category. On the other hand, the value of Tot col. ranged from 27,863 to 68,771 CFU/100 mL. Looking into the results from future scenarios, it can be inferred that water quality status is significantly affected by all the drivers—climate change, land-use/land cover change, and population changes, and water quality will deteriorate further in 2050 when compared with the current situation. Furthermore, predominantly, the fifth scenario (e) has the highest negative impact on the water quality, which is the effect of population growth, land-use land cover changes, and climate change with RCP 8.5. Here, the average percentage increase in the BOD and Tot. coli by 2050 is 161.3% and 215.4%, respectively, when compared with the situation in 2018. We have further analyzed the impact of individual drivers and pressures on river water quality deterioration, and the results are presented in Table 3. Results show that the river water quality in terms of both BOD and Tot. coli is determined by the individual drives in the decreasing order of population growth > land-use land cover change > climate change. This result can be explained as the amount of wastewater being generated will grow proportionately with the population increase without any countermeasures. On the other hand, the drivers—land-use land cover change and climate change-induced extreme weather conditions, will trigger the changes in River discharge due to various complex environmental interactions in the river catchment. This change in River discharge and other local factors will induce a cascading change in water quality parameters due to the modifications in oxidation-reduction process and overall contaminate transport. The results depict the significance of these key drivers of change which are likely to affect water resources of a region, and the development of local level climate resilient plan must consider these drivers for effective policy planning. Further, as the river quality simulations indicate a high degree of water contamination, this is likely to pose potential health risks like gastroenteritis if consumed accidentally (microbial contamination), and death of aquatic organisms such as fish (because of high BOD).

Table 3. Summary for order of contribution of different drivers on water quality deterioration in Santa Rosa sub-watershed.

Parameter	Average % Increase (2018 to 2050)	% Contribution from Population Growth	% Contribution from LULC Change	% Contribution from Climate Change
BOD	161	68	20	12
Tot. coli	215	69	18	13

5. Possible Management Options Based on the Scientific Evidence

Finally, based on the hydrological simulation output, we propose the following feasible management solutions for the water resource management in the study area.

(a) Coastal area zoning—The Government of India has a very comprehensive Integrated Coastal Zone Management (ICZM) program, which acts as a spatial decision-support tool using both ground-based data and satellite data. However, its field implementation is not satisfactory due to various factors such as rapidly changing environmental systems in these coastal environments owing to both natural and anthropogenic factors [51]. If scientific forecasting of different environmental components such as water quality and biodiversity based on the "what if" situation are incorporated in these programs, the zoning for different activities can be made more effective and sustainable [52]. For example, in the case of Brahmani River, people indiscriminately using the downstream floodplain area for aquaculture, and bring brackish or saltwater towards inland [53]. This disturbs the sea water-fresh water equilibrium, a main cause for saltwater intrusion, deterioration of mangrove species, and permanently damaging agricultural fields. Hence government should demarcate the area with clear job allocation and there should be a strict penalty if someone fails to comply.

(b) Building sufficient wastewater management infrastructure—Good river water quality is very important for maintaining the ideal environment for a healthy coastal ecosystem as shown by Kumar [10]. So, monitoring and regulating all wastewater flowing into the river is vital. At present, there is not enough infrastructure to cater to the wastewater being generated from both the domestic as well as industrial sector. Hence, we must have precise information about wastewater being generated both for the current situation and in the coming future. This will help to design the infrastructure of sufficient capacity.

(c) Public awareness—Local people must be made aware of the benefits of a healthy coastal ecosystem in providing economic benefits, protection from climate change, and overall human well-being as reported by several scientific reports and publications [54,55]. They should understand that a participatory approach is a key to achieve the sustainable management of the coastal environment. Since water is the limiting factor for socio-economic dimension for people living in Kendrapara district, it is very important for the people to understand about different water management options available and how they can play a critical role in this.

6. Conclusions

Despite diverse international and national initiatives and programs, coastal zone management in India still faces many obstacles at institutional, societal and financial levels. This study shows that by including different factors such as population growth, climate change and land-use land cover changes, average value of the BOD and Tot. coli in the river water body will be increased 161.3% and 215.4%, respectively, by year 2050 when compared with the situation in 2018. Due to the dynamic nature of coastal areas and the major impacts of demographic and climatic changes (historically and in the future), it is essential that climate uncertainty be included in water resource management policy planning. Active involvement of local stakeholders in scientific research on water resources and environmental change is essential to help them understand their future requirements and risks, and design effective mitigation measures. This process of climate resilient planning will help in developing ecosystems and social resilience in the region against increasing climate vulnerability and uncertainty. The situation of Indian coastal zone management requires an all-inclusive, integrated multidisciplinary and transdisciplinary approach (both bottom-up and top-down) for addressing water resource management issues. Especially, vulnerable coastal zones such as inland and coastal water bodies being an important part of the complex coastal ecosystem require a more sound water resource management, and if not efficiently planned, can be the limiting factor for the ecological, social and economic growth of the region that may largely affect human well-being.

Our study provides clear insights about the future water quality under diverse "what if" conditions, and provides broader perspectives and meaningful considerations for appropriately designing diverse policy relevant to sustainable management solutions keeping climate uncertainty in its core. Hydrological simulation for water quality of Brahmani River in the sensitive coastal district Kendrapara in Odisha provided a clear overview of the existing as well as plausible alternative future scenarios. Our study illustrates that the Participatory Coastal Land-Use Management (PCLM) approach is an effective approach ready for implementation for ensuring sustainable management of coastal water resources. The PCLM approach, as an all-inclusive integrated framework for water resource management, helped in retrofitting models by regular interactions and feedbacks from diverse stakeholders having stakes in water resource use and management in the district to ensure stakeholder needs are well addressed and included, and they also understand and own simulated results. The key enabling condition for effectiveness of PCLM approach is an all-inclusive participation of stakeholders that supports the planning, implementation, and monitoring of the approach in a more holistic manner.

Author Contributions: Conceptualization: P.K.; methodology: P.K.; formal analysis: P.K.; investigation: P.K., R.D. and R.K.; writing—original draft preparation: P.K.; writing—review and editing: P.K., R.D., S.D., R.K., B.A.J., S.H., R.A., N.S., O.S., S.C. and B.K.M. All authors have read and agreed to the published version of the manuscript.

Funding: Data collection and field survey for this research work is supported by the project from Asia Pacific Network for Global Change Research (APN) under Collaborative Regional Research Programme (CRRP) with grant number CRRP2018-03MY-Hashimoto, https://doi.org/10.13039/100005536. Fund for publication fee for this work is supported by IPCC Strategy Research Fund 2020 in-house grant from Institute for Global Environmental Strategies (IGES).

Institutional Review Board Statement: Not applicable.

Informed Consent Statement: Not applicable.

Data Availability Statement: Not applicable.

Conflicts of Interest: The authors declare no conflict of interest.

References

1. Nayak, S. Coastal zone management in India- present status and future needs. *Geo Spat. Inf. Sci.* **2017**, *20*, 174–183. [CrossRef]
2. Michael, H.A.; Post, V.E.A.; Wilson, A.M.; Werner, A.D. Science, society, and the coastal groundwater squeeze. *Water Resour. Res.* **2017**, *53*, 2610–2617. [CrossRef]
3. Krishnamurthy, R.R.; Dasgupta, R.; Chatterjee, R.; Shaw, R. Managing the Indian coast in the face of disasters and climate change: A review and analysis of India's coastal zone management policies. *J. Coast. Conserv.* **2014**, *18*, 657–672. [CrossRef]
4. Steyl, G.; Dennis, I. Review of coastal-area aquifers in Africa. *Hydrogeol. J.* **2010**, *18*, 217–225. [CrossRef]
5. DasGupta, R.; Shaw, R. An indicator based approach to assess coastal communities' resilience against climate related disasters in Indian Sundarbans. *J. Coast. Conserv.* **2015**, *19*, 85–101. [CrossRef]
6. UN Water. *Water Security and the Global Water Agenda*; UNU-INWEH: Hamilton, Canada, 2013; p. 38. ISBN 9789280860382.
7. Mohanty, P.K.; Panda, U.S.; Pal, S.R.; Mishra, P.K. Monitoring and management of environmental changes along the Orissa coast. *J. Coast. Res.* **2008**, *24*, 13.e27. [CrossRef]
8. Kumar, P.; Dasgupta, R.; Johnson, B.; Saraswat, C.; Basu, M.; Kefi, M.; Mishra, B.K. Effect of Land use changes on water quality in an ephemeral coastal plain: Khambhat City, Gujarat, India. *Water* **2019**, *11*, 724. [CrossRef]
9. Identifying the Relationships Between Water Quality and Land Cover Changes in the Tseng-Wen Reservoir Watershed of Taiwan. *Int. J. Environ. Res. Public Health* **2013**, *10*, 478–489. [CrossRef] [PubMed]
10. Kumar, P. Numerical quantification of current status quo and future prediction of water quality in eight Asian Mega cities: Challenges and opportunities for sustainable water management. *Environ. Monit. Assess.* **2019**, *191*, 319. [CrossRef]
11. Lakshmi, A.; Rajagopalan, R. Socio-economic implications of coastal zone degradation and their mitigation: A case study from coastal villages in India. *Ocean Coast. Manag.* **2000**, *43*, 749–762. [CrossRef]
12. World Resource Institute (WRI). *Bigger Problem, Better Solutions, World Resource Institute Annual Report 2018–2019*; WRI: Washington DC, USA, 2019; p. 52.
13. Kumar, P.; Tsujimura, M.; Nakano, T.; Minoru, T. The effect of tidal fluctuation on ground water quality in coastal aquifer of Saijo plain, Ehime prefecture, Japan. *Desalination* **2012**, *286*, 166–175. [CrossRef]
14. Dabrowski, J.M.; De Klerk, L.P. An Assessment of the Impact of Different Land Use Activities on Water Quality in the Upper Olifants River Catchment. *Water SA* **2013**, *39*, 231–241. [CrossRef]
15. Saraswat, C.; Mishra, B.K.; Kumar, P. Integrated urban water management scenario modeling for sustainable water governance in Kathmandu Valley, Nepal. *Sustain. Sci.* **2017**, *12*, 1037–1053. [CrossRef]
16. Assessment and management of coastal multi-hazard vulnerability along the Cuddalore-Villupuram, east coast of India using geospatial techniques. *Ocean Coast. Manag.* **2011**, *54*, 302–311. [CrossRef]
17. Zeilhofer, P.; Lima, E.B.N.R.; Lima, G.A.R. Spatial patterns of water quality in the Cuiaba river basin, Central Brazil. *Environ. Monit. Assess.* **2006**, *123*, 41–62. [CrossRef] [PubMed]
18. Ding, J.; Jiang, J.; Fu, L.; Liu, Q.; Peng, Q.; Kang, M. Impacts of Land Use on Surface Water Quality in a Subtropical River Basin: A Case Study of the Dongjiang River Basin, Southeastern China. *J. Water* **2015**, *7*, 4427–4445. [CrossRef]
19. United Nation Sustainable Development Goals. *The 2030 Agenda for Sustainable Development*; A/RES/70/1; United Nation Sustainable Development Goals: New York, NY, USA, 2015; p. 41.
20. Dong, L.; Liu, J.; Du, X.; Dai, C.; Liu, R. Simulation-based risk analysis of water pollution accidents combining multi-stressors and multi-receptors in a coastal watershed. *Ecol. Indic.* **2018**, *92*, 161–170. [CrossRef]
21. Greggio, N.; Giambastiani, B.M.S.; Mollema, P.; Laghi, M.; Capo, D.; Gabbianelli, G.; Antonellini, M.; Dinelli, E. Assessment of the main geochemical processes affecting surface water and groundwater in a low-lying coastal area: Implication for water management. *Water* **2020**, *12*, 1720. [CrossRef]

22. Yang, H.J.; Jeong, H.J.; Bong, K.M.; Jin, D.R.; Kang, T.W.; Ryu, H.S.; Han, J.H.; Yang, W.J.; Jung, H.; Hwang, S.H.; et al. Organic matter and heavy metal in river sediments of southwestern coastal Korea: Spatial distributions, pollution, and ecological risk assessment. *Mar. Pollut. Bull.* **2020**, *159*, 111466. [CrossRef]

23. Oude Essink, G.H.P. Improving fresh groundwater supply problems and solutions. *Ocean Coast Manag.* **2001**, *44*, 429–449. [CrossRef]

24. Murali, R.M.; Dhiman, R.; Choudhary, R.; Seelam, J.K.; Ilangovan, D.; Vethamony, P. Decadal shoreline assessment using remote sensing along the central Odisha coast, India. *Environ. Earth Sci.* **2015**, *74*, 7201–7213. [CrossRef]

25. Jha, D.K.; Devi, M.P.; Vidyalakshmi, R.; Brindha, B.; Vinithkumar, N.V.; Kirubagaran, R. Water quality assessment using water quality index and geographical information system methods in the coastal waters of Andaman sea, India. *Mar. Pollut. Bull.* **2015**, *100*, 555–561. [CrossRef]

26. Keesstra, S.; Mol, G.; de Leeuw, J.; Okx, J.; de Cleen, M.; Visser, S. Soil-related sustainable development goals: Four concepts to make land degradation neutrality and restoration work. *Land* **2018**, *7*, 133. [CrossRef]

27. Kadaverugu, R.; Dhyani, S.; Dasgupta, R.; Kumar, P.; Hashimoto, S.; Pujari, P. Mutiple values of Bhitarkanika mangrove for human well-being: Synthesis of contemporary scientific knowledge for mainstreaming ecosystem services in policy planning. *J. Coast. Conserv.* **2021**, *25*, 32. [CrossRef]

28. Kathiresan, K. Mangrove forests of India. *Curr. Sci.* **2018**, *114*, 5–976. [CrossRef]

29. Gupta, M.; Fletcher, S. The application of a proposed generic institutional framework for integrated coastal management to India. *Ocean Coast. Manag.* **2001**, *44*, 757–786. [CrossRef]

30. Panigrahi, J.K.; Mohanty, P.K. Effectiveness of the Indian coastal regulation zones provisions for coastal zone management and its evaluation using SWOT analysis. *Ocean Coast. Manag.* **2012**, *65*, 34–50. [CrossRef]

31. Barik, J.; Mukhopadhyay, A.; Ghosh, T.; Mukhopadhyay, S.K.; Chowdhury, S.M.; Hazra, S. Mangrove species distribution and water salinity: An indicator species approach to Sundarban. *J. Coast. Conserv.* **2018**, *22*, 361–368. [CrossRef]

32. Johnson, B.A.; Macandog, D.B.; Kumar, P.; Dasgupta, R.; Kawai, M. *Participatory Coastal Land-Use Management (PCLM) Guidebook: Participatory Approaches and Geospatial Modeling for Increased Resilience to Climate Change at the Local Level*; IGES publication: Kanagawa, Japan, 2021; p. 88.

33. ENVIS Centre of Odisha's State of Environment. Orienvis.nic.in. Available online: http://webcache.googleusercontent.com/search?q=cache:24-b22YoyscJ:orienvis.nic.in/index1.aspx%3Flid%3D1%26mid%3D2%26langid%3D1%26linkid%3D1+&cd=2&hl=en&ct=clnk&gl=jp (accessed on 12 August 2019).

34. CGWB. Ground Water Information Booklet of Kendrapara District, Odisha. 2013. Available online: http://cgwb.gov.in/District_Profile/Orissa/kendrapara.pdf (accessed on 6 February 2021).

35. Rahman, M.d.R.; Thakur, P.K. Detecting, mapping and analysing of flood water propagation using synthetic aperture radar (SAR) satellite data and GIS: A case study from the Kendrapara District of Orissa State of India. *Egypt. J. Remote Sens. Space Sci.* **2018**, *21*, S37–S41. [CrossRef]

36. Naik, P.K.; Pati, G.; Srivastava, D.S.; Tiwari, S.K.; Hota, R.N. Hydrogeology of Kendrapara District with Special Reference to Salinity Hazard. *Univ. Spec. Publ. Geol.* **2005**, *4*, 137–143.

37. DLIC. District Irrigation Plan of Kendrapara. 2016. Available online: http://www.dowrodisha.gov.in/DIP/2015-20/KENDRAPARA.pdf (accessed on 6 February 2021).

38. Census. Census of India. Website: Office of the Registrar General & Census Commissioner, India. 2011. Available online: http://censusindia.gov.in/ (accessed on 5 February 2019).

39. DasGupta, R.; Hashimoto, S.; Okuro, T.; Basu, M. Scenario-based land change modelling in the Indian Sundarban delta: An exploratory analysis of plausible alternative regional futures. *Sustain. Sci.* **2019**, *14*, 221–240. [CrossRef]

40. Seiber, J.; Purkey, D. *WEAP—Water Evaluation and Planning System User Guide for WEAP 2015*; Somerville, Stockholm Environment Institute: Stockholm, Sweden, 2015.

41. Ministry of Home Affairs, Government of India, Census Digital Library. Available online: https://censusindia.gov.in/DigitalLibrary/MapsCategory.aspx (accessed on 5 March 2020).

42. United Nations, Department of Economic and Social Affairs, Population Division (UN DESA). *World Urbanization Prospects: The 2014 Revision, 2015, (ST/ESA/SER.A/366)*; United Nations Publication: New York, NY, USA, 2015; p. 517.

43. Odisha Water Resource Board, Digital Library. Available online: http://www.dowrodisha.gov.in/ (accessed on 13 February 2020).

44. Indian Meteorological Department. Ministry of Earth Sciences, Government of India, Digital Library. Available online: https://mausam.imd.gov.in/ (accessed on 5 January 2020).

45. Intergovernmental Panel on Climate Change (IPCC) Data Distribution Center. Available online: https://www.ipcc-data.org/ (accessed on 20 April 2020).

46. Landsat Missions. USGS. Available online: https://www.usgs.gov/core-science-systems/nli/landsat/landsat-8?qt-science_support_page_related_con=0#qt-science_support_page_related_con (accessed on 25 June 2020).

47. Eastman, J.R. *TerrSet Geospatial Monitoring and Modeling System Manual*; Clark University: Worcester, MA, USA, 2016.

48. American Public Health Association (APHA). *Standard Methods for the Examination of Water and Wastewater*, 22nd ed.; Rice, E.W., Baird, R.B., Eaton, A.D., Clesceri, L.S., Eds.; American Public Health Association (APHA), American Water Works Association (AWWA) and Water Environment Federation (WEF): Washington, DC, USA, 2012; p. 1496.

49. The United Nations World Water Development Report 2017. *Wastewater: The Untapped Resource*; UNESCO: Paris, France, 2017.

50. Kadaverugu, R.; Dhyani, S.; Dasgupta, R.; Kumar, P.; Hashimoto, S. Dynamics of land use land cover changes and its impact on coastal environment. Under Review.
51. Portman, M.E.; Esteves, L.S.; Le, X.Q.; Khan, A.Z. Improving integration for integrated coastal zone management: An eight country study. *Sci. Total Environ.* **2012**, *439*, 194–201.d. [CrossRef] [PubMed]
52. Badham, J.; Elsawah, S.; Guillaume, J.H.A.; Hamilton, S.H.; Hunt, R.J.; Jakeman, A.J.; Peirce, S.A.; Snow, V.O.; Sebens, M.B.; Fu, B.; et al. Effective modeling for integrated water resource management: A guide to contextual practices by phases and steps and future opportunities. *Environ. Model. Softw.* **2019**, *116*, 40–56. [CrossRef]
53. Das, M.K.; Samanta, S.; Sajina, A.M.; Sudheesan, D.; Naskar, M.; Bandopadhyay, M.K.; Paul, S.K.; Bhowmick, S.; Srivastava, P.K. Fish diversity, community structure and ecological integrity of river Brahmani. *J. Inland Fish Soc. India* **2016**, *48*, 1–13.
54. Kumar, P.; Avtar, R.; Dasgupta, R.; Johnson, B.A.; Mukherjee, A.; Ahsan, M.N.; Nguyen, D.C.H.; Nguyen, H.Q.; Shaw RMishra, B.K. Socio-hydrology: A key approach for adaptation to water scarcity and achieving human well-being in large riverine islands. *Prog. Disaster Sci.* **2020**, *8*, 100134. [CrossRef]
55. Easman, E.S.; Abernethy, K.E.; Godley, B.J. Assessing public awareness of marine environmental threats and conservation efforts. *Mar. Policy* **2018**, *87*, 234–240. [CrossRef]

MDPI
St. Alban-Anlage 66
4052 Basel
Switzerland
Tel. +41 61 683 77 34
Fax +41 61 302 89 18
www.mdpi.com

Sustainability Editorial Office
E-mail: sustainability@mdpi.com
www.mdpi.com/journal/sustainability